Physics of Electrolytes

Physics of Electrolytes

Volume 2

Thermodynamics and
Electrode Processes in
Solid State Electrolytes

edited by

J. HLADIK

Department of Physics,
University of Dakar,
Senegal, Africa

1972

ACADEMIC PRESS · London · New York

CHEMISTRY

ACADEMIC PRESS INC. (LONDON) LTD.
24/28 Oval Road,
London NW1

United States Edition published by
ACADEMIC PRESS INC.
111 Fifth Avenue
New York, New York 10003

Library of Congress Catalog Card Number: 70–170742
ISBN: 0–12–349802–3

PRINTED IN GREAT BRITAIN BY
ROYSTAN PRINTERS LIMITED
Spencer Court, 7 Chalcot Road
London NW1

CONTRIBUTORS

G. B. BARBI, *High Temperature Chemistry Group, Euratom CCR, Ispra, Italy.*

D. A. DE VOOYS, *Van't Hoff Laboratorium, State University, Sterrenbos 19, Utrecht, Netherlands.*

J. DUPUY, *Laboratoire d'Emission Electronique, Faculté des Sciences, Villeurbanne, France.*

R. T. FOLEY, *Chemistry Department, The American University, Washington DC 20016, USA.*

R. J. FRIAUF, *Département of Physics and Astronomy, University of Kansas, Lawrence, Kansas 66044, USA.*

K. S. GOTO, *Department of Metallurgical Engineering, Tokyo Institute of Technology, Tokyo, Japan.*

K. O. HEVER,† *School of Chemistry, City of Leicester Polytechnic, Leicester, England.*

J. HLADIK, *Département de Physique, Faculté des Sciences, Université de Dakar, Senegal, West Africa.*

A. R. KAUL, *Laboratory of General Chemistry, University of Moscow, USSR.*

J. H. KENNEDY, *University of California, Santa Barbara, California, USA.*

W. PLUSCHKELL, *Max-Plank Institut für Metallforschung, Institut für Metallkunde, Stuttgart, West Germany,*

H. RICKERT, *Lehrstuhl für Physikalische Chemie, Universität Dortmund, West Germany.*

J. SCHOONMAN, *Laboratorium voor Kernfijsica en Vaste Stof, State University, Princetonplein 1, Utrecht, Netherlands.*

T. TAKAHASHI, *Department of Applied Chemistry, Faculty of Engineering, Nagoya University, Nagoya, Japan.*

Y. D. TRETYAKOV, *Laboratory of General Chemistry, University of Moscow, USSR.*

A. C. H. VAN PESKI, *Dalco Incorporated, De Beaufortlaan 28, Soestduinen, Netherlands.*

W. L. WORRELL, *School of Metallurgy and Materials Science, University of Pennsylvania, Philadelphia, Pennsylvania 19104, USA.*

†Deceased

FOREWORD

Electrochemistry deals with chemical reactions in response to electrical phenomena and vice versa. It has always been based on a mutual interaction between people studying the phenomenological laws of electricity and those concerned with the atomistic concepts of matter. The basis of the introduction of electrochemistry as a scientific discipline was the discovery of electrolytes. For a long period only liquid electrolytes were known, the reason for this being that the the mobility of ions in liquid systems are by and large several orders of magnitude higher than in solids.

Electrical potential differences in electrochemical systems amount to several volts and reflect the magnitude of the chemical affinities. In order to build up and keep stable these differences in electrical potential by a separation of electrical charges in a finite time, at least a minimum electrical conductivity is necessary for the system under consideration. In solid systems this minimum conductivity is in general obtained only at elevated temperatures. Thus, much of solid state electrochemistry turns out also to be high temperature electro-chemistry. Generally both ionic and electronic conductivity occurs in solids. Since by definition solid electrolytes must have ionic transference numbers close to unity, one may state that solid electrolytes occur only among those ionic compounds of which the components differ in Pauling's electronegativity scale by two or more. This rule of thumb is in line with experimental facts; by and large only halides and oxides exhibit component activity regions in which the transference number of ionic species comes close to unity, and there seems to be little chance of finding sulphides, nitrides or carbides sufficiently stable to permit their use as solid electrolytes.

Historically, the application of solid electrolytes is much newer than the application of salt melts or aqueous ionic solutions as electrolytes. Nevertheless, at the beginning of this century such physical chemists as Nernst, Warburg, Haber and Tubandt knew of the electrolytic nature of certain ceramics or glasses. A systematic approach towards ionic conductivity in solids, however, was not possible before the advent of point defect thermodynamics in crystals, which must be regarded as a prerequisite for the understanding of transport properties in solids. This field was investigated in principle in the late twenties and early thirties by Frenkel, Wagner, Schottky and Jost, among others, and has the same importance for solid state electrochemistry as has the field of electrolyte structure and ion interactions for the electrochemistry of aqueous or liquid systems.

During the following years C. Wagner presented his classical work on the oxidation of metals and solid state reactions in which he assumed that these heterogeneous reactions are electrochemical in character. This meant that he postulated an independent transport of ions and electrons (and/or holes) through the reaction product layer to find the reactants at the phase boundary. It is well known today that the formulation of the transport equations in the reaction product layer and their integration, taking into account the proper boundary conditions, lead directly to the parabolic rate law of tarnishing and solid state reactions. Integration of the same transport equations, taking into account the boundary conditions of an open circuit galvanic cell, yields the electromotive force of the galvanic cell, in which the Gibbs free energy of the virtual cell reaction is the essential term beside the transference numbers. In view of this it becomes clear why research in the fields of solid state electrochemistry and solid state reactions is so mutually beneficial.

In the last decade there were two main stimuli for solid state electrochemistry at elevated temperatures. Firstly, there appeared a paper by Kiukkola and Wagner in the *Journal of the American Electrochemical Society* of 1957 which pointed out the potentialities of solid electrolytes for thermodynamic investigations at moderate and high temperatures and which reintroduced doped zirconia as a solid electrolyte. This and similar electrolytes have subsequently gained great popularity among chemists and physical chemists for thermodynamic and kinetic investigations. Secondly, modern techniques in space and in reactor plants require new methods for the conversion of different sorts of energy into electrical energy, a subject right within the province of solid state electrochemists. Here the interest has been stimulated also by the discovery of the ionic solids Ag_3SI, $RbAg_4I_5$ and related compounds, which conduct anomalously well at room temperature. Research on fuel cells and new batteries for smogless propulsion should also be mentioned in this context.

Where are the main activities in solid state electrochemistry nowadays, and in which direction can the future research in this field be expected to go? Looking through journals for original publications, it is still the application of the traditional solid electrolytes for thermodynamic investigations including coulometric titrations which seems to have priority. Moreover, solid electrolytes have recently been used extensively to determine transport coefficients of condensed systems without introducing new principles. Another point worth mentioning is the search for new solid electrolytes or a testing of new combinations of known electrolytes in order to extend the range of applicability in high temperature electrochemistry.

What is the future direction then for solid state electrochemistry? Regarding the analogy with electrochemical research on liquid systems, one may expect considerable efforts from solid state electrochemists in the field of electrodes, which is concerned with the structure and the kinetic aspects of solid/liquid and solid/solid interfaces. The state of knowledge in this area is still deficient, however by improving it, not only will interfaces be understood much better, but also the application of electrochemical probes for various questions in solid state kinetics will be promoted. In any case, solid state electrochemistry will prove to be an exciting scientific field in the years to come, and its influence on technology can hardly be overrated.

June, 1972.

H. SCHMALZRIED
Institute for Theoretical Metallurgy, Technical University, Clausthal, West Germany.

CONTENTS OF VOLUME 1

CONTENTS

C. THERMODYNAMICS

D. ELECTRODE PROCESSES

E. ENERGETICS AND PHOTO-ELECTROCHEMICAL PHENOMENA

C. THERMODYNAMICS

12. ELECTROCHEMICAL KNUDSEN CELLS FOR INVESTIGATING THE THERMO-DYNAMICS OF VAPOURS

Hans Rickert

Lehrstuhl für Physikalische Chemie, Universität Dortmund, West Germany.

Summary

A "Knudsen" cell can be combined with solid state galvanic cells $Pt/Ag/AgJ/Ag_2S/Pt$ or $Pt/Ag/AgJ/Ag_2Se/Pt$. In these cells silver iodide is practically a pure ionic conductor. The sulphur (or selenium) vapour in the Knudsen cell is in equilibrium with the Ag_2S (or Ag_2Se) of a galvanic cell. A flow of current through a cell, with the positive pole at the Ag_2S (or Ag_2Se) is a measure of the rate of outflow of vapour out of the Knudsen cell, while the e.m.f. is a measure of the chemical potential of the sulphur (or selenium) vapour. In this way the rate of outflow of vapour can be measured as a function of its chemical potential or thermodynamic activity. Thus new theoretical means are available for the determination of the individual kinds of sulphur or selenium molecules. By the combination of an electrochemical Knudsen cell with a mass spectrometer it is possible to separate parent from fragment ions in the mass spectrometer, so that exact thermodynamic data of sulphur and selenium molecules are obtained.

I. Introduction

In this chapter the combination of a normal Knudsen cell with a solid state galvanic cell—thus forming an electrochemical Knudsen cell (Ratchford and Rickert, 1962)—will be described. An analysis of the partial pressure of the vapour in a normal Knudsen cell is possible only if the exact composition of the vapour is known, or better if there is only one molecular species in the gas phase. The advantage of the electrochemical Knudsen cell is that the total pressure of the vapour in the Knudsen vessel and thus the evaporation rate of the Knudsen cell can be varied within a large range, and that the total rate of effusion can be measured as a function of the activity of the species of vapour in the Knudsen cell. There is a further improvement if an electrochemical Knudsen cell is combined with a mass spectrometer (Detry et al., 1967). In a mass spectrometer with a simple Knudsen cell it is often difficult to distinguish between parent ions and fragments and thus it is often impossible to make an exact analysis of the vapour in the Knudsen cell. The combination of an electrochemical Knudsen cell and a mass spectrometer makes it possible to distinguish between parent ions and those ions formed by fragmentation of larger molecules in the mass spectrometer. With the combination of an electrochemical Knudsen cell and a mass spectrometer the exact composition of sulphur (Detry et al., 1967) and selenium (Detry et al., 1971) vapour and the thermodynamic properties of the molecular species in these vapours were investigated. This will also be described in this chapter. For most discussions the investigation of the thermodynamics of sulphur vapour using the galvanic cell $Pt/Ag/AgJ/Ag_2S/Pt$ will be used as an example.

II. The Knudsen Cell

Knudsen (1909) demonstrated with mercury the dynamic measurement of equilibrium vapour pressures at very low pressures. The Knudsen cell is shown schematically in Fig. 1. It consists of a small vessel containing a condensed phase with its equilibrium vapour. Outside the Knudsen cell is a vacuum, the Knudsen cell itself having a small hole connecting the interior of the Knudsen cell to the vacuum. Through this small hole the vapour effuses out of the Knudsen cell. From gas–kinetic considerations it follows that j, the rate of effusion in mol per cm^2 per sec is given by

$$j = \frac{P}{\sqrt{2\pi MRT}} \cdot \tag{1}$$

Where p is the pressure in the Knudsen cell, R the gas constant, T the absolute temperature, M the molecular weight of the effusing vapour molecules.

If there are different vapour molecules eqn (1) holds for every species and the rate of effusion is given by the sum of them all. From this it is clear that evaluation of the vapour pressure is only simple if the composition of the vapour is known, that is if one knows whether the vapour consists of atoms, molecules or a definite mixture of them. It is especially easy if there is only one molecular species. For complex vapours additional information is necessary. This can be given by combining a Knudsen cell with a solid state galvanic cell, thus forming an electrochemical Knudsen cell, and then combining the electrochemical Knudsen cell with a mass spectrometer. Before discussing the electrochemical Knudsen cell, thermodynamic and kinetic properties of a solid state galvanic cell will be described.

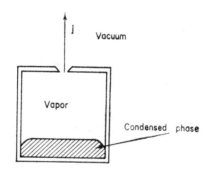

$$j = \frac{p}{\sqrt{2 \pi MRT}}$$

FIG. 1. Knudsen cell.

III. The Solid State Galvanic Cell Pt/Ag/AgJ/Ag₂S/Pt

For investigating the thermodynamics of sulphur the solid state galvanic cell

$$Pt/Ag/AgJ/Ag_2S/Pt \tag{I}$$

was used in the electrochemical Knudsen cell. The galvanic cell (I) is usually constructed by pressing tablets of the components and mounting these face to face, but other geometries are possible and were used in the electrochemical Knudsen cell. The cell (I) can be used in the temperature range 200–450° C. The cell has been used previously by Reinhold (1934) and

Wagner (1953) for thermodynamic studies, and by Kobayashi and Wagner (1957) and Rickert (1961) for kinetic investigations. The galvanic cell (I) has both thermodynamic and kinetic properties, which are very important for using the cell in an electrochemical Knudsen cell; both types of properties will therefore be discussed in the following.

IV. Thermodynamic Properties of the Solid State Cell
Pt/Ag/AgJ/Ag$_2$S/Pt

In open circuit, and also under current flow conditions where polarization effects are negligible, the e.m.f. of this cell is related to the chemical potential μ_{Ag} of silver in the Ag$_2$S sample according to

$$\mu_{Ag} - \mu_{Ag}^0 = - EF \qquad (2)$$

μ_{Ag}^0 is the chemical potential of silver in the standard state, which is pure metallic silver. Thus it is possible, with the aid of cell (I), to measure the chemical potential of silver in Ag$_2$S, and from the relationship

$$\mu = \mu^0 + RT \ln a \qquad (3)$$

one can also obtain the thermodynamic activity a_{Ag} of silver in Ag$_2$S. It follows fron eqns (2) and (3) that the activity of silver in Ag$_2$S is given by

$$a_{Ag} = \exp\left(- \frac{EF}{RT}\right). \qquad (4)$$

With this cell it is also possible to measure the chemical potential of sulphur μ_s since μ_{Ag} and μ_s are interrelated by the Gibbs–Duhem equation. Since the range of non-stoichiometry of Ag$_2$S is small, one may write

$$2(\mu_{Ag} - \mu_{Ag}^0) = \mu_s' - \mu_s \qquad (5)$$

where μ_s' is the chemical potential of sulphur in Ag$_2$S which is in equilibrium with silver. Let the chemical potential of silver in the sulphide in equilibrium with sulphur ($\mu_s = \mu_s^0$) be μ_{Ag}^*, then from eqn (5) it follows

$$2(\mu_{Ag}^* - \mu_{Ag}^0) = \mu_s' - \mu_s^0. \qquad (6)$$

Let the e.m.f. of cell (I) for Ag$_2$S in equilibrium with sulphur be E^*, that is

$$\mu_{Ag}^* - \mu_{Ag}^0 = - E^*F. \qquad (7)$$

By combination of eqns (2), (3), (6) and (7) it follows

$$\mu_s - \mu_s^0 = 2(E - E^*)F. \qquad (8)$$

This equation gives the dependence of the chemical potential of sulphur from the e.m.f. of cell (I).

It follows from eqns (3) and (8)

$$a_s = \exp\left(\frac{2(E - E^*)F}{RT}\right).$$ (9)

In this way the measurement of e.m.f. yields a value for the activity of sulphur atoms in the silver sulphide of cell (I). The thermodynamic activity of sulphur is taken as "one" for equilibrium with liquid sulphur and thus indicating the used standard state for sulphur. It is important in what follows to note that the chemical potential or the activity of sulphur and that the partial pressures of the sulphur molecules in the gas phase can be measured if the Ag_2S in cell (I) is equilibrated with the gas phase and the e.m.f. of the cell is measured. Then the sulphur activity in Ag_2S is the same as the sulphur activity in the gas phase. Under such conditions the e.m.f. of cell (I) is a direct measure of the chemical potential of sulphur or the thermodynamic activity of sulphur in the gas phase. The partial pressures of the different sulphur molecules in the gas phase depend typically upon the thermodynamic activity of sulphur. In sulphur vapour there are the molecular species S_2, S_3, S_4, S_5, S_6, S_7 and S_8 in the low temperature region (Luft, 1955; Berkowitz and Chupka, 1964; Detry et al., 1967) that is as long as liquid sulphur can also be present, in contradiction to previous assumptions where one has assumed that only S_2, S_4, S_6 and S_8 were present (Braune et al., 1951) Even if in the low temperature range sulphur atoms are not present in measurable amounts in the vapour we may use sulphur atoms for obtaining thermodynamic relationships. Consider the equilibrium in the gas phase

$$xS(g) = S_x(g)$$ (10)

where x may be 2, 3, 4, 5, 6, 7 and 8, that is the equilibrium between sulphur atoms and S_x-molecules. From this it follows that under conditions of equilibrium the chemical potential of the sulphur atoms μ_s is related to that of the molecular species S_x by

$$x\mu_S = \mu_{S_x}.$$ (11)

Equation (11) is also true for all sulphur molecules in equilibrium with liquid sulphur, and in this case we designate a corresponding chemical potential with the superscript "0" to represent the chosen standard state:

$$x\mu_S^0 = \mu_{S_x}^0.$$ (12)

From eqns (11) and (12) it follows

$$x(\mu_S - \mu_S^0) = \mu_{S_x} - \mu_{S_x}^0.$$ (13)

From the definition of the chemical potential, we can write for μ_{S_x}:

$$\mu_{S_x} = \mu_{S_x}^0 + RT \ln \frac{p_{S_x}}{p_{S_x}^0}$$ (14)

where R is the gas constant, T the absolute temperature, p_{S_x} the partial vapour pressure of the sulphur molecules S_x and $p_{S_x}^0$ the equilibrium partial vapour pressures above liquid sulphur.

In eqn (14) the ideal gas law is assumed to be valid for sulphur vapour, which is reasonable because of the very small vapour pressures encountered. Using eqns (3), (13) and (14) we can derive the dependence of the partial pressure pS_x of the different sulphur molecules from the activity of sulphur atoms in the gas phase

$$p_{S_x} = p_{S_x}^0 \, a_S^{\,x}.$$ (15)

From eqns (15) and (9) it follows

$$p_{S_x} = p_{S_x}^0 \exp\left(\frac{2x(E - E^*)F}{RT} \right).$$ (16)

Equation (16) expresses the equilibrium partial pressure of sulphur molecules S_x over Ag_2S in terms of the corresponding equilibrium partial pressure over liquid sulphur and the e.m.f. of cell (I). Thus we have the dependence of the partial pressure p_{S_x} of the sulphur molecules S_x from the e.m.f. of cell (I) when Ag_2S of this cell is in equilibrium with the vapour. This dependence will be very important in our further discussions.

V. Kinetic Properties of the Galvanic Cell (I)

In the kinetic applications the electric current passing through cell (I) is a measure of the rate of addition to, or removal of silver from, the silver sulphide. In the steady state, that is when the silver–sulphur ratio in the sulphide remains constant, and when no other side reactions occur, the silver loss from the silver sulphide corresponds to sulphur evolved by the Ag_2S phase. The reverse occurs when sulphur is absorbed by the Ag_2S phase. Thus it is possible to measure by an electric current the evaporation rate of sulphur from the silver tablet, and if the cell is placed into a Knudsen cell the evaporation rate of sulphur out of the Knudsen cell. We may discuss the achievement of steady state conditions during the evaporation of sulphur from silver sulphide in detail. If the negative pole of the current carrying circuit is connected to the silver and the positive pole to the platinum sheet which is in contact with the Ag_2S tablet, the current will result in the withdrawal of silver from the Ag_2S the silver ions migrating through the AgJ and the electrons through the platinum sheet. This is indicated by arrows in Fig. 2. The current I can be controlled by a series of resistances and measured by means of an ammeter. Two processes are now occuring simultaneously. First, silver is removed from the Ag_2S by the passage of the current. This alone causes a decrease of the silver potential in the Ag_2S and a corresponding increase of the sulphur potential. Secondly, sulphur evaporates with an increasing velocity from the Ag_2S surface as the sulphur potential rises. This evaporation process of sulphur alone would cause the sulphur potential of the Ag_2S to decrease and so under the present conditions a steady state is set up. In the stationary state, which is automatically achieved, an amount of sulphur being evaporated is equivalent to the amount of silver withdrawn from the sulphide by the passage of the current. The current in the steady state thus yields a direct measurement of the vaporization rate of the sulphur, i.e.

$$J = I/F. \tag{17}$$

Where J is the vaporization rate of sulphur in equivalents, I is the electrical current drawn through cell (I), F is the Faraday constant. The stationary state is clearly indicated when the chemical potential of the silver or sulphur

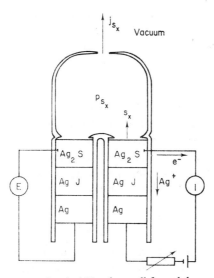

FIG. 2. Electrochemical Knudsen cell for sulphur vapour.

reaches a constant value, which is equivalent to saying that the e.m.f. of the cell is constant with time. The attainment of equilibrium in the Ag_2S is very rapid as compared with the establishment of the steady state. Because of the large diffusion coefficient of the silver ions (10^{-5} cm^2 s^{-1}), the high electronic conductivity and the narrow range of stoichiometry, which gives an effective diffusion coefficient of 10^{-2} cm^2 s^{-16}, potential differences in an Ag_2S tablet which is 1 mm thick are eliminated in less than one second.

The achievement of the steady state depends on the magnitude of the current which is passed as well as on the thickness of the tablet. The time taken was about one minute when large currents were passed, but required several hours for the smallest currents used (Ratchford and Rickert, 1962; Detry et al., 1967).

VI. The Electrochemical Knudsen Cell

The general considerations will be presented using once again the example of silver sulphide and sulphur. In the electrochemical Knudsen cell shown in Fig. 2, a Knudsen cell is combined with two solid-state galvanic cells of type (I). The two solid-state galvanic cells which are completely equal in construction have two different tasks. One cell is used under open circuit conditions. Under these conditions the silver sulphide of the cell will reach equilibrium with the sulphur in the gas phase of the electrochemical Knudsen cell and thus the e.m.f. of the cell is a measure of the activity of sulphur in the gas phase according to eqn (9). The other cell is used under current flow. A current I in this cell with a positive pole at the Ag_2S tablet is a measure of the rate at which silver is removed from the Ag_2S. Under steady state conditions this rate equals the rate of effusion of sulphur from the opening of the Knudsen cell. The special significance of this combination of a Knudsen cell with a galvanic solid-state cell is that measurements can be carried out not only at a definite chemical potential and total vapour pressure of sulphur, but that these quantities can be varied by changing the rate of effusion merely by adjusting the current through the cell. Thereby new theoretical possibilities become available for the determination of the partial pressures of the individual molecular species of sulphur. The rate of effusion j_x of a definite molecular species S_x in moles per unit surface area and time is given by the well-known relation:

$$j_x = \frac{p_{S_x}}{(2\pi M_{S_x} RT)^{1/2}} \tag{18}$$

where the equilibrium partial pressure p_{S_x} of the molecules S_x is a function of the absolute temperature T and the chemical potential μ_S of the sulphur in the gas and under equilibrium conditions in the silver sulphide of the cell used under open circuit conditions. M_{S_x} designates the molecular weight of the S_x molecules and R is the gas constant. Substitution of eqn (15) in eqn (18) gives

$$j_x = \frac{p^0_{S_x}}{(2\pi M_{S_x}RT)^{1/2}} \, a_S{}^x \tag{19}$$

and with eqn (16)

$$j_x = \frac{p^0_{S_x}}{(2\pi M_{S_x}RT)^{1/2}} \exp\left[2x(E - E^*)F/RT\right]. \tag{20}$$

In the special case of equilibrium with liquid sulphur eqn (20) takes the form:

$$j_x{}^0 = \frac{p^0_{S_x}}{(2\pi M_{S_x}RT)^{1/2}}. \tag{21}$$

According to eqns (19) and (20) the rate of effusion of the different molecular species of sulphur from the orifice of the electrochemical Knudsen cell is given by the equilibrium partial pressures $p^0_{S_x}$ over liquid sulphur, the activity of mono-molecular sulphur a_{S_x} or the e.m.f. E of the solid-state galvanic cell (I), respectively. The electrochemical Knudsen cell yields the rate of effusion in electrical units as electrical current densities i_x, since

$$i_x = 2x\,j_x\,F. \tag{22}$$

The total current density i is given by

$$i = I/q = \Sigma\,i_x. \tag{23}$$

where q is the cross-sectional area of the Knudsen cell opening and I is the current through the cell (I). In eqn (19) and 20) the number of atoms x in the S_x molecules appear in the exponential term. For this reason the proportion of the different molecular species in the effusing vapour is strongly influenced by the chemical potential of sulphur μ_s or the activity a_s or the voltage E of the cell (I) respectively. Therefore one can expect that in different potential regions, different molecular species will predominate, whereby an excellent opportunity exists to determine separately the partial pressures of the individual molecular species. The relationship between the partial current i_x of one molecular species S_x and the e.m.f. E of cell (I) is given from eqns (20) and (22) by

$$\ln i_x = \ln 2x\,F\,j_x{}^0 + 2x(E - E^*)F/RT. \tag{24}$$

From eqn (24) one can see that a plot of $\ln i_x$ versus E is linear with a slope of $2x\,F/RT$, from which one can recognize the molecular species in question.

If one molecular species predominates, the electrical current measured shows the dependence on the e.m.f. according to eqn (24) and together with eqn (20) one can get the partial pressure of these molecules and from this thermodynamic data. When there are many molecular species such as in the case of sulphur and selenium vapours there is an advantage to be gained in combining the electrochemical Knudsen cell with a mass spectrometer. It enables separation in the mass spectrometer of parent ions from fragment ions and thus a complete analysis of the vapour is possible (Detry *et al.*, 1967). This will be described in the next section.

VII. Combination of an Electrochemical Knudsen Cell with a mass Spectrometer for the Study of the Thermodynamics of Sulpher Vapour

The combination of the electrochemical Knudsen cell (Ratchford and Rickert, 1962) with a mass spectrometer (Detry *et al.*, 1967) makes it possible to obtain accurate thermodynamic values for all vapour species, and thus to elucidate the thermodynamics of sulphur vapour. In particular, it enables one to distinguish between molecular fragments and primary particles as well as to measure ionization cross-sections in the spectrometer ion source.

Apart from theoretical considerations (Luft, 1955), our own (Detry *et al.*, 1971) and independent experimental studies (Berkowitz and Chupka, 1964) with the mass spectrometer have shown that the species S_3, S_5 and S_7 appear in the sulphur vapour as well as the species S_2, S_4, S_6 and S_8 (Braune *et al.*, 1951). In the electrochemical Knudsen cell sulphur is vaporized from Ag_2S by means of one of the cells I. An electric current, I, is passed through one cell with the positive pole of an external current supply on the Ag_2S. In this way the vaporization rate, and hence the partial pressure of sulphur in the cell, can be varied. For each steady state value of the current there are three measurable quantities:

(a) The electrical current, I. This gives a measure of the sulphur evaporation rate:

$$i = \frac{I}{q} = \frac{\Sigma I_{S_x}}{q} = 2F \, \Sigma \, x j_{S_x} \qquad (25)$$

where i is the electrical current density, q is the orifice cross-sectional area of the Knudsen cell, I_{S_x} is the partial current corresponding to the evaporation of the S_x species, F is the Faraday constant, x is the number of atoms in the sulphur molecules and j_{S_x} is the effusion rate of the S_x molecules in moles per unit area and unit time. The relationship to the partial pressure is given by eqn (18).

(b) The e.m.f., E, of the other cell (I). E gives a measure of the activity,

a_s, of sulphur in the Knudsen cell according to eqn (9). The sulphur partial pressure p_{S_x} depends, for each vapour species S_x in a characteristic manner on the sulphur activity according to eqn (15).

(c) The intensity $I_{S_x}^+$ of the S_x molecule after separation in the mass spectrometer: this intensity usually consists of two parts—one of which derives from primary S_x molecules and the other from fragments of larger molecules formed during the ionization. The intensity $(I_{S_x}^+)_{prim}$ of the ions formed directly from S_x molecules by electron collision is related to the S_x partial pressure in the Knudsen cell by the equation

$$p_{S_x} = \frac{(I_{S_x}^+)_{prim}\, T}{A_{S_x}} \qquad (26)$$

where A_{S_x} is the sensitivity of the mass spectrometer for the S_x molecules.

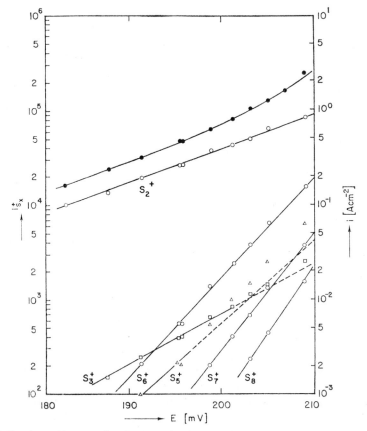

FIG. 3. Ion intensities $I_{S_x}^+$ for the S_x molecules (in arbitrary units) and the total current density, i, as a function of the e.m.f., E, at 300°C.

The method described makes it now possible to separate primary from fragment ions. On a plot of $\ln I_{S_x}^+$ versus e.m.f. the straight line for the primary ions of type S_x should have a characteristic slope $2x\,F/RT$, according to eqns (9), (15) and (26). The intensities of the fragment ions, on the other hand, have the same potential dependence as that of the S_x molecules from which they are formed.

In order to calculate the partial pressures from the intensities it is necessary to know the instrumental sensitivity A_{S_x}.

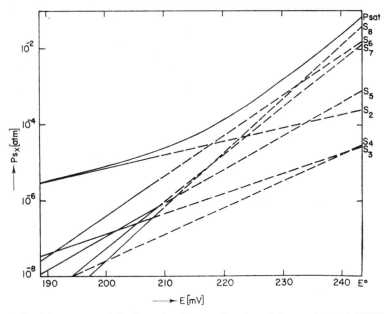

FIG. 4. Partial pressures of the S_x molecules as a function of the e.m.f., E, at 300°C.

In a region of potential where one molecular species is of predominant importance, the total current of the electrochemical cell corresponds to the effusion out of the Knudsen cell and A_{S_x} can be obtained from eqns (9), (18), (22) and (26). The current is a direct measure for the pressure. If one has more than one molecular species together, the total current still gives information since it is equal to the sum of all the partial currents I_{S_x}. Measurements of this type have been carried out in the temperature range 200–400°C. Figures 3 and 4 show, as an example, some measurements at 300°C. In Fig. 3 the ion intensities for the sulphur molecules and the total current density are shown as a function of the e.m.f. In Fig. 4 the partial pressures which have been calculated from these results are shown. By extrapolation of these straight lines up to the saturation point, $E = E^*$, one can obtain the saturation partial pressure $p_{S_x}^0$. Addition of these partial pressures gives the total pressure (200°C: 3.03×10^{-3} atm; 250°C: 1.96×10^{-2} atm; 300°C: 8.09×10^{-2} atm; 350°C: 2.43×10^{-1} atm; 400°C: 6.11×10^{-1} atm). These values are in good agreement with those obtained by the direct measurements of West and Menzies (1929) (Table I). In order to avoid the inaccuracies inherent in obtaining the total pressure by the addition of each of the partial pressures, the relative partial pressures obtained in the present study were combined with the total pressure of West and Menzies. The absolute partial pressures obtained in this way are shown in Table I.

TABLE I.   Partial pressures in saturated sulphur vapour and total pressure

pressure, atm	Temperature, °C				
	200	250	300	350	400
p_{S_2}	1.40×10^{-6}	2.60×10^{-5}	2.68×10^{-4}	1.90×10^{-3}	9.40×10^{-3}
p_{S_3}	1.70×10^{-7}	3.38×10^{-6}	3.66×10^{-5}	2.68×10^{-4}	1.34×10^{-3}
p_{S_4}	1.65×10^{-7}	3.04×10^{-6}	3.25×10^{-5}	2.15×10^{-4}	1.04×10^{-3}
p_{S_5}	1.56×10^{-5}	1.72×10^{-4}	9.64×10^{-4}	4.20×10^{-3}	1.43×10^{-2}
p_{S_6}	5.50×10^{-4}	3.60×10^{-3}	1.60×10^{-2}	5.25×10^{-2}	1.37×10^{-1}
p_{S_7}	3.28×10^{-4}	2.63×10^{-3}	1.27×10^{-2}	4.55×10^{-2}	1.26×10^{-1}
p_{S_8}	1.89×10^{-3}	1.02×10^{-2}	3.64×10^{-2}	9.70×10^{-2}	2.14×10^{-1}
p_{tot}	2.97×10^{-3}	1.66×10^{-2}	6.62×10^{-2}	2.00×10^{-1}	4.98×10^{-1}

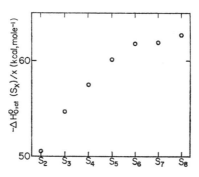

FIG. 5. Enthalpies of formation $H_0°/x$ of the different sulphur molecules S_x out of the atoms devided by the number of atoms x in the molecules.

The thermodynamic properties of the gaseous sulphur molecules were calculated from the equilibrium constants k of the reactions listed in Table II with the well-known relations

$$\Delta G^\circ = - RT \ln k = \Delta H_T^\circ - T\Delta S^\circ \tag{29}$$

$$\mathrm{d}\ln k / \mathrm{d}(1/T) = - \Delta H_T^\circ / R. \tag{30}$$

From the data in Table II one gets the entropies S_T° of the different sulphur molecules their enthalpies of formation out of sulphur atoms and their heats

TABLE II.   Standard enthalpies and entropies of reaction for a number of reactions with gaseous sulphur molecules

Reaction	Temperature °K	ΔH_T° kcal/mol	ΔS_T°
$2S(rh) \rightarrow S_2(g)$	460–670	$28 \cdot 08 \pm 0 \cdot 35$	$32 \cdot 6$
$2S_3(g) \rightarrow 3S_2(g)$	566–669	$26 \cdot 6 \pm 2 \cdot 0$	$(37 \cdot 6 \pm 3 \cdot 0)^*$
$S_4(g) \rightarrow 2S_2(g)$	615	$28 \cdot 2 \pm 2 \cdot 0$	$(36 \cdot 7 \pm 2 \cdot 5)^*$
$2S_5(g) \rightarrow 5S_2(g)$	565–620	$95 \cdot 5 \pm 4 \cdot 0$	$(111 \cdot 4 \pm 8 \cdot 0)^*$
$3/4S_8(g) \rightarrow S_6(g)$	435–625	$6 \cdot 26 \pm 0 \cdot 33$	$7 \cdot 6 \pm 0 \cdot 3$
$7/8S_8(g) \rightarrow S_7(g)$	435–625	$5 \cdot 77 \pm 0 \cdot 31$	$7 \cdot 2 \pm 4 \cdot 2$
$S_8(g) \rightarrow 4S_2(g)$	460–625	$96 \cdot 8 \pm 2 \cdot 2$	$110 \cdot 2 \pm 4 \cdot 2$

* calculated values from statistical mechanical formulae.

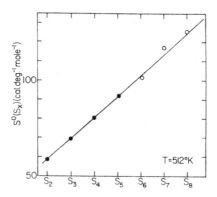

FIG. 6. Standard entropies of the sulphur molecules S_x at 521°K.

of sublimation ΔH_6°, which were calculated using the known heat of sublimation for S_2 (Stull and Sinke, 1956). All these values are given in Table III. Included are the values for the enthalpies of formation devided by the number of atoms, thus giving the bond energy per atom in the molecules. These enthalpies are illustrated in Fig. 5.

Figure 6 shows the entropies of the sulphur molecules for $T = 512°K$.

TABLE III. Standard entropies S_T°, enthalpies of formation ΔH_0° out of the atoms, enthalpies of formation $\Delta H_0^\circ/x$ out of the atoms divided by the number of atoms in the molecule and heats of sublimation $\Delta H_{0,\,sub}^\circ$

Molecule	S_T° cal/degree mol	$\Delta H_{0,\,form}^\circ$ kcal/mol	$\Delta H_{0,\,form}^\circ/x$ kcal/g. atom	$\Delta H_{0,\,sub}^\circ$ kcal/mol
S_2	59·6 (565°K)(ref.[14])	101·0	50·5	30·86 $\pm$ 0·35
S_3	(71·8 $\pm$ 1·5)*(618°K)	164·5	54·8	33·3 $\pm$ 1·3
S_4	(83·9 $\pm$ 2·5)*(615°K)	230·2	57·6	33·5 $\pm$ 2·2
S_5	(94·9 $\pm$ 4·0)*(592°K)	300·6	60·1	29·1 $\pm$ 2·5
S_6	101·6 $\pm$ 3·3 (512°K)	370·8	61·8	24·8 $\pm$ 2·7
S_7	116·9 $\pm$ 3·8 (512°K)	433·3	61·9	28·1 $\pm$ 2·7
S_8	126·8 $\pm$ 4·2 (530°K)	501·4	62·7	26·0 $\pm$ 2·7

* The standard entropies S_6, S_7, S_8 were derived from experimental ΔS_T° differences in entropy using the known entropy of Se (Stull and Sinke, 1956). Those of S_3, S_4, and S_5 were calculated from statistical mechanical formulae.

VIII. Experimental Details

The electrochemical Knudsen cells were made out of Pyrex glass. The Knudsen cell vessel had a diameter of about 25 mm. The opening had a profile pictured in Figs 2 and 7, so that no resistance to effusion could develop in the opening. Figure 7 shows the actual construction of the electrochemical Knudsen cell and the arrangement of the solid-state galvanic cells in the Knudsen cell. In order to ensure that the silver sulphide tablet was vacuum tight in the cavity the silver iodide was momentarily melted so that it made intimate contact with the glass and was free from pores. The cell in this heated condition was placed in a furnace and then, without cooling it down to room temperature, put into a mass spectrometer. In the mass spectrometer the furnace could be adjusted so that the effusing molecules could come directly into the ionization chamber. The silver sulphide was prepared from 99·97 pure silver and especially purified sulphur supplied by

Degussa, Frankfurt and Merck, Darmstadt, respectively. Measurements were carried out with different electrochemical Knudsen cells which had a cross sectional area of the openings of about 0·1–1 mm². Two different types of mass spectrometers were used, but gave the same results.

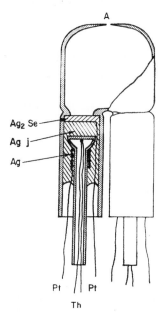

FIG. 7. Construction of the electrochemical Knudsen cell, A: opening, Pt: Platinum-leads, Th: Thermocouple.

IX. Measurements on Selenium Vapour

The partial pressures of the different selenium molecules in selenium vapour and their thermodynamic properties were investigated as well (Detry *et al.*, 1971) with a combination of an electrochemical Knudsen cell and a mass spectrometer. It is well known that there are different molecules present in selenium vapour. But since previously (Preunner and Brockmöller, 1913; Illarionov and Lapina, 1957) only Se_2, Se_4, Se_6 and Se_8 were assumed to be present (and at temperatures higher than 1000°C Se_1), our own investigations (Detry *et al.*, 1968, 1971) and others (Goldfinger and Jeunehomme, 1959; Berkowitz and Chupka, 1966; Fujisaki *et al.*, 1966) have shown that there are Se_3, Se_5 and Se_7 as well in the vapour. With the combination of an electrochemical Knudsen cell and a mass spectrometer it was possible to get exact thermodynamic data for all molecules.

TABLE IV. Partial pressures in saturated selenium vapour and the total pressure

Temperature	200°C	250°C	300°C	350°C	400°C	450°C
$p^*(Se_2)$	$3\cdot51\times10^{-8}$	$9\cdot39\times10^{-7}$	$1\cdot26\times10^{-5}$	$1\cdot07\times10^{-4}$	$6\cdot54\times10^{-4}$	$3\cdot08\times10^{-3}$
$p^*(Se_3)$	$4\cdot02\times10^{-11}$	$2\cdot36\times10^{-9}$	$5\cdot56\times10^{-8}$	$7\cdot36\times10^{-7}$	$6\cdot63\times10^{-6}$	$4\cdot43\times10^{-5}$
$p^*(Se_4)$	$9\cdot69\times10^{-11}$	$6\cdot14\times10^{-9}$	$1\cdot45\times10^{-7}$	$1\cdot90\times10^{-6}$	$1\cdot70\times10^{-5}$	$1\cdot09\times10^{-4}$
$p^*(Se_5)$	$3\cdot03\times10^{-7}$	$5\cdot24\times10^{-6}$	$4\cdot07\times10^{-5}$	$2\cdot02\times10^{-4}$	$7\cdot97\times10^{-4}$	$2\cdot53\times10^{-3}$
$p^*(Se_6)$	$1\cdot48\times10^{-6}$	$2\cdot11\times10^{-5}$	$1\cdot33\times10^{-4}$	$5\cdot57\times10^{-4}$	$1\cdot86\times10^{-3}$	$4\cdot89\times10^{-3}$
$p^*(Se_7)$	$3\cdot98\times10^{-7}$	$6\cdot58\times10^{-6}$	$4\cdot34\times10^{-5}$	$1\cdot88\times10^{-4}$	$6\cdot49\times10^{-4}$	$1\cdot80\times10^{-3}$
$p^*(Se_8)$	$3\cdot57\times10^{-8}$	$7\cdot25\times10^{-7}$	$5\cdot18\times10^{-6}$	$2\cdot20\times10^{-5}$	$8\cdot01\times10^{-5}$	$2\cdot36\times10^{-4}$
$\Sigma p^*(Se_x)$	$2\cdot25\times10^{-6}$	$3\cdot46\times10^{-5}$	$2\cdot35\times10^{-4}$	$1\cdot08\times10^{-3}$	$4\cdot06\times10^{-3}$	$1\cdot27\times10^{-2}$

In this case the solid state galvanic cell

$$Pt/Ag/AgJ/Ag_2S/Pt \qquad \text{(II)}$$

was used in the Knudsen cell. Apart from this the technique was the same than that used in the investigation of sulphur vapour.

TABLE V.   Standard entropies S_T° for the different selenium molecules at 300°K, 600°K and 1000°K

	S [cal/grad mol]		
	300°K	600°K	1000°K
Se_2*	60·28	66·31	70·86
Se_3	75·31	84·71	91·77
Se_4	90·75	104·26	114·36
Se_5	92·23	109·52	122·58
Se_6	103·69	125·12	141·22
Se_7	116·29	141·60	160·70
Se_8	127·10	156·59	178·74

* Stull and Sinke (1956).

TABLE VI.   Heats of sublimation $\Delta H_{0,sub}^\circ$ of the selenium molecules

Molecule	$-\Delta H_{0,\,sub}^\circ$ kcal mol^{-1}
Se_2	34·92
Se_3	44·1
Se_4	45·6
Se_5	34·4
Se_6	33·5
Se_7	36·2
Se_8	38·4

Table IV shows the partial pressures in saturated selenium vapour and the total pressure. Table V gives the entropies of the different selenium molecules at 300°K, 600°K and 1000°K. For these data statistical mechanical formulae were used as well as is discussed in a separate paper (Detry *et al.*, 1971). The heats of sublimation are given in Table VI and the enthalpies of formation out of atoms in Table VII.

The enthalpies of formation of the molecules divided by the number of atoms x in molecule are illustrated in Fig. 8.

TABLE VII.   Heats of formation of the selenium molecules out of atoms and the same value divided by the number of atoms

Molecule Se_x	$\Delta_{atom} H_0^{\circ}$ kcal/mol	$\Delta_{atom} H_0^{\circ}/x$ kcal/mol
Se_2	72·94	36·47
Se_3	117·7	39·2
Se_4	170·3	42·6
Se_5	235·3	47·1
Se_6	290·1	48·4
Se_7	341·5	48·8
Se_8	393·1	49·1

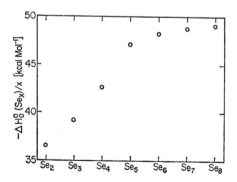

FIG. 8. Enthalpies of formation $\Delta H_0^{\circ}/x$ of the different selenium moelcules Se_x out of the atoms divided by the number of atoms x in the molecules.

References

Berkowitz, J. and Chupka, W. A. (1964). *J. Chem. Phys.* **40,** 287.

Berkowitz, J. and Chupka, W. A. (1966). *J. chem. Phys.* **45,** 4289.

Braune, H., Peter, S. and Neveling, V. (1951). *Z. Naturf.* **6a,** 32.

Detry, D., Drowart, J., Goldfinger, P., Keller, H. and Rickert, H. (1967). *Z. phys. Chem.* (Frankf. Ausg.) **55,** 314.

Detry, D., Drowart, J., Goldfinger, P., Keller, H. and Rickert, H. (1968). Bunsentagung, Augsburg.

Detry, D., Drowart, J., Goldfinger, P., Keller, H. and Rickert, H. (1971). *Z. phys. Chem.* (Frankf. Ausg.) **75,** 273.

Fujisaki, Westmore, J. B. and Tickner, A. W. (1966). *Can. J. Chem.* **44,** 3036.

Goldfinger, P. and Jeunehomme, M. (1959). *In* "Advances in Mass-Spectrometry", pp. 534–546. Pergamon Press, London.

Illarionov, V. V. and Lapina, L. M. (1957). *Dokl. Akad. Nauk SSSR* **114,** 1021.

Knudsen, M. (1909). *Ann. Phys.* (Leipzig) **29,** 179.

Kobayashi, H. and Wagner, C. (1957). *J. chem. Phys.* **26,** 1609.

Luft, N. M. (1955). *Mh. Chem.* **86,** 474.

Preuner, G. and Brockmöller, J. (1913). *Z. phys. Chem.* **81,** 129.

Ratchford, R. J. and Rickert, H. (1962). *Z. Elektrochem.* **66,** 497.

Reinhold, H. (1934). *Z. Elektrochem.* **40,** 361.

Rickert, H. (1961). *Z. Electrochem.* **65,** 463.

Saure, H. and Block, J. (1970). Bunsentagung, Heidelberg.

Stull, D. R. and Sinke, G. C. (1956). "Thermodynamic Properties of the Elements", Advances in Chemistry Series, No. 18, American Chem. Soc. Washington D.C.

Wagner, C. (1953). *J. Chem. Physics* **21,** 1819.

West, W. A. and Menzies, A. W. C. (1929). *J. phys. Chem.* **33,** 1880.

13. OXYGEN CONCENTRATION CELLS

Kazuhiro S. Goto[†] and Wolfgang Pluschkell[‡]

*†Tokyo Institute of Technology,
Department of Metallurgical Engineering, Tokyo, Japan*

*‡Max-Planck-Institut für Metallforschung,
Institut für Metallkunde, Stuttgart, West Germany.*

I. Introduction

An oxygen concentration cell consists of a solid oxide electrolyte separating two electrode compartments (I) and (II) with definite oxygen partial pressures p_{O_2}' and p_{O_2}''. Two metal electrodes are in contact with the interfaces of the electrolyte. Such a cell is shown schematically in Fig. 1. If one con-

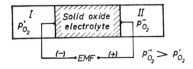

Fig. 1. Schematic diagram of an oxygen concentration cell.

siders the cell as a thermodynamic system then one of the combined statements of the first and second laws of thermodyanamics for a reversible process is

$$dF = V\,dP - S\,dT - dw.$$

The free energy change dF, on passing the one Faraday electrical charge $\tilde{F}$ across the cell is equal to the reversible electrical work $\tilde{F}dE$, as long as the intensive properties, temperature T and total pressure P, are constant so that

$$dF = -dw = -\tilde{F}dE \quad \text{or} \quad \Delta F = -E\tilde{F},$$

where $\tilde{F}$ is the Faraday constant and ΔF the free energy change when one Faraday of electricity is passed at constant E. When the virtually reversible process consists only of the transference of oxygen through the electrolyte, the electrical potential E for the case where the intensive properties are constant and the electrodes of the same metal is given by

$$E = \frac{RT}{4\tilde{F}} \ln \frac{p_{O_2}''}{p_{O_2}'},$$

where R is the gas constant. The positive pole is at the interface with the higher oxygen partial pressure p_{O_2}''. If p_{O_2}'' is known and taken as a reference, the unknown value p_{O_2}' can be computed from this equation.

Because the principles of the cells are relatively simple and since they provide much information, a considerable amount of research has been carried out on oxygen concentration cells with various solid electrolytes since the pioneering work of Kiukkola and Wagner (1957). This work provided the impetus for many investigations in the fields of metallurgy, fuel cells, fused salts, glass, and the more general field of physical chemistry at elevated temperature.

One of the most commonly used electrolytes in oxygen concentration cells consists of zirconia, ZrO_2, doped with 10–20 mol % CaO, Y_2O_3, MgO, or other rare earth metal oxides. It is now well known that these solid solutions contain large concentrations of oxygen vacancies which are needed to accommodate the valence difference between the Zr^{4+} and the doped cations. The electrical properties of these solid solutions were investigated for the first time by Nernst (1899–1900) and Reynolds (1902) about seventy years ago. Baur and Preis (1937) used this electrolyte to construct fuel cells. Haber (1908) and his coworkers and Treadwell (1916) were led to study oxygen concentration cells using Thüringer hard glass and porcelain. Further development was to follow from more basic studies on the conduction mechanism of solid electrolytes and their relation to the measured electromotive forces. Wagner (1933) derived an expression for the electromotive force of a galvanic cell involving a mixed conducting electrolyte as a function of the sum of the ionic transference numbers. He also reported the predominant oxygen ion conductivity in ZrO_2–Y_2O_3 solid solutions due to a very high concentration of oxygen vacancies in the lattice (Wagner, 1943).

In their basic investigation, Kiukkola and Wagner (1957) used a solid solution of 15 mol % CaO in ZrO_2 as a solid electrolyte and succeeded in obtaining accurate standard free energies of formation of several metal oxides. Independently, Peters and Möbius (1958) and Peters and Mann (1959) carried out similar experiments with solid electrolytes of ThO_2 doped with La_2O_3 and ZrO_2 doped with Y_2O_3. These comments should be taken to serve only as a brief survey of the investigations on solid oxide electrolytes which initiated the rapid development of oxygen concentration cells in the past decade.

The present authors intend to limit this article to a short comment on the expected electromotive force of cells with various electrolytes, on possible experimental errors, and on typical examples of applications of these types of galvanic cells to thermodynamic and kinetic studies in metallurgy.

II. Electromotive Forces of Oxygen Concentration Cells

In general, the e.m.f. of an oxygen concentration cell depends on the different chemical potentials of oxygen at the two electrodes and on the transport properties of the solid electrolyte under consideration. The transference numbers of ions, of positive holes and of excess electrons may therefore appear in the equation for the e.m.f. In the following sections some limiting cases of practical importance are discussed. For simplicity a quasi-thermostatic treatment will be used as the basis for a discussion of the e.m.f. of oxygen concentration cells. A treatment using irreversible thermodynamics leads to substantially the same results (Wagner, 1966).

A. The Transference Number of Oxygen is Unity

It is known that in certain solid solutions of zirconia or thoria, oxygen is the only substantially mobile species over a wide range of oxygen pressures and temperatures. For this reason the half cell reactions connected with the virtual transport of 4 Faradays from the right to the left side of the electrolyte (Fig. 1) can easily be formulated. At the cathode gaseous oxygen of chemical potential μ_{O_2}'' is ionized and incorporated into the lattice of the electrolyte:

II

$$O_2(p_{O_2}'') + 4e' \to 2\overset{\leftarrow}{O}^{2-}.$$

At the anode the reverse reaction takes place and oxygen of chemical potential μ_{O_2}' is released so that

I

$$2\overset{\leftarrow}{O}^{2-} \to O_2(p_{O_2}') + 4e'.$$

The arrows over the oxygen symbols indicate the direction of migration. The sum of I and II describes the overall cell reaction, namely, the transport of one mole of oxygen gas from potential μ_{O_2}'' to potential μ_{O_2}':

I + II

$$O_2(p_{O_2}'') \to O_2(p_{O_2}').$$

The free energy change for this reaction is given by

$$\Delta F = \mu_{O_2}' - \mu_{O_2}'' = -4\tilde{F}E,$$

when the intensive properties are kept constant. Insertion of $\mu_{O_2} = \mu^{\circ}_{O_2} + RT \ln p_{O_2}$ leads to the equation

$$E = \frac{RT}{4\tilde{F}} \ln \frac{p_{O_2}''}{p_{O_2}'} \tag{1}$$

which was mentioned in the introduction of this chapter. Equation (1) is always valid regardless of whether the electrolyte consists of a solid oxide solution of two, three or more cationic constituents provided that only the transference number of oxygen ions is unity.

B. The Transference Number of Cation A^{2+} is Unity in Oxide AO

The motion of the cation A^{2+} in an oxide AO, e.g., MgO, may be described relative to the oxygen ion sublattice which is thought to be fixed. The transference number $t_{A^{2+}}$ is then unity. The half cell reactions coupled with the transport of four Faradays through the cell are as follows:

II

$$2 \overrightarrow{A}^{2+} + 4 e' + O_2(p_{O_2}'') \rightarrow 2(AO)$$

and

I

$$2(AO) \rightarrow 2 \overrightarrow{A}^{2+} + 4 e' + O_2(p_{O_2}').$$

Since the activity of the oxide AO is unity throughout the electrolyte, the overall cell reaction only consists of the transport of one mole oxygen gas from the right to the left cell compartment:

I + II

$$O_2(p_{O_2}'') \rightarrow O_2(p_{O_2}').$$

The e.m.f. developed is given by eqn (1) derived in the preceding section.

C. The Sum of the Transference Numbers of Cations is Unity in the Double Oxide AB_2O_4

Recently the double oxides as solid electrolytes in oxygen concentration cells have been the subject of considerable interest. In the following discussion it is assumed that only cations in the double oxide AB_2O_4 contribute to electrical conduction. The sum of the transference numbers of A^{2+}- and B^{3+}-ions is taken to be unity:

$$t_{A^{2+}} + t_{B^{3+}} = 1.$$

This assumption seems to be justified in the case of some spinels and related compounds which do not contain transition metal ions. Some details are discussed in Section III.D.

An oxygen concentration cell may be set up with an homogeneous electrolyte AB_2O_4. If a certain amount of electrical charge is passed through the cell in order to test its reversibility, the double oxide will partially decompose into its constituent oxides AO and B_2O_3. The formation of AO or B_2O_3 on the right or left side of the electrolyte depends on the polarity of the applied voltage and on the magnitude of the individual transference numbers. To evaluate the e.m.f., the following cell arrangement is considered:

$$
\begin{array}{cc}
\text{I} & \text{II}
\end{array}
$$

$$
\text{Pt} - {p_{O_2}}'\ \left|\begin{array}{c}\text{AB}_2\text{O}_{4'}\\ \text{AO}\end{array}\right|\ \text{AB}_2\text{O}_4\ \left|\begin{array}{c}\text{AB}_2\text{O}_{4'}\\ \text{B}_2\text{O}_3\end{array}\right|\ {p_{O_2}}'' - \text{Pt}
$$

It is assumed that the left-hand layer of the electrolyte contains isolated particles of pure oxide AO and the right hand layer particles of pure B_2O_3 and that consequently there are oppositely directed activity gradients of AO and B_2O_3 across the intermediate layer of homogeneous oxide AB_2O_4. The virtual half cell reactions for the transport of four Faradays through the cell and the overall reaction are then given by the following equations (Pluschkell, 1969):

II

$$
O_2({p_{O_2}}'') + 4e' + \tfrac{4}{3}\,\bar{\imath}_{\text{B}^{3+}}\,(\overrightarrow{\text{B}^{3+}}) + 2\,\bar{\imath}_{\text{A}^{2+}}\,(\overrightarrow{\text{A}^{2+}}) \rightarrow
$$

$$
2\,\bar{\imath}_{\text{A}^{2+}}\,(\text{AB}_2\text{O}_4) + (\tfrac{2}{3}\,\bar{\imath}_{\text{B}^{3+}} - 2\,\bar{\imath}_{\text{A}^{2+}})\,(\text{B}_2\text{O}_3)
$$

I

$$
(2\,\bar{\imath}_{\text{A}^{2+}} + \tfrac{2}{3}\,\bar{\imath}_{\text{B}^{3+}})\,(\text{AB}_2\text{O}_4) \rightarrow
$$

$$
\tfrac{4}{3}\,\bar{\imath}_{\text{B}^{3+}}\,(\overrightarrow{\text{B}^{3+}}) + 2\,\bar{\imath}_{\text{A}^{2+}}\,(\overrightarrow{\text{A}^{2+}}) + 2\,\bar{\imath}_{\text{A}^{2+}}\,(\text{AB}_2\text{O}_4)
$$

$$
+ (\tfrac{2}{3}\,\bar{\imath}_{\text{B}^{3+}} - 2\,\bar{\imath}_{\text{A}^{2+}})\,(\text{AO}) + 4e' + O_2({p_{O_2}}')
$$

I + II

$$
O_2({p_{O_2}}'') + (\tfrac{2}{3}\,\bar{\imath}_{\text{B}^{3+}} - 2\,\bar{\imath}_{\text{A}^{2+}})\,(\text{AB}_2\text{O}_4) \rightarrow
$$

$$
(\tfrac{2}{3}\,\bar{\imath}_{\text{B}^{3+}} - 2\,\bar{\imath}_{\text{A}^{2+}})\,(\text{AO} + \text{B}_2\text{O}_3) + O_2({p_{O_2}}')
$$

$$
\Delta F = -4\tilde{F}E = RT \ln\frac{{p_{O_2}}'}{{p_{O_2}}''} - (\tfrac{2}{3}\,\bar{\imath}_{\text{B}^{3+}} - 2\,\bar{\imath}_{\text{A}^{2+}})\,\Delta F^{\circ}_{\text{AB}_2\text{O}_4}
$$

$$
E = \frac{RT}{4\tilde{F}} \ln\frac{{p_{O_2}}''}{{p_{O_2}}'} + \frac{1}{4\tilde{F}}(\tfrac{2}{3}\,\bar{\imath}_{\text{B}^{3+}} - 2\,\bar{\imath}_{\text{A}^{2+}})\,\Delta F^{\circ}_{\text{AB}_2\text{O}_4} \tag{2}
$$

The positive sign between the two terms of formula (2) is to be replaced by a negative one, if AO is thought to be precipitated in the right layer and B_2O_3 in the left layer of the electrolyte.

The transference numbers are average values for transport over the whole homogeneity range of the double oxide. A more detailed derivation is given in the comprehensive paper of Wagner (1966).

The discussion so far has shown that the transport of one mole of oxygen gas from cell compartment II to cell compartment I is accompanied by the decomposition or formation of $(\frac{2}{3} t_{B^{3+}} - 2 t_{A^{2+}})$ moles of double oxide AB_2O_4. The e.m.f. developed depends not only on the different oxygen potentials of cell compartments I and II, but also on the transport properties of the electrolyte and its free energy of formation. As a rough estimate if it is assumed that $\Delta F_{AB_2O_4}$ is about 5 kcal mol^{-1} and $t_{B^{3+}} \sim 0$, then the second term of eqn (2) takes on a value of about 0·1 V, which is by no means a minor correction.

Thus, in order to establish well-defined conditions it is recommended that a two phase mixture, i.e., either AB_2O_4, AO or AB_2O_4, B_2O_3 be used as solid electrolyte. The finely dispersed oxide AO or B_2O_3 fixes the activities of the components throughout the electrolyte and prevents decomposition by charge transport. When this condition is fulfilled formula (2) reduces to eqn (1), since the second term vanishes.

D. Electrolytes with Substantial Electronic Conductivity

In the preceding sections only ionic conduction has been taken into account. Oxides which are usually thought to be typical ionic conductors may also exhibit substantial electronic conductivity under certain conditions of oxygen pressure and temperature (Rapp, 1967).

Wagner (1933) has derived a general formula for the e.m.f. of oxygen concentration cells:

$$E = \frac{1}{4\tilde{F}} \int_{\mu_{O_2}'}^{\mu_{O_2}''} t_{\text{ion}} \, d\mu_{O_2} \qquad (3)$$

This equation is valid for an idealized cell in which local equilibrium at the phase boundaries is virtually established. If t_{ion} is smaller than unity, an internal short circuit is present and cathodic as well as anodic reactions have finite rates. If mixed conduction is significant, such deviations from local equilibrium may be minimized by using a thick electrolyte.

Eqn (1) follows from formula (3) when $t_{\text{ion}} = t_{O^{2-}} = 1$. In the case of a mixed conductor the integration of eqn (3) is possible for the case where the partial pressure p_{O_2}'' does not differ very much from p_{O_2}'. For this condition one obtains:

$$E = \bar{t}_{\text{ion}} \frac{RT}{4\tilde{F}} \ln \frac{p_{O_2}''}{p_{O_2}'} = (1 - \bar{t}_e) \frac{RT}{4\tilde{F}} \ln \frac{p_{O_2}''}{p_{O_2}'}. \qquad (4)$$

The transference numbers of ions and electronic species are average values for the oxygen pressure range investigated. A general solution of the integral formula (3) requires the formulation of t_{ion} as a function of oxygen pressure. This problem has been solved by Schmalzried (1962, 1963) and the following sections are based on his work.

The total conductivity of a mixed conductor is the sum of the partial conductivities of ion σ_{ion}, excess electrons $\sigma_{e'}$ and electron holes $\sigma_{e\cdot}$, and hence

$$\sigma_{total} = \sigma_{ion} + \sigma_{e'} + \sigma_{e\cdot}. \tag{5}$$

From the definition of transference numbers it follows that

$$t_{ion} = \frac{\sigma_{ion}}{\sigma_{total}} = \left[1 + \frac{\sigma_{e\cdot}}{\sigma_{ion}} + \frac{\sigma_{e'}}{\sigma_{ion}} \right]^{-1}. \tag{6}$$

In order to combine the partial conductivities with oxygen partial pressure, the following models are discussed:

(a) If the chemical potential of oxygen in the surrounding gas atmosphere is decreased, oxygen is removed from the lattice of the oxide. By this reaction, oxygen vacancies $V_O^{\cdot\cdot}$ and excess electrons e' may be formed:

$$O_O \rightleftarrows V_O^{\cdot\cdot} + 2e' + \tfrac{1}{2}O_2(p_{O_2}). \tag{7}$$

If equilibrium between the gas phase and the oxide is established and there are no interactions between the charged defects, the mass action law applied to reaction (7) yields in terms of concentrations instead of activities:

$$(e') = K_1 (V_O^{\cdot\cdot})^{-\frac{1}{2}} p_{O_2}^{-\frac{1}{4}}. \tag{8}$$

Two different cases may be considered:

(i) According to reaction (7), the concentration of oxygen vacancies is one half of the concentration of electrons. This is only true, if the concentrations of all other defects are sufficiently small. With this condition, it follows from eqn (8):

$$(e') = K_1' p_{O_2}^{-1/6}. \tag{9}$$

In general, the mobility of electrons exceeds by far that of the ionic species. The transference number of electrons is therefore close to unity in an oxide with equivalent concentrations of oxygen vacancies and excess electrons (n-type semi-conductor). Such electrolytes are not useful for purposes of galvanic cell measurements.

(ii) In some solid solutions (stabilized zirconia and thoria base solutions) the concentration of oxygen vacancies can be fixed at a comparatively high level by chemical doping. The conductivity of the electrolyte is then predominantly ionic in character. Only at very small oxygen pressures do electrons also contribute to the electrical conduction. Since the concentration of oxygen vacancies is fixed by doping, eqn (8) reduces to

$$(e') = K_1'' p_{O_2}^{-\frac{1}{4}}. \tag{10}$$

The partial conductivity $\sigma_{e'}$ is proportional to the concentration and to the mobility of excess electrons. If it is assumed that the mobility $u_{e'}$ is not a function of oxygen pressure it follows that

$$\sigma_{e'} = K_1'' \tilde{F} u_{e'} p_{O_2}^{-\frac{1}{4}}. \tag{11}$$

(b) The other limiting case is that in which at large oxygen potentials in the gas phase an excess of oxygen is accommodated in the lattice of the oxide. The defects produced are interstitial oxygen ions O_i'' or cation vacancies V_{Me}''. To maintain electrical neutrality electron holes are formed:

$$\tfrac{1}{2} O_2(p_{O_2}) \rightleftarrows O_i'' + 2\,e^{\bullet} \tag{12}$$

or

$$\tfrac{1}{2} O_2(p_{O_2}) \rightleftarrows V_{Me}'' + 2\,e^{\bullet} + \text{MeO}. \tag{13}$$

When the relations $(O_i'') = \tfrac{1}{2}(e^{\bullet})$ or $(V_{Me}'') = \tfrac{1}{2}(e^{\bullet})$ are valid, the concentration of electron holes is proportional to $p_{O_2}^{1/6}$ in both cases. By the same argument as in section (a) (i), the electrolyte is then predominantly a p-type semiconductor and thus unsuited for oxygen concentration cells.

If the concentration of oxygen vacancies of the electrolyte is large as a result of chemical doping, the free lattice sites are occupied according to the reaction

$$\tfrac{1}{2} O_2(p_{O_2}) + V_O^{\bullet\bullet} \rightleftarrows O_O + 2\,e^{\bullet}. \tag{14}$$

Since $(V_O^{\bullet\bullet})$ is nearly constant, one then obtains

$$(e^{\bullet}) = K_2'' p_{O_2}^{\frac{1}{4}}. \tag{15}$$

The partial conductivity by electron holes is described by

$$\sigma_{e^{\bullet}} = K_2'' \tilde{F} u_{e^{\bullet}} p_{O_2}^{\frac{1}{4}} \tag{16}$$

when the same procedure and suppositions as that of the preceding section are used.

Eqn (6) may be expressed in terms of the oxygen partial pressures. If the special values of p_{O_2} at which the ionic conductivity of the oxide equals the conductivity by excess electrons or electron holes are designated by $p_{e'}$ and $p_{e^{\bullet}}$, respectively, the insertion of eqns (11) and (16) into (6) gives

$$t_{ion} = \left[1 + \left(\frac{p_{O_2}}{p_{e^{\bullet}}} \right)^{\frac{1}{4}} + \left(\frac{p_{O_2}}{p_{e'}} \right)^{-\frac{1}{4}} \right]^{-1}. \tag{17}$$

It should be mentioned that in certain electrolytes only electron or defect electron conductivity comes into play at a given oxygen pressure. One of the terms of eqn (17) is then to be omitted. In Fig. 2 the ionic transference number is shown schematically as a function of oxygen pressure. At $p_{e'}$ and $p_{e\cdot}$ the ionic transference number is equal to 0·5. From the figure t_{ion} is unity over a wide range of oxygen partial pressures only if $p_{e'}$ is much smaller than $p_{e\cdot}$.

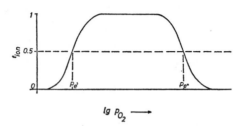

FIG. 2. Ionic transference number in a mixed conducting oxide as a function of oxygen partial pressure.

The insertion of (17) into (3) followed by integration gives

$$E = \frac{RT}{\tilde{F}} \left[\ln \frac{p_{e\cdot}^{\frac{1}{4}} + p_{O_2}'^{\frac{1}{4}}}{p_{e\cdot}^{\frac{1}{4}} + p_{O_2}''^{\frac{1}{4}}} + \ln \frac{p_{e'}^{\frac{1}{4}} + p_{O_2}''^{\frac{1}{4}}}{p_{e'}^{\frac{1}{4}} + p_{O_2}'^{\frac{1}{4}}} \right] \tag{18}$$

provided that $p_{e'} \ll p_{e\cdot}$.

In the following section, special limiting cases which arise from eqn (18) are discussed.

(a) If the sequence of oxygen partial pressures is

$$p_{O_2}'' > p_{O_2}' \gg p_{e\cdot} \gg p_{e'}$$

or

$$p_{e\cdot} \gg p_{e'} \gg p_{O_2}'' > p_{O_2}'$$

the e.m.f. according to eqn (18) is zero. In the first case the electrolyte behaves like an electron hole, in the second case like an excess electron-semiconductor.

(b) The condition

$$p_{e\cdot} \gg p_{O_2}'' > p_{O_2}' \gg p_{e'}$$

leads to the important eqn (1). The tuning between the range of measurements and the electrolyte selected is optimal since the transference number t_{ion} is always unity.

(c) If the sequence of p_{O_2}' and $p_{e'}$ in example (b) is exchanged (since p_{O_2}' becomes smaller than $p_{e'}$), the e.m.f. according to eqn (18) has a constant value which depends only on the reference pressure p_{O_2}'' and is given by

$$E = \frac{RT}{4\tilde{F}} \ln \frac{p_{O_2}''}{p_{e'}}. \tag{19}$$

The linear dependence of the e.m.f. on $\ln p_{O_2}$ with decreasing p_{O_2}' (case b) and the asymtotic approach to the value given by eqn (19) for case c is shown in Fig. 3. The e.m.f. cannot be zero since a thin layer of the electrolyte always remains as an ionic conductor.

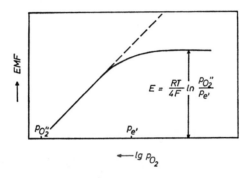

FIG. 3. The e.m.f. of an oxygen concentration cell which has an electrolyte that exhibits n-type semiconduction at low oxygen pressure.

(d) In the case

$$p_{O_2}'' \gg p_{e\bullet} \gg p_{O_2}' \gg p_{e'}$$

electron holes influence the measurements. The e.m.f. described by

$$E = \frac{RT}{4\tilde{F}} \ln \frac{p_{e\bullet}}{p_{O_2}'}. \tag{20}$$

as a function of $\ln p_{O_2}$ is given by a straight line parallel to the curve which describes the "ideal" relation (1). It is shifted to smaller values as shown in Fig. 4. If p_{O_2}' and $p_{e'}$ are interchanged so that

$$p_{O_2}'' \gg p_{e\bullet} \gg p_{e'} \gg p_{O_2}'$$

the e.m.f. is again constant and given by

$$E = \frac{RT}{4\tilde{F}} \ln \frac{p_{e\bullet}}{p_{e'}}. \tag{21}$$

It may be remarked here that the formulae of this section have been derived for some simple and very special situations. Care should therefore be exercised whenever these equations are used to interpret experimental results.

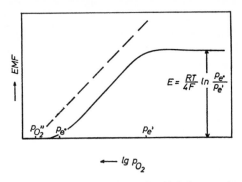

The equation shown in the figure:

$$E = \frac{RT}{4F} \ln \frac{P_{e'}}{P_{e'}}$$

$\longleftarrow lg\, P_{O_2}$

FIG. 4. The e.m.f. of an oxygen concentration cell which has an electrolyte that exhibits p-type semiconduction at high and n-type semiconduction at low oxygen pressures.

III. Special Electrolytes

A. Stabilized Zirconia

On heating, pure zirconia (ZrO_2) exhibits a phase transformation from a monoclinic to a tetragonal structure which is accompanied by a volume shrinkage (Ryshkewitch, 1960) of about 9%. The transformation seems to be of the athermal diffusionless type with a pronounced hysteresis between 950 and 1200°C (Fehrenbacher and Jacobsen, 1965).

In certain concentration ranges, solid solutions of ZrO_2 with CaO, MgO or Y_2O_3 are cubic and have a CaF_2-type lattice structure. These solutions are called "stabilized", since the additives decrease the rate of phase transformation. The concentration range of the cubic phase in the ZrO_2–CaO system is from 10 to 20 mol % CaO at 1200°C according to Hund (1952). Nearly the same values have been reported by Carter and Roth (1967) and by Tien and Subbarao (1963) for 1400°C and Cocco (1959) for 1600°C. Duwez et al. (1959) have reported that the solubility limits at 16 and 20 mol % CaO are temperature independent. Although special regions of the phase diagrams of zirconia with other oxides have been intensively investigated, most of the systems and phase relations are not as yet well known.

The cubic solutions of ZrO_2 with 10–20 mol % CaO accommodate the valence difference of the cations according to the following reaction:

$$(CaO) \rightarrow Ca_{Zr}'' + V_O^{\cdot\cdot} + ZrO_2 \tag{22}$$

Ca_{Zr}'' indicates a Ca^{2+} ion on a Zr^{4+} ion lattice site, the charge difference being -2. An equal amount of oxygen vacancies $V_O^{\bullet\bullet}$ with a charge of $+2$ are formed to maintain electrical neutrality. Equations similar to (22) may be written down for the dissolution of MgO or Y_2O_3. Oxygen vacancies were suggested by Wagner (1943) in order to explain the abnormal high ionic conductivity of these solutions which did not show any detectable decomposition of the electrolyte. Hund (1952) was able to verify this suggestion by comparing X-ray- and pyknometer-densities with density-computations based on the dissolution reaction (22).

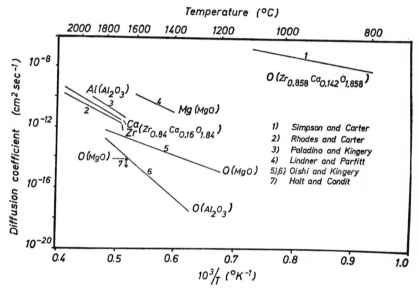

FIG. 5. Component diffusion coefficients in MgO, Al_2O_3 and stabilized zirconia.

The component–diffusion coefficients of oxygen (Simpson and Carter, 1966), zirconium and calcium (Rhodes and Carter, 1966) in stabilized zirconia as a function of temperature are shown in Fig. 5. Since the cations are relatively immobile, oxygen ions are very rapidly transported via oxygen vacancies introduced by the CaO doping. The electrical conductivity of stabilized zirconia together with other oxides is plotted in Fig. 6 for comparison. Carter and Roth (1967) measured the conductivity of the cubic phase as a function of CaO content and found a flat maximum at about 14 mol % CaO. The effect of decreasing ionic conductivity with further increasing dopant concentration is not clearly understood. It has been attributed to vacancy ordering (Tien and Subbarao, 1963) or vacancy clustering (Strickler and Carlsson, 1965) or to interactions between oppositely charged

552   K. Goto and W. Pluschkell

defects (Kroger, 1964, 1966). It has been further observed that the activation energy increases with increasing CaO content according to $Q = 6640\,(1 + 22 \cdot 9\,N_{CaO})$ cal/mol (Carter and Roth, 1967). Tien and Subbarao (1963) have explained this effect by assuming that the large Ca^{2+} ions impede the motion of oxygen ions.

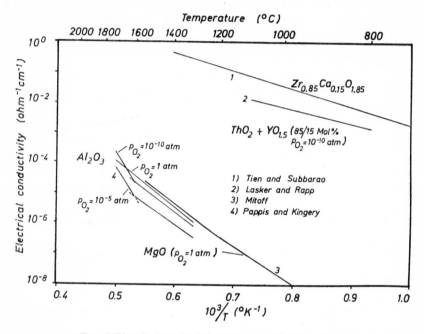

FIG. 6. Electrical conductivity of some oxide electrolytes.

Kingery and coworkers (1959) converted their oxygen diffusion data into conductivity units with the aid of the Nernst–Einstein equation (Lidiard, 1957):

$$\sigma = \frac{n_i\, z_i^2\, e^2}{f\, t_i\, kT}\, D_{i,\,Tr} \tag{23}$$

(σ total conductivity, n_i concentration, z_i valence, t_i transference number, $D_{i,\,Tr}$ tracer diffusion coefficient of species i; f correlation factor, in this special case $0 \cdot 65$). Since the computed values agreed fairly well with the measured conductivity and activation energy provided that t_i was taken as unity, Kingery et al. (1959) concluded that oxygen is the only mobile species. The fit would have been better if the correlation factor had not been neglected. It is now generally accepted that the transference number of oxygen in stabilized zirconia is substantially unity. The simple relation (1) is therefore valid for the e.m.f. of oxygen concentration cells set up with this electrolyte. With air as a reference gas the e.m.f. of the cell is described by eqn (1) with decreasing p_{O_2}'. At very low oxygen pressures an increasing departure is observed. This is due to the presence of excess electrons in one side layer of the electrolyte as discussed in Section II.D.c. It is thus of the utmost importance to determine the special oxygen partial pressure $P_{e'}$ as a function of temperature, since its magnitude determines the limits of accurate performance of the cell. Figure 7 shows the estimates made by various investigators.

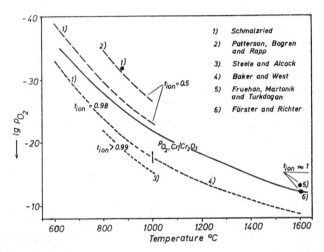

FIG. 7. Ionic transference numbers in $Zr_{0.85}Ca_{0.15}O_{1.85}$ as a function of temperature and oxygen partial pressure.

Schmalzried (1963) performed e.m.f.-measurements on the cell combination

$$Pt - Ca, CaO \mid stab\ ZrO_2 \mid air - Pt$$

for the temperature range 600–1000°C. Since p_{O_2}' of the Ca, CaO-equilibrium and p_{O_2}'' are known and because it was assumed that there is no substantial error when $p_{e\bullet} \gg 0.21$ atm, the value of $p_{e'}$ could be computed from eqn (18) and the measured e.m.f. Fig. 7 represents $p_{e'}$ for which $t_{ion} = 0.5$, and 0.98 as a function of temperature. The results of Steele and Alcock (1965) are in fair agreement; those of Baker and West (1966) fit closely at higher temperatures.

Rapp and coworkers (Patterson et al., 1967) determined $p_{e'}$ by conductivity measurements for equal oxygen pressures on both sides of the electrolyte. The results are likely to be reliable since in this method absolute gas-tightness of the electrolyte is not necessary. The close agreement with the value determined by Schmalzried (1963) for the cell

$$Pt - Ca, CaO \mid stab\ ZrO_2 \mid Ni, NiO-Pt$$

should thus be emphasized. Obviously very different oxygen pressures applied to the cell can cause substantial errors owing to oxygen transfer (see Section IV.D). The same tendency is exhibited in the measurements of Baker and West (1966) who determined $p_{O_2} = 10^{-8.4}$ atm at 1600°C for the "onset" of electronic conduction on a cell with H_2O/H_2 gas electrodes. Turkdogan and coworkers (Fruehan $et\ al.$, 1969) and Förster and Richter (1969) have reported that $t_{ion} = 1.0$ at oxygen partial pressures as low as 3×10^{-13} and 10^{-12} atm at the same temperature. In both cases Cr/Cr_2O_3 mixtures were used to fix p_{O_2}''. Recently, Steiner (1969) measured the conductivity due to electrons and electron holes in zirconia and thoria base solid solutions by an improved polarization technique (Hebb, 1952; Wagner, 1957). He reported $p_{e'} = 1.4 \times 10^{-28}$ atm for $ZrO_2 + 16.5$ mol % CaO and $p_{e'} = 5.8 \times 10^{-31}$ atm for $Zr_{0.9}Y_{0.1}O_{1.95}$ at 1000°C. The first value is in fair agreement with the results of Rapp and coworkers (Patterson $et\ al.$, 1967).

It is not unreasonable in the case of stabilized zirconia that

$$(Zr_{0.85}Ca_{0.15}O_{1.85})\ t_{ion} > 0.98,$$

when P_{O_2}' is larger than the oxygen partial pressure given by the Cr/Cr_2O_3-equilibrium (Schwerdtfeger, 1967; Tretjakow and Schmalzried, 1965). According to Kröger (1966) relatively small concentrations of impurities may seriously increase the electronic conductivity of the electrolyte. This possibility should be taken into account when materials of different origin are compared.

B. Thoria Base Solid Solutions

The lattice structure of pure thoria ThO_2 is of the CaF_2-type at all temperatures up to the melting point of about 3300°C. Conductivity measurements under various oxygen pressures and experiments on oxygen concentration cells by Lasker and Rapp (1966) have revealed that pure ThO_2 is a typical mixed conductor. At 1000°C and oxygen pressures greater than 10^{-5} atm, the transference number $t_{e\cdot}$ exceeds 0.5. At $p_{O_2} < 10^{-19}$ atm the onset of excess electron conductivity has been observed. In the intermediate range, the charge transport is predominantly by ions. Since a $\sigma_{e\cdot} \propto p_{O_2}^{\frac{1}{4}}$-relation has been found (Bauerle, 1966; Lasker and Rapp, 1966; Rudolph, 1959; Steiner, 1969), the pure substance is characterized by a relatively high concentration of oxygen defects as discussed in Section II.D. It is not quite clear whether the oxygen defects are produced by an intrinsic equilibrium reaction or are introduced by small amounts of aliovalent impurities. The latter possibility seems to be the more likely (Bauerle, 1966; Lasker and Rapp, 1966; Rapp, 1967).

Solid solutions of ThO_2 with Y_2O_3, La_2O_3 or CaO compensate the valence difference of the cations by the formation of oxygen vacancies (Hund and Dürrwächter, 1951; Hund and Mezger, 1952; Subbarao et al., 1965). The dissolution reaction may be formulated in a similar fashion to that given by eqn (22):

$$(Y_2O_3) \rightarrow 2\,Y_{Th}{}' + V_O^{\bullet\bullet} + 2(ThO_2) \tag{24}$$

The fluorite-type structure is stable up to 20–25 mol % $YO_{1.5}$ at 1400°C and up to 45–50 mol % $YO_{1.5}$ at 2200°C (Subbarao et al., 1965).

The large amount of oxygen vacancies introduced by the dopant effectively increases the ionic conductivity. For this reason and as shown in Fig. 8 from the investigation of Lasker and Rapp (1966) the oxygen pressure range with ionic transference numbers equal to unity is broadened.

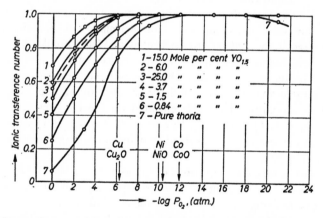

FIG. 8. Ionic transference numbers in pure thoria and thoria base solid solutions with yttria as a function of oxygen partial pressure at 1000°C (after Lasker and Rapp, 1966).

In summary the following conclusions can be drawn:

(1) At relatively high p_{O_2} the ThO_2–$YO_{1.5}$ electrolytes exhibit p-type semiconduction proportional to $p_{O_2}^{\frac{1}{4}}$ which is in agreement with the dilute solution model developed in Section II.D. The reference pressure p_{O_2}'' of galvanic cells should not be fixed in this range. Oxygen pressures corresponding to the Co, CoO-equilibrium guarantee ionic transference numbers very close to unity.

(2) The ionic conductivity of the $ThO_2-YO_{1.5}$ solid solutions exhibits a flat maximum at about 15 mol % $YO_{1.5}$ (Lidiard, 1957; Rapp, 1967). This behaviour is similar to that found for stabilized zirconia and indicates a pronounced interaction between the ionic defects.

(3) The observation that no conductivity contribution by excess electrons could be detected in $ThO_2-YO_{1.5}$ solutions down to oxygen pressures of the Cr, Cr_2O_3 equilibrium is of particular importance. Rapp and coworkers (Patterson *et al.*, 1967) estimate from polarization measurements the following maximum oxygen pressure values for which $t_{ion} > 0.99$ in $Th_{0.85} \times Y_{0.15}O_{1.925}$: $10^{-34.3}$ atm at 1000°C, $10^{-39.5}$ atm at 900°C, and $10^{-44.7}$ atm at 800°C. In oxygen concentration cells thoria base solid solutions are obviously superior to stabilized zirconia provided that the oxygen pressures are extremely low.

C. Oxides of the type A_mO_n

Solid electrolytes based on zirconia or thoria have been most widely used in oxygen concentration cells for high accuracy thermochemical measurements. These materials nevertheless exhibit some inherent disadvantages: First, they are subjected to an awkward thermal shock sensitivity. This is a disadvantage if the cell has to be heated up very rapidly. Second, they become *n*- or *p*-type semiconductors in certain ranges of oxygen partial pressures. For these reasons there have been some attempts to use other more favourable materials as solid electrolytes in oxygen concentration cells. The investigations have concentrated on the commercially available oxides of MgO, Al_2O_3 and SiO_2. In Section II.B it was shown that relation (1) is valid for these oxides provided that the ionic transference number is equal to unity.

Since the electronic configurations of the ions Al^{3+}, Mg^{2+}, Si^{4+} and O^{2-} correspond to rare gases, the oxides are expected to be ionic conductors. Optical measurements have provided evidence that the energy gap between the valence- and the conduction-band is very large and is about 7.5 eV in the case of MgO (Nelson, 1955). According to theoretical considerations of Yamashita and Kurosawa (1954), ionic defects of the Schottky type are more likely in alkali-earth-oxides than Frenkel defects. The free energy of formation of a Schottky pair in MgO lies between 4 and 6 eV.

These data indicate that the concentrations of electronic and also of ionic defects are small in the pure oxides. Hence, as shown in Fig. 6, the electrical conductivity of MgO and also Al_2O_3 is approximately 10^6 times smaller than that of zirconia or thoria base solid solutions. Because of the high internal resistance galvanic cells made with these oxides are particular sensitive to the stray fields of the furnace used. It is also evident that the transport properties are extremely sensitive to small amounts of impurities which may act as donors or acceptors in the matrix and so cause transference numbers which cannot be controlled.

Al_2O_3

Numerous investigations on the electrical conductivity of alumina are summarized in the paper of Pappis and Kingery (1961). The results compiled scatter by some powers of ten, showing clearly that extrinsic effects have masked the properties of the pure substance. The measurements of Pappis and Kingery on Al_2O_3-single crystals seem to be the most reliable at present and are included in Fig. 6. The conductivity is a function of oxygen pressure and has an almost temperature-independent minimum at about 10^{-5} atm. Pappis and Kingery interpret this behaviour as being one in which, at low p_{O_2} values, excess electrons and oxygen vacancies or interstitial aluminium ions and, at high p_{O_2}, electron holes together with aluminium vacancies are formed. This results in n- or p-type semiconduction at the low and high p_{O_2} values respectively. For temperatures esceeding 1600°C intrinsic electronic conductivity is observed with an activation energy of about 5·5 eV. According to this interpretation the energy gap between valence- and conduction-band should be 11eV which is not unreasonable.

Diffusion experiments on Al_2O_3 have shown as in Fig. 5 that the mobility for aluminium ions (Paladino and Kingery, 1962) is larger than for oxygen ions (Oishi and Kingery, 1960). The application of the Nernst–Einstein relation (23) to the diffusion- and conductivity-data gives $t_{Al}3 + = 0·03$ for the transference number of aluminium ions in Al_2O_3 (Paladino and Coble, 1963). This value is in fair agreement with some direct measurements (Fischer and Achermann, 1966; Kingery and Meiling, 1961). Several investigators have tested the use of alumina as a useful solid electrolyte in oxygen concentration cells by measuring the e.m.f. of alumina cells with fixed and known oxygen pressures p_{O_2}' and p_{O_2}'' and computing t_{ion} with the help of eqn (4) (Baker and West, 1966; Catoul et al., 1967; Fischer and Ackermann, 1968; Fischer and Janke, 1969; Schmalzried, 1963).

On the basis of the preceding discussion, one is led to expect considerable smaller e.m.f. values than those predicted by eqn (1). Above 1300°C this has always been the case. The reported values are however confused. Since Schmalzried (1963) and Matsumura (1966) reported an almost wholly electronic conductivity above 1300°C, other investigators (Baker and West, 1966; Catoul et al., 1967; Fischer and Ackermann, 1968) have found values of t_{ion} between 0·6 and 0·9. The extensive study of Fischer and Janke (1969a) shows that $t_{ion} \sim 1·0$, if $p_{O_2} > 3 \times 10^{-16}$ atm at 1000°C and $p_{O_2} > 10^{-13}$ atm at 1200°C. At 1600°C they measured $t_{ion} = 0·56$ unaffected by oxygen pressure.

Since large ionic transference numbers are always measured on poly-crystalline alumina, it may be concluded that those grain boundaries which have a quasi-supercooled liquid structure are favourable paths for enhanced diffusion of oxygen- or aluminium ions. On the whole, alumina is not a useful electrolyte in oxygen concentration cells, since its transport properties are strongly influenced by so many factors which are not easy to control.

MgO

As shown in Fig. 5 the diffusion coefficient of magnesium (Lindner and Parfitt, 1957) in single crystals of MgO is roughly three powers of ten larger than that of oxygen (Oishi and Kingery, 1960). This difference is still larger, if the results of Holt and Condit (1966) are taken into account. These in-vestigators carried out diffusion runs using an O^{18} exchange-technique and afterwards bombarded the bevelled specimens with protons so as to induce the nuclear reaction $O^{18}\,(p, n)\,F^{18}$. The positron radiation of F^{18} can be made visible by autoradiography. The experiments revealed that the oxygen transport in bulk material is slower than hitherto reported. Along grain boundaries and dislocations the mobility of oxygen is enormously acceler-ated. Second phase impurities deposited along grain boundaries aided the oxygen transport.

The electrical properties of MgO single crystals were investigated by Mitoff (1959, 1962), who used an oxygen concentration cell to evaluate the transference numbers of ions and electrons. Measurements on the cell

$$Pt - air \mid MgO \mid p_{O_2}'' = 1\ atm - Pt$$

showed that the transference number t_{ion} is nearly unity at 1000°C. This result is in agreement with that of Schmalzried (1960) who obtained the same value at 1100°C. According to Mitoff, the transference number t_{ion} decreases with increasing temperature and reaches about 0·1 at 1500°C. MgO seems therefore to be a predominant ionic conductor at low and a pre-dominant electronic conductor at high temperatures as long as the impurity concentration is small. The conductivity of MgO as a function of temperature shows a slight curvature as revealed in Fig. 6. This is a consequence of the different activation energies of the two processes.

Mitoff has moreover stated that the conductivity is also a function of oxygen partial pressure and exhibits a minimum at about 10^{-5} atm. This minimum becomes flatter and is shifted to lower conductivity values when the concentration of aliovalent impurity ions like iron is reduced. According to Baker and West (1966) MgO is a mixed conductor with a transference number t_{ion} of 0·75 at 1300°C ($p_{O_2} = 4 \times 10^{-13}$ atm) and 0·35 at 1550°C ($p_{O_2} = 10^{-11}$ atm). In marked contrast to these statements several investigators have reported ionic transference numbers near to unity up to temperatures as high as 1600°C (Luzgin *et al.*, 1963; Ohtani and Sanbongi, 1963; Palguer and Neuimin, 1962; Sanbongi *et al.*, 1966). Fischer and Janke (1969b) tested a mixture of MgO with 4·5 % SiO_2. If SiO_2 is dissolved into MgO to some extent, possibly by the reaction

$$(SiO_2) \rightarrow Si_{Mg}^{\cdot\cdot} + V_{Mg}'' + 2(MgO), \qquad (25)$$

the ionic conductivity should be increased due to chemically induced vacancies in the Mg^{2+}-sublattice. However the solubility seems to be small and most of the SiO_2 is incorporated into the coexisting phase forsterite, Mg_2SiO_4. The measurements of Fischer and Janke (1969b) show that electronic conductivity lowers the e.m.f. when the pressure is smaller than 10^{-11} atm at 1600°C.

Large ionic transference numbers similar to those of alumina are measured when the electrolyte consists of a fine grained sinter disc. In this case transport along grain boundaries makes a considerable contribution to the total conductivity. Moreover impurities with different, but stable valence may introduce point defects into the matrix and so increase the ionic part of the conductivity. In this case the e.m.f. is also influenced by the transference numbers of the individual ionic species. In summary, MgO is not a suitable electrolyte for accurate oxygen potential measurements but may be helpful for qualitative experiments on high temperature systems (Steele, 1968).

SiO_2 *and rare earth metal oxides*

According to Schmalzried (1963), SiO_2 is a pure ionic conductor between $p_{O_2} = 1$ atm and $p_{O_2} = 10^{-16}$ atm at 1000°C. The material does not seem to be very useful in oxygen concentration cells since its resistance is very large and it readily reacts with other substances at higher temperatures. Furthermore, at low p_{O_2} the reduction to SiO gas comes into play.

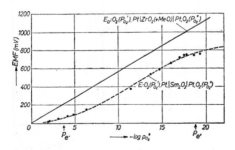

FIG. 9. E.m.f. of an oxygen concentration cell with Sm_2O_3 as solid electrolyte at 845°C (after Tare and Schmalzried, 1964).

The conduction mechanism of rare earth metal oxides was investigated by Noddack and Walch (1959). They reported complete electronic conductivity. But e.m.f. measurements of Schmalzried (1963) and Tare and Schmalzried (1964) on Nd_2O_3, Sm_2O_3, Gd_2O_3, Dy_2O_3, Yb_2O_3, Sc_2O_3 and Y_2O_3 as solid electrolytes showed that there exist wide ranges of oxygen pressures for which the ionic transference number of these oxides is substantially unity. Their results on Sm_2O_3 are given as an example in Fig. 9. The dotted line of the figure corresponds to the theoretical considerations of Section II.D with $p_{e'} = 10^{-19.2}$ atm and $p_{e^{\bullet}} = 1$ atm.

D. Double Oxides

As discussed in the foregoing section, binary oxides like MgO, Al_2O_3 and SiO_2 are not very useful in oxygen concentration cells since their electrical conductivity is extremely low. Conditions are much more favourable in the case of some ternary oxide compounds which exhibit a wide homogeneity range indicative of a marked disorder in the lattice. In Fig. 10 the conductivity of forsterite Mg_2SiO_4 is shown as function of composition and temperature (Pluschkell and Engell, 1965). The stoichiometric compound dissolves an excess of about 1 mol % SiO_2, possibly by the formation of magnesium ion vacancies and silicon ions on interstitial positions. The chemi-

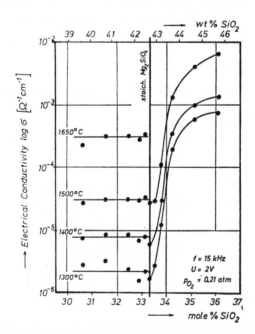

Fig. 10. Electrical conductivity of forsterite Mg_2SiO_4. (after Pluschkell and Engell, 1968).

cally induced disorder results in an increase of the conductivity by roughly a factor of 250.

In the ternary compound AB_2O_4 an excess of the constituent oxide B_2O_3 may be dissolved for instance as follows:

$$(B_2O_3) \rightarrow V_A'' + V_O^{\cdot\cdot} + (AB_2O_4) \tag{26}$$

In this case, the transference number of oxygen cannot be expected to be unity since A^{2+} ions via vacancies can also contribute to charge transport. Moreover, Schmalzried and Tretjakow (1966) have shown that in a number of spinels and related compounds the oxygen sublattice is nearly the same throughout the homogeneity range of the compound. The dissolution of a constituent oxide produces point defects mainly in the cationic sublattice such that the foregoing reaction may be written as

$$4(B_2O_3) \rightarrow 2B_A^{\cdot} + V_A'' + 3(AB_2O_4) \tag{27}$$

or

$$4(B_2O_3) \rightarrow 2B_1^{\cdot\cdot\cdot} + 3\,V_A'' + 3(AB_2O_4) \tag{28}$$

In any case the conductivity is mainly determined by the cations since the oxygen ions are fixed.

Whether there is also a certain contribution by oxygen ions in complicated silicate structures is not well established (Scholze, 1955). But if the oxygen ions are closely packed, the assumption

$$t_A2+ + t_B3+ = 1 \tag{29}$$

seems to be justified. As in Section II.C, it is recommended that a heterogeneous mixture of AB_2O_4 with one of the co-existing phases, possibly AO or B_2O_3 be used so as to get reproducible results in e.m.f. measurements.

Spinel ($MgAl_2O_4$)

Hans and coworkers (Catoul et al., 1967) performed some successful measurements on an oxygen concentration cell of the type

$$Fe - [O]_{Fe} \,|\, MgAl_2O_4 \,|\, air - Pt$$

They claim sufficient stability against thermal shock and corrosion resistance of the spinel used.

The following cell combinations have been investigated in detail by Fischer and Janke (1969b):

$$Ir - H_2O/H_2 \,|\, MgAl_2O_4 \,|\, air - PtRh \quad (1100\text{--}1700°C)$$

$$PtRh - [O]_{Fe} \,|\, MgAl_2O_4 \,|\, air - PtRh \quad (1600°C)$$

The composition of the electrolyte used was obviously within the homogeneity range of the spinel (Muan and Osborn, 1965). The e.m.f. measured was a little low in comparison with that given by eqn (1). Fischer and Janke conclude that there should be a considerable electronic conductivity when $p_{O_2}' < 10^{-10}$ atm at 1600°C. A direct proof by conductivity measurements has not as yet been given.

Forsterite (Mg$_2$ SiO$_4$)

In Fig. 11 the results of e.m.f. measurements at 1400°C with forsterite as solid electrolyte are shown (Pluschkell, 1969). In order to fix the activities of MgO and SiO$_2$, the electrolyte consisted in the first case of a heterogeneous mixture of Mg$_2$ SiO$_4$ + MgO, in the second case of Mg$_2$ SiO$_4$ + MgSiO$_3$. The comparison between eqn (1) and the measured values reveals that forsterite is predominantly an ionic conductor. In the case of an excess of MgO the e.m.f. exhibits an increasing deviation when p_{O_2}' becomes smaller than about 16^{-8} atm. Since the conductivity of specimens of the same composition increases with decreasing oxygen pressure the effect can be attributed to excess electrons (Pluschkell and Engell, 1968).

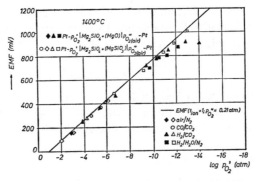

FIG. 11. e.m.f. of an oxygen concentration cell with forsterite Mg$_2$SiO$_4$ as solid electrolyte. (after Pluschkell, 1969).

Mullite (3Al$_2$O$_3$. 2SiO$_2$)

Porcelain when heated to high temperatures consists mainly of the silicate phase mullite. Haber (1908), Treadwell (1916) and Treadwell and Terebesi (1933) have used this material in oxygen concentration cells and have reported a satisfactory performance.

E.m.f. measurements of Schmalzried (1963) at 1000°C resulted in values about 40 mV too low in comparison with those calculated from eqn (1). The extent of the homogeneity range of mullite is still a matter of controversy and it is thus not possible to decide if the electrolyte used was heterogeneous or not.

Marincek and coworkers (Schuh *et al.*, 1968) conclude from their results that the transference number of oxygen is unity at 1450°C and $p_{O_2}' > 10^{-12}$ atm. According to Fischer and Janke (1969a) the electrolyte should be heated up to 1500°C prior to measurements in order to get reproducible results. As a result of this heat treatment the unstable sillimanite is decomposed. Figure 12 from the publication of Fischer and Janke (1969a) shows quite clearly that mullite is superior to stabilized zirconia with regard to electronic conductivity. Successful e.m.f. measurements with this electrolyte have also been performed by Löscher (1969, 1970).

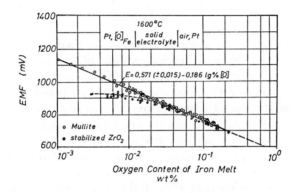

FIG. 12. Oxygen potential in liquid iron at 1600°C. (after Fischer and Janke, 1969a).

Although the influence of reduction to SiO gas on the e.m.f. has until now not been investigated, mullite seems to be a promising material for oxygen concentration cells up to steel-making temperatures.

IV. Possible Experimental Errors in Electromotive Force Measurements

A. Temperature Gradients

In a cell with a solid electrolyte two temperature gradients should be taken into account. One is a temperature gradient across the solid electrolyte, and the other is a temperature difference at the interface itself when the contact area between the metal and the electrolyte is large. The non-isothermal e.m.f. of an oxygen concentration cell with oxygen partial pressures of

p_{O_2}'' and p_{O_2}' at the two electrodes with temperatures of T_2 and T_1 is given (Wagner, 1970) by:

$$E(T_2, p_{O_2}''; T_1, p_{O_2}') = \frac{1}{4\tilde{F}} [\mu_{O_2}^\circ(T_2) - \mu_{O_2}^\circ(T_1)]$$

$$+ \frac{1}{4\tilde{F}} [RT_2 \ln p_{O_2}'' - RT_1 \ln p_{O_2}']$$

$$+ \frac{1}{2\tilde{F}} \int_{T_1}^{T_2} \left(\bar{S}_{O^{2-}}^{(ZrO_2)} + \frac{Q_{O^{2-}}^{*(ZrO_2)}}{T} \right) dT$$

$$- \frac{1}{\tilde{F}} \int_{T_1}^{T_2} \left(\bar{S}_{e'}^{(Pt)} + \frac{Q_{e'}^{*(Pt)}}{T} \right) dT \qquad (30a)$$

In this equation, $\mu_{O_2}^\circ$ is the standard chemical potential of oxygen at 1 atm, $\bar{S}_{O^{2-}}^{(ZrO_2)}$ and $\bar{S}_{e'}^{(Pt)}$ are the partial molar entropies of O^{2-} ions in $ZrO_2 \cdot CaO$ and of electrons in platinum, respectively. $Q_{O^{2-}}^{*(ZrO_2)}$ and $Q_{e'}^{*(Pt)}$ are heats of transfer of O^{2-} ions in $ZrO_2 \cdot CaO$ and of electrons in platinum, respectively.

The first and the second terms are oxygen chemical potential differences divided by $4\tilde{F}$, while the third and the last terms are characteristic properties of the solid electrolyte and the electrode metals.

Because the partial molar entropies and the heats of transfer of oxygen anions and electrons are nearly constant over a wide range of oxygen partial pressures (Cusack and Kendall, 1958; Ruka et al., 1968) the above equation can be reduced to

$$E = \frac{1}{4\tilde{F}} [\mu_{O_2}(T_2, p_{O_2}'') - \mu_{O_2}(T_2, p_{O_2}')] + \alpha(T_2 - T_1) \qquad (30b)$$

where α is a constant related to the partial entropies and heats of transfer. The coefficient α depends on the nature of the electrolyte and the metal electrodes which are in contact with it. Goto et al. (1969a) and Fischer (1967), respectively, have measured the electromotive force of $ZrO_2 \cdot CaO$ and $ThO_2 \cdot CaO$ and $ZrO_2 \cdot Y_2O_3$ with platinum electrodes under isobaric but non-isothermal conditions. The coefficient α was found to be about 0.095 ± 0.005 mV/°C for the zirconia-lime and 0.050 ± 0.005 mV/°C for the thoria-lime electrolytes. The negative electrode was at the higher temperature for both systems. Figure 13 is an isoelectromotive force line diagram for isobaric oxygen ($p_{O_2} = 0.21$ atm) and with different temperatures T_1 and T_2 at the electrodes. From the above measurements, one can estimate the experimental error due to temperature differences at the two electrodes.

For example, if there is a temperature difference of 10°C and the oxygen partial pressures are the same the electromotive force would be about 5·3 mV for a cell containing zirconia–lime with platinum electrodes. Of this 0·95 mV would be due to the thermoelectromotive force and 4·35 mV to the oxygen chemical potential difference. If there is a temperature difference at the interface of a platinum electrode itself, the oxygen will be transferred from the lower temperature position to the higher and this would result in a mixed potential. The errors in e.m.f. measurements for this case are no greater than in the case of different temperatures across the electrolyte.

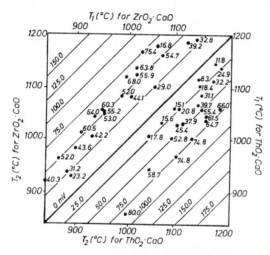

FIG. 13. Iso-electromotive force line diagrams for thoria-line (lower half) and zirconia-lime (upper half) electrolytes. The results are for isobaric oxygen (0·21 atm) and for temperatures T_1 and T_2 at the two platinum electrodes. (The negative electrode is at the higher temperature). (after Goto *et al.*, 1969a).

The temperature of a platinum electrode will fluctuate when an unstable jet of cold gas is made to impinge on it. Many investigators have reported that the e.m.f. is modified by the flow rate of a cold gas at an electrode. For a temperature change of 1·0°C, the e.m.f. would differ by 0·53 mV, which is usually big enough to disturb e.m.f. measurements made with a potentiometer and a sensitive galvanometer. The flow rate of the cold reference gas has thus to be carefully controlled.

The above discussion is limited to the case where the oxygen partial pressure is not affected by the temperature difference. In the case of metal/metal-oxide reference mixtures the equilibrium oxygen pressure also depends on the temperature. This effect can readily be computed from tabulated thermochemical data (Elliott and Gleiser, 1960).

B. Oxygen Pressure Gradient Parallel to Metal–solid Electrolyte Interface

In many cases of industrial applications, a long ZrO_2–CaO tube with a relatively thin wall is immersed into a liquid metal in order to determine the oxygen content. The e.m.f. measurements may be influenced by an oxygen pressure gradient due to nonequilibrium conditions between the gas and the liquid phase. As shown in Fig. 14 oxygen will be transferred from the upper to the lower part by a circulating current. The oxygen pressure gradient on the metal/tube interface is thus smaller than in the bulk liquid metal. If the reference electrode of platinum in air is in contact with the electrolyte wall at the lower position as in many cases of "oxygen probes" used for quick oxygen analysis in liquid metals, the electromotive force will to a large extent be determined by the oxygen pressure $p_{O_2}{}^+$ at the opposite interface of the reference electrode (in the half circle indicated by a broken line in Fig. 14).

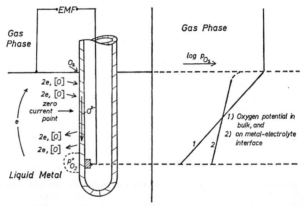

FIG. 14. Schematic diagram of oxygen transfer by a circulating current when there is an oxygen pressure gradient parallel to the metal/electrolyte interface.

The absolute value of $p_{O_2}{}^+$ is determined in part by the electrical resistance of the circuit, the transfer rate of oxygen in the liquid metal, and the oxygen pressure gradient in the bulk liquid metal. Although it is difficult to evaluate the error, the oxygen content determined by "oxygen probes" would be greater than the real value when the oxygen pressure in the gas phase is higher than that in the liquid metal. A measured mixed potential can sometimes be interpreted under certain special conditions but in general the interpretation is ambiguous.

C. Interference with a Gas Phase

Typical cell constructions with and without separate electrode compartments are shown in Fig. 15. If the two electrode compartments are not separated the gas mixture cannot be neutral with respect to the oxygen pressure at the two electrodes. When a single condensed phase electrode is used without any buffering reaction, the gaseous oxygen which penetrates the interface from the bulk gas phase can cause serious errors in some types of e.m.f. measurements. The extremely close contact required between solid metals and the solid electrolytes in potentiostatic and galvanostatic measurements (Pastorek and Rapp, 1969) for diffusivity of oxygen in solid metals is an example of this effect. Similar precautions are required in the determination (Tare and Schmalzried, 1966) of the redox reaction rates of iron and wüstite.

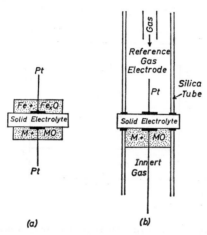

FIG. 15. Schematic diagrams of typical cell constructions (a) without (b) with separated electrode compartments.

When powder mixtures of metal and its oxide are used as electrodes, the direct contact of platinum with the solid electrolyte not only keeps the internal resistance to a minimum but also keeps the cell reversible for longer times. If different metals are used for the two electrodes, the thermo-electromotive force of the thermocouple so established should be added to or subtracted from the measured e.m.f.

D. Errors Due to Oxygen Transfer through the Solid Electrolyte

Since oxygen concentration cells are not in thermodynamic equilibrium, the driving force for oxygen transfer through the solid electrolyte causes equalizing phenomena of oxygen pressures in the two electrode compartments.

The transfer of oxygen through the electrolyte can be accomplished by an internal current due to electronic conduction, by Knudsen diffusion of gaseous oxygen through pores with various labyrinth factors, by chemical diffusion of oxygen through grain boundaries, and by molecular diffusion through micro cracks.

When the oxygen anion is the main charge carrier in the electrolyte BO_2 and the transference number of electrons is small, the oxygen transfer due to the internal current is shown schematically in Fig. 16. Under steady state conditions, the oxygen mass fluxes should all have the same value through the pored AO_2 oxide, the solid electrolyte without pores, and the oxide DO_2 with pores.

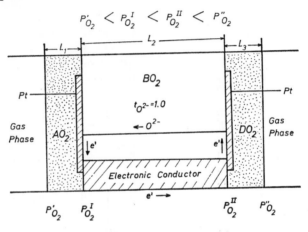

FIG. 16. The transfer of oxygen across an electrolyte BO_2 which has a predominantly anionic but small electronic conductivity.

The internal current divided by $nF(n = 4$ for one mole of O_2) must be equal to the oxygen diffusion per unit time through the pores in AO_2 and DO_2 as follows:

$$D_{AO_2}^{eff}\left(\frac{p_{O_2}^{I} - p_{O_2}'}{L_1}\right) = \left(\frac{1}{\sigma_{ion}} + \frac{1}{\sigma_e}\right)^{-1} \frac{(RT)^2}{(n\tilde{F})^2 L_2} \ln\frac{p_{O_2}^{II}}{p_{O_2}^{I}}$$

$$= D_{DO_2}^{eff}\left(\frac{p_{O_2}'' - p_{O_2}^{II}}{L_3}\right) \quad (31)$$

where D_i^{eff} is the effective diffusion constant of gaseous oxygen through pores in oxide i, σ_{ion} and σ_e are the specific ionic and electronic conductivities, and L_i is the thickness of the phase i.

From the equation it is deduced that when the absolute values of σ_{ion} and σ_e increase at high temperatures that $p_{O_2}^{II}$ approaches $p_{O_2}^{I}$ in value so that the voltage is reduced. One should thus take into consideration not only the transference numbers but also the absolute values of ionic and electronic conductivities. If there are micropores smaller than the mean free path of oxygen ($\sim 1\mu m$) in the solid electrolyte, the oxygen will be transferred by Knudsen diffusion, the rate of which is given by

$$\frac{dn_{O_2}}{dt} = \gamma\xi\,\frac{d}{6L_2}\left(\frac{8RT}{\pi M_{O_2}}\right)^{\frac{1}{2}}(p_{O_2}{}^{II} - p_{O_2}{}^{I})\frac{1}{RT} \tag{32}$$

where γ is the volume porosity, ξ is the labyrinth factor, dn_{O_2} is the number of moles of O_2 transferred in time dt, and d is the average diameter of pores.

Tien and Subbarao (1963) studied the grain growth in $Ca_{0.16}\,Zr_{0.84}$ $O_{1.84}$ sintered from powders of ZrO_2 and $CaCO_3$. The porosity of the sinter was 40% after heating for 15 min at 1400°C, 10% at 1700°C after 2 h. The grain size was about 25 μm after 2 h at 1900°C. A close inspection of the micrographs published by Tien and Subbarao shows that pores are independently dispersed both at grain boundaries and in the crystal matrix. The present authors therefore believe that the experimental error due to Knudsen diffusion is usually very small, as long as the porosity is smaller than 10%. The labyrinth factor ξ is then essentially zero in the above equation.

Möbius and Hartung (1965) measured the permeability of oxygen through stabilized zirconia, sintered corundum, mullite and quartz glass. The penetration rates of oxygen ($cm^3\,O_2/s$) through one centimetre thick polycrystalline oxide with an area of about 4·0–8·0 cm^2 were found to be $2 \times 10^{-5}\,cm^3/s$ for ZrO_2 . MgO at 1500°C, $3 \times 10^{-7}\,cm^3/s$ for sintered corundum at 1500°C, $1\cdot5 \times 10^{-7}\,cm^3/s$ for mullite at 1500°C, and $< 10^{-7}cm^3/s$ for quartz glass at 1100°C, against an argon flow in the oxide tubes and air on the outside.

Hayes et al. (1961, 1963) have reported that oxygen but not nitrogen can penetrate polycrystalline oxides of aluminium and sintered corundum. The logarithm of the permeability is a simple linear relation of the reciprocal temperature. The diffusion mechanism is not clear. Similar results have been reported by other investigators (Robert and Roberts, 1967; Smith et al., 1966; Ullmann, 1968).

Wagner (1968) measured the solubility of water vapour in ZrO_2 . Y_2O_3 and obtained the diffusion constant of protons in the dense oxide, which determines the transfer rate of H_2O. According to him, water vapour is dissolved in the solid electrolyte so that

$$H_2O(g) + V_O^{\cdot\cdot} = 2H_i^{\cdot} + O_O \tag{33}$$

where $H_i^{\cdot}$ is an interstitial proton. The diffusion constant $D_{H_i^{\cdot}}$ so obtained was $1\cdot57 \times 10^{-6}\,cm^2/s$ at 994°C. The permeability of H_2O gas was estimated at 994°C to be $1\cdot568 \times 10^{-7}\,NTP\,cm^3$ of H_2O gas per second per cm^2 of the solid electrolyte with 1 cm thickness (8 wt % Y_2O_3). The result is for partial pressures of H_2O which are 1 atm and 0 atm on both sides of the solid electrolyte.

When the oxygen or water vapour pressure difference between the two electrode compartments is too large and oxygen and water vapour will penetrate the electrolyte and cause errors in the e.m.f. measurements. Leakage may in addition be caused by sudden micro-crack formation, as observed in Al_2O_3 at high temperatur (Hayes et al., 1963; Pluschkell, 1970). For this case, the oxygen transfer rate can be estimated by means of the Knudsen diffusion equation given above.

E. Errors Due to Reaction Products between Electrode Materials and Solid Electrolytes

If an interoxide compound, ABO_3 is formed between AO used in an electrode compartment and the solid electrolyte BO_2 with pure oxygen anion conduction, the cell can be expressed as,

$$p_{O_2}', Pt \,|\, A + AO \underset{p_{O_2}^{I}}{\overset{I}{|}} ABO_3 \underset{p_{O_2}^{II}}{\overset{II}{|}} BO_2 \,(t_{O^{2-}} = 1 \cdot 0)\,|\, Pt, p_{O_2}''$$

where ABO_3 is assumed to be the only possible compound between AO and BO_2.

If BO_2 is a pure oxygen anion conductor and ABO_3 is a mixed conductor, one may use eqn (22) in the paper by Wagner (1966) to describe the e.m.f. of the above cell. If there are no anions other than oxygen anions, this equation becomes

$$E = \frac{1}{2\tilde{F}}[\mu_O'' - \mu_O'] - \frac{1}{2\tilde{F}}\int_I^{II} t_{A^{2+}}\, d\mu_{AO}$$

$$- \frac{1}{4\tilde{F}}\int_I^{II} t_{B^{4+}}\, d\mu_{BO_2} - \frac{1}{2F}\int_I^{II} t_e\, d\mu_O \qquad (34)$$

where μ_O, μ_{AO}, and μ_{BO_2} are the chemical potentials per g-atom oxygen, mole AO, and mole BO_2, respectively, and I and II refer to the interfaces of A, AO/ABO_3 and ABO_3/BO_2.

(a) When the electronic conduction in ABO_3 is dominant, the e.m.f. is given by

$$E = \frac{RT}{4\tilde{F}} \ln \frac{p_{O_2}''}{p_{O_2}''}. \qquad (35)$$

The oxygen pressure at the interface II, $p_{O_2}{}^{II}$ can then be estimated for several cases:

(i) $t_e \gg t_{O^{2-}} \gg t_{A^{2+}}$ or $t_{B^{4+}}$.

Under these conditions, the chemical potential of atomic oxygen is almost constant throughout the ABO_3 phase so that $p_{O_2}{}^{II} = p_{O_2}{}'$ and therefore,

$$E = \frac{RT}{4\tilde{F}} \ln \frac{p_{O_2}{}''}{p_{O_2}{}'} .$$

(ii) $t_e \gg t_{A^{2+}} \gg t_{B^{4+}}$ or $t_{O^{2-}}$.

Under these conditions, the chemical potential of metal A is almost constant throughout the ABO_3 phase, i.e., equal to μ_A°. If the equilibrium of $A + \frac{1}{2}O_2 + BO_2 = ABO_3$ is assumed, one can estimate $p_{O_2}{}^{II}$ as

$$RT \ln p_{O_2}{}^{II} = 2 \Delta F^\circ_{ABO_3} + RT \ln p_{O_2}{}'$$

and the expression for the e.m.f. is then given by

$$E = \frac{RT}{4\tilde{F}} \ln \frac{p_{O_2}{}''}{p_{O_2}{}^{II}} = \frac{RT}{4\tilde{F}} \ln \frac{p_{O_2}{}''}{p_{O_2}{}'} - \frac{\Delta F^\circ_{ABO_3}}{2\tilde{F}} . \tag{36}$$

(iii) $t_e \gg t_{B^{4+}} \gg t_{A^{2+}}$ or $t_{O^{2-}}$ (which are not very likely to occur in a real system).

The e.m.f. can be shown in a similar way to be

$$E = \frac{RT}{4\tilde{F}} \ln \frac{p_{O_2}{}''}{p_{O_2}{}^{II}} = \frac{RT}{4\tilde{F}} \ln \frac{p_{O_2}{}''}{p_{O_2}{}'} + \frac{\Delta F^\circ_{ABO_3}}{4\tilde{F}} . \tag{37}$$

(b) When the electronic conduction in ABO_3 is negligible eqn (34) becomes

$$E = \frac{1}{2\tilde{F}} [\mu_O{}'' - \mu_O{}'] - \frac{1}{2\tilde{F}} \int_I^{II} t_{A^{2+}} \, d\mu_{AO} - \frac{1}{4\tilde{F}} \int_I^{II} t_{B^{4+}} \, d\mu_{BO_2}.$$

(i) If $t_{O^{2-}}$ (in ABO_3) $= 1$, one has

$$E = \frac{1}{2\tilde{F}} [\mu_O{}'' - \mu_O{}'] = \frac{RT}{4\tilde{F}} \ln \frac{p_{O_2}{}''}{p_{O_2}{}'} . \tag{38}$$

(ii) If $t_{A^{2+}}$ (in ABO_3) $= 1$, it follows that

$$E = \frac{RT}{4\tilde{F}} \ln \frac{p_{O_2}{}''}{p_{O_2}{}'} - \frac{\mu_{AO}{}^{II} - \mu_{AO}^\circ}{2\tilde{F}} . \tag{39}$$

At the ABO_3/BO_2 interface, the relations $\mu_{ABO_3} = \mu^\circ_{ABO_3}$ and $\mu_{BO_2} = \mu^\circ_{BO_2}$ are valid.

The equilibrium condition for the virtual reaction

$$AO + BO_2 = ABO_3$$

is

$$\mu_{AO}{}^{II} + \mu^\circ_{BO_2} = \mu^\circ_{ABO_3}.$$

With the definition

$$\Delta F^\circ_{ABO_3} = \mu^\circ_{ABO_3} - \mu^\circ_{AO} - \mu^\circ_{BO_2}.$$

One thus obtains an equation for the e.m.f. which is given by

$$E = \frac{RT}{4\tilde{F}} \ln \frac{p_{O_2}{}''}{p_{O_2}{}'} - \frac{\Delta F^\circ_{ABO_3}}{2\tilde{F}}. \tag{40}$$

(iii) If $t_{A^{2+}} + t_{4B^+} = 1$, one has

$$E = \frac{1}{2\tilde{F}} [\mu_O{}'' - \mu_O{}'] - \frac{\mu_{AO}{}^{II} - \mu^\circ_{AO}}{2\tilde{F}} t_{A^{2+}} - \frac{\mu^\circ_{BO_2} - \mu_{BO_2}{}^I}{4\tilde{F}} t_{B^{4+}}. \tag{41}$$

For the same suppositions as in II.C, one finds for the e.m.f. that

$$E = \frac{RT}{4\tilde{F}} \ln \frac{p_{O_2}{}''}{p_{O_2}{}'} + \frac{1}{4\tilde{F}} (t_{B^{4+}} - 2 t_{A^{2+}}) \Delta F^\circ_{ABO_3}. \tag{42}$$

This expression is slightly different to that derived by Schmalzried (1966).

At the end of these considerations two examples are given: If a $ZrO_2 . CaO$ electrolyte is immersed into Fe–O melts, iron oxide (Fe_xO) is absorbed and the iron oxide layer becomes a mixed conductor according to Hoffmann and Fischer (1958, 1962). In coulometric experiments the colour of the $ZrO_2 . CaO$ electrolyte becomes black (Casselton, 1968) with the onset of mixed conduction (Douglas and Wagner, 1966) when the oxygen pressure is extremely reduced by an externally applied high d.c. potential.

F. Reversibility Test of the e.m.f.

The reversibility of the measured e.m.f. with respect to an externally applied current must be tested experimentally. Only the reversible e.m.f. can give exact thermodynamic properties at elevated temperatures. It should be said here however that it is in general difficult to determine whether the e.m.f. is a mixed potential or not.

Figure 17 shows schematically the electrical potential change for the galvanostatic condition when an external current is supplied to an oxygen concentration cell. In the figure, IR_0 is the potential drop without any polarization due to the circuit resistance R_0. When the current is off, the e.m.f. will decay to the equilibrium value as a result of depolarization phenomena. Qualitatively potential–current hysteresis curves of an oxygen concentration cell are shown in Fig. 18. In all cases, the depolarization rate, which determines the shape of the hysteresis curves, depends not only on the micromechanism of the polarization phenomena but also on the current density and its time of duration.

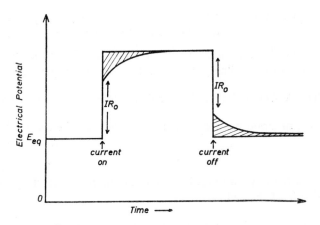

FIG. 17. The electric potential change across an oxygen concentration cell when an external current is supplied under galvanostatic conditions.

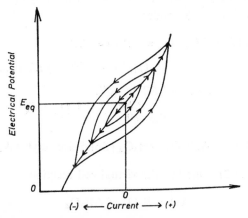

FIG. 18. Hypothetical potential–current hysteresis curve for an oxygen concentration cell. The shape of the curves is determined by the mechanism of the polarization and depolarization and by the current density and its duration.

Although there is some difference among O_2 + Pt, H_2/H_2O + Pt (Weiss-bart and Ruka, 1961, 1962), and CO/CO_2 + Pt (Karpachev et al., 1964), the reversibility of gas electrodes of the type $Pt/ZrO_2 \cdot CaO$ is known to be very good. Condensed phase reactions caused by an external current seriously decrease the reversibility of the e.m.f. Such reactions can be initiated by the current used to obtain the zero point of a galvanometer.

V. Applications to Thermodynamic Studies

A. Standard Free Energy of Formation of Binary and Ternary Compounds of Metal–Oxygen Systems

The following cell has been used extensively to determine the standard free energy of formation of binary compounds of metal–oxygen systems:

Cell(I)   A, $AO(p_{O_2}')$ | Solid electrolyte | Reference electrode (p_{O_2}'').

The cell reaction upon passing $2\tilde{F}$ across the cell may be formulated as

$$A + \tfrac{1}{2}O_2 \,(p_{O_2}'') = AO$$

where p_{O_2}'' is the oxygen partial pressure at the reference electrode on the right hand side of the cell.
Thus,

$$\Delta F[A + \tfrac{1}{2}O_2 \,(p_{O_2}'') = AO] = -2E\tilde{F}.$$

Because the standard state of oxygen is one atm, the standard free energy of formation can be calculated from the measured e.m.f. and the oxygen pressure at the reference electrode;

$$\Delta F_f^\circ [A + \tfrac{1}{2}O_2 \,(p_{O_2} = 1 \text{ atm}) = AO]$$
$$= \Delta F[A + \tfrac{1}{2}O_2 \,(p_{O_2}'') = AO] + \tfrac{1}{2}RT \ln p_{O_2}'' \qquad (43)$$
$$= -2E\tilde{F} + \tfrac{1}{2}RT \ln p_{O_2}''.$$

Alternatively, one may set up the cell

Cell(II)   A, B_2O_3, AB_2O_4 | Solid electrolyte | A, AO

so that on passing $2\tilde{F}$, one has the virtual cell reaction

$$AO + B_2O_3 = AB_2O_4$$

whereupon it follows that

$$\Delta F^\circ[AO + B_2O_3 = AB_2O_4] = -2E\tilde{F}. \qquad (44)$$

One can thus obtain the standard free energy of the above reaction. This is sometimes called the standard free energy of formation of the binary metal oxide compound.

TABLE I

Metal–oxygen liquid solutions investigated by galvanic cells

System	Literature-references
Na–O	Isaacs *et al.*, 1968; Kolodney *et al.*, 1965; Meyers and Hunter, 1968
Mn–O	Schwerdtfeger, 1967
Fe–O	Baker and West, 1966; Fischer and Ackermann, 1965, 1966; Förster and Richter, 1969; Fruehan *et al.*, 1969; Sanbongi *et al.*, 1966; Schwerdtfeger, 1967
Fe–O–S	Fischer and Ackermann, 1966
Fe–O–Si	Fruehan *et al.*, 1969; Schwerdtfeger, 1967
Fe–O–Mn	Schwerdtfeger, 1967
Fe–O–V	Fischer and Ackermann, 1966
Fe–O–Cr	Fischer and Ackermann, 1966
Fe–O–Co	Fischer and Ackermann, 1966; Fischer *et al.*, 1970
Fe–O–Ni	Fischer and Ackermann, 1966; Fischer *et al.*, 1970
Fe–O–Cu	Abraham, 1969; Fischer *et al.*, 1970
Steels	Engell *et al.*, 1967; Fischer, 1968; Fischer and Ackermann, 1965; Fitterer, 1966, 1967; Fitterer *et al.*, 1969; Ihida *et al.*, 1969; Luzgin *et al.*, 1963; Marincek, 1967a, b; Matsushita and Goto, 1966; Ohtani and Sanbongi, 1963; Pargeter, 1968; Sanbongi *et al.*, 1966; Ulrich and Borowski, 1968
Co–O	Fischer and Ackermann, 1966
Co–O–S	Fischer and Ackermann, 1966
Co–O–Ni	Fischer *et al.*, 1970
Ni–O	Fischer and Ackermann, 1966
Ni–O–S	Fischer and Ackermann, 1966
Cu–O	Abraham, 1969; Diaz and Richardson, 1968; El-Naggar *et al.*, 1967; Fischer and Ackermann, 1966; Fruehan and Richardson, 1969; Kozuka *et al.*, 1968; Marincek, 1967a, b; Osterwald, 1965; Pluschkell and Engell, 1965; Rickert and Wagner, 1966
Cu–O–Ag	Fruehan and Richardson, 1969
Cu–O–Sn	Fruehan and Richardson, 1969
Cu–O–Co	Abraham, 1969
Cu–O–Ni	Abraham, 1969
Ag–O	Diaz *et al.*, 1966; Diaz and Richardson, 1968; Fischer and Ackermann, 1966
Sn–O	Belford and Alcock, 1965; Fischer and Ackermann, 1966
Pb–O	Alcock and Belford, 1964; Fischer and Ackermann, 1966

By using the above result the standard free energies of formation of NiO, Cu_2O, MnO, SnO, PbO, MoO_2, Cr_2O_3, WO_2, $WO_{2\cdot72}$, $WO_{2\cdot90}$ and WO_3, ZnO, and NbO, NbO_2, and Nb_2O_5 have been determined (Alcock and Zador, 1967; Barbi, 1964; Belford and Alcock, 1965; Fischer and Ackermann, 1966; Gerasimov *et al.*, 1960, 1962; Goto and Matsushita, 1963; Hiraoka *et al.*, 1968; Jeffes and Sridhar, 1967; Kashida *et al.*, 1968; Kiukkola and Wagner, 1957; Kozuka *et al.*, 1967, 1968; Ksenofontova *et al.*,

1962; Laurentev *et al.*, 1961; Moriyama *et al.*, 1969; Osterwald, 1968; Palguev and Neuimin, 1960; Pluschkell and Engell, 1965; Rapp, 1963; Rickert and Wagner, 1966; Rizzo *et al.*, 1967; Schwerdtfeger, 1967; Steele and Alcock, 1965; Tretjakow and Schmalzried, 1965; Wilder, 1966, 1969; Worrell, 1966). As solid electrolytes, only zirconia base or thoria base solid solutions have been used because their conduction mechanisms are already known. The mixture of iron and wüstite or of nickel and nickel oxide has been widely used in the reference electrode since the equilibrium oxygen pressure had been accurately determined and their buffering capacity is large. As shown in Fig. 15, circulating air or pure oxygen gas in conjunction with a platinum electrode has also been used as the reference electrodes. The temperature range between 500°C and about 1500°C has been studied and the accuracy of the e.m.f. measurements has in most cases been found to be $\pm 1 \cdot 0 \, \text{mV}$.

This corresponds to $\pm 92 \, \text{cal/mol}$ of oxygen gas ($\tilde{F} = 23 \cdot 060 \, \text{cal/Volt}$. equiv.). For Cu_2O, PbO (*s.* or *l.*), NiO, CoO, SnO (*s.* or *l.*), $Fe_{0 \cdot 947}O$, MoO_2, and MnO, the free energy deduced from e.m.f. measurements agrees to within $200 \, \text{cal/mol}$ oxygen with that determined by other methods (Elliott and Gleiser, 1960). The disagreement is larger for Cr_2O_3, niobium oxides, and higher tungsten oxides.

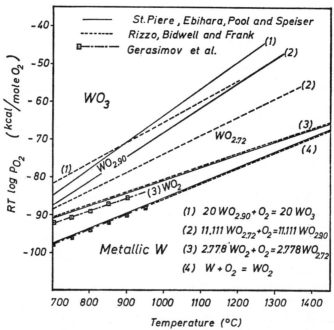

FIG. 19. Standard free energy changes of reaction of the W–O system.

The problem of determining selected values of standard free energies of formation of binary metal oxides is well illustrated in the case of the W–O system. The stable oxides in W–O system are WO_2 (brown, monoclinic), $WO_{2.72}$ or $W_{18}O_{49}$ (violet, monoclinic), $WO_{2.90}$ or $W_{20}O_{58}$ (blue, monoclinic) and WO_3 (yellow, monoclinic). These all have a small homogeneity range (Rieck, 1967). In addition, W_3O as reported in some of the literature ("Gmelins Handbuch", 1933; Hagg and Schonberg, 1954; Rieck, 1967), is an extremely unstable phase.

Gerasimov et al., (1962) measured the e.m.f. of the cell, W–WO_2, or WO_2–$WO_{2.72}$ mixture | ZrO_2 . CaO | Fe–Fe_xO (reference electrode) and found

$$E = 76 \cdot 8 - 0 \cdot 06T(°K) \pm 0 \cdot 5 \, mV \; (for \; W–WO_2)$$

$$E = -6 \cdot 68 + 0 \cdot 045T(°K) \pm 0 \cdot 5 \, mV \; (for \; WO_2–WO_{2.72}).$$

By combining the measured e.m.f. with ΔF_f° for Fe_xO, they were able to obtain the ΔF_f° values for the tungsten oxides.

Rizzo et al., 1967 carried out more extensive e.m.f. measurements on cells with separated electrode compartments. The experiments were repeated for three different kinds of reference electrodes. They obtained the following results,

$$\tfrac{1}{2}W + \tfrac{1}{2}O_2 = \tfrac{1}{2}WO_2, \qquad \Delta F_f^\circ = -68{,}660 + 20 \cdot 31T(°K) \pm 300 \, cal.$$

$$\frac{1}{0 \cdot 72}WO_2 + \tfrac{1}{2}O_2 = \frac{1}{0 \cdot 72}WO_{2.72},$$
$$\Delta F^\circ = -59{,}640 + 15 \cdot 01T(°K) \pm 300 \, cal$$

$$\frac{1}{0 \cdot 18}WO_{2.72} + \tfrac{1}{2}O_2 = \frac{1}{0 \cdot 18}WO_{2.90},$$
$$\Delta F^\circ = -67{,}860 + 24 \cdot 21T(°K) \pm 300 \, cal$$

$$\frac{1}{0 \cdot 10}WO_{2.90} + \tfrac{1}{2}O_2 = \frac{1}{0 \cdot 10}WO_3,$$
$$\Delta F^\circ = -66{,}765 + 26 \cdot 76T(°K) \pm 500 \, cal$$

St. Pierre et al. (1962) determined the equilibrium gas compositions from the weight loss and gain, using two different kinds of gas mixtures of H_2–H_2O and of CO–CO_2. The following expressions were obtained from Fig. 9 of their report

$$W + O_2 = WO_2, \qquad \Delta F_f^{\,0} = -125{,}920 + 39 \cdot 8T(°C)$$
$$2 \cdot 778 \, WO_2 + O_2 = 2 \cdot 778 \, WO_{2.72}, \qquad \Delta F^\circ = -114{,}070 + 32 \cdot 56T(°C)$$
$$11 \cdot 111 \, WO_{2.72} + O_2 = 11 \cdot 111 \, WO_{2.90}, \qquad \Delta F^\circ = -132{,}970 + 64 \cdot 87T(°C)$$
$$20 \, WO_{2.90} + O_2 = 20 \, WO_3, \qquad \Delta F^\circ = -134{,}000 + 70 \cdot 0T(°C).$$

The results found by the above three investigations are compared in Fig. 19.

The thermodynamics of WO_2 and $WO_{2.72}$ have been studied by Coughlin (1954—calculation), Bousquet and Perachon (1964—gas equilibrium), Barbi (1964—electromotive force), Griffis (1959—gas equilibrium), Ackermann and Rank (1963—effusion), and Vasiléva et al. (1960—gas equilibrium). All the results except those by Bouquet and Perachon, by Barbi and by Vasiléva et al. agree to within 300 cal/half mol of O_2. There is excellent agreement among the results reported by St. Pierre et al. (1962), Rizzo et al. (1967), Gerasimov et al. (1962) and also by Kashida et al. (1968). In the St. Pierre et al. study, no difference in $\Delta F°$ values was obtained despite the use of different kinds of gas mixtures. The gas composition ranges used were $p_{H_2O}/p_{H_2} = 0\cdot3-0\cdot9$ and $p_{CO_2}/p_{CO} = 0\cdot4-0\cdot6$. They have examined the accuracy of their experimental method by comparing their results with measurements on the $CO-CO_2-Fe-Fe_xO$ and $H_2-H_2O-Fe-Fe_xO$ equilibria. Excellent agreement with the values of Darken and Gurry (1945) were found in all cases. Rizzo et al. (1967) obtained identical $\Delta F°$ values on using $Cu-Cu_2O$, $Fe-Fe_xO$ and air reference electrodes. The electromotive forces were strictly constant, at least for two weeks and at most for one month.

With many precautions, Griffis (1959) has carried out the H_2-H_2O equilibrium experiments, circulating the gas mixture over the condensed phases of the W–O system. The calibration experiments with $Fe-H_2O-H_2-Fe_3O_4$ for his new apparatus (Griffis, 1959) were found to be in agreement with those of his old apparatus (Griffis, 1958) for which incorrect values were obtained for the W–O system. The present authors are under the impression that the errors in gas equilibrium methods (Griffis, 1958; Bousquet and Perachan, 1964; Vasiléva et al., 1960) are larger than those in the weight-loss-and-gain method by St. Pierre et al. (1962) and that this is responsible for the paradoxical results reported by Griffis. The old work by Vasiléva et al. (1950) is subject to large errors. St. Pierre et al. (1962) calculated $\Delta H°_{298}$ for $W + \frac{3}{2}O_2 = WO_3$ and obtained $-201\cdot190$ cal/mol of WO_3, which is in excellent agreement with the calorimetric value of $-201\cdot460$ by Mah (1959).

In conclusion, the equation for $\Delta F°$ given either by St. Pierre et al. (1962) or by Rizzo et al. (1967) are judged to give the more reliable values for the standard free energy changes from metallic tungsten to WO_2 and to $WO_{2.72}$.

There is large disagreement however if one compares the $\Delta F°$ values for oxidation of $WO_{2.72}$ to $WO_{2.90}$ and to WO_3. The small amount of oxygen required to oxidize $WO_{2.72}$ to $WO_{2.90}$ and to WO_3 seriously lowers the buffering capacity of the electrode (Kashida et al., 1968; Rizzo et al., 1967). In addition, if platinum is not in direct contact with the electrolyte, the resistance of the $WO_2-WO_{2.72}$ layer may decrease the sensitivity of the e.m.f. measurements. On the other hand, St. Pierre et al. (1962) used the range of gas ratio of p_{CO_2}/p_{CO} of 14 to 20 for $WO_{2.72}-WO_{2.90}$ and 18 to 40 for $WO_{2.90}-WO_3$. There is no serious experimental difficulty in the preparation and supply of such gas to the specimen at temperatures between 700°C and 1050°C. Carbon deposition for the chosen gas composition range can also be ruled out. The $\Delta F°$ values of the reactions concerned with $WO_{2.90}$ and WO_3 reported by St. Pierre *et al.* (1962) seem to be more accurate than those obtained from oxygen concentration cell measurements.

The effect of the polymorphic transitions (Rieck, 1967) of WO_3 (monoclinic → orthorohmbic, etc.) on the $\Delta F°$ values is small and masked by experimental errors. There are large differences in the temperatures for the three phase equilibria WO_2–$WO_{2.72}$–$WO_{2.90}$ and WO_2–$WO_{2.90}$–WO_3 determined from the intersections of the $\Delta F°(T)$ curves reported by different investigators.

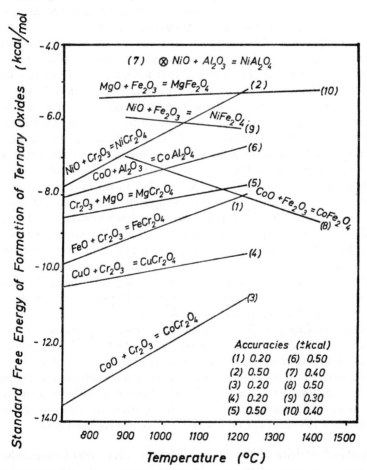

FIG. 20. Standard free energy of formation of ternary oxides determined by oxygen concentration cells. (after Tretjakow and Schmalzried, 1965).

The standard free energies of formation of ternary oxides have been determined by cell (II) with a three condensed phase equilibrium in an electrode compartment. This is an application of the oxygen concentration cell which has been exploited by Schmalzried (1960) and his coworkers (Benz and Schmalzried, 1961; Taylor and Schmalzried, 1964; Tretjakow and Schmalzried, 1965) and others (Inoue and Sanbongi, 1969; Levitskil and Rezukhina, 1966; Rezukhina et al., 1961a, b, 1963; Tretjakow, 1963). Because there is only a very limited number of $\Delta F°$ values for ternary oxides, the determinations by oxygen concentration cells are very important. The results by Tretjakow and Schmalzried (1965) are shown in Fig. 20. The slopes of the lines in the figure give the entropy change of the reactions. A discussion of why some of the slopes are negative and the others are positive has been given by Tretjakow and Schmalzried (1965).

Since Tretjakow and Schmalzried used air as the reference electrode, there must have been a large oxygen pressure difference between the two electrode compartments. The e.m.f. values for $Cr + Cr_2O_3$ against the air reference electrode are given by $1866\cdot4 - 0\cdot432T(°K) \pm 2\,mV$. A $\Delta F_f°$ value of 67,800 cal/half mol O_2 at 1000°K obtained from this equation is to be compared with the value of $-69,715$ cal given by Elliott and Gleiser (1960). The difference of about 2 kcal is very large even though there was excellent agreement in all cases among the values for the binary oxides, Fe_xO, CoO, NiO, and Cu_2O. Because the e.m.f. is so large, errors of similar magnitude are expected in the case of the ternary oxides. In the case of a Fe, Cr_2O_3, $FeCr_2O_4$ mixture against an air reference electrode for example, the temperature dependence of the e.m.f. is given by $1,669–0\cdot460T \pm 0\cdot2\,mV$.

A further difficulty is the attainment of an equilibrium oxygen pressure in the electrode compartment, because the equilibrium oxygen pressure is determined only by contacting lines of three condensed phases.

Accurate $\Delta F°$ values for the ternary oxides may nevertheless be obtained provided that a cell with a thoria-based electrolyte operated at low oxygen gas pressures and with separate electrode compartments is used to measure the e.m.f. as a function of temperature. In such an experiment the metallic electrode in contact with the electrolyte should be tamped down by a large amount of a mixture of the three condensed phases. An equal volume ratio of these should be used to so as to take care of the small buffering capacity. The investigations on ternary oxides by Russian scientists (Levitskii and Rezukhina, 1966; Rezukhina et al., 1961a, b, 1963) are well summarized by Toropov and Barzakovskii (1966) and will not be repeated here.

B. Chemical Activity of the Components of Alloys and Oxide Solutions

The chemical activity of the constituents in alloys and oxide solutions can be determined from the following simple relations:

(A–B) alloy, AO | solid electrolyte | A, AO (reference)

$$E = \frac{RT}{2\widetilde{F}} \ln a_A \text{ (standard state is pure A)} \tag{45}$$

B, (AO–BO$_2$) solution | solid electrolyte | B, BO$_2$ (reference)

$$E = -\frac{RT}{4\widetilde{F}} \ln a_{BO_2} \text{ (standard state is pure BO}_2) \tag{46}$$

(E is positive when the left electrode is the negative pole.) The free energies of formation of the oxides and the solid electrolyte must satisfy the relation:

$$-\Delta F_f^\circ \text{ of electrolyte} \gg -\Delta F_f^\circ \text{ of AO} \gg -\Delta F_f^\circ \text{ of BO}_2.$$

It should be borne in mind, however, that since the alloys are always saturated with oxygen, ternary instead of binary solutions are being studied. The following solid solutions have been investigated: Fe–Au, Fe–Pd, Fe–Pt (Kubik and Alcock, 1967), Fe–Ni (Rapp and Maak, 1962), Co–Pd, Pt–Ni, Pd–Ni (Bidwell and Speiser, 1965; Schwerdtfeger and Muan, 1967), Au–Ni (Sellars and Maak, 1966), Sn–Fe (Kozuka et al., 1968), Ta–W, Ta–Mo, Nb–Mo (Singhall and Worrell, 1970); the liquid solutions investigated are: Sn–Pb (Goto and St. Pierre, 1963), Fe–Cr (Fruehan, 1969), Cu–Zn (Wilder and Galin, 1969) and Ni–Pb (Cavanaugh and Elliott, 1964). These liquid and solid solutions comprise the alloy systems which have been studied.

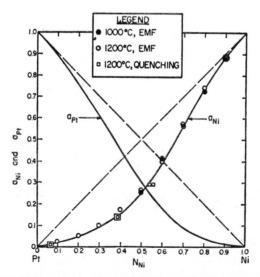

FIG. 21. Activity–composition curves for Ni and Pt in Pt–Ni alloys in the temperature range 1000–1200°C ("Quenching" means gas–solid equilibrium method). (after Schwerdtfeger and Muan, 1965).

Liquid solutions of the oxides: $SnO-SiO_2$ (Kozuka et al., 1968), and $PbO-SiO_2$ (Benz and Schmalzried, 1961; Charette and Flengas, 1969; Esin et al., 1957; Kozuka and Samis, 1970; Goto and Matsushita, 1963); and solid solutions of: $Fe_3O_4-CoFe_2O_4$ (Carter, 1960) and $Fe_3O_4-MgFe_2O_4$ (Gordeev and Tretjakov, 1963), $Fe_3O_4-Mn_3O_4$ (Ulrich et al., 1966), $NiO-MgO$ (Seetharaman and Abraham, 1968a), $NiO-MnO$ (Seetharaman and Abraham, 1968(b), $FeO-MnO$ (Engell, 1962; Prasad et al., 1969), $FeO-MgO$ (Engell, 1962), and $CoO-MnO$ (Seetharaman and Abraham, 1969) have also been investigated.

Figure 21 shows values for the activity of nickel in Ni–Pt alloys determined by Schwerdtfeger and Muan (1965) by two different methods: an e.m.f. method and the $CO-CO_2-NiO-(Ni + Pt)$ equilibrium method were used. The activity of the noble metal was calculated by means of the Duhem–Margules equation. A detailed discussion of the experimental errors is given in the report. Figure 22 shows schematically the structure of two cells used for the SnO activity measurement and for the determination of the free energy of formation of SnO by Kozuka et al. (1968).

Cells with a double-crucible structure such as in Fig. 22 have been widely used to determine the activity in liquid solutions.

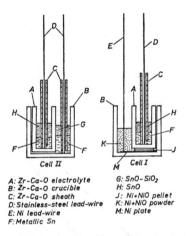

Cell II   Cell I

A: Zr-Ca-O electrolyte G: SnO-SiO₂
B: Zr-Ca-O crucible H: SnO
C: Zr-Ca-O sheath J: Ni+NiO pellet
D: Stainless-steel lead-wire K: Ni+NiO powder
E: Ni lead-wire M: Ni plate
F: Metallic Sn

FIG. 22. Schematic diagram of two galvanic cells used for the determination of SnO activities and of the standard free energy of formation of SnO. (after Kozuka et al., 1968).

Various liquid oxides, such as $Na_2O-B_2O_3$, $CaO-SiO_2$, $FeO-SiO_2$, $PbO-SiO_2$, $PbO-B_2O_3$ and $CaO-Al_2O_3-SiO_2$ have been used as electrolytes in oxygen concentration cells. Jeffes and Sridhar (1967) have compared values for the activity of PbO determined by galvanic cells with solid (Kozuka and Samis, 1970; Goto and Matsushita, 1963), of liquid $PbO-SiO_2$ (Bockris and Mellors, 1956; Esin *et al.*, 1957; Sridhar and Jeffes, 1967) electrolytes with those obtained by other methods (Callow, 1951; Richardson and Webb, 1955). They have discussed the possible causes of discrepancy in the various values which are shown in Fig. 23. One source of error is that associated with the construction of the cell electrodes. It is known for example that the measured e.m.f. of the cell (Ito and Yanagase, 1960),

$$Pb(1) \mid PbO\text{--}SiO_2 (1) \mid Pt(s), O_2(g)$$

depends on the Pt oxygen electrode construction. Figure 24 shows the various forms of electrode used by Sridhar and Jeffes (1967) ((a) and (b)); by Lepinskikh and Esin (1961) (c); and by Ranford and Flengas (1965) (d). A constant but increased e.m.f. is always obtained when the end of the Pt wire (see (a) and (b) in Fig. 24) is retracted so as to be just inside the alumina sheath. This implies that there is a steep potential gradient in the melt and the Pt wire picks up the potential for a low oxygen pressure. To obtain reproducible e.m.f.'s a uniform oxygen supply at the Pt/oxide interface is therefore absolutely necessary.

We shall in what follows discuss the conditions under which an oxygen concentration cell with a liquid oxide electrolyte is expected to exhibit reversibility. The investigation by Jeffes and Sridhar (1967) has shown that the kinetics of the reactions at the electrodes have always to be considered. A cell used by Sanbongi and Omori (1959) and which uses $CaO\text{--}SiO_2$ as an electrolyte, will be suitable for the purpose of the present discussion. The cell can be specified by

$$\left. \begin{array}{c} \text{Liquid Fe--Si} \\ (27\% \text{ Si}) \end{array} \right| CaO\text{--}SiO_2 \ (1) \left| \begin{array}{c} \text{MgO} \\ \text{or C} \end{array} \right| CaO\text{--}SiO_2 \ (2) \left| \begin{array}{c} \text{Liquid Fe--Si} \\ (27\% \text{ Si}) \end{array} \right.$$

The e.m.f. of the cell is given by eqn (1) since the ionic transference numbers in the silicate, bortate and aluminate melts of alkali and alkali earth metal oxides are all unity provided that the Fe_xO content is smaller than several percent. The solid MgO or carbon serves only as a separating wall with a fixed conductivity. A current flow is inevitable even when a potentiometer is used since a balance point has to be established. One such problem is that related to the change of surface composition of electrode metals caused by a current flow. For this reason the interpretation of the measured e.m.f. and its reversibility has been a source of controversy.

Let us now consider some possible electrode reactions and the effect these have on the cell reversibility. Suppose electrons are supplied to the electrode on the left and removed from the one on the right by means of an external current. Since only calcium cations move in the electrolyte, the current flow

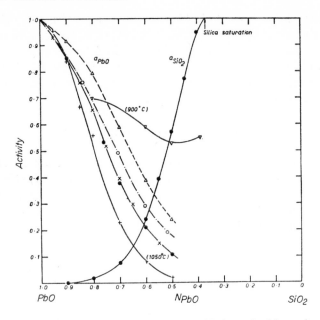

FIG. 23. Activities of PbO in PbO–SiO$_2$ melts at 1000°C determined by various methods. (after Jeffes and Sridhar, 1967).

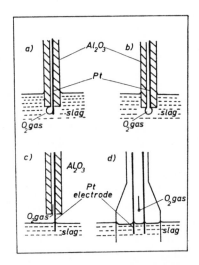

FIG. 24. Configurations of Pt–O$_2$ electrodes (a) and (b) after Sridhar and Jeffes, (c) after Lepinskik and Esin, (d) after Ranford and Flengas. (after Jeffes and Sridhar, 1967).

in the cell will be accomplished as indicated schematically by the diagram below.

$$
\begin{array}{ccccc}
\text{I} & \text{II} & \text{III} & \text{IV} \\
\text{Fe–27\% Si} \mid & \text{CaO–SiO}_2 \text{ (l)} \mid & C \mid & \text{CaO–SiO}_2 \text{ (2)} \mid & \text{Fe–27\% Si}
\end{array}
$$

The wall of graphite or MgO serves to separate the two liquid oxides which, unlike the Fe–Si alloys, differ in composition. Possible electrode reactions at the four interfaces are then given by,

at interface I, $2\,\overleftarrow{Ca}_{ad}{}^{2+} + 4e' \rightarrow 2\,Ca_{ad}$

$2\,Ca_{ad} + O_{2,\,ad} \rightarrow 2(CaO)_{ad}$

$$\rule{5cm}{0.4pt} \tag{47}$$

for a net reaction $2\,\overleftarrow{Ca}_{ad}{}^{2+} + 4e' + O_{2,\,ad} \rightarrow 2(CaO)_{ad}$

at interface II, $2\,O_{ad}{}^{2-} \rightarrow O_{2,\,ad} + \overrightarrow{4e'}.$

Oxygen $(O_{2,\,ad})$ is accumulated and electrons $(4e')$ are transferred across the graphite.

at interface III, $2\,\overleftarrow{Ca}_{ad}{}^{2+} + O_{2,\,ad} + 4e' \rightarrow 2(CaO)_{ad}.$

at interface IV, $2\,Ca_{ad} \rightarrow 2\,\overleftarrow{Ca}_{ad}{}^{2+} + 4e'$

$2(CaO)_{ad} \rightarrow 2\,Ca_{ad} + O_{2,\,ad}$ (48)

$$\rule{5cm}{0.4pt}$$

for a net reaction $2(CaO)_{ad} \rightarrow 2\,\overleftarrow{Ca}^{2+} + 4e' + O_{2,\,ad}$

For the specified current flow, these reactions show that the oxygen pressure increases at the Fe–Si/CaO–SiO$_2$ interface IV. The pressure decreases at interface I. Constant oxygen pressures at the interfaces can be maintained by means of suitable fast buffer reactions. In the case of CaO–SiO$_2$ melts, such buffer reactions are given by

$$2\,O_{ad}{}^{2-} + Si_{ad} + O_{2,\,ad} \rightleftarrows SiO_4{}^{4-} \tag{49}$$

$$3\,O_{ad}{}^{2-} + 2\,Si_{ad} + 2\,O_{2,\,ad} \rightleftarrows Si_2O_7{}^{6-}, \text{ etc.,} \tag{50}$$

where the equilibrium constants are defined respectively by

$$K_3 = \frac{a_{SiO_4{}^{4-}}}{a_{O_{ad}{}^{2-}}{}^2\, a_{Si_{ad}}\, p_{O_2}} \quad \text{and} \quad K_4 = \left(\frac{a_{Si_2O_7{}^{6-}}}{a_{O_{ad}{}^{2-}}{}^3\, a_{Si_{ad}}{}^2\, p_{O_2}{}^2} \right)^{\frac{1}{2}}.$$

K_3 and K_4 are identical when the following relations are satisfied:

$$a_{SiO_4}{}^{4-} = a_{SiO_2} \cdot a_{O^{2-}}{}^2$$

$$a_{Si_2O_7}{}^{6-} = a_{SiO_2}{}^2 \cdot a_{O^{2-}}{}^3.$$

It follows that

$$K_3 = K_4 = \frac{a_{SiO_2}}{a_{Si, ad} \cdot p_{O_2}}. \tag{51}$$

Even when there is a current flow, the interface oxygen pressure will be given by eqn (51), provided only that the buffer reactions rate is large. Good reversibility of the e.m.f. is then to be expected. When this kinetic criterion is satisfied and the oxygen pressure refers to the same Fe–Si composition, then

$$E = \frac{RT}{4\tilde{F}} \ln \frac{a_{SiO_2, 2}}{a_{SiO_2, 1}}. \tag{52}$$

This result follows from eqn (1). A similar discussion holds for the Al_2O_3 activity measurements on cells using Fe–Al alloys and CaO–Al_2O_3–SiO_2 melts as electrolytes. The current flow changes not only the concentration of the absorbed species but also the electrolyte composition close to the Fe–Si/CaO–SiO$_2$ interface. This composition change of the liquid electrolyte causes a change in the equilibrium oxygen pressures and also a certain value of diffusion potential. The diffusion rates of CaO or SiO_2 in liquid electrolytes are large. Schwerdtfeger (1966) has used SiO_2–saturated liquid electrolytes so as to reduce the diffusion potential to a minimum. The reaction rates of the electrolytes and the graphite or MgO separating wall has to be small if the slag composition is not to change during the e.m.f. measurements. When the oxygen pressures at the two electrodes differ, equalizing processes can still take place even in the absence of an external current. The metal electrode at the lower oxygen pressure is then continuously oxidized either by transport of the dissolved neutral metal or by dissolved oxygen as O_2 or $O°$. The counter diffusion of cations of different valences with diffusion of oxygen anions to keep electrical neutrality in the oxide electrolytes is also a possibility. The diffusion constant and relative concentration in the melt determines the dominant mechanism. The rate of this oxidation reaction should be small so that the electrode surface composition can be maintained.

The agreement between the values for the activity of silica determined from e.m.f. measurements, with those obtained by other methods can be taken as indirect proof that the kinetic criteria discussed above are satisfied. Sanbongi and Omori (1959) have made such a comparison and have found a fair agreement for values of the activity of SiO_2 in the CaO–SiO_2 system at 1600°C. They have also established the following experimental results for any composition of slag:

(i) the e.m.f. is the same whether graphite or magnesia is used for the separating wall.

(ii) there is no change in the e.m.f. when the concentration of silicon in the Fe–Si alloy of either side of electrodes is changed from 27·3% to 42·1% by weight. These results also suggest that the kinetic criteria are satisfied. In the second paragraph of p. 204 of the report by Jeffes and Sridhar (1967), they stated that the use of MgO results in an erratic e.m.f. owing to its electronic conduction. The e.m.f. is not affected when MgO or graphite is used only as a separating wall and is not in direct contact with the Fe–Si electrodes. This is provided the slag is an ionic conductor and the oxygen is transferred from and to the gas phase as discussed above. MgO or graphite is not used as an electrolyte in the experiment.

C. Coulometric Titration

Deviations from stoichiometric composition in transition metal oxides and oxide compounds have been determined by chemical, electrochemical analysis or by gravimetry or volumetry. A further, very elegant method is that of coulometric titration *in situ*. In this method oxygen is supplied to or removed from one compartment of the oxygen concentration cell by applying a non-equilibrium voltage across the cell. The amount of oxygen transferred through the electrolyte—in most cases stabilized zirconia—is given by

$$n_O = \frac{I \cdot t}{2\tilde{F}}.$$ (53)

The final equilibrium oxygen pressure regulates the distribution of the titrated oxygen between the specimen and the surrounding gas atmosphere. Because of this, the activity of oxygen can be obtained as a function of the composition of homogeneous phases such as metals and nonstoichiometric oxides. The following solutions have been investigated by this titration method:

$O_{Cu,s}$ (Pastorek and Rapp, 1969), $O_{Cu,1}$ (Gerlach *et al.*, 1968), $O_{Pb,1}$ (Alcock and Belford, 1964), $O_{Sn,1}$ (Belford and Alcock, 1965), $Fe_{1-\Delta}O$ (Rizzo *et al.*, 1968), $Co_{1-\Delta}O$ (Sockel and Schmalzried, 1968), $Ni_{1-\Delta}O$ (Sockel and Schmalzried, 1968; Tretjakow and Rapp, 1969), $Fe_{3-\Delta}O_4$ (Sockel and Schmalzried, 1968), $TiO_{2-\Delta}$, $NbO_{2-\Delta}$ (Zador, 1968), $Co_xFe_{3-x}O_4$ (Sockel and Schmalzried, 1968), $(Mg, Fe)_{3+\Delta}O_4$ (Schmalzried and Tretjakow, 1966), $LiFe_5O_{8-\Delta}$ (Tretjakow and Rapp, 1969).

In some titration experiments open cell constructions have been used (see for example Fig. 15). In this case an interaction between the surrounding "inert" gas atmosphere or the "vacuum" cannot be avoided. For this reason, Rizzo *et al.* (1968) for example, have had to make corrections of between 3 and 20% to their measurements of the oxygen partial pressure—composition isotherms of wüstite. The direct contact between the specimen and the electrolyte may introduce additional errors as a result of the formation of solid solutions or the precipitation of new phases, whenever a bad contact results in local high current densities.

In Fig. 25 a closed titration cell is shown (Tretjakow and Rapp, 1969).

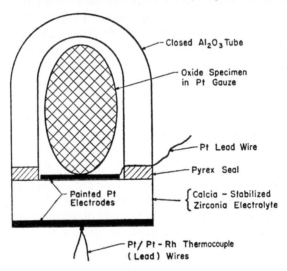

FIG. 25. A closed cell for coulometric titration involving an oxide specimen. (after Tretjakow and Rapp, 1969).

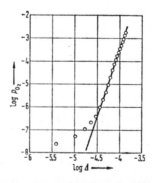

FIG. 26 Oxygen partial pressure p_{O_2} as a function of metal deficit in $Ni_{1-\Delta}O$ at 1200°C ($\log p_{O_2} = -7.79$ at equilibrium Ni, NiO). (after Sockel and Schmalzried, 1968).

Pyrex glass provides a gas tight seal on heating. The specimen is not in contact with the electrolyte and the oxygen supplied to or removed from it is transferred over the gas phase inside the container. At low oxygen pressures reactions are possible between the sample and the platinum. A separating alumina plate is then helpful (Sockel and Schmalzried, 1968). Satisfactory performance of the cell is limited by the leakage of oxygen through the container wall by pore diffusion or transport combined with electronic conduction. Figure 26 from the investigation of Sockel and Schmalzried (1968) reproduces the titration curve for $Ni_{1-\Delta}O$ at 1200°C. In this diagram, the slope of the line corresponds to 1 : 6, as it should be, if in nickel oxide the metal vacancies are doubly ionized.

In titration experiments the current density should not be in excess of about 100 $\mu A/cm^2$. This is necessary in order to avoid the possible precipitation of blocking layers in the buffer- and reference-mixtures or on the specimen. Förster (personal communication) and coworkers tried to deoxidize liquid steel by titration through a wall of stabilized zirconia at current densities of more than 1 A/cm^2. After the experiment a black layer of reduced zirconia was found to have formed on the side of the electrolyte in contact with the steel bath. A similar decomposition and blackening of stabilized zirconia at the cathode was observed by Weininger and Zemany (1954) and Casselton (1968), when the electrolyte was subjected to very high current densities.

D. Oxygen Content in Liquid Metals and Alloys

The determination of the oxygen content in liquid metals by galvanic cells (see Table I, p. 575) has received considerable interest in the last few years not only because of its scientific interest but also because of possible industrial applications. The method can give a rapid estimate of the oxygen content from simple e.m.f. measurement provided that only a calibration curve of $E = f(\% \, O, T,$ alloying elements) has been determined. Chemical analysis gives the total oxygen content of the melt, while the Galvanic method leads to an estimate of the free oxygen which has not combined with deoxidizers. Deoxidation practice will be improved, once reliable cell measurements can be made.

Gaseous oxygen is dissolved by a liquid metal according to the reaction

$$\tfrac{1}{2}O_2 \, (p_{O_2}') \rightleftarrows \underline{O} \tag{54}$$

In equilibrium, the oxygen partial pressure of the coexisting gas phase is given by the relation

$$(p_{O_2}')^{\frac{1}{2}} = K^{-1} \cdot a_O = f_O \, [\% \, O] \exp \frac{\Delta F°}{RT}. \tag{55}$$

In this equation, a_O is the oxygen activity, f_O the oxygen activity coefficient, and $\Delta F°$ the standard free energy of reaction. The formula applies to the case of an infinitely dilute binary solution of Me–O where the standard state of the oxygen is taken as 1 atm and a weight per cent scale is used for the oxygen content. If eqn (55) is substituted into eqn (1), it follows that

$$E = \frac{RT}{2\tilde{F}} \ln \frac{(p_{O_2}'')^{\frac{1}{2}}}{[\% \, O]} - \frac{RT}{2\tilde{F}} \ln f_O - \frac{\Delta F°}{2\tilde{F}}. \tag{56}$$

The activity coefficient f_O is unity when the oxygen content is zero. One therefore obtains from eqn (56) that

$$\Delta F° = \lim_{[\% \, O] \to 0} \left[RT \ln \frac{(p_{O_2}'')^{\frac{1}{2}}}{[\% \, O]} - 2 E\tilde{F} \right]. \tag{57}$$

Thus, by measuring values of $[\% \, O]$ and E for binary Me–O melts, calculating the expression in brackets and plotting it versus $[\% \, O]$ one may readily obtain $\Delta F°$ by extrapolation of the aforementioned plot. In most cases it has been observed that this plot results in a horizontal line which extends up to the point where there is a saturation concentration of oxygen and the consequent formation of the coexisting oxide. In this case Henry's law is obeyed over the total concentration range.

One may get $\ln f_O$ for the more general case where f_O is not unity, by substituting the value of $\Delta F°$, obtained by extrapolation, into equation (56) which is rearranged to give

$$\ln f_O = \ln \frac{(p_{O_2}'')^{\frac{1}{2}}}{[\% \, O]} - \frac{\Delta F°}{RT} - \frac{2E\tilde{F}}{RT}. \tag{58}$$

For dilute multi-component solutions the following relation (Wagner, 1952) holds:

$$f_O = f_O° \cdot f_O^X \cdot f_O^Y \dots . \tag{59}$$

The coefficients f_O^X and f_O^Y characterize the influence of the elements X and Y on the activity coefficient of oxygen f_O.

Littlewood has published a comprehensive paper on the direct determination of oxygen in liquid metals by the e.m.f. method (Littlewood, 1966). For this reason only a table of the systems investigated so far together with some comments on new developments will be given here.

The main difficulty in oxygen probe measurements on liquid metals is the high thermal shock sensitivity of the solid electrolytes which are based on ZrO_2 and ThO_2. It has been reported that partially stabilized zirconia tubes are more resistant to thermal shock than fully stabilized tubes (Fruehan and Richardson, 1969). Unfortunately, however, the gas permeability of the electrolyte increases with increasing content of the monoclinic phase (Karaulov *et al.*, 1968). Fitterer (1966, 1967) and Fitterer *et al.* (1969) have overcome the thermal shock difficulty by constructing a cell in which a small disc of stabilized zirconia is fused on to one end of a fused silica tube. A metal cap on the tip of the device is used so as to avoid slag contamination. This arrangement is useful when cell components have rapidly to be immersed in hot melts. The insulation due to the silica tube increases the accuracy of the e.m.f. measurements. The reason for this is that the rapid transport of oxygen through the wall of the immersed zirconia tube, which would otherwise take place, is prevented. The effects which arise because of non-equilibrium conditions between the gas and the metal phase have been discussed in Section IV.B.

According to eqns (56) and (59) alloying elements may influence the e.m.f. of an oxygen probe even though the oxygen content of the melt has not changed. This effect is only negligible in the case of rimming steels which contain small amounts of deoxidizers. In general the interaction of alloying elements and their influence on the activity coefficient f_O should be taken into account.

If the process of precipitation deoxidation is investigated by e.m.f. measurements, one should be careful with regard to possible reactions between the electrolyte and the deoxidation products. This seems to be of special importance when pure MgO is used as solid electrolyte (Matsushita and Goto, 1966).

As to the most suitable reference electrode, there is no agreement among the investigators. A gas electrode with air or pure oxygen exhibits the advantage of a temperature independent reference pressure p_{O_2}''. The dis-

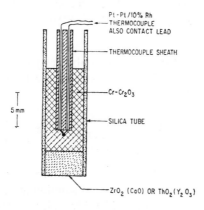

Fig. 27. Schematic diagram of an oxygen probe. (after Fruehan *et al.*, 1969).

advantage here is that in most cases the difference in chemical potential $\Delta\mu_{O_2}$ between the two cell compartments becomes quite large so that failure due to leakage is likely. An added difficulty is that the flow rate of the gas has to be carefully controlled so as to avoid errors due to temperature gradients (see Section IV.A).

A suitable Me–MeO mixture may be selected in order to minimize $\Delta\mu_{O_2}$. In this case the temperature coefficient of the cell e.m.f. is relatively low. But mixtures of the type Me–MeO or MeO–Me$_2$O$_3$ are readily polarizable (Diaz and Richardson, 1968) and there is the possibility of reactions with the solid electrolyte. Figure 27 shows the cell arrangement used by Fruehan *et al.*, 1969 for oxygen potential measurements in liquid steel. According to this investigation the electrolyte ZrO$_2$ + 4% CaO works satisfactorily down to oxygen pressures of the Cr–Cr$_2$O$_3$ equilibrium. This corresponds to about 20 ppm oxygen dissolved in liquid iron at 1600°C. It should be noted however that when the oxygen pressure is as low as 10^{-16} atm (6 ppm [O]) only the ThO$_2$ + 7% Y$_2$O$_3$ electrolyte exhibits a stable accurate voltage (Fig. 28).

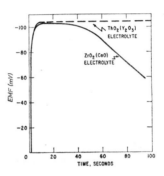

FIG. 28. Electromotive force tracing for a iron melt containing 10·1 wt% Si at 1600°C. The curve shows the unreliability of the ZrO$_2$ (CaO) electrolyte at oxygen activities below 0·001% O. (from Fruehan *et al.*, 1969).

Tretjakow and Muan (1969) tested a double cell which involved a combination of ZrO$_2$- and ThO$_2$-solid electrolytes in the temperature range from 700 to 1100°C. The cell is given by

$$\text{Pt} - p_{O_2}' \left| \begin{array}{c} \text{ThO}_2 + 8 \text{ mol}\% \\ \text{Y}_2\text{O}_3 \end{array} \right| \begin{array}{c} \text{Co/CoO} \\ \text{buffer and} \\ \text{contact} \end{array} \left| \begin{array}{c} \text{ZrO}_2 + 15 \text{ mol}\% \\ \text{CaO} \end{array} \right| \text{air} - \text{Pt}$$

This cell utilizes the favourable properties of stabilized zirconia at high oxygen pressures and that of the thoria base solid solution at very low oxygen pressures. The cell operates over a considerably wider range of oxygen pressure than those described previously.

VI. Application to Kinetic Studies

A. Determination of Oxygen Diffusion Constants in Metals

An oxygen concentration cell can also be used to determine the diffusion constants of dissolved oxygen in metals when the redox reaction rate at the metal/electrolyte interface is much faster than the diffusion of oxygen in the metal.

Let us consider the following cell arrangement:

$$\begin{matrix} \text{I} & & \text{II} \\ p_{O_2}{}', M + \underline{O} & | \ \text{solid electrolyte} \ | & \text{reference electrode}, p_{O_2}{}''. \end{matrix}$$

When a voltage is applied to the cell, oxygen may be removed from or delivered at the metal phase. D_O values can be calculated from the resultant current or the potential change under potentiostatic or galvanostatic conditions when this process is determined by oxygen diffusion in the metal. The e.m.f. change which follows when $p_{O_2}{}'$ is suddenly changed in the gas phase of the left hand compartment of the cell may also be used to evaluate D_O.

We assume at the start of this section that the phase boundary reaction at interface I determines the reaction rate. Although this situation is not realized in practice, the following comments are given for the sake of completeness.

The original e.m.f., E_0 with zero current is $RT/4F \ln p_{O_2}{}''/p_{O_2}{}'$. If the potential E_1 is applied, the interface oxygen pressure will change instantaneously to $p_{O_2}{}'''$ at the left side interface. The redox reaction at interface I, $O^{2-} = O + 2e'$ can be split into the following elementary reactions:

$$O^{2-} = O_{ad}{}^{2-} \qquad \text{(Chemisorption or desorption)}$$

$$O_{ad}{}^{2-} = O_{ad} + 2e' \quad \text{(Charge exchange at the interface)}$$

$$O_{ad} = O \qquad \text{(Chemisorption or desorption)}$$

If the redox reaction rate is controlled by the third reaction, namely by the rate of absorption or desorption of neutral oxygen atoms, then the rate of oxygen transfer will be given by the equation

$$\dot{n}_O = kA([O_{ad}] - [O_{ad}{}^e]) \tag{60}$$

where $\dot{n}_O$ is number of gram atoms of oxygen transferred per unit time, k is an isothermal rate constant, A is area of reaction interface, $[O_{ad}]$ is the concentration of absorbed oxygen at the interface and $[O_{ad}{}^e]$ is the oxygen concentration in equilibrium with the metal phase. When there is no polarization at the reference electrode side and the Ohmic potential drop is negligible, the following expression is valid for the change in potential difference

$$E_1 - E_0 = \frac{RT}{2\tilde{F}} \ln \left(\frac{p_{O_2}{}'}{p_{O_2}{}'''} \right)^{\ddagger} = \frac{RT}{2\tilde{F}} \ln \frac{[O_{ad}] \text{ at } p_{O_2}{}'}{[O_{ad}] \text{ at } p_{O_2}{}'''}. \qquad (61)$$

The current density at the interface is determined by a reaction rate given by

$$\frac{I}{2\tilde{F}A} = k([O_{ad}] \text{ at } p_{O_2}{}' - [O_{ad}] \text{ at } p_{O_2}{}'''). \qquad (62)$$

On inserting eqn (62) into (61), one obtains

$$E_1 - E_0 = \frac{RT}{2\tilde{F}} \left\{ \ln[O_{ad}] \text{ at } p_{O_2}{}' - \ln\left([O_{ad}] \text{ at } p_{O_2}{}' - \frac{I}{2k\,\tilde{F}A} \right) \right\}. \qquad (63)$$

According to this equation, both the electrical potential and the current are nearly independent of time when the volume of metal is large. This result is contradicted by the measurements of various investigators (Osterwald and Schwarzlose, 1968; Pastorek and Rapp, 1969; Rickert, 1967; Rickert and Steiner, 1966; Rickert and El Miligy, 1968; Rickert et al., 1966) which clearly show that either the potential or the current changes with time. This is true even for a semi-infinite metal phase geometry.

In the section which now follows the situation will be discussed where the reaction rate is determined by a diffusion controlled process. This seems to be the dominant process in most cells.

In a previous discussion on liquid oxide electrolytes, it was suggested that the redox reactions at metal–oxide interfaces are very rapid. This conclusion is supported in the following section on the decay curves of oxygen concentration cells.

When the oxygen transfer rate is controlled by oxygen diffusion in metals, the current I is related to the diffusion rate of oxygen at $x = 0$ (the interface) by

$$\frac{I}{2\tilde{F}} = - \frac{AD_O}{V_m} \left(\frac{\partial N_O}{\partial x} \right)_{x=0} \qquad \text{(gram atoms O/sec)} \qquad (64)$$

where N_O is the atom fraction of oxygen and the other symbols have their usual meaning.

Explicit expressions for $N_O(x, t)$ can be obtained by solving Fick's second law of diffusion

$$\left(\frac{\partial N_O}{\partial t} \right) = D_O \frac{\partial^2 N_O}{\partial x^2}. \qquad (65)$$

Solutions for various initial conditions and cell geometries are available (Crank, 1956). The time dependence of the current I in terms of D_0 can then be calculated from eqn (64) since

$$\left(\frac{\partial N_0}{\partial x}\right)_{x=0} \quad \text{follows from } N_0(x, t).$$

The diffusion equation can be solved for both the potentiostatic and the galvanostatic condition. Examples of these limiting cases will be discussed.

Rickert (1967) and his coworkers (Rickert and Steiner, 1966; Rickert and El Miligy, 1968; Rickert et al., 1966) have exploited this method for cells of linear and of cylindrical geometry. Osterwald and Schwarzlose (1968) and Pastorek and Rapp (1969) have also used this method. The oxygen diffusion constant has been determined for solid copper, solid silver, liquid silver, and liquid copper.

The following example, which is based on a publication by Pastorek and Rapp (1969), will serve to bring out the details of the method. The cell considered is:

$$\boxed{-\text{FeO, Fe}_3\text{O}_4 \mid \text{ZrO}_2 . \text{CaO} \mid \text{Cu-O} \mid \text{ZrO}_2 . \text{CaO} \mid \text{FeO, Fe}_3\text{O}_4-}$$

$$\text{Electrode 1} \qquad\qquad \text{Electrode 2} \qquad\qquad \text{Electrode 3}$$

The cell voltage E can be related to the activity of oxygen a_0 at the electrode/electrolyte interfaces by,

$$E - IR_0 = \frac{RT}{2\bar{F}} \ln \frac{a_0'(\text{Cu})}{a_0''(\text{FeO, Fe}_3\text{O}_4)} \tag{66}$$

where R_0 is the ionic electrical resistance of the electrolyte.

Potentiostatic experiment

In such an experiment the potentiometer is set for an electromotive force E_1 such that the copper is in equilibrium with an oxygen activity somewhat higher than that for which $\text{FeO-Fe}_3\text{O}_4$ would coexist. At the start of the experimental run the e.m.f. is increased to E_2. A value is chosen for which Cu_2O does not form. The new oxygen activity, which is established at both copper/electrolyte interfaces, follows instantaneously with the voltage increase. Since current flows from both $\text{FeO-Fe}_3\text{O}_4$ electrodes (but in opposite directions) a symmetrical concentration is produced in the solid copper. This concentration distribution changes with time and can be calculated. A solution of eqn (65) for the initial conditions

$$N_0 = N_0(E_1) \quad \text{for} \quad 0 < x < L, \qquad t = 0$$

and the boundary conditions

$$N_O = N_O(E_2) \quad \text{for} \quad x = 0 \quad \text{and} \quad x = L, \qquad t > 0$$

$$N_O = N_O(E_2) \quad \text{for all} \quad x, \qquad t = \infty$$

is given (Crank, 1956) by

$$N_O(x, t) = N_O(E_2) - [N_O(E_2) - N_O(E_1)] \times$$

$$\sum_{n=1,3,5\ldots}^{\infty} \frac{4}{n\pi} \sin\left(\frac{n\pi x}{L}\right) \exp\left(\frac{-n^2 \pi^2 D_O t}{L^2}\right). \qquad (67)$$

In this equation $N_O(E_1)$ is the equilibrium atom fraction of oxygen in the copper at an oxygen activity $a_O(E_1)$ when a voltage E_1 is applied to the cell. L is the thickness of the copper sample.

The instantaneous current I for the two cells ($\frac{1}{2}$ for each cell) which is given by

$$I = \frac{16\tilde{F}AD_O}{V_{Cu}} [N_O(E_2) - N_O(E_1)] \sum_{n=1,3,5\ldots}^{\infty} \exp\left(\frac{-n^2 \pi^2 D_O t}{L^2}\right) \qquad (68)$$

then follows on substitution of eqn (67) differentiated with respect to x into (64). Neglecting the ohmic potential drop, and assuming the validity of Henry's law, eqn (66) yields

$$\frac{N_O(E_2)}{N_O(E_1)} = \exp\frac{2\tilde{F}}{RT} [(E_2 - E_1)]. \qquad (69)$$

Equation (68) can be written as

$$I = B \sum_{n=1,3,5\ldots}^{n} \exp\left(\frac{-n^2 t}{\tau}\right) \qquad (70)$$

where

$$B = [N_O(E_2) - N_O(E_1)] \frac{16\tilde{F}A D_O}{L V_{Cu}}$$

and

$$\tau = 1^2/\pi^2 D_O.$$

$$I \approx B \exp(-t/\tau) \quad \text{for} \quad t/\tau > 0.5.$$

A plot of log I versus t is thus asymptotic to a line with a slope of $-1/2.303\tau$ and an intercept of log B. D_O is obtained from the slope. $N_O(E_1)$ can be calculated from this value for D_O and the intercept value of log B on substituting these into eqn (69). The saturation solubility of oxygen in copper can also be calculated on applying Sieverts law. Figure 29 shows the relation between log I and time obtained by Pastorek and Rapp (1969) in their determination of D_O for solid copper.

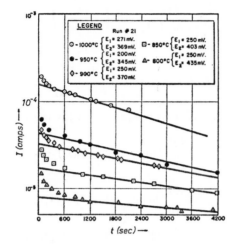

FIG. 29. Linear relations between logarithmic current and time at 800–1000°C under potentiostatic conditions. (after Pastorek and Rapp, 1969).

Galvanostatic experiment

In a galvanostatic experiment the solid copper is brought to equilibrium with an oxygen activity which corresponds to that for which $FeO–Fe_3O_4$ coexist. This is done by short-circuiting the cell. A predetermined constant current is then passed through the left side of the cell. A constant flux of oxygen atoms is thus made to enter the left-hand face of the cylindrical copper electrode. The cell voltage E is measured as a function of time up until Cu_2O is formed at the electrode. The solution of the diffusion equation which is required is that for one dimensional diffusion of oxygen into a semi-infinite copper specimen where the initial condition is

$$N_O = N_O(t_0) \quad \text{for} \quad x \geqslant 0, \qquad t = 0$$

and boundary conditions are

$$N_O = N_O(t_0) \quad \text{for} \quad x = \infty, \qquad t > 0$$

$$\left(\frac{\partial N_O(x, t)}{\partial x} \right)_{x=0} = - \frac{I V_{Cu}}{2 \tilde{F} A D_O} \quad \text{for} \quad t > 0.$$

A solution is given (Crank, 1956) by

$$N_O(x, t) = N_O(t_0) + \frac{V_{Cu} \, I t^{\frac{1}{2}}}{\tilde{F} A \, \pi^{\frac{1}{2}} D_O^{\frac{1}{2}}} \exp\left(-\frac{x^2}{4 D_O t}\right)$$

$$- \frac{I V_{Cu} \, x}{2 \tilde{F} A \, D_O} \operatorname{erfc}\left(\frac{x}{2(D_O t)^{\frac{1}{2}}}\right) \cdots \tag{71}$$

At $x = 0$,

$$N_O(t) = N_O(t_0) + \frac{V_{Cu} \, I t^{\frac{1}{2}}}{\tilde{F} A \, \pi^{\frac{1}{2}} D_O^{\frac{1}{2}}} \cdots . \tag{72}$$

A combination of eqn (72) with eqn (66) yields the following expression:

$$E(t) - I R_0 = \frac{RT}{2\tilde{F}} \left\{ \ln f_0 + \ln\left[N_O(t_0) + \frac{V_{Cu} \, I t^{\frac{1}{2}}}{\tilde{F} A \, \pi^{\frac{1}{2}} D_O^{\frac{1}{2}}} \right] \right\}$$

$$- \frac{RT}{4\tilde{F}} \ln \left[p_{O_2}''(\text{FeO–Fe}_3\text{O}_4) \right]. \tag{73}$$

Except for short times (5 sec or less at 1000°C), $V_{Cu} I t^2 / \tilde{F} A \, \pi^{\frac{1}{2}} D_{O_2}^{\frac{1}{2}} \gg N_O(t_0)$, so that a plot of $[E(t) - I R_0]$ versus $\ln I t^{\frac{1}{2}}$ is linear and with slope $(RT/2\tilde{F})$. A plot of $(p_{O_2}''^{\frac{1}{2}}/f_0) \exp 2\tilde{F}(E - I R_0)/RT$ versus $I t^{\frac{1}{2}}$ is also linear.

The slope of such a plot is $V_{Cu}/FA \, \pi^{\frac{1}{2}} D_O^{\frac{1}{2}}$ and there is an intercept at $N_O(t_0)$. The results obtained by Pastorek and Rapp (1969) show such a linear dependence. This is taken as direct experimental proof that the rate is controlled by the diffusion rate. As discussed above, the potential in the case of a semi-infinite metal phase is independent of time when the rate-controlling step is an interface reaction.

Danckwerts (1951) has solved Fick's second law for the case where the rate-controlling mechanism is mixed both by diffusion and chemical reaction. Using the previous equations based upon the assumption that only diffusion is rate controlling, one may obtain smaller D_O values than the true value when the actual mechanism is a mixed one. Despite this however, the diffusion constant calculated from the diffusion equations are not always smaller then those D_O obtained by other methods.

In order to measure D_O accurately one has to overcome two experimental difficulties. Firstly, there has to be no polarization at the interface of the reference electrode and secondly, a close contact at the metal/electrolyte interface is required.

Iskoe and Worrell (personal communication) have measured the degree of polarization of the reference electrode as a function of the applied current density and temperature. The polarization for a Cu, Cu_2O reference electrode was found to be much smaller than that for FeO–Fe_3O_4. According to their work, a close uniform interface between the reference electrode and the solid electrolyte is very important.

Another way to determine the diffusion constant in metals by the e.m.f. method is to measure the time-dependent e.m.f. without current flow (Goto *et al.*, 1968b, 1969c; Masson and Whiteway, 1967; Sano *et al.*, 1969) after the oxygen pressure of gas phase is rapidly changed from one pressure p_{O_2}' to another constant pressure p_{O_2}'''. A solution of Fick's second law can be contructed to describe the process and D_O can then be obtained as in the previous examples. Since the principle is the same, details of the calculation will not be repeated here.

B. Depolarization and Decay Curves

This short section will be devoted to a qualitative discussion of the shapes of e.m.f. decay curves. One is led to conclude that the redox reaction, $O^{2-} = O + 2e'$ is faster than the diffusion rate of oxygen in the liquid metal.

Suppose the cell is given by

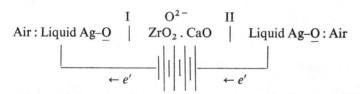

$$
\begin{array}{ccc}
\text{I} & O^{2-} & \text{II} \\
\text{Air : Liquid Ag–O} & \text{ZrO}_2 . \text{CaO} & \text{Liquid Ag–O : Air}
\end{array}
$$

$\leftarrow e'$ $\leftarrow e'$

A current flow will initiate polarization changes on both sides of the electrolyte and after the current is cut off the e.m.f. will decay with time to its initial zero value if such a cell is used.

When the rate of the depolarization process is controlled by the interface redox reaction, then as discussed in the previous section, the e.m.f. after the current cut-off can be expressed as follows (Goto *et al.*, 1970):

$$
E = \frac{RT}{2\bar{F}} \ln \frac{[(p_{O_2}^{\frac{1}{2}})_{t=0}{}^{\text{I}} - (p_{O_2}^{\frac{1}{2}})_e] \exp - kt + (p_{O_2}^{\frac{1}{2}})_e}{[(p_{O_2}^{\frac{1}{2}})_{t=0}{}^{\text{II}} - (p_{O_2}^{\frac{1}{2}})_e] \exp - kt + (p_{O_2}^{\frac{1}{2}})_e}. \tag{74}
$$

In this equation $(p_{O_2})_{t=0}{}^{\text{I}}$ and $(p_{O_2})_{t=0}{}^{\text{II}}$ are the oxygen pressures at the left I and the right II interfaces immediately after the current cut-off. $(p_{O_2})_e$ is the oxygen pressure in equilibrium with the dissolved oxygen in liquid silver, and k is the isothermal rate constant. The electromotive force, E in eqn (74) is zero at $t = \infty$. If the diffusion is slow an applied current will cause a non-steady state concentration gradient of oxygen in the liquid silver. The distribution of oxygen is then given by eqn (71) for the case of a semi-infinite medium under a galvanostatic condition. The equation in the previous section is complicated. The result may however be simplified by slightly changing the shape of the concentration distribution. For the purpose of the present argument it will be assumed that

$$\frac{[O] - [O]_\infty}{[O]_{t=0}{}^i - [O]_\infty} = \exp - K_1 x \quad \text{(for the left cell compartment)}$$

or $\exp - K_2 x$ (for the right cell compartment).

The e.m.f. is then given (Goto et al., 1970) by

$$E = \frac{RT}{2\tilde{F}} \log \left\{ 1 + \frac{2([O]_{t=0} - [O]_\infty)}{\pi^{\frac{1}{2}}[O]_\infty} [(4DK_1{}^2)^{-\frac{1}{2}} \cdot t^{-\frac{1}{2}} - 2(4DK_1{}^2)^{-\frac{3}{2}} t^{-\frac{3}{2}} \right.$$
$$\left. + 12(4DK_1{}^2)^{-5/2} t^{-5/2} - \ldots] \right\}$$
$$- \frac{RT}{2\tilde{F}} \log \left\{ 1 + \frac{2([O]_{t=0}{}^{II} - [O]_\infty)}{\pi^{\frac{1}{2}}[O]_\infty} [(4DK_2{}^2)^{-\frac{1}{2}} t^{-\frac{1}{2}} \right. \tag{75}$$
$$\left. - 2(4DK_2{}^2)^{-\frac{3}{2}} t^{-\frac{3}{2}} + 12(4DK_2{}^2)^{-5/2} t^{-5/2} - \ldots] \right\}$$

$$\text{(when } (4Dt)^{\frac{1}{2}} K_1, \quad (4Dt)^{\frac{1}{2}} K_2 \gg 1 \cdot 0)$$

where $[O]_{t=0}{}^I$ and $[O]_{t=0}{}^{II}$ are the interface oxygen contents in the left and the right cell compartments immediately after the current cut-off; K_1 and K_2 are constants determined by the concentration distribution at $t = 0$ and D is the diffusion constant of oxygen in liquid silver. E again tends to zero as t tends to infinity. Goto et al., (1970) obtained the following results: (1) the degree of polarization is increased both by increasing current and by increasing the time, (2) an increase in the metal–electrolyte interface area decreases the degree of the polarization, (3) the rate of depolarization is accelerated when the liquid silver is stirred. From eqn (74),

$$\log_{10} \left[\frac{2EF}{10^{2 \cdot 3} RT} - 1 \right]$$

is proportional to the time when the depolarization is controlled by the first order reaction rate of absorption or desorption of oxygen to or from the interface. However, Fig. 30 shows that the relation is not linear. According to eqn (75), the gradient dE/dt is zero at $t = \infty$. Thus the gradient of the curves in Fig. 30 must also be zero at $t = \infty$.

The solution given above for the diffusion equation is not valid for small values of t and in fact the gradient is $-\infty$ as $t \to 0$. From the results of the above discussion it can again be concluded that diffusion is the rate controlling mechanism.

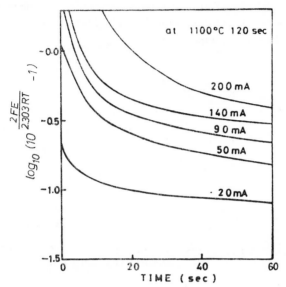

FIG. 30. Relation between time and

$$\log \left[10 \frac{2EF}{2 \cdot 3RT} - 1 \right]$$

calculated from e.m.f. decay curves. (after Goto et al., 1970).

C. Rate of Gas–Solid Reactions

Oxygen concentration cells can be used to determine not only the diffusion constant of oxygen but also the rate constants of slow surface reactions such as the oxidation of iron to wüstite by a $CO–CO_2$ gas mixture.

In the experiment a thin metal film is fixed to the gas-tight polished plate of $ZrO_2 . CaO$, which separates two compartments with different oxygen pressures. If the metal is then oxidized the oxygen pressure at the metal/electrolyte interface is maintained at that of the two phase equilibrium until the metal is completely consumed. An apparent rate constant for the oxidizing reaction can be calculated from the time taken to achieve a constant e.m.f. when the rate-controlling process is that of a slow surface reaction while all the diffusion steps are rapid.

Tare and Schmalzried (1966) have determined the rate constant for oxidation of iron foils to wüstite using this method. Figure 31 shows the e.m.f.

change with time for an iron foil $5\,\mu m$ thick which was fixed onto a $ZrO_2 . CaO$ electrolyte and then oxidized by $CO–CO_2$ gas at 960°C. In Fig. 31, part A is the time necessary for dissolution of oxygen into the iron foil until the nucleation of wüstite. Part B is the time to oxidize through all the thickness of the iron foil. Part C reflects the change in the composition of wüstite. The e.m.f. is constant with time when the oxide and the gas phase are in complete equilibrium.

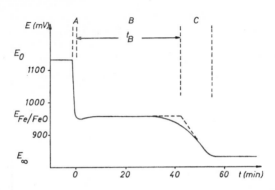

FIG. 31. Electromotive force *vs* time curve for the oxidation of an iron foil to wüstite. $T = 960°C, \Delta = 5\mu, k = 3 \cdot 7 \times 10^{-8}$ (mol/cm² s. atm). (after Tare and Schmalzried, 1966.)

The rate of the surface reaction between wüstite and the oxidizing $CO–CO_2$ gas mixture may be formulated for a low oxygen coverage (Grabke, 1965) as

$$\frac{dn_0}{dt} = k_1 p_{CO_2} - k_2 p_{CO} \Gamma_{O(ad)} \tag{76}$$

where n_0 = number of moles of oxygen which react per cm² and $\Gamma_{O(ad)}$ = surface concentration of absorbed oxygen atoms, k_1 and k_2 are the rate constants. Using the equilibrium condition $dn_0/dt = 0$ and integrating over t_B, one obtains k_1 defined by the above equation, if the relation

$$\int dn_0 = \Delta . \rho_{Fe}/M_{Fe} . (1 - x)$$

is taken into account:

$$k_1 = \frac{\Delta . \rho_{Fe}}{t_B . M . (1 - x)} \left[p_{CO_2} \left(1 - \frac{p_{CO}}{p_{CO_2}} \left(\frac{p_{CO_2}}{p_{CO}} \right)_{Fe/FeO} \right) \right]^{-1} \tag{77}$$

Δ = thickness of iron foil, ρ = density, M = atomic weight of iron, $(1 - x)$ = composition of $Fe_{1-x}O$ in equilibrium with iron. The values of k_1 so obtained were in the range of $1 \cdot 4 \times 10^{-8}$ to $6 \cdot 5 \times 10^{-8}$ (mol/cm² · sec atm) at 850°C–975°C. These values are about a factor of two higher than those obtained by Grabke (1965). The disagreement in the measured values seems to be due in part to a non-ideal contact between the iron and ZrO_2. CaO and in part to the rough wüstite surface which grows on the iron foil. Both effects would result in larger values for the constant. In Fig. 31 there is a minimum value for the e.m.f. between A and B. This minimum can be considered as supersaturation to form wüstite nuclei on the surface of the iron foil in a CO_2–CO mixture. The value of the supersaturation ranged from about 5 to 8%.

Although the principle is very attractive, it is extremely difficult in experiments to establish an absolute dense contact with the electrolyte. Only one report has as yet been published on this method (Tare and Schmalzried, 1966).

D. Oxygen Pressure Change in Liquid Oxide Phases

(a) Experimental procedure and results

As already discussed, the e.m.f. of an oxygen concentration cell is determined by the oxygen pressures at the interfaces between the solid electrolyte and the metallic electrodes. Figure 32 (Sasabe et al., 1970) shows schematically a cell assembly which can be used to measure the oxygen pressure change at the bottom of liquid oxides, when the oxygen pressure in the gas phase is changed from one constant value to another. The platinum wire insulated by an alumina tube is in contact with the ZrO_2. CaO electrolyte only at bottom of the melt. When the cell assembly is stabilized at a certain experimental temperature, the oxygen pressure in the gas phase changes rapidly from 0·21 atm down to about 10^{-4} atm. The e.m.f. of the cell changes

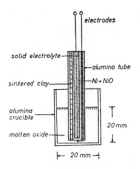

FIG. 32. Schematic diagram of a cell assembly used to measure oxygen pressure changes in liquid oxide phases. (after Sasabe, Goto and Someno, 1970).

linearly after an initial period of about 300 s and its value depends on the oxygen pressure change at the bottom of the liquid oxides. Binary liquid solutions of various compositions of $PbO-SiO_2$, $PbO-GeO_2$ and Na_2O-GeO_2 have been studied at temperatures between 900°C and 1300°C. These melts do not dissolve solid $ZrO_2 \cdot CaO$ electrolyte during prolonged experimental runs.

(b) An interpretation of oxygen pressure change in liquid oxides

There is no direct observation on solubilities of oxygen gas or metallic lead in the melts studied here. However, one can assume that the distribution coefficient between gas and liquid phases is nearly one when the same units are used for the concentration of oxygen or metal vapours in gas and liquid phases, e.g. mol/cm^3 of gas or melt. This unit distribution coefficient may be regarded as a first approximation which does not include the effects of liquid oxide composition and of temperature. The correction is however in most cases less than a factor of 10.

The assumption of a unit distribution coefficient is applicable in some other cases, such as solubilities of various alkali metal vapours in halide melts (Breding, 1964; Janz, 1967), of HF gas in molten $NaF-ZrF_4$ mixtures (Schaefer et al., 1959), of oxygen in molten carbonates (Schenke et al., 1966), and metal vapours in silicate melts (Meyer and Richardson, 1961–62; Richardson and Billington, 1955–56).

One can estimate the solubility of oxygen gas to be about 1.3×10^{-5} mol O_2/cm^3 of melt at 1000°K and under pure O_2 gas at 1 atm. This result is on the assumption of a unit distribution coefficient. The concentration of neutral Pb° in $PbO-SiO_2$ melts can be estimated from the equilibrium of the following reaction:

$$Pb(vapour) + \tfrac{1}{2}O_2(gas) = PbO(in\ PbO-SiO_2)$$

The estimated value is 6.5×10^{-17} mol Pb/cm^3 of melt at 1000°K in a $PbO-SiO_2$ melt with $a_{PbO} = 0.5$ under pure oxygen gas at 1 atm. Since it is not known whether oxygen dissolves as mono- or di-atomic molecules both types will be considered in the following discussion.

The oxygen pressure may equally well be described by

$$O_2(gas) = [O_2]\quad \text{or}\quad 2[O°]\ (dissolved)$$

by

$$[Pb°] + \tfrac{1}{2}O_2(gas) = Pb^{2+} + O^{2-}$$

or

$$Pb^{2+} + \tfrac{1}{2}O_2(gas) = Pb^{4+} + O^{2-}.$$

These reactions are not independent but related by

$$[Pb^\circ] + \tfrac{1}{2}[O_2] \quad \text{or} \quad [O^\circ] = Pb^{2+} + O^{2-}$$

and

$$Pb^{2+} + \tfrac{1}{2}[O_2] \quad \text{or} \quad [O^\circ] = Pb^{4+} + O^{2-}.$$

The concentrations of Pb^{2+} and O^{2-} are virtually constant. The oxygen pressure change will be controlled by the diffusion rate of that species among $[O^\circ]$, $[O_2]$, $[Pb^\circ]$, and Pb^{4+} which has the largest concentration. The diffusion of a species which has a small concentration range can hardly affect the oxygen pressure in the melt, even when its diffusion constant is large. Diffusion processes take place simultaneously with rapid homogeneous chemical reactions.

One may estimate the value of the diffusion constant for the neutral species with the largest content from the measured e.m.f. change, if the concentration ratio, for example, $[O^\circ]/[O^\circ]^i$ is calculated from Fick's second law. For the boundary conditions of the present problem (neglecting the change in diffusion path by dipping the cell):

$$\frac{[O^\circ]}{[O^\circ]^i} = \frac{4}{\pi} \sum_{n=0}^{\infty} \frac{(-1)^n}{2n+1} \exp - \frac{D(2n+1)^2 \pi^2}{4L^2} t \cos (2n+1) \frac{\pi x}{2L}$$

$$\text{(for large } t) \quad (78)$$

where $[O^\circ]$ is the concentration of the monoatomic oxygen in the melt at time t and at a distance x from the bottom and $[O^\circ]^i$ is the initial concentration, L is the depth of oxide melt and D is the diffusion constant of $[O^\circ]$ (see Fig. 32). From the relation $[O^\circ] = K p_{O_2}^{\frac{1}{2}}$, the change of e.m.f. at time t, $\Delta E_t (=$ the initial e.m.f. minus e.m.f. at time $t)$ can be obtained from

$$\Delta E_t = \frac{RT}{2\tilde{F}} \ln \frac{p_{O_2}^{\frac{1}{2}(\text{in melt})}}{p_{O_2}^{\frac{1}{2}(\text{initial})}} = \frac{RT}{2\tilde{F}} \ln \frac{[O^\circ]}{[O^\circ]^i}. \quad (79)$$

For $x = 0$ (the bottom), $R = 1\cdot98$ cal/deg mol and $F = 23{,}070$ cal/equiv, the e.m.f. change is found to be

$$\Delta E_t = 0\cdot1023 \, T \, \frac{Dt}{L^2} - 0\cdot00785 \, T. \quad (80)$$

The measured e.m.f. changes always satisfy a linear relation with time after about 300 s. Since it is impossible to determine which species determines the pressure change, the calculated diffusion constant will be referred to as a "diffusion constant of the neutral species" and will be designated by D°.

(c) A comparison of "the diffusion constant of the neutral species" with those of cations

In Fig. 33 $D°$ values are given for various compositions of Na_2O–GeO_2 and PbO–GeO_2 melts. Values of $D°$ for PbO–SiO_2 melts are about the same and are not plotted. For comparison, Fig. 33 shows the tracer diffusion constants of some cations in silicate melts. Some of these were calculated for $f = 1·0$ and $t_i = 1·0$ on the basis of the Nernst–Einstein equation (23) given in the previous section (III.A).

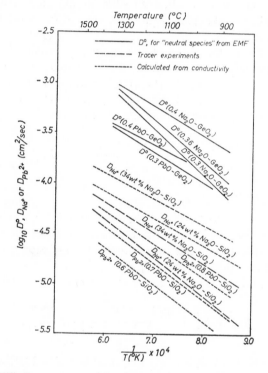

Fig. 33. Diffusion constants of "neutral species" and cations in oxide melts.

From Fig. 33, it can be seen that the "diffusion constants of the neutral species" are much larger than those for charged particles. The difference is larger than the experimental or calculation errors for charged particles. Mulfinger and Scholze (1962) measured the diffusion constant of dissolved helium gas in various liquid silicates. The diffusion constant was found to be in the range of 10^{-3}–10^{-4} cm^2/sec at 1480°C–1200°C and decreased with the silica content. The results given in Fig. 33 are in the same range and the diffusivity also decreases with the increase of acid oxides. The only other possible mechanism by which the oxygen pressure in an oxide melt can be changed is one for which there is a counter diffusion of cations of different valences coupled with a diffusion of anions to maintain charge neutrality. For example, if Pb^{3+} or Pb^{4+} with O^{2-} diffuses from the melt surface toward the bottom against a counter diffusion of Pb^{2+}, the oxygen pressure in the melt will be increased. This mechanism is, however, considered slow and is probably controlled by the slow diffusion of O^{2-} anions. A comparison of diffusion constants in Fig. 33 suggests that the rate of oxygen pressure change in oxide melts is controlled by diffusion of the dominant neutral species.

(d) *The validity of the Nernst–Einstein equation when applied to the case of oxide melts*

The validity of the equation is an interesting problem in liquid oxide melts and depends upon errors in correlation factors, in number of charge carriers, in transference numbers, in electrical conductivity and in tracer diffusion constant. Many investigators (Adachi *et al.*, 1965; Bockris and Mellors, 1956; Bockris *et al.*, 1948, 1955, 1956; Dancy and Derge, 1966; Goto, 1969; Goto *et al.*, 1969d; Ito and Yanagase, 1960; Kato and Minowa, 1969; Mackenzie, 1956; Saito *et al.*, 1969) have studied the physical properties of liquid oxide systems. In general the experimental accuracy has not always been as good as in the case of fused halides or aqueous solutions. This difference is due essentially to the high experimental temperatures required for liquid oxide systems. From a combined study of the electrolysis of liquid oxides (Bockris and Mellors, 1956; Bockris *et al.*, 1952; Dancy and Derge, 1966; Dickson and Dismukes, 1962; Dukelow and Derge, 1960; Ito and Yanagase, 1958) and Hittorf transference numbers (Bockris and Mellors, 1956; Dancy and Derge, 1966), one can conclude that cationic transference numbers are unity in silicate and borate melts of alkali and alkali earth metal oxides, while electronic conduction is significant in the melts containing transition metal oxides such as FeO$_x$ of more than several per cent. Serious disagreement in the conductivity measurements of liquid oxides can arise as a result of the frequency and conductivity dependence of the cell constant. The conductivity of a melt might at a particular composition be so high that the very small resistance of the liquid would be masked by the temperature dependent resistance of the connecting wires and electrodes. The frequency dependence of the polarization resistance of liquid oxides has not yet been clarified. The errors in the estimate of the number of charge carriers depends not only on the reliability of the density measurements but also on the decomposition constants of the "basic" oxide (e.q. Na_2O, PbO, CaO etc.) However it has been said that almost all basic oxides are decomposed to yield free cations and anions. Some portions of free oxygen anions are consumed by the formation of silicate or borate anion polymers. A comparison of the values of $D_{Ca^{2+}}$ determined by three investigations (Niwa, 1957; Saito and Maruya, 1958; Towers et al., 1953) shows that the relative error in the diffusion constant of cations when measured by a tracer technique is about 20%. The determination of the correlation factor is difficult in the case of liquid oxides. A value of 0·7 will be used as a first approximation. This value is based on that obtained for various solids (Compaan and Haven, 1956; Shewmon, 1963).

A reasonable estimate of the range of the relative errors involved in the calculation of a tracer diffusion constant by the Nernst–Einstein equation is that which ranges between about $+30\%$ and -50%. This estimate is based on the following relative errors: $\sigma = \pm20\%, f = -30\% \ (f = 0·7), n_i = \pm5\%$ (errors only in density measurements), $t_i = \pm5\%$, and $T = \pm0·3\%$. The errors refer to measurements on liquid oxides at temperatures between 700°C and 1400°C. The present authors have calculated the self diffusion constants of Pb^{2+} and Na^+ in silicate melts, using the electrical conductivity and the density measurements of Bockris and Mellors (1956) together with a value of the thermal expansivity (White, 1955) given by $1·0 \times 10^{-4} [°C]^{-1}$. The results on Pb^{2+} are shown in Fig. 33. Gupta and King (1967) have used the capillary-reservoir technique (Fig. 33) to measure the tracer diffusion constant of Na^+ in Na_2O–SiO_2 melts at 850–1500°C. They have also calculated D_{Na^+} using the conductivity measured by Hutchins (1959) (Fig. 33). Based upon $t_{Fe^{2+}} = 0·9$ Simnad et al. (1956) have calculated $D_{Fe^{2+}}$ and the activation energy for a SiO_2 saturated FeO melt at 1250°C. Their values of $5·8 \times 10^{-5}$ cm^2/s and 16 kcal/mol are to be compared with the measured values of $7·9 \times 10^{-5}$ cm^2/s and 40 kcal/mol obtained by Yang and Derge (1959). A similar calculation in the case of $40\% \ CaO$–$20\% \ Al_2O_3$–$40\% \ SiO_2$ at 1250°C gave a value of $1·9 \times 10^{-7}$ for $D_{Ca^{2+}}$ and an activation energy of 25 kcal/mol and these are to be compared to the measured values of $1·3 \times 10^{-7}$ cm^2/s and 70 kcal/mol obtained by Towers and Chipman (1957). Simnad et al. (1956) have measured the mobility of Fe^{2+} in a $34\% \ SiO_2$–FeO melt at 1250°C under an electrical potential gradient and as stated above have calculated $D_{Fe^{2+}}$ to be $5·8 \times 10^{-5}$ cm^2/s. Their estimate of the Fe^{2+} concentration based on the Nernst–Einstein equation was 0·0057 moles of Fe^{2+} cm^3 for a value of $t_{Fe^{2+}} = 0·9$ and a mobility of 9×10^{-4} cm^2/s volt. A value of 0·034 moles of Fe^{2+}/cm^3 was obtained from a calculation which used only the density and composition. The above comparison between the calculated and measured tracer diffusion constants suggests that the range of $+30\%$ and -50% for the possible error is reasonable.

E. Oxygen Pressure Distribution in a Gas Phase During a Reaction

There exists a concentration gradient in the gaseous species at the gas phase close to the reaction surface whenever a heterogeneous reaction between a gas phase and a condensed phase is controlled by a slow rate of mass transport in the gas phase. Figure 34 shows schematically an oxygen concentration cell (Goto *et al.*, 1968a) which can be used to measure the oxygen pressure distribution in the gas phase close to the reaction surface of liquid iron during

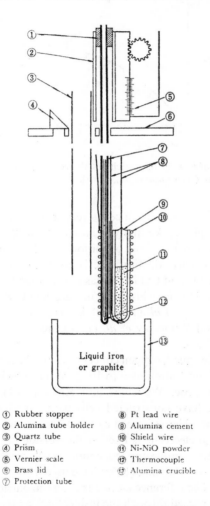

① Rubber stopper	⑧ Pt lead wire
② Alumina tube holder	⑨ Alumina cement
③ Quartz tube	⑩ Shield wire
④ Prism	⑪ Ni-NiO powder
⑤ Vernier scale	⑫ Thermocouple
⑥ Brass lid	⑬ Alumina crucible
⑦ Protection tube	

FIG. 34. Schematic diagram of an oxygen concentration cell which can be used to measure the oxygen pressure distribution in the vicinity of a reaction surface. (after Goto *et al.*, 1968a).

decarburization by the reaction $CO_2 + C_{(Fe-liq.)} = 2 CO$. It is possible to measure both the temperature and the electromotive force of the oxygen concentration cell as a function of distance from the liquid iron surface on which the decarburization reaction takes place. The distribution of CO_2 across such a cell calculated from the measured e.m.f. and the temperature is shown in Fig. 35.

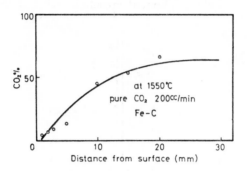

FIG. 35. Distribution of CO_2 in the vinicity of the reaction surface of liquid Fe–C alloys during decarburization. (after Goto *et al.*, 1968a).

It has been observed that such a cell is very sensitive to rapid concentration changes in the gas phase. An analysis based on the $\%CO_2$ distribution and the observed rates under various conditions has led Goto *et al.* (1969b) to conclude that the decarburization rate is controlled predominantly by the transport rates of CO_2 and CO in the gas phase when the iron contains more than 0·2% carbon. This conclusion is in agreement with the results of Ito and Sano (1964), Gunji *et al.* (1964), Baker *et al.* (1964), Löscher (personal communication) and Whiteway *et al.* (1968) but not with those of Swisher and Turkdogan (1967).

When properly designed, oxygen concentration cells can be used to measure oxygen pressures from 1 atm down to 10^{-30} atm in flowing high temperature gases. For this purpose, Weissbart and Ruka (1961), Goto (1962), and Littlewood (1964), have constructed cells with very long electrolyte tubes with one closed end. The reference electrode is at the inside bottom of the tube and a platinum electrode is fixed at the outside bottom. High temperature gases for analysis are supplied continuously to the outside platinum electrode. Möbius and Hartung (1965) have used a solid electrolyte tube with both ends opened. The reference electrode is on the outside and the gas which is to be analysed is passed through the inside. Such an oxygen gauge has been used to measure the permeability of various oxides at high temperatures. "Oxygen gauges" are now commonly used to monitor oxygen pressures in flowing high temperature gases both in the laboratory (Derge, 1969; El-Naggar *et al.*, 1967; Goto and Matsushita, 1967; Möbius and Hartung, 1965; Whiteway *et al.*, 1968; Yuan and Kröger, 1969) and in industry (Matsushita *et al.*, 1966).

F. The Intra-Particle Oxygen Pressure Change during a Reaction

In chemical engineering analyses (Barner *et al.*, 1963; Bogdandy, 1958; Bogdandy and Engell, 1967; Edström and Bitsianes, 1955; Hara, 1966; Landler and Komarek, 1966; Mori, 1960; Moriyama, 1965; Nakamura, 1964; Spitzer *et al.*, 1966; Tokuda, 1965) of the reduction of sintered oxide particles, only the apparent rate determined by a thermobalance has been used so far. However, if the intra-particle gas composition during the reduction is known, this information would increase the accuracy of the analyses. The oxygen pressure in the pores of tamped oxide powder has been measured (Kashida *et al.*, 1968) during reduction by reducing gas mixtures. Oxygen concentration cells as shown in Fig. 36 have the reducing gas flowing on the outside of the cell assembly. Reduction only takes place from the left side. Figure 37 shows the change of the electromotive force and the gas

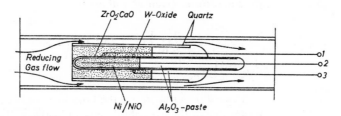

FIG. 36. Cell assembly used to measure the intra-particle gas composition during the reduction by hydrogen gas. (after Kashida *et al.*, 1968).

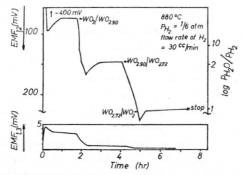

FIG. 37. Change of the electromotive force and the gas composition in pores of WO_3 with time at 880°C. (after Kashida *et al.*, 1968).

composition in pores during reduction of WO_3 at 880°C by H_2–A mixture ($p_{H_2} = 0.1667$ atm).

The lower part of the figure shows the electromotive force between the two measuring electrodes 1 and 3 located at 15 and 35 mm from the WO_3 surface exposed to the H_2–A gas mixture. The oxygen pressure at 15 mm was always slightly smaller than that at 35 mm. It is clear that the intraparticle gas composition is constant over a fairly long time. This corresponds to the equilibrium value of WO_3 and $WO_{2.90}$, $WO_{2.90}$ and $WO_{2.72}$, $WO_{2.72}$ and WO_2, or WO_2 and W (St. Pierre et al., 1962). The results for other oxide powders also show that the gas composition in pores does not change smoothly during reduction or oxidation but rather that it is constant for a certain time at the two phase equilibrium composition values. This is irrespective of whether or not the reaction takes place topochemically. Similar changes in e.m.f. have been observed by Tare and Schmalzried (1966) in the case of oxidation of iron as in Fig. 31. The magnitude of "supersaturation" shown in Fig. 37 has been measured for the reduction of porous oxide phases.

Pluschkell and Yoshikoshi (1970) have measured the growth rate of the iron phase on single crystals of wüstite by means of a high temperature microscope. The growth rate is controlled by the reaction rate at the ternary phase reaction zone. The rate equation is given as follows:

$$r = \frac{b}{h} k \, V_{Fe} \frac{1}{1 + K_0 \left(\frac{p_{H_2O}}{p_{H_2}}\right)} p_{H_2} \left[1 - \left(\frac{p_{H_2}}{p_{H_2O}}\right)_{eq.} \left(\frac{p_{H_2O}}{p_{H_2}}\right)\right] t \qquad (81)$$

where r is radius of the reduced iron phase, b is the broadness of the reaction zone, h is the height of the edge of the iron phase, k is an isothermal reaction rate constant, V_{Fe} is the molar volume of iron, K_0 is the oxygen absorption coefficient, p_{H_2} and p_{H_2O} are the partial pressures of hydrogen and water vapour, and (p_{H_2}/p_{H_2O}) eq. is the equilibrium gas ratio for the coexistence of iron and wüstite. From this equation, it is clear that during the initial stages of the reduction the growth rate of the iron phase (which corresponds to the rate of consumption of the reactant gases) is proportional to the total area of the three phase reaction zone. Pluschkell and Yoshikoshi have shown quantitatively that the reduction rate is rapidly accelerated by the formation of the iron phase. An "autocatalytic effect" for metal phases in reduction processes has been reported by Pease and Taylor (1921, 1922) in the case of Cu–CuO and by Benton and Emmett (1924) for Ni–NiO. The super-saturation observed as in Fig. 37 can be interpreted as follows. At the start of the reaction, the area of the reaction zone of the ternary phase co-existence is so small that the reactant gas which is supplied cannot be consumed. When the total area of the reaction zone is increased following the growth of the new phase all of the transferred reactant gas is consumed. The gas is thus maintained at an equilibrium concentration. When a catalytic action is defined as one for which there is either an increase in the number of reaction sites or a decrease in the activation energy or both, as the result of a changing reaction mechanism, the growth of the new phase can be said to be accelerated by an "auto-catalytic action". It seems clear that absolute values of the "super-saturation" are not reproducible at high temperature solid surfaces.

From the above discussions, one may assume that oxide phases formed at all the steps of the $WO_3 \rightarrow WO_{2.90} \rightarrow WO_{2.72} \rightarrow WO_2 \rightarrow W$ sequence are also auto-catalytic in both directions of reduction and oxidation.

VII. Conclusions

The study of the action of concentration cells is a very specialized problem in electrochemistry. There have nevertheless been a great number of investigations which have concerned themselves with this problem in the last decade. The study of concentration cells has been shown to have a wide applicability to thermodynamic and kinetic studies. In the present chapter, the authors have tried to collect and summarize all the information on oxygen concentration cells which use solid electrolytes. Fuel cell studies have however been excluded. The chapter may be summarized as follows. In Section II the theoretical electromotive forces generated in oxygen concentration cells with solid electrolytes have been discussed. Cases where the conduction mechanisms are purely anionic, purely cationic, or of the mixed type have been treated. Special limiting cases of the e.m.f. equation, which is a function of oxygen pressures and transference numbers of species in electrolytes, have been discussed. Section III deals with the information collected in the case of special solid electrolytes. The use of ZrO_2 and thoria base solutions, oxides such as MgO, Al_2O_3, SiO_2 and of some ternary oxides has also been treated. Special emphasis has been given to the future use of ternary oxide compounds as solid electrolytes. In Section IV the possible experimental errors which arise in e.m.f. measurements at high temperatures have been discussed. Estimates of the magnitude of various kinds of errors have been given. Section V deals with examples of the application to thermodynamic studies. Methods of measuring the standard free energy of reactions and of estimating the chemical activity of components in alloys and in oxide solutions have been presented. A discussion of the coulometric titration of oxygen in metals and in non-stoichiometric oxides and of the oxygen content in alloys has also been included. Specific discussion has been given on: (1) the selected values of standard free energies of reactions which involve WO_3, $WO_{2.90}$, $WO_{2.72}$, WO_2 and W, (2) liquid oxide electrolytes for oxygen concentration cells,

(3) the coulometric titration of non-stoichiometric oxides, and (4) the problems connected with the determination of the oxygen content in liquid alloys. In Section VI, examples of applications to kinetic studies have been given. The diffusivity of oxygen in liquid or solid copper and silver has been measured by potentiostatic, galvanostatic and other methods. The effect of depolarization and the shapes of decay curves of the e.m.f. have been discussed. The way in which the reaction rate of oxidation of pure iron to wüstite can be determined from the time-dependent e.m.f. has been given. The mechanism by which the oxygen pressure changes in liquid oxide phase and the applicability of the Nernst–Einstein equation to such a system has also been discussed. Some comments on the rate-controlling step of decarburization of liquid iron-carbon alloys have been given. The change of intra-particle gas composition when WO_3 was reduced to metallic tungsten through intermediate steps has also been discussed.

The future study of concentration cells is likely to be concerned with their behaviour at temperatures up to 2000°C. The experimental errors discussed in Section IV will then be serious. If the transport properties of solid carbides, sulphides, and nitrides are elucidated, it seems that it will be possible in future to construct carbon, sulphur, and nitrogen concentration cells. However, in any case, the conduction mechanism of electrolytes and true virtual electrode reactions must be clarified in order to apply the e.m.f. methods to various studies.

Acknowledgment

The authors would like to express their sincere appreciation to Professor Carl Wagner of Max–Planck–Institut für Physikalische Chemie at Göttingen, who has given numerous constructive comments on the manuscript.

References

Abraham, K. P. (1969). *Trans. Indian Inst. Metals* 5–7.
Ackermann, R. J. and Rank, E. G. (1963). *J. phys. Chem.* **67**, 2596–2601.
Adachi, A., Ogino, K., Suetaki, T. and Saito, T. (1965). *Tetsu-to-Hagane* **51**, 1857–1860.
Alcock, C. B. and Belford, T. N. (1964). *Trans. Faraday Soc.* **60**, 822–835.
Alcock, C. B. and Zador, S. (1967). *Electrochim. Acta* **12**, 673–677.
Baker, R. and West, J. M. (1966). *J. Iron Steel Inst.* **204**, 212–216.
Baker, L. A., Warner, N. A. and Jenkins, A. E. (1964). *Trans. metall. Soc. A.I.M.E.* **230**, 1228–1235.
Barbi, G. B. (1964). *J. phys. Chem.* **68**, 1025–1029.
Barner, H. E., Manning, F. S. and Philbrook, W. O. (1963). *Trans. metall. Soc. A.I.M.E.* **227**, 897–904.

Baur, E. and Preis, H. (1937). *Z. Electrochem.* **43**, 727–732.

Bauerle, J. E. (1966). *J. chem. Phys.* **45**, 4162–4166.

Belford, T. N. and Alcock, C. B. (1965). *Trans. Faraday Soc.* **61**, 443–453.

Benton, A. F. and Emmett, P. H. (1924). *J. Am. chem. Soc.* **46**, II, 2728–2737.

Benz, R. and Schmalzried, H. (1961). *Z. phys. Chem.* **29**, 77–82.

Bidwell, L. R. and Speiser, R. (1965). *Acta. metall.* **13**, 61–70.

Bockris, J. O'M. and Mellors, G. W. (1956). *J. phys. Chem.* **60**, 1321–1328.

Bockris, J. O'M., Kitchener, J. A., Ignatowicz, S. and Tomlinson, J. W. (1952). *Trans. Faraday Soc.* **48**, 15–91, *Discuss. Faraday Soc.* (1948). **4**, 265–281.

Bockris, J. O'M., Kitchener, J. A. and Davies, A. E. (1952). *Trans. Faraday Soc.* **48**, 536–548.

Bockris, J. O'M., Mackenzie and Kitchener, J. A. (1955). *Trans. Faraday Soc.* **51**, 1734–1748.

Bockris, J. O'M., Tomlinson, J. W. and White, J. D. (1956). *Trans. Faraday Soc.* **52**, 299–310.

Bogdandy, L. von (1958). *Arch. EisenhüttWes.* **29**, 603–609.

Bogdandy, L. von and Engell, H. J. (1967). "Die Reduktion der Eisenerze" Springer und Stahleisan Verlag.

Bousquet, J. and Perachon, G. (1964). *Compt. Rend.* **258**, 934–936.

Breding, M. A. (1964). *In* "Molten salt chemistry", 367–425, M. Blander, ed. Interscience Publishers.

Callow, R. J. (1951). *Trans. Faraday Soc.* **37**, 370–376.

Carter, R. E. (1960). *J. Am. Ceram. Soc.* **43**, 448–452.

Carter, R. E. and Roth, W. L. (1967). *In* "Electromotive force measurements in high temperature systems", 125–144. Inst. Min. Metall. London.

Casselton, R. E. W. (1968). *In* "Electromotive force measurements in high temperature systems", 151–157. Inst. Min. Metall., London.

Catoul, P., Tyon, P. and Hans, A. (1967). *Centre Nat. Rech. Met.* **11**, 57–62.

Cavanaugh, C. R. and Elliott, J. F. (1964). *Trans. metall. Soc.* **230**, 633–638.

Charette, G. G. and Flengas, S. N. (1969). *Can. Met. Quarterly* **7**, 191–200.

Cocco, A. (1959). *Chimica Ind.* **41**, 882–886.

Compaan, K. and Haven, Y. (1956). *Trans. Faraday Soc.* **52**, 786–801.

Coughlin, J. (1954). *U. S. Bur. Mines, Bull.* No. 542.

Crank, J. (1956). "The mathematics of diffusion" Oxford University Press.

Cusack, N. and Kendall, P. (1958). *Proc. phys. Soc.* **72**, 870–898.

Danckwerts, P. V. (1951). *Trans. Faraday Soc.* **47**, 1014–1023.

Dancy, E. A. and Derge, G. J. (1966). *Trans. metall. Soc. A.I.M.E.* **236**, 1642–1648.

Darken, L. S. and Gurry, R. W. (1945). *J. Am. chem. Soc.* **67**, 1398–1412.

Derge, G., Private communication, Carnegie-Mellon University Pittsburgh, Penn. U.S.A. (1969).

Diaz, C. M. and Richardson, F. D. (1968). *In* "Electromotive force measurements in high temperature systems", 29–41. (Ed. C. B. Alcock, Inst. Min. Metall., London).

Diaz, C. M., Masson, C. R. and Richardson, F. D. (1966). *Trans. Instn. Min. Metall.* **C75**, C183–C185.

Dickson, W. R. and Dismukes, E. D. (1962). *Trans. metall. Soc. A.I.M.E.* **224**, 505–511.

Douglas, D. L. and Wagner, C. (1966). *J. electrochem. Soc.* **113**, 670–676.

Dukelow, D. A. and Derge, G. (1960). *Trans. metall. Soc. A.I.M.E.* **218**, 136–141.

Duwez, P., Odell, F. and Brown, F. H. (1952). *J. Am. Ceram. Soc.* **35**, 107–113.

616   K. Goto and W. Pluschkell

Edström, J. O. and Bitsianes, G. (1955). *Trans. metall. Soc. A.I.M.E.* **203**, 760–765.
Elliott, J. F. and Gleiser, M. (1960). "Thermochemistry of Steelmaking". Addison-Wesley, New York.
El-Naggar, M. M. A., Horsley, G. B. and Parlee, N. A. D. (1967). *Trans. metall. Soc. A.I.M.E.* **239**, 1994–1996.
Engell, H.-J. (1962). *Z. phys. Chem.* **35**, 192–195.
Engell, H.-J., v.d.Esche, W. and Schulte, E. (1967). *Hoesch-Berichte* 146–155.
Esin, O. A., Sryvalin, I. T. and Khlynov, V. V. (1957). *Zh. neorg. Khim.* **2**, 2429–2435. *Russ. J. inorg. Chem.* (1957). **2**, 237–248.
Fehrenbacher, L. L. and Jacobsen, L. A. (1965). *J. Am. Ceram. Soc.* **48**, 157–161.
Fischer, W. (1967). *Z. Naturf.* **22a**, 1575–1581.
Fischer, W. A. (1968). *Berg-u. hüttenm. Mh.* **113**, 141–148.
Fischer, W. A. and Ackermann, W. (1965). *Arch. EisenhüttWes.* **36**, 643–648, 695–698.
Fischer, W. A. and Ackermann, W. (1966). *Arch. EisenhüttWes.* **37**, 43–47, 697–700, 779–781, 959–961.
Fischer, W. A. and Ackermann, W. (1968). *Arch. EisenhüttWes.* **39**, 273–276.
Fischer, W. A. and Janke, D. (1969a). *Arch. EisenhüttWes.* **40**, 707–716.
Fischer, W. A. and Janke, D. (1969b). *Arch. EisenhüttWes.* **40**, 837–841.
Fischer, W. A., Janke, D. and Ackermann, W. (1970). *Arch. EisenhüttWes.* **41**, 361–367.
Fitterer, G. R. (1966). *J. Metals* **18**, 961–966.
Fitterer, G. R. (1967). *J. Metals* **19**, 92–96.
Fitterer, G. R., Cassler, C. D. and Vierbicky, V. L. (1969). *J. Metals* **21**, 46–52.
Förster, E. and Richter, H. (1969). *Arch. EisenhüttWes.* **40**, 475–478.
Fruehan, R. J. (1969). *Trans. metall. Soc. A.I.M.E.* **245**, 1215–1218.
Fruehan, R. J. and Richardson, F. D. (1969). *Trans. metall. Soc. A.I.M.E.* **245**, 1721–1726.
Fruehan, R. J., Martonik, L. J. and Turkdogan, E. T. (1969). *Trans. metall. Soc. A.I.M.E.* **245**, 1501–1509.
Gerasimov, Ya. I., Vasiléva, I. A., Chusova, T. P., Geiderikh, V. A. and Timofeeva, M. A. (1960). *Dokl. Akad. Nauk SSSR* **134**, 1350–1352.
Gerasimov, Y. I., Vasiléva, I. A., Chusova, T. P., Geiderikh, V. A. and Timofeeva, M. A. (1962). *Russ. J. phys. Chem.* (English Translation) **36**, 180–183.
Gerlach, J., Osterwald, J. and Stichel, W. (1968). *Z. Metallk.* **59**, 576–579.
"Gmelins Handbuch der anorganischen Chemie," (1933). System No. 54, p. 107, Verlag Chemie GMBH, Berlin, Germany.
Gordeev, I. V. and Tretjakov, Y. D. (1963). *Zh. neorg. Khim.* **8**, 1814–1819. *Russ. J. inorg. Chem.* (1963). **8**, 943–947.
Goto, K. (1962). Ph., D. Thesis at Ohio State University Columbus, Ohio, U.S.A.
Goto, K. (1969). Second Int. Symp. on ESR at Pittsburgh. A collection of papers I.
Goto, K. and Matsushita, Y. (1963). *Tetsu-to-Hagane* **49**, 1936–1937.
Goto, K. and Matsushita, Y. (1967). *Trans. electrochem. Soc. Japan* **35**, 1–7.
Goto, K. and St. Pierre, G. R. (1963). *Tetsu-to-Hagane* **49**, 1873–1879.
Goto, K., Kawakami, M. and Someno, M. (1968a). *Trans. Iron Steel Inst. Japan* **8**, 346–347.
Goto, K., Ito, T. and Someno, M. (1969a). *Trans. metall. Soc. A.I.M.E.* **245**, 1664–1665.
Goto, K., Sasabe, M. and Someno, M. (1968b). *Trans. metall. Soc. A.I.M.E.* **242**, 1757–1759.

Goto, K., Kawakami, M. and Someno, M. (1969b). *Trans. metall. Soc. A.I.M.E.* **245**, 293–301.

Goto, K., Sasabe, M., Kawakami, M. and Someno, M. (1969c). *J. Iron Steel Inst. Japan* **55**, 1007–1029.

Goto, K., Someno, M., Sasabe, M. and Saito, H. (1969d). A collection of technical papers of second international symposium on ESR technology. Part II.

Goto, K., Someno, M., Sano, N. and Nagata, K. (1970). *Trans. metall. Soc. A.I.M.E.* **244**, 23–29.

Grabke, H. J. (1965). *Ber. Bunsenges. phys. Chem.* **69**, 48–57.

Griffis, R. C. (1958). *J. electrochem. Soc.* **105**, 398–402.

Griffis, R. C. (1959). *J. electrochem. Soc.* **106**, 418–422.

Gunji, K., Katase, Y. and Aoki, H. (1964). *Tetsu-to-Hagane* **50**, 1828–1830, (1966). **52**, No. 11, PS 36 and (1967). **53**, 764–766.

Gupta, Y. P. and King, T. B. (1967). *Trans. metall. Soc. A.I.M.E.* **239**, 1701–1707.

Haber, F. (1908). *Z. anorg. Chem.* **57**, 154–184.

Hagg, G. and Schonberg, N. (1954). *Acta. crystallogr.* **7**, 351–352.

Hara, Y. (1966). *Nippon. Kink. Gakk.* **30**, 1192–1193.

Hayes, D., Budworth, D. W. and Roberts, J. P. (1961). *Trans. Br. Ceram. Soc.* **60**, 494–504.

Hayes, D., Budworth, D. W. and Roberts, J. P. (1963). *Trans. Br. Ceram. Soc.* **62**, 507–523.

Hebb, M. (1952). *J. chem. Phys.* **20**, 185–190.

Hiraoka, T., Sano, N. and Matsushita, Y. (1968). *Tetsu-to-Hagane* **55**, 470–474.

Hoffmann, A. and Fischer, W. A. (1958). *Z. phys. Chem.* **31**, 30–38.

Hoffmann, A. and Fischer, W. A. (1962). *Z. phys. Chem.* **35**, 95–108.

Holt, J. B. and Condit, R. H. (1966). *Mat. Sc. Res.* **3**, 13–29.

Hund, F. (1952). *Z. phys. Chem.* **199**, 142–151.

Hund, F. and Dürrwächter, W. (1951). *Z. anorg. allg. Chem.* **265**, 67–75.

Hund, F. and Mezger, R. (1952). *Z. phys. Chem.* **201**, 268–277.

Hutchins, J. R. III, (1959). Sc. D. Thesis MIT U.S.A.

Ihida, M., Tsuchida, S. and Kawai, Y. (1969). *Tetsu-to-Hagane* **55**, 279.

Inoue, H. and Sanbongi, K. (1969). *Tetsu-to-Hagane* **55**, 521.

Isaacs, H. S., Minushkin, B. and Salzano, F. J. (Nov. 1968). *In* "Proc. Int. Conf. Sodium Technology and Large Fast Reactor Design", Argonne.

Ito, K. and Sano, K. (1964). *Tetsu-to-Hagane* **50**, 873–877 and (1965). **51**, 1252–1259.

Ito, H. and Yanagase, T. (1958). *J. electrochem. Soc. Jap.* **26**, E. 126–128.

Ito, H. and Yanagase, T. (1960). *Trans. Japan Inst. Metals* **1**, 115–120.

Janz, G. J. (1967). *In* "Molten salts handbook", 102 Academic Press.

Jeffes, J. H. E. and Sridhar, R. (1967). *In* E.m.f. measurements in high temperature systems". 199–213. (Ed. C. B. Alcock) Inst. Mining and Metall, London.

Karaulov, A. G., Grebenyuk, A. A., Gul'ko, N. V. Rudyak, I. N. and Zoz, E. I. (1968). *Ogneupory* 45–51.

Karpachev, S. V., Filayev, A. T. and Palguev, S. F. (1964). *Electrochim Acta.* **9**, 1681–1685.

Kashida, K., Goto, K. and Someno, M. (1968). *Trans. metall. Soc. A.I.M.E.* **242**, 82–87.

Kato, M. and Minowa, S. (1969). *Trans. I.S.I.J.* **9**, 31–38, 39–46.

Kingery, W. D. and Meiling, G. E. (1961). *J. appl. Phys.* **32**, 556.

Kingery, W. D., Pappis, J., Doty, M. E. and Hill, D. C. (1959). *J. Am. Ceram. Soc.* **42**, 393–398.

Kiukkola, K. and Wagner, C. (1957). *J. electrochem. Soc.* **104**, 379–387.

Kolodney, M., Minushkin, B. and Steinmetz, H. (1965). *Electrochem. Techn.* **3**, 244–249.

Kozuka, Z. and Samis, C. S. (1970). *Trans. metall. Soc. A.I.M.E.* **1**, 871–876.

Kozuka, Z., Siahaan, O. P. and Moriyama, J. (1967). *Nippon Kink. Gakk.* **31**, 1272–1278.

Kozuka, Z., Siahaan, O. P. and Moriyama, J. (1968). *Trans. Japan Inst. Metals* **9**, 200–205.

Kozuka, Z., Shidahara, Y., Sugimoto, E., Watanabe, N. and Moriyama, J. (1968). *J. Min. metall. Inst. Japan* **84**, 1657–1662.

Kozuka, Z., Sazuki, K., Oishi, T. and Moriyama, J. (1968). *Nippon Kink. Gakk.* **32**, 1132–1137.

Kröger, F. A. (1964). "The Chemistry of Imperfect Solids" p. 275, North Holland Publ. Co., Amsterdam.

Kröger, F. A. (1966). *J. Am. Ceram. Soc.* **49**, 215–218.

Ksenofontova, R. F., Vasiléva, I. A. and Lomonosov, M. V. (1962). *Dokl. Akad. Nauk SSSR* (English Translation) **143**, 1105–1107.

Kubik, A. and Alcock, C. B. (1967). *In* "Electromotive force measurements in high-temperature systems", 43–49 and *Metal Science J.* (1967). **1**, 19–24.

Landler, P. F. J. and Komarek, K. L. (1966). *Trans. metall. Soc. A.I.M.E.* **236**, 138–149.

Lasker, M. F. and Rapp, R. A. (1966). *Z. phys. Chem.* **49**, 198–221.

Laurentev, V. I., Gerasimov, Y. I. and Rezakhina, T. N. (1961). *Dokl. Akad. Nauk SSSR* **136**, 1372–1375.

Lepinskik, B. M. and Esin, O. A. (1961). *Zh. neorg. Khim.* **6**, 1223–1224. *Russ. J. inorg. Chem.* (1961). **6**, 625–627.

Levitskii, V. A. and Rezukhina, T. N. (1966). *Inorg. Materials* **2**, 122–126.

Lidiard, A. B. (1957). *In* "Handbuch der Physik", Bd. XX p. 324. Springer Verlag. Berlin.

Lindner, R. and Parfitt, G. (1957). *J. chem. Phys.* **26**, 182–185.

Littlewood, R. (1964). *Steel Times* September, 423–425.

Littlewood, R. (1966). *Can. Met. Q.* **5**, 1–17.

Löscher, W. (1969). *Arch. EisehnüttWes.* **40**, 479–487.

Löscher, W. (1970). *Hoesch Berichte* 30–37.

Luzgin, V. P., Viškarev, A. F. and Javojskij, V. I. (1963). Isvestija vysšich, ucebnych zavedenij Černaja metallurgija **6**, 44–50.

Mackenzie, J. D. (1956). *Chem. Rev.* **56**, 455–470.

Mah, A. D. (1959). *J. Am. chem. Soc.* **81**, 1582–1583.

Marincek, B. (1967a). *Schweizer Arch.* **33**, 143–148.

Marincek, B. (1967b). *Helv. Chim. Acta.* **50**, 988–990.

Masson, C. R. and Whiteway, S. G. (1967). *Can. Met. Q.* **6**, 199.

Matsumura, T. (1966). *Can. J. Phys.* **44**, 1685–1698.

Matsushita, M. and Goto, K. (1966). *In* "Thermodynamics" Int. Atom. Energy Agency, Vol. 1, 111–129. Vienna.

Matsushita, Y., Goto, K., Cho, A., Isobe, K., Tate, M. and Sasao, H. (1966). *Tetsu-to-Hagane* **52**, 393–394.

Meyer, H. W. and Richardson, F. D. (1961–62). *Trans. Instn. Min. Metall.* **71**, 201–214.

Meyers, J. E. and Hunter, R. A. (Nov. 1968). *In* "Proc. Int. Conf. Sodium Technology and Large Fast Reactor Design", Argonne.
Möbius, H.-H. and Hartung, R. (1965). *Silikattechnik* **16**, 276–281.
Mori, K. (1960). Private communication, Lecture Note, Ibaragi Univ. Japan.
Moriyama, A., Yagi, J. and Muchi, I. (1965). *Nippon Kink. Gakk.* **29**, 528–534.
Moriyama, J., Sato, N., Asao, H. and Kozuka, Z. (1969). Memoirs, F. Eng. Kyoto, Univ. Vol. **31**, 253–267.
Mitoff, S. P. (1959). *J. chem. Phys.* **31**, 1261–1269.
Mitoff, S. P. (1962). *J. chem. Phys.* **36**, 1383–1389.
Muan, A. and Osborn, E. F. (1965). "Phase Equilibria among Oxides in Steelmaking". Addison-Wesley Publ. Co.
Mulfinger, H.-O. and Scholze, H. (1962). *Z. Glaskde* **35**, 495–500.
Nakamura, Y. (1964). Private communication, Dr. Sci. Thesis, Univ. of Tokyo.
Nelson, J. (1955). *Phys. Rev.* **99**, 1902.
Nernst, W. (1899–1900). *Z. Elektrochem.* **6**, 41–43.
Niwa, K. (1957). *Nippon Kink. Gakk.* **21**, 304–308.
Noddack, W. and Walch, H. (1959). *Ber. Bunsenges. phys. Chem.* **63**, 269–274.
Ohtani, M. and Sanbongi, K. (1963). *Tetsu-to-Hagane* **49**, 22–29.
Oishi, Y. and Kingery, W. D. (1960). *J. chem. Phys.* **33**, 480–486.
Osterwald, J. (1965). Habilitationsschrift, TU Berlin, Germany.
Osterwald, J. (1968). *Z. Metallk.* **59**, 576–579.
Osterwald, J. and Schwarzlose, G. (1968). *Z. phys. Chem.* **62**, 119–126.
Paladino, A. E. and Coble, R. E. (1963). *J. Am. Ceram. Soc.* **46**, 133–136.
Paladino, A. E. and Kingery, W. D. (1962). *J. chem. Phys.* **37**, 957–962.
Palguev, S. F. and Neuimin, A. D. (1960). Trudy Inst. Electrokhim., Akad. Nauk SSSR, Ural, Filial., No. I, 111. *Chem. Abstr.* **55**, (1961). 19556.
Palguev, S. F. and Neuimin, A. D. (1962). *Soviet Phys. Solid State* **4**, 629–632.
Pappis, J. and Kingery, W. D. (1961). *J. Am. Ceram. Soc.* **44**, 459–464.
Pargeter, J. K. (1968). *J. Metals* **20**, 10, 27–31.
Pastorek, R. L. and Rapp, R. A. (1969). *Trans. metall. Soc. A.I.M.E.* **245**, 1711–1720.
Patterson, J. W., Bogren, E. C. and Rapp, R. A. (1967). *J. electrochem. Soc.* **114**, 752–758.
Pease, R. N. and Taylor, H. S. (1921). *J. Am. Chem. Soc.* **43**, II, 2179–2188 and (1922). **44**, II, 1637–1647.
Peters, H. and Mann, G. (1959). *Z. Elektrochem.* **63**, 244–248.
Peters, H. and Möbius, H. (1958). *Z. phys. Chem.* **209**, 298–309.
St. Pierre, G. R., Ebihara, W. T., Pool, M. and Speiser, R. (1962). *Trans. metall. Soc. A.I.M.E.* **224**, 259–264.
Pluschkell, W. (1969). *Arch. EisenhüttWes.* **40**, 401–403.
Pluschkell, W. (1970) unpublished experimental work.
Pluschkell, W. and Engell, H.-J. (1965). *Z. Metallk.* **56**, 450–452.
Pluschkell, W. and Engell, H.-J. (1968). *Ber. dt. keram. Ges.* **45**, 388–394.
Pluschkell, W. and Yoshikoshi, H., private communication, to be published in Arch. EisnehüttWes. (1970). **41**,
Prasad, K. K. Seetharaman, S. and Abraham, K. P. (1969). *Trans. Indian Inst. Metals* **22**, 7–9.
Ranford, R. E. and Flengas, S. N. (1965). *Can. J. Chem.* **43**, 2879–2887.
Rapp, R. A. (1963). *Trans. metall. Soc. A.I.M.E.* **227**, 371–374.
Rapp, R. A. (1967). *In* "Thermodynamics of Nuclear Materials", 559–584, Vienna.
Rapp, R. A. and Maak, F. (1962), *Acta. metall* **10**, 63–69.

Reynolds, H. (1902). Dissertation, University of Göttingen, Germany.

Rezukhina, T. N., Lavrent'ev, V. I., Levitskii, V. A. and Kuznetsov, F. A. (1961a). *Russ. J. phys. Chem.* **35**, 671–672.

Rezukhina, T. N., Levitskii, V. A. and Kazimirova, N. M. (1961b). *Russ. J. phys. Chem.* **35**, 1305–1306.

Rezukhina, T. N., Levitskii, V. A. and Ozhegov, P. (1963). *Russ. J. Phys. Chem.* **37**, 358–359.

Rhodes, W. H. and Carter, R. E. (1966). *J. Am. Ceram. Soc.* **49**, 244–249.

Richardson, F. D. and Billington, J. C. (1955–56). *Trans. Instn. Min. Metall.* **65**, 273–297.

Richardson, F. D. and Webb, L. E. (1955). *Trans. Instn. Min. Metall.* **64**, 529–564.

Rickert, H. (1967). *In* "Electromotive force measurements in high-temperature systems", 59–90.

Rickert, H. and El Miligy, A. (1968). *Z. Metallk.* **59**, 635–641.

Rickert, H. and Steiner, R. (1966). *Z. phys. Chem.* **49**, 127–137.

Rickert, H. and Wagner, H. (1966). *Electrochim. Acta.* **11**, 83–91.

Rickert, H., Wagner, H. and Steiner, R. (1966). *Chemie-Ingr-Tech.* **38**, 618–622.

Rieck, G. D. (1967). "Tungsten and its Compounds". Pergamon Press.

Rizzo, F. E., Bidwell, L. R. and Frank, D. F. (1967). *Trans. metall. Soc. A.I.M.E.* **239**, 593–595.

Rizzo, H. F., Gordon, R. S. and Cutler, J. B. (1968). *In* "Mass Transport in Oxides" Nat. Bur. Standards, Spec. Publ. **296**, 129–142.

Robert, E. W. and Roberts, J. P. (1967). *Bull. Soc. fr. Ceram.* **77**, 3–13.

Rudolph, J. (1959). *Z. Naturf.* **A14**, 727–737.

Ruka, R. J., Bauerle, J. E. and Dykstra, L. (1968). *J. electrochem. Soc.* **115**, 497–501.

Ryshkewitch, E. (1960). "Oxide Ceramics". Academic Press, New York and London.

Saito, T. and Maruya, K. (1958). *Bull. Res. Inst. Min. Metall.* **14**, 306–314.

Saito, H., Goto, K. and Someno, M. (1969). *Tetsu-to-Hagane* **55**, 539–549.

Sanbongi, K. and Omori, Y. (1959). Science Reports of Res. Inst., Tohoku Univ. **A11**, 244–263 and 339–359, and **A13**, (1961) 174–182, and 238–246.

Sanbongi, K., Ohtani, M., Omori, Y. and Inoué, H. (1966). *Trans. I.S.I.J.* **6**, 76–79.

Sano, N., Honma, S. and Matsushita, Y. (1969). *Trans. I.S.I.J.* **9**, 404–408.

Sasabe, M., Goto, K. and Someno, M. (1970). *Trans. metall. Soc. A.I.M.E.* **1**, 811–817.

Schaefer, J. H., Crimes, W. R. and Watson, G. M. (1959). *J. phys. Chem.* **63**, 1999–2002.

Schenke, M., Broers, G. H. and Ketelaar, J. A. A. (1966). *J. electrochem. Soc.* **113**, 404.

Schmalzried, H. (1960a). *Z. phys. Chem.* **25**, 178–192.

Schmalzried, H. (1960b). *J. chem. Phys.* **33**, 940.

Schmalzried, H. (1962). *Ber. Bunsenges. phys. Chem.* **66**, 572–576.

Schmalzried, H. (1963). *Z. phys. Chem.* **38**, 87–102.

Schmalzried, H. (1966). *In* "Thermodynamics" Vol. 1. 97–109. complied by Intntl. Atomic Energy Agency, Vienna, Austria.

Schmalzried, H. and Tretjakow, J. D. (1966). *Ber. Bunsenges, phys. Chem.* **70**, 180–189.

Scholze, H. (1955). *Ber. dt. keram. Ges.* **32**, 381–385.

Schuh, B., Korousic, B. and Marincek, B. (1968). *Schweizer. Arch.* **34**, 380–387.

Schwerdtfeger, K. (1966). *Trans. metall. Soc. A.I.M.E.* **236**, 32–35.

Schwerdtfeger, K. (1967). *Trans. metall. Soc. A.I.M.E.* **239**, 1276–1281.

Schwerdtfeger, K. and Muan, A. (1965). *Acta. metall.* **13**, 509–515.

Seetharaman, S. and Abraham, K. P. (1968a). *Indian J. Technol.* **6**, 123–124.

Seetharaman, S. and Abraham, K. P. (1968b). *Trans. Instn. Min. Metall.* **77**, c209–211.

Seetharaman, S. and Abraham, K. P. (1969). *Scripta Metall.* **3**, 911–926.

Sellars, C. M. and Maak, F. (1966). *Trans. metall. Soc. A.I.M.E.* **236**, 457–464.

Shewmon, P. G. (1963). *In* "Diffusion in solids", p. 100. McGraw-Hill Book Co. Inc.

Singhall, S. and Worrell, W. L. (1970). Private communication, to be published in *Acta. Metall.*

Simnad, M. T., Yang, L. and Derge, G. (1956). *Trans. metall. Soc. A.I.M.E.* **214**, 690–692.

Simpson, L. A. and Carter, R. E. (1966). *J. Am. Ceram. Soc.* **49**, 139–144.

Smith, A. W., Meszaros, F. W. and Amata, C. D. (1966). *J. Am. Ceram. Soc.* **49**, 240–244.

Sockel, H. G. and Schmalzried, H. (1968). *Ber. Bunsenges. phys. Chem.* **72**, 745–754.

Spitzer, R. H., Manning, F. S. and Philbrook, W. O. (1966). *Trans. metall. Soc. A.I.M.E.* **236**, 726–742.

Sridhar, R. and Jeffes, J. H. E. (1967). *Trans. Instn. Min. Metall.* **76**, C44–60.

Steele, B. C. H. (1968). *In* "Electromotive force measurements in high temperature systems", p. 3–25, Inst. Min. Metall. London.

Steele, B. C. H. and Alcock, C. B. (1965). *Trans. metall. Soc. A.I.M.E.* **233**, 1359–1365.

Steiner, R. (1969). Dissertation, TU Karlsruhe, Germany, to be published in *Ber. Bunsenges. phys. Chem.*

Strickler, D. W. and Carlsson, W. G. (1965). *J. Am. Ceram. Soc.* **48**, 286–289.

Subbarao, E. C., Sutter, P. H. and Hrizo, J. (1965). *J. Am. Ceram. Soc.* **21**, 443–446.

Swisher, J. H. and Turkdogan, E. T. (1967). *Trans. metall. Soc. A.I.M.E.* **239**, 602–610.

Tare, V. B. and Schmalzried, H. (1964). *Z. phys. Chem.* **43**, 30–32.

Tare, V. B. and Schmalzried, H. (1966). *Trans. metall. Soc. A.I.M.E.* **236**, 444–446.

Taylor, R. W. and Schmalzried, H. (1964). *J. phys. Chem.* **68**, 2444–2449.

Tien, T. Y. and Subbarao, E. C. (1963a). *J. chem. Phys.* **39**, 1041–1047.

Tien, T. Y. and Subbarao, E. C. (1963b). *J. Am. Ceram. Soc.* **46**, 489–492.

Tokuda, M. (1965). Private communication, Dr. Eng. Thesis, Univ. of Tokyo.

Toropov. N. A. and Barzakovskii, V. P. (1966). *In* "High-temperature chemistry of silicates and other oxide systems", 63–90. Consultant Bureau, New York.

Towers, H. and Chipman, J. (1957). *Trans. metall. Soc. A.I.M.E.* **209**, 769–773.

Towers, H., Paris, M. and Chipman, J. (1953). *J. Metals* **5**, 1455–1459.

Treadwell, W. D. (1916). *Z. Elektrochem.* **22**, 411–421.

Treadwell, W. D. and Terebesi, L. (1933). *Helv. chim. Acta.* **16**, 922–939.

Tretjakow, J. D. and Muan, A. (1969). *J. electrochem. Soc.* **116**, 331–334.

Tretjakow, J. D. and Schmalzried, H. (1965). *Ber. Bunsenges. phys. Chem.* **69**, 396–402.

Tretjakow, J. D. and Rapp, R. A. (1969). *Trans. metall. Soc. A.I.M.E.* **245**, 1235–1241.

Tretjakow, J. D. *et al.* (1963). *J. Inorg. Chem. USSR* **8**, 145 and 1814.

Ullmann, H. (1968). *Z. phys. Chem.* **237**, 71–80.

Ulrich, K.-H. and Borowski, K. (1968). *Arch. EisenhüttWes.* **39**, 259–263.

Ulrich, K.-H., Bohnenkamp, K. and Engell, H.-J. (1966). *Z. phys. Chem.* **51**, 36–49.

Vasiléva, I. A., Gerasimov, Ya. I. and Simanov, O. P. (1950). *Russ. J. Phys. Chem.* **24**, 1811–1815.

Vasiléva, I. A., Gerasimov, Ya. I. and Simanov, Yu. P. (1960). *Russ. J. Phys. Chem.* **34**, 863–866.

Wagner, C. (1933). *Z. phys. Chem.* **B21**, 25–41.

Wagner, C. (1943). *Naturwissenshaften* **31**. 265–268.

Wagner, C. (1952). "Thermodynamics of Alloys". Addison-Wesley Publishing Company.

Wagner, C. (1957). *In* "Intern, Comm. Electrochem. Thermodyn. and Kinetics", p. 361ff. London.

Wagner, C. (1966). *In* "Advances in Electrochemistry and Electrochemical Engineering" **4**, 1–46.

Wagner, C. (1968). *Z. Bunsenges. phys. Chem.* **72**, 778–781.

Wagner, C. (1970). Private communication.

White, J. L. (1955). Ph. D. Thesis, Univ. California.

Whiteway, S. G., Peters, R. J. W., Jamieson, W. D. and Masson, C. R. (1968). *Can. Metall. Q.* **7**, 211–215.

Weininger, J. L. and Zemany, P. D. (1954). *J. chem. Phys.* **22**, 1469–1470.

Weissbart, J. and Ruka, R. (1961). *Rev. Sci. Instr.* **32**, 593–595.

Weissbart, J. and Ruka, R. (1962). *J. electrochem. Soc.* **109**, 723–726.

Wilder, T. C. (1966). *Trans. metall. Soc. A.I.M.E.* **236**, 1035–1040.

Wilder, T. C. (1969). *Trans. metall. Soc. A.I.M.E.* **245**, 1370–1372.

Wilder, T. C. and Galin, W. E. (1969). *Trans. metall. Soc. A.I.M.E.* **245**, 1287–1290.

Worrell, W. L. (1966). "Thermodynamics" Vol. 1, I.A.E.A. paper No. SM 66/66. Vienna.

Wu. Y. C., Goto, K. and Matsushita, Y. (1964). *Tetsu-to-Hagane* **50**, 470–473.

Yamashita, J. and Kurosawa, T. (1954). *J. phys. Soc. Jap.* **9**, 944–953.

Yang, C. L. and Derge, G. (1959). *J. chem. Phys.* **30**, 1627.

Yuan, D. and Kröger, F. A. (1969). *J. electrochem. Soc.* **116**, 594–600.

Zador, S. (1968). *In* "Electromotive force measurements in high temperature systems", 145–150. Inst. Min. Met., London.

14. ELECTROCHEMICAL CELLS WITH NON-OXYGEN CONDUCTIVITY

Y. D. Tretyakov and A. R. Kaul

Laboratory of General Chemistry, University of Moscow, USSR.

I. Introduction

Electrochemical cells incorporating solid electrolytes proved to be applicable in thermodynamic and kinetic investigations. They turned out to be useful for studying thermodynamic properties of simple binary and ternary oxides, halides, phosphides, carbides, metal alloys and solid solutions of salts. The coulometric titration technique is an excellent and up to now unique way of investigating the thermodynamics of compounds with a very narrow homogenity range.

The application of galvanic cells with solid electrolytes is based on the following considerations. The analysis of the crystal system $M_A X_B$ with some differences between chemical potentials of constituents (as shown in Fig. 1) shows that when the crystal is in equilibrium with the gaseous phase, and platinum electrodes contacting the opposite ends of the crystal from the direction of the chemical potential change, the arising e.m.f., according to Wagner (1933), is

$$E = -\frac{1}{2Z_xF}\int_{\mu_{x_2}'}^{\mu_{x_2}''}t_i d\mu_{x_2} = \frac{1}{Z_MF}\int_{\mu_M'}^{\mu_M''}t_i d\mu_M. \tag{1}$$

Here μ_{x_2}', μ_{x_2}'' are chemical potentials of non-metal components on the opposite sides of the crystal in equilibrium with the platinum electrodes; μ_M' and μ_M''—chemical potentials of metal components on the corresponding sides of the crystal; Z_x and Z_M—anion and cation valency respectively; F—Faraday's constant; t_i—ionic transport number. It is obvious from eqn (1) that chemical potentials of both crystal components on one side may be calculated if chemical potentials on the other side and the $t_i = f(\mu_{x_2}$ or $\mu_M)$ dependence are known and equilibrium e.m.f. value is measured. The simplified version of eqn (1) when $t_i = 1$ is

$$E = \frac{\mu_{x_2}' - \mu_{x_2}''}{2Z_xF} = \frac{\mu_M'' - \mu_M'}{Z_MF}. \tag{2}$$

The design and application of solid-state galvanic cells would present no difficulties if there were a great number of crystals of pure ionic conductivity with a wide μ_{x_2} (μ_M) and temperature range. By applying quasichemical approximation, it is easy to show that an ionic crystal has a limited pure ionic conductivity range (if any). It may be shown on the $M^{IV}X_2^{II}$ crystal sample exhibiting intrinsic disorder of the Frenkel type in the anionic sublattice. The latter may be expressed by the following equation

$$0 = X_i'' + V_x^{\cdot\cdot} \tag{3}$$

where X_i'' and $V_x^{\cdot\cdot}$ are interstitial anion and anion vacancy respectively

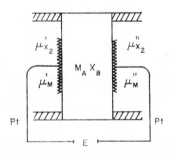

FIG. 1. Crystals M_AX_B separating two gaseous spaces with different chemical potentials of X_2 and M.

with double effective electronic charge. Intrinsic electronic disorder of the crystal and its interaction with the gaseous phase may be expressed as follows:

$$0 = h^{\cdot} + e'$$ (4)

$$0 = \tfrac{1}{2}X_2 + V_x^{\cdot\cdot} + 2e'$$ (5)

where e' and $h^{\cdot}$ are electrons and electron holes. When the law of mass action is applied to equilibrium states of processes (3), (4), (5), one obtains:

$$[X_i''][V_x^{\cdot\cdot}] = K_s$$ (3a)

$$n \cdot p = K_i$$ (4a)

$$[V_x^{\cdot\cdot}]n^2 p_{x_2}^{\frac{1}{2}} = K_{x_2}$$ (5a)

Equations (3a)–(5a) combined with the electroneutrality condition

$$n + 2[X_i''] = p + 2[V_x^{\cdot\cdot}]$$ (6)

form a set, the solution of which results in any defect concentration as $f(P_{x_2})$ at fixed temperature and equilibrium constants due to this temperature (see Fig. 2). Taking into consideration the fact that the electron and hole mobility is higher than that of ions, one may find pure ionic conductivity within the P_{x_2} range limited by dotted lines. If $P_{x_2} > P_{x_2}^{*}$ excess electron hole conductivity arises, while if $P_{x_2} < P_{x_2}^{**}$ excess electron conductivity is observed. Both make the use of simplified Wagner's equation (2) impossible.

Oxygen-conductive solid electrolytes of the $ZrO_2(CaO)$, $ThO_2(Y_2O_3)$ type are of considerable use, although they are not considered here. It is solid electrolytes with cationic and nonoxygen-anionic conductivity that have attracted great interest recently. Their application extends the range of the e.m.f. method to acquire thermodynamic data.

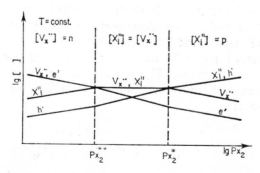

FIG. 2. Concentration of intrinsic defects in a $M_A X_B$ crystal as a function of partial pressure of non-metal component.

The following requirements should be observed when constructing the galvanic cells:

1. Chemical potentials $\mu_{x_2}(\mu_M)$ of the studied electrode and the reference electrode should be in the predominantly ionic (electrolytic) range of the solid electrolyte. The criterion of electrolytic application (Vetcher and Vetcher, 1967a) is

$$\bar{t}_i = \frac{1}{\mu_{x_2}'' - \mu_{x_2}'} \int_{\mu_{x_2}'}^{\mu_{x_2}''} t_i \, d\mu_{x_2} \geqslant 0.99. \tag{7}$$

2. The chemical potential $\mu_{x_2}(\mu_M)$ at the electrolyte—Pt-electrode—electrode interface should be equal to the chemical potential $\mu_{x_2}(\mu_M)$ of the electrode, as if it were taken separately. This above condition is often broken for the following reasons:

(a) chemical interaction between the three constituents of the above-mentioned interface,

(b) polarization effects occurring in the electrolyte and the electrode studied,

(c) non-electrochemical transport of volatile components through the gaseous phase and electrolyte (owing to porosity of the latter).

The effects of point 2 can be considerably reduced or ruled out in these cells with a separated gaseous space which help to avoid contact between the electrolyte, Pt-electrode and electrode studied (Tretyakov, 1968). Where the Pt electrodes are chemically active with the electrode studied (or electrolyte) they should be replaced by other electron conductors (Ir, Rh, Mo, W, Ta). Difficulties arising from the electron conductivity effect in solid electrolytes ($\bar{t}_i < 0.99$), can be reduced if the galvanic cell works in a compensation regime, i.e. the chemical potential $\mu_{x_2}(\mu_M)$ of the reference electrode is constantly controlled so that the measured e.m.f. is about zero. According to eqn (1), if $E = 0$ then μ_{x_2}' of the electrode studied is equal to μ_{x_2}'' of the reference electrode and is t_i—independent, so that t_i may considerably deviate from 1 (Komarov and Tretyakov, 1970b).

As was mentioned above the number of ionic crystals with pure ionic conductivity is very limited. Theoretically this property characterises systems with remarkable intrinsic disorder of the Frenkel type in anion or cation sublattices:

$$0 = X_i'' + V_x^{\cdot\cdot}; \quad K_F' = [X_i''] \cdot [V_x^{\cdot\cdot}] \tag{8}$$

$$0 = M_i^{\cdot\cdot\cdot} + V_M''''; \quad K_F'' = [M_i^{\cdot\cdot\cdot}] \cdot [V_M'''']. \tag{9}$$

Predominantly anionic conductivity may be expected in (9) and cationic in (10). It is obvious that ionic crystals containing ions of easily variable valency should not be regarded as potential solid electrolytes because the $K_F' \gg K_{x_2}$ condition cannot be preserved (K_{x_2} is the equilibrium constant of a quasi-chemical reaction of the type (5) or

$$\tfrac{1}{2}X_2 = V_M'' + 2h^{\cdot} + 2X_x{}^{\times}, \tag{10}$$

where $X_x{}^{\times}$ is the non-metal ion in its regular site). The highest level of ionic conductivity in ionic crystals may be also achieved by doping altervalent impurity in the maximum quantity which does not break the one-phase system.

II. Anion-Conductive Solid Electrolytes

A. Electrolytic Properties of CaF₂

Among anion-conductive solid electrolytes halides are the most widespread, CaF_2 occupying the first place among them. Ure (1957) measured transport numbers and diffusion coefficient of Ca in pure CaF_2 and in CaF_2 doped by NaF and YF_3. He came to a conclusion that CaF_2 is predominantly ion conductive, namely F^--ion conductive. Earlier Zintl and Udgard (1939) showed that doping CaF_2 with YF_3 leads to the formation of interstitial F^--ion structures. It is now acknowledged that the dominant disorder type of CaF_2 is that of anti-Frenkel's

$$0 \rightarrow F_i' + V_F. \quad [F_i'] \cdot [V_F^{\cdot}] = K_{a-F}. \tag{11}$$

The temperature dependence of defect concentration as shown by Ure (1957) is presented in Fig. 3. The high mobility of F^--ion may be due to both

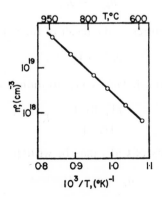

Fig. 3. Temperature dependence of anti-Frenkel defect concentration (after Ure, 1957).

vacancy and interstitial mechanisms. The latter is more probable (according to Ure) and F^--ion mobility through interstitials is found to be

$$U = (1\cdot 34 \times 10^7)\, T^{-1} \exp\left[-(19 \pm 4) \times 10^3/T\right] \text{cm}^2 \sec^{-1} \text{volt}^{-1} \quad (12)$$

The transport number of cations in CaF_2 is very small: 3×10^{-8} at 850°C (Ure, 1957), 8×10^{-7} at 1000°C and 1×10^{-6} at 1100°C (Matzke and Lindner, 1964). Consequently CaF_2 is considered as anion conductive (cation transport may be practically ignored) when used as a solid electrolyte in galvanic elements. Possible electron (when α_{F_2}[†] is low) and hole (when α_{F_2} is high) conductivity as a result of processes

$$0 \rightarrow \tfrac{1}{2}F_2 + V_F^{\cdot} + e' \quad (13)$$

$$\tfrac{1}{2}F_2 \rightarrow F_i' + h^{\cdot} \quad (14)$$

is another limitation in the use of CaF_2 as a solid electrolyte. Wagner (1968) has made a theoretical attempt to find out the conditions of the onset of electronic conductivity in CaF_2. According to data of Wöhler and Rodewald (1909), Mollwo (1934), Botinck (1958) and Lichter and Bredig (1965) CaF_2 crystals become non-stoichiometric when dissolving an excess of Ca, if the activity of F_2 is low, α_{Ca} is high. The formation of the non-stoichiometric phase is due to the following mechanism of the onset of defects

$$Ca(g) + 2F_F^{\times} = CaF_2 + 2e_F^{\times} \quad (15)$$

where $e_F^{\times}$ is an electron substituted for an anion also known as a colour centre. Applying the law of mass action to reaction (15) one may show that the number of colour centres per cm^3 is proportional to the square root of calcium activity

$$n_{e_F \times} \sim \alpha_{Ca}^{\tfrac{1}{2}}. \quad (16)$$

In addition to colour centres as the result of the reaction

$$e_F^{\times} \rightarrow V_F^{\cdot} + e' \quad (17)$$

mobile electrons are formed. Their concentration is related to $[e_F^{\times}]$ as

$$\frac{[V_F^{\cdot}][e']}{[e_F^{\times}]} = K. \quad (18)$$

The electron conductivity of CaF_2 may be written as

$$\sigma_e = n_{e'} \times U_{e'} \times e, \quad (19)$$

[†] Here and later "α" is the thermodynamic activity of a component.

where $n_{e'}$ and $U_{e'}$ are the number per cm^3 and mobility of excess electrons respectively, and e is the electronic charge. In view of

$$n_{e_F \times} \times U_{e_F \times} = n_{e'} U_{e'} \tag{20}$$

and keeping in mind (16) one finds

$$\sigma_e = n^\circ_{ie_F \times} U_{e_F \times} e \, \alpha_{Ca}^{\frac{1}{2}} \tag{21}$$

where $n^\circ_{e_F \times}$ is the number of colour centres per cm^3 at $\alpha_{Ca} = 1$. Thus the transference number of electrons t_e is found to be

$$t_e = \frac{n^\circ_{e_F \times} U_{eF \times} e \, \alpha_{Ca}^{\frac{1}{2}}}{\sigma_{ion}} \tag{22}$$

where σ ion is the ionic conductivity of pure CaF_2. Equation (22) makes it possible to calculate the values of α_{Ca} at which electronic conductivity should not be neglected. Applying data of concentration and mobility of colour centres in CaF_2 in equilibrium with saturated Ca-vapour (Lichter and Bredig, 1965) and conductivity data of Ure (1957), Wagner (1968) concluded that $t_e \leqslant 10^{-2}$ at $\alpha_{Ca} \leqslant 1 \times 10^{-5}$ and 6×10^{-6} at $600°$ and $840°$ C respectively. It is easy to prove when ΔG data for CaF_2 formation is used, that the lower limiting value for CaF_2 as a solid electrolyte corresponds to $P_{F_2} = 10^{-58.6}$ atm at $600°C$ and $10^{-43.0}$ atm at $840°C$. Recent d.c. polarization, a.c. conductivity and open circuit e.m.f. measurements by Hinze (1968) give an overall indication that the electrolytic domain for CaF_2 extends down to the stability limit determined by the equation

$$\log P_{F_2} = 9 \cdot 0 - \frac{6 \cdot 34 \times 10^4}{T}. \tag{23}$$

The apparent practical difficulties associated with conductivity and/or e.m.f. measurements on CaF_2 at high fluorine pressures and elevated temperatures seem to have thus far discouraged the experimental measurements required to determine the high P_{F_2} electrolytic domain boundary for this material. However, in view of the extreme electronegativity difference between $Ca^{··}$ and F' ions and the correspondingly large forbidden band gap for CaF_2 it would not be surprising if this boundary were found to occur above one atmosphere fluorine pressure (Patterson, 1970). From the above one may conclude that CaF_2 as a solid electrolyte with pure anionic conductivity is much better than ordinary anion-conductive solid electrolytes of the ZrO_2 (CaO), $ZrO_2(Y_2O_3)$, $ThO_2(Y_2O_3)$, $ThO_2(CaO)$ type. It is characterized by a wider P_{x_2} and T range, where $t_i \simeq 1$.

It follows—from data of Ure (1957)—that the level of conductivity of pure CaF_2 is sufficiently high to permit its use in galvanic cells. The dissolution of impurities, when the cation is more than bivalent, results in the increase of conductivity dye to increased concentration of interstitial F-ions, for example

$$YF_3(\to CaF_2) = Y_{Ca}{}^{\cdot} + F_i' + 2_F{}^{\times} \qquad (24)$$

YF_3-doped CaF_2, however, cannot be recommended for practical use as the increase of cation mobility is followed by a narrowing of the range of pure anionic conductivity (see Table I).

This may be explained in terms of Schottky-type disorder in cationic sublattice

$$0 = V_{Ca}'' + 2V_F{}^{\cdot} \qquad (25)$$

$$K_S = [V_{Ca}''][V_F{}^{\cdot}]^{2\cdot} \qquad (25a)$$

In accord with eqn (24) introduction of YF_3 increases the F_i' concentration. In view of eqns (11) and (25a) an increase of $[V_{Ca}'']$ may be expected as a result of CaF_2 doping and Ca^{++} transport intensification through the vacancy mechanism. Information concerning the doping of CaF_2 with impurities including the monovalent cation is inconsistent. It was stated by Ure (1957) that CaF_2-doping with NaF increases its conductivity as well as the transport number of Ca^{++} ions ($t_{Ca}^{1000°C}$ in CaF_2 "pure" $\sim 8 \times 10^{-7}$; $t_{Ca}^{660°C}$ in CaF_2 (0·06 mol % NaF) $\sim 3 \times 10^{-4}$). On the other hand Short and Roy (1964) consider that the self-diffusion rate of Ca at 760°C in CaF_2 (3 mol % NaF) is 0·7 of that in pure CaF_2. Their conclusion seems to be more consistent if the following model is accepted:

$$NaF(\to CaF_2) = Na_{Ca}' + V_F{}^{\cdot} + F_F{}^{\times}.$$

TABLE I. Calcium Diffusion Coefficients in CaF_2 doped with YF_3 at 1100°C†

Solution comp.	D, $cm^2\,sec^{-1}$
CaF_2 "pure"	$8\cdot6 \times 10^{-11}$
CaF_2 (20% YF_3)	$1\cdot9 \times 10^{-10}$
CaF_2 (40% YF_3)	$1\cdot6 \times 10^{-9}$

† Data of Short and Roy (1964).

If (14) and (25a) are taken into consideration the increase of $[V_F^{\cdot}]$ leads to the decrease of conductivity and t_{Ca}:

$$t_{Ca} \simeq \frac{\sigma \text{ cat}}{\sigma \text{ an}} \sim \frac{[V_{Ca}'']}{[F_i']} = \frac{K_S[V_F^{\cdot}]}{[V_F^{\cdot}]^2 K_{a-F}} = \frac{K}{[V_F^{\cdot}]}.$$

Nevertheless it is clear that the doping of CaF_2 with NaF does not perfect the electrolytic properties of CaF_2, whichever analysis is the more consistent.

High chemical activity of CaF_2 which strongly increases with temperature makes the use of it as a solid electrolyte more limited compared with oxide electrolytes. At $T > 1000°C$ its use is practically impossible. Existing information about the oxygen influence on electrolytic properties of CaF_2 is rather conflicting. The simplest thermodynamic evaluation shows the displacement of equilibrium of the reaction

$$CaF_2 + \tfrac{1}{2}O_2 \leftrightarrows CaO + F_2 \tag{26}$$

to the left even at $1500°$ K $(\Delta G_{1500}° = 116.4 \text{ kcal})$. However, Ure (1957) demonstrated that annealing of CaF_2 in air at $1000°$ C considerably modifies its conductivity. In view of the fact that ionic radius of oxygen is very close to that of fluorine $[r_{O^{2-}} = 1.36 \text{ Å}; r_{F^-} = 1.33 \text{ Å}; \text{(Bokiy, 1960)}]$, σ modifications may be connected with the formation of defect structures:

$$\tfrac{1}{2}O_2 \to O_i^{\times} \to O_i'' + 2h^{\cdot} \tag{27a}$$

$$\tfrac{1}{2}O_2 + V_F^{\cdot} \to O_F' + 2h^{\cdot} \tag{27b}$$

In their study of penetration of oxygen into CaF_2 single crystals Phillips and Hanlon (1963) showed that solubility of O_2 at $T > 800°$ C is about 0.2 mol %. One might, therefore, expect the onset of considerable hole conductivity after heating CaF_2 in an oxygen-containing atmosphere, making its use as a solid electrolyte impossible. At the same time, many experiments (Benz and Wagner, 1961; Taylor and Schmalzried, 1964; Rezukhina et al., 1966) showed that CaF_2 preserves its anionic conductivity in pure oxygen at $T < 1000°C$. Moreover, the doping of the electrolyte with $CaO(0.2 \text{ wt } \%)$ does not influence t_i (Vetcher, 1970). It was shown that the presence of moisture in the surrounding atmosphere changes the electrolytic properties of CaF_2 for the worse.

The most significant limitation on the use of CaF_2 as an electrolyte at high temperatures is its interaction with the electrodes studied. The following galvanic cells are examples (Vetcher and Vetcher, 1967b):

$$CaO \mid CaF_2 \mid ZnF_2 \tag{28}$$

$$CaO \mid CaF_2 \mid CdF_2 \tag{29}$$

$$CaO \mid CaF_2 \mid AlF_3. \tag{30}$$

Their e.m.f. corresponds to thermodynamic values only at the start of the experiment and then drops rapidly. This phenomenon is evidently due to the formation of the compounds $CaZnF_4$ (Schnering and Bleckmann, 1965) and $CaAlF_5$ (Holm, 1965) from CaF_2, and ZnF_2, AlF_3 respectively; CaF_2 and CdF_2 result in unlimited solid solutions (Weller, 1966). The study (Vetcher and Vetcher, 1967b) of galvanic cell

$$Pt, O_2 \mid CaZnF_4, CaF_2, ZnO \mid CaF_2 \mid ZnF_2, ZnO \mid O_2, Pt \qquad (31)$$

proves the interaction between electrode and electrolyte. At the beginning the e.m.f. is 200 mV at 700°C and then drops to zero in two days. It is not surprising as the formation of $CaZnF_4$ at the right-hand side of the cell makes both electrodes identical. The instability of e.m.f. in Benz and Wagner's (1961) thermodynamic study of the CaO–SiO_2 system in the cells

$$Pt, O_2 \mid CaO \mid CaF_2 \mid CaSiO_3, SiO_2 \mid O_2, Pt \qquad (32)$$

$$Pt, O_2 \mid CaO \mid CaF_2 \mid Ca_3Si_2O_7, CaSiO_3 \mid O_2, Pt \qquad (33)$$

$$Pt, O_2 \mid CaO \mid CaF_2 \mid Ca_2SiO_4, Ca_3Si_2O_7 \mid O_2, Pt \qquad (34)$$

is probably due to the interaction of CaF_2 with CaO and SiO_2 resulting in the formation of fluorosilicates $3CaO \cdot 2SiO_2 \cdot CaF_2$ and $11CaO \cdot 4SiO_2 \cdot CaF_2$. Electrode and electrolyte interaction narrows the T-range in which galvanic cells may be used for thermodynamic studies. However, as has already been stated, this difficulty may be avoided in a galvanic cell with separated gaseous space, which discontacts electrolyte and electrodes. The recommended cell arrangement analogous to that already used with oxide electrolyte (Komarov and Tretyakov, 1970a) is given in Fig. 4.

Galvanic cells with F^--conductive solid electrolyte were used in the study of thermodynamics of fluorides, carbides, borides, phosphides, sulphides, metal alloys and intermetal compounds as well as simple and

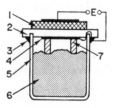

Fig. 4. Design of cell with separate-gaseous spaces; (1) reference electrode, (2) solid electrolyte, (3) high temperature vacuum seal, (4) Pt electrode, (5) crucible of inert material, (6) electrode studied, (7) inert block to press Pt electrode.

binary oxides. Basic results of reported researches and their experimental features are discussed below.

B. Thermodynamic Study of Fluorides

The first attempts to study thermodynamics of fluorides with the help of electrochemical cells reversible to F^--ions were simultaneously made by Heus and Egan (1966) and Lofgren and McIver (1966). The e.m.f.s of galvanic cells

$$Mg, MgF_2 \mid CaF_2 \mid Th, ThF_4 \tag{35}$$

$$Th, ThF_4 \mid CaF_2 \mid Al, AlF_3 \tag{36}$$

$$U, UF_3 \mid CaF_2 \mid Al, AlF_3 \tag{37}$$

$$Th, ThF_4 \mid CaF_2 \mid Ni, NiF_2 \tag{38}$$

$$Al, AlF_3 \mid CaF_2 \mid PbF_{2(S)}, Pb_{(1)} \tag{39}$$

$$Al, AlF_3 \mid CaF_2 \mid CoF_2, Co \tag{40}$$

were measured by Heus and Egan (1966). The overall reaction of the first cell is

$$\tfrac{1}{2}Mg + \tfrac{1}{4}ThF_4 = \tfrac{1}{2}MgF_2 + \tfrac{1}{4}Th \tag{35a}$$

where all reagents and products are in standard states in case there is no significant interaction between electrodes and electrolyte. The freed energy of this reaction is

$$\Delta G = \tfrac{1}{2}G^\circ_{MgF_2} - \tfrac{1}{4}\Delta G^\circ_{ThF_4} = - FE \tag{35b}$$

where $\Delta G^\circ_{MgF_2}$ and $\Delta G^\circ_{ThF_4}$—standard molar free energies of formation of fluorides. Similar relationships may also be formulated for other cells. Table II shows the e.m.f. of the cells studied at $T = 600°C$.

TABLE II.   E.m.f. values of Cells Including Fluorides.

Cell	e.m.f., V
(35)	0.310 ± 0.005
(36)	0.310 ± 0.003
(37)	0.070 ± 0.003
(38)	1.980 ± 0.005
(39)	1.580 ± 0.003
(40)	1.611

Taking Mg/MgF_2 mixture as a standard electrode ($\Delta G^\circ_{MgF_2} = -232 \cdot 4$ kcal at 600°C) and using the e.m.f. values of Table II Heus and Egan calculated ΔG° of formation of other fluorides:

$\Delta G^\circ_{ThF_4} = -436 \cdot 2$ kcal; $\Delta G^\circ_{UF_3} = -310 \cdot 5$ kcal; $\Delta G^\circ_{AlF_3} = -305 \cdot 7$ kcal;

$\Delta G^\circ_{NiF_2} = -126 \cdot 8$ kcal; $\Delta G^\circ_{PbF_2} = -130 \cdot 8$ kcal; $\Delta G^\circ_{CoF_2} = -129 \cdot 6$ kcal

Bones *et al.* (1967), Lofgren and McIver (1966) measured the e.m.f. of the cells

$$Ni, NiF_2 \mid CaF_2 \mid Al, AlF_3 \tag{41}$$

$$Ni, NiF_2 \mid CaF_2 \mid Mg, MgF_2 \tag{42}$$

$$Ni, NiF_2 \mid CaF_2 \mid Fe, FeF_2 \tag{43}$$

$$Al, AlF_3 \mid CaF_2 \mid U, UF_3 \tag{44}$$

$$UF_3, UF_4 \mid CaF_2 \mid Mg, MgF_2 \tag{45}$$

$$UF_3, UF_4 \mid CaF_2 \mid U, UF_3 \tag{46}$$

as well as (37). The equilibrium e.m.f.-values of these cells agree well with calorimetric data by Rudzitis *et al.* (1964) and the above mentioned results of Heus and Egan (1966). The cell designs used in these experiments are presented in Figs 5a, b, c and Fig. 6. These cells do not differ greatly from common arrangements with an oxygen-conductive solid electrolyte. Where electrodes were not particularly volatile (cells 36, 37) it was possible to use the cell with combined gaseous space (see Fig. 5a). Excessive volatility of electrodes (for example those containing Mg, Pb) made it necessary to

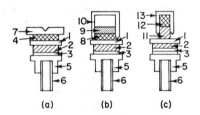

(a)   (b)   (c)

FIG. 5a, b, c. Fluoride cells used by Heus and Egan (1966). (1) CaF_2 single crystal, (2) reference electrode (Al + AlF$_3$ pellet), (3) aluminium disc, (4) Th + ThF$_4$ or U + UF$_3$ pellets, (5) Mo electrode, (6) mullite tube, (7) Mo electrode, (8) Pb + PbF$_2$ filings, (9) Pb, (10) Mo cup, (11) CaF$_2$ pellet, (12) Mg + MgF$_2$ pellet, (13) Ta—cup.

separate gaseous spaces of the right and left electrodes (Figs 5b, c and 6). Mo or Ta contacts are recommended (Heus and Egan, 1966) as suitable because of their inertness to fluorides. Composition of the surrounding inert atmosphere (Ar), is very important, and it should be carefully freed from O_2. It was achieved by passing argon over Ti-granules at 600°C or heated Ca-chips. Insufficient purification of argon is evidently the reason for overestimated and unstable e.m.f. values of the cell (38) reported by Heus and Egan (1966). The practice of using a Ni/NiO electrode showed the difficulty of obtaining equilibrium e.m.f. values in the cells with this electrode especially when T is below 800°C (Kaul *et al.*, 1970). In cell (38) oxygen traces present in an inert atmosphere form an oxide film on the Ni surface. The result of this is the decrease of chemical potential of Ni and the increase of that of fluorine leading to a higher e.m.f. than that corresponding to the equilibrium of Ni–NiF$_2$.

There is information (Lofgren and McIver, 1966) that single crystals of CaF$_2$ are more suitable as an electrolyte than polycrystaline pressed pellets, because penetration of the fluorides under study into them is possible.

Markin and Bones (1962) made an interesting and so far unique attempt of coulometric titration in galvanic cells with CaF$_2$ solid electrolyte. They investigated the cell:

$$\text{Al, AlF}_3 \mid \text{CaF}_2 \mid \text{UF}_{3-x} \qquad (46a)$$

where single phase uranium fluoride served as the right electrode. Electrochemical transport of F$^-$-ions from the right electrode to the left allowed evaluation of the chemical potential of F versus F/U ratio and the lower limit of UF$_3$ homogenity range (it corresponds to UF$_{2.97}$).

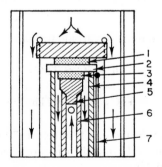

FIG. 6. Fluoride cell used by Bones *et al.* (1967) and Lofgren and McIver (1966). (1) and (3) electrodes, (2) CaF$_2$ electrolyte, (4) alumina tube, (5) iron pellet support, (6) stainless steel tube, (7) Pt–Pt, Rh thermocouple.

C. Thermodynamic Study of Carbides

Thermodynamic data in the thorium–carbon system have been obtained by Aronson (1964), Egan (1964), Aronson and Sadofsky (1965) and Satow (1967a, b) who used galvanic cells of the type:

$$Th, ThF_4 \mid CaF_2 \mid ThF_4, ThC_2, C \qquad (47)$$

$$Th, ThF_4 \mid CaF_2 \mid ThF_4, ThC_2, ThC \qquad (48)$$

$$Th, ThF_4 \mid CaF_2 \mid ThF_4, ThC_{0.7}, Th_\alpha \qquad (49)$$

$$Th, ThF_4 \mid CaF_2 \mid ThF_4, ThC_{1-x} \qquad (50)$$

$$Th, ThF_4 \mid CaF_2 \mid ThF_4, ThC_{2-x}. \qquad (51)$$

In each case the reaction in the left-hand half-cells is $Th + 4F^- = ThF_4 + 4e'$ and in the right-hand half-cells the reactions are

$$ThF_4 + 2C + 4e' = ThC_2 + 4F^- \qquad (47a)$$

$$ThF_4 + ThC_2 + 4e' = 2ThC + 4F^- \qquad (48a)$$

$$ThF_4 + \tfrac{1}{6}ThC_{0.7} + 4e' = \tfrac{7}{6}Th_\alpha + 4F^- \qquad (49a)$$

$$ThF_4 + \frac{1-x-y}{y} ThC_{1-x} + 4e' = \frac{1-x}{y} ThC_{1-x-y} + 4F^- \qquad (50a)$$

$$ThF_4 + \frac{2-x-y}{y} ThC_{2-x} + 4e' = \frac{2-x}{y} ThC_{2-x-y} + 4F^-. \qquad (51a)$$

It is obvious that the reactions of carbide formation

$$Th + 2C = ThC_2 \qquad (47b)$$

$$ThC_2 + Th = 2ThC \qquad (48b)$$

lead to the onset of equilibrium e.m.f. in cells (47,) (48). In cell (49) this the result of the dissolution of Th metal in monocarbide $ThC_{0.7}$ with the formation of solid solution Th_α

$$\tfrac{1}{6}ThC_{0.7} + Th = \tfrac{7}{6}Th_\alpha. \qquad (49b)$$

In (50) and (51) the dissolution reactions of additional thorium in non-stoichiometric mono- and dicarbides at the limits of homogenity ranges

$$\frac{1 - x - y}{y}\, ThC_{1-x} + Th = \frac{1 - x}{y}\, ThC_{1-x-y} \qquad (50b)$$

$$\frac{2 - x - y}{y}\, ThC_{2-x} + Th = \frac{2 - x}{y}\, ThC_{2-x-y} \qquad (51b)$$

are overall cell ones. The thermodynamic data calculated from the e.m.f. of cells (47)–(51) are summarized in Table III.

The conclusions of Aronson (1964) and Satow (1967a) contradict each other in their evaluation of the homogenity range of ThC_2. According to Aronson (1964) the range is very narrow, though the exact data is not given. The results of Satow (1967a) seem more convincing. He claims the existence of monophase thorium dicarbide within the $Thc_{1.72}$–$ThC_{1.97}$ range at 720–950°C. The limits of homogenity of thorium monocarbide are held to be $ThC_{0.70}$–$ThC_{0.95}$ (Satow, 1967b) in agreement with Aronson and Sadofsky's (1965) data—$ThC_{0.66}$–$ThC_{0.96}$.

Behl and Egan (1966) used the cells

$$U, UF_3 \mid CaF_2 \mid UF_3, UC_2, C \qquad (52)$$

$$U, UF_3 \mid CaF_2 \mid UF_3, U_2C_3, C \qquad (53)$$

to derive $\Delta G°$ of formation of U_2C_3 and UC_2 from the elements. It is of interest that the equilibrium in the right-hand electrode of cell (52) is meta-

TABLE III. Standard Molar Free Energy, Enthalpy and Entropy of Formation of Thorium Carbides

Carbide	$\Delta G°_{1173°K}$, kcal/g. mol	$\Delta H°$, kcal/g. mol	$\Delta S°$, cal/°K × g. mol	Investigators
ThC_2	-27.7 ± 2	-35 ± 4	-6 ± 3	Aronson (1964)
	$-28.7 \pm 0.3†$	-37.1	-7.2	Egan (1964)
	$-28.3 \pm 2.5†$	-35.2 ± 2.5	-5.9 ± 2	Satow (1967a)
ThC	-23.8 ± 2	-29 ± 4	-4 ± 2	Aronson (1964)
	$-25.7 \pm 2.5†$	-32.9 ± 2.5	-6.6 ± 2	Satow (1967a)

† The values shown are extraploated from e.m.f. data at lower temperatures, working on the assumption that $\Delta H°$ and $\Delta S°$ are independent of temperature.

stable and is maintained such for an indefinitely long time at 700–900°C. As in cells (47)–(51) it took a long time to reach equilibrium and care was taken to remove the traces of oxygen and moisture from the inert atmosphere. The e.m.f. measurements of cells (52), (53) in the interval of 700–900°C showed that for reactions

$$U_\beta + 2C_{(g)} = UC_{2(S)} \tag{52a}$$

$$U_\gamma + 2C_{(g)} = UC_{2(S)} \tag{52b}$$

$$2U_\beta + 3C_{(g)} = U_2C_{3(S)} \tag{53a}$$

$\Delta G° = -15·820–8·2T$ (kcal); $\Delta G° = -18·980–5·2T$ (kcal); $\Delta G° = -43,860–7T$ (kcal) respectively.

D. Thermodynamic Study of Sulphides

The application of e.m.f. cells including an F^--conductive solid electrolyte to the study of sulphides is confined to investigation (Aronson, 1967) of cells:

$$Th, ThF_4 \mid CaF_2 \mid ThF_4, ThS, Th_\alpha \tag{54}$$

$$Th, ThF_4 \mid CaF_2 \mid ThF_4, Th_2S_3, ThS \tag{55}$$

$$Th, ThF_4 \mid CaF_2 \mid ThF_4, Th_7S_{12}, Th_2S_3 \tag{56}$$

$$Th, ThF_4 \mid CaF_2 \mid ThF_4, ThS_2, Th_7S_{12} \tag{57}$$

where Th_α is thorium metal, saturated with sulphur in equilibrium with ThS. The cell reactions are:

$$Th + yThS = (1 + y)Th_\alpha \tag{54a}$$

$$Th + Th_2S_3 = 3ThS \tag{55a}$$

$$Th + Th_7S_{12} = 4Th_2S_3 \tag{56a}$$

$$Th + 6ThS_2 = Th_7S_{12} \tag{57a}$$

Table IV represents the free energy values of reactions (54a)–(57a) calculated from the equation $\Delta G = -4EF$ as well as ΔS and ΔH values for the cells where e.m.f. versus temperature dependence is unambiguous.

The small value of $\Delta G_{(54a)}$ is evidence of a slight decrease of thorium activity when it is saturated with sulphur and consequently of a very limited range of sulphur solubility in thorium metal. Combining the heat of formation of Th_2S_3 from elements obtained by Eyring and Westrum (1953) and ΔH°-values of reactions (54a)–(56a) Aronson (1967) calculated ΔH° of formation of other thorium sulphides. The results of his calculations as well as independent estimated data are listed in Table V.

It should be noted, however that the cells used by Aronson (1967) differ from the cells given above in having a thin region interposed between electrolyte and electrode studied. For example, cell

$$Th, ThF_4 \mid CaF_2 \mid CaF_2, ThF_4, Th_2S_3, ThS \mid ThF_4, Th_2S_3, ThS \qquad (55b)$$

was substituted for cell (55). The interlayer consisted of a homogenated mixture of CaF_2, ThF_4 and sulphides. The author observed that the stability of the cells was increased by the extra interfacial area.

TABLE IV.   Thermodynamic Data on Thorium Sulphides

Reaction	$-\Delta G$ at 900°C, (kcal/g. atom Th)	$-\Delta S$, (cal/°C. g. atom Th)	$-\Delta H$, (kcal/g. atom Th)
(54a)	$1 \cdot 1 \pm 1$	$1 \cdot 9 \pm 2$	$3 \cdot 3 \pm 3$
(55a)	$19 \cdot 8 \pm 2$	$5 \cdot 9 \pm 2$	$26 \cdot 7 \pm 4$
(56a)	$39 \cdot 7 \pm 2$	$6 \cdot 3 \pm 3$	$47 \cdot 1 \pm 4$
(57a)	$95 \cdot 2 \pm 4$	—	—

TABLE V.   The Heats of Formation of Thorium Sulphides

Compound	$-\Delta H^\circ_{1173°K}$,[†] (kcal/g. atom S)	$-\Delta H^\circ_{1173°K}$,[‡] (kcal/g. atom S)	
ThS	$109 \cdot 0$	119	→
Th_7S_{12}	$96 \cdot 2$	94	
ThS_2	$87 \cdot 7$	84	

† Data of Aronson (1967) from e.m.f. measurements.

‡ Estimated data of Eastman et al. (1950).

E. Thermodynamic Study of Thorium Borides and Phosphides

Information about the free energy of formation of thorium borides was obtained solely through the e.m.f. technique. Aronson and Auskern (1966) measured the e.m.f. of the following galvanic cells:

$$Th, ThF_4 \mid CaF_2 \mid ThF_4, ThB_6, B \tag{58}$$

$$Th, ThF_4 \mid CaF_2 \mid ThF_4, ThB_6, ThB_4 \tag{59}$$

$$Th, ThF_4 \mid CaF_2 \mid ThF_4, ThB_4, Th_\alpha \tag{60}$$

where

$$Th + 6B = ThB_6 \tag{58a}$$

$$Th + 2ThB_6 = 3ThB_4 \tag{59a}$$

$$Th + yThB_4 = (1 + y)Th_\alpha \tag{60a}$$

are the respective overall cell reactions. Here Th_α is thorium metal, saturated with boron and equilibrated with ThB_4. The free energy change in reactions (58a)–(60a) was calculated from equilibrium e.m.f. data at 850°C through the equation $\Delta G = -4EF$ and gave $-54 \cdot 4 \pm 2$, $-47 \cdot 1 \pm 2$, $-1 \cdot 4 \pm 1$ (kcal/g. atom Th), respectively. Hence, at 850°C

$$\Delta G^\circ_{ThB_6} = -54 \cdot 4 \pm 2 \text{ kcal/g. atom Th};$$

$$\Delta G^\circ_{ThB_4} = -52 \cdot 0 \text{ kcal/g. atom Th.}$$

The small e.m.f. values of cell (60) (9–20 mV) indicate low solubility of boron in thorium metal. Non-linear though monotonical dependence of e.m.f. on temperature made it impossible to obtain ΔS and ΔH of formation of borides. It is noteworthy that stable and reproducible e.m.f. values of cell (58) were obtained only at 800–900°C. At higher temperatures considerable drop of e.m.f. with time was observed. This was probably due to the metastability of the equilibrium (58a) which was maintained over some days because of weak interdiffusion of components. At higher temperatures there evidently arises a ThB_4 phase coexisting with ThB_6 at temperatures lower than 1400°C.

Thermodynamic treatment of thorium phosphides by Gingerich and Aronson (1966) in cells

$$Th, ThF_4 \mid CaF_2 \mid ThF_4, ThP, Th_3P_4 \tag{61}$$

$$Th, ThF_4 \mid CaF_2 \mid ThF_4, ThP_{0.55}, Th_\alpha \tag{62}$$

cannot be considered very successful because the e.m.f. values were not always reproducible, especially in cell (62). The authors claim the reactivity of phosphides to be the reason. The reaction

$$Th_3P_4 \rightarrow 3ThP + \tfrac{1}{2}P_{2(g)} \tag{61a}$$

produces P_2-vapour of such a density as to destroy the Pt-lead and Ta-foil contacts by the end of the experiment which takes some days. Nevertheless, Gingerich and Aronson (1966) evaluated the partial free energy change of thorium in the reaction:

$$Th + yThP_{0.55} = (1 + y)Th_\alpha \tag{62b}$$

as $\Delta \bar{G}_{Th(1173°K)} = -0.82 \pm 0.7$ kcal/g. atom Th, while in the reaction

$$Th + Th_3P_4 = 4ThP \tag{61b}$$

at 1173° K they found that

$$\Delta G = -53.7 \pm 2.8 \text{ (kcal/g. atom Th)}$$

$$\Delta S = -9.3 \pm 3.7 \text{ (cal/°K g. atom Th)}$$

$$\Delta H = -64.6 \pm 7.1 \text{ (kcal/g. atom Th)}.$$

In their studies of carbides, borides, sulphides and phosphides in galvanic cells reversible to the F^--ion Aronson and co-workers used an equilibrium mixture of Th/ThF_4 as the reference electrode. The possibility of the interaction between ThF_4 and the CaF_2-electrolyte and its after-effects were touched upon at the "International Symposium on Thermodynamics of Nuclear Materials", I.A.E.A. Vienna, 1966. It was noted (Aronson, Nowotny) that the interaction is so slight that it cannot influence the e.m.f. values. At the same time considerable solubility of ThF_4 in CaF_2 was reported (Zintl and Udgard, 1939; Heus and Egan, 1966). The possible effect of this cannot be estimated until the mechanism of Th^{4+} penetration into the CaF_2 lattice is known. When it corresponds to the reaction:

$$ThF_4(\rightarrow CaF_2) = Th_{Ca}^{\cdot\cdot} + 2F_i' + 2F_F^{\times} \tag{63}$$

one can expect considerable increase of cationic conductivity, on condition that concentration of doped defects exceeds that of intrinsic ionic defects.

F. Thermodynamic Study of Intermetal Compounds and Metal Alloys

The e.m.f. method, using an anion-conductive solid electrolyte, was applied to the study of intermetal compounds by Samokhval and Vetcher (1968). When measuring the e.m.f. in the galvanic cell

$$Mo|graphite|Al,CaAlF_5,CaF_2|CaF_2|AlSb,Sb,CaAlF_5,CaF_2|graphite|Mo \qquad (64)$$

they found the free energy of formation of AlSb from elements. It is claimed that equilibrium mixture of Al, $CaAlF_5$, CaF_2 was used as reference electrode to prevent the reaction

$$2Al + AlF_3 \rightarrow 3AlF \qquad (64a)$$

which is possible in the presence of AlF_3. The reactions on the left-hand and right-hand electrodes are:

$$Al + CaF_2 + 3F^- = CaAlF_5 + 3e' \qquad (64b)$$

$$CaAlF_5 + Sb + 3e' = CaF_2 + AlSb + 3F^- \qquad (64c)$$

and consequently the overall cell reaction is

$$Al_{(S)} + Sb_{(S)} = AlSb_{(S)} \qquad (64d)$$

Samokhval and Vetcher (1968) pointed out that it was possible to obtain stable e.m.f. values using half-cells (Fig. 7) prepared by compressing sintered and homogenized samples of the mixture studied with graphite and CaF_2 powder. This helped to reduce evaporation of antimony in the course of the experiment. The free energy of reaction (64d) calculated from equilibrium e.m.f.-values is given by the equation

$$\Delta G = -12 \cdot 86 + 3 \cdot 44 \times 10^{-3} \, T \pm 0 \cdot 10 \, (kcal/mol)/778-896°K/.$$

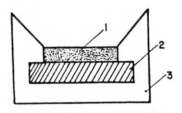

FIG. 7. Design of half cell used by Samokhval and Vetcher (1968). (1) graphite contact (2) electrode studied, (3) CaF_2 electrolyte.

Hence

$$\Delta H^{\circ}_{AlSb} = -12 \cdot 86 \pm 0 \cdot 19 \, \text{kcal/mol} \quad (800°K)$$

$$\Delta S^{\circ}_{AlSb} = -3 \cdot 44 \pm 0 \cdot 22 \, \text{cal/}°\text{K.mol} \quad (800°K).$$

It is interesting to find that reproducible e.m.f. values in cell (64) were obtained even at 500°C. Electrodes being homogenized at 800°C, the equilibrium was quickly achieved (in the course of 6–8 h) which should be attributed to speedy diffusion of Al in the studied medium. At temperatures above 600°C the e.m.f. values of cell (64) correspond to those of cell

$$Al_{(S)} \, | \, \text{melt LiCl} + KCl, Al^{3+} \, | \, AlSb_{(S)}, Sb_{(S)} \quad (65)$$

studied by Vetcher *et al.* (1965). Slight divergence in the data at $T < 600°C$ can be explained by experimental errors in cell (65), since both Al^+—and Al^{3+}—ions are likely to be present in the liquid electrolyte.

The study of solid solutions of AlSb–GaSb in the galvanic cell

$$AlSb, Sb, CaAlF_5, CaF_2 \, | \, CaF_2 \, | \, (Al, Ga)Sb, Sb, CaAlF_5, CaF_2 \quad (66)$$

at 850°K by Samokhval *et al.* (1969) is of considerable interest. The overall cell process is

$$AlSb = AlSb \text{ (in solid solution).} \quad (66a)$$

The values of AlSb activity in the solution obtained through the equation

$$\ln \alpha_{AlSb} = -\frac{\Delta \bar{G}_{AlSb}}{RT} = \frac{3EF}{RT} \quad (66b)$$

are given in Table VI. The table also presents CaSb activity data calculated with the help of the Gibbs–Duhem equation as well as the data of integral free energy of Gibbs. It follows from Table VI that solid solutions of AlSb–GaSb demonstrate a negative deviation from the ideal behaviour in the presence of excess antimony and consequently are characterized by a tendency to ordering and not by a tendency to lamination as was stated earlier (Romanenko and Ivanov–Omskij, 1959; Nikitina and Romanenko, 1964). A successful attempt (Samokhval and Vetcher, 1970) to study alloys of highly reactive metals such as Ti and Al in cells reversible to the F^--ion is of special interest. E.m.f. measurements in a galvanic cell

$$Al_{(S,1)}, CaAlF_5, CaF_2 \mid CaF_2 \mid Al_xTi_{1-x(S)}, CaAlF_5, CaF_2 \qquad (67)$$

carried out at 890–1000° C, and $0.05 \leqslant X \leqslant 0.5$ made it possible to conclude that solution in the range of compound Ti_3Al is the only stable phase in this interval of concentrations (see Fig. 8). Considerable decrease in entropy of alloys rich in Ti is due to the ordering effect, the nature of which is not yet clear.

TABLE VI.   Thermodynamic Data of GaSb–AlSb Solid Solutions†

Content of AlSb	e.m.f., mV	α_{AlSb}	α_{GaSb}	$-\Delta G$, (kcal/mol)
0·1	250	$3·7 \times 10^{-5}$	$7·2 \times 10^{-1}$	1·7
0·2	195	$3·5 \times 10^{-4}$	$4·8 \times 10^{-1}$	2·7
0·3	159	$1·5 \times 10^{-3}$	$3·0 \times 10^{-1}$	3·8
0·4	128	$5·4 \times 10^{-3}$	$1·5 \times 10^{-1}$	4·3
0·5	97·6	$1·9 \times 10^{-2}$	$5·45 \times 10^{-2}$	4·7
0·7	47·9	$1·4 \times 10^{-1}$	$3·0 \times 10^{-3}$	4·3
0·9	13·4	$5·8 \times 10^{-1}$	$1·7 \times 10^{-4}$	1·75

† Given by Samokhval *et al.* (1969).

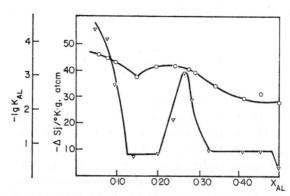

FIG. 8. Partial entropy ($\triangledown$) and log of activity coefficient of Al ($\bigcirc$) in Ti–Al alloys at 960°K (after Samokhval and Vetcher, 1970).

G. Thermodynamic Study of Oxides

The first attempt to study thermodynamics of oxygen-containing compounds by the galvanic cells including the F^--conductive solid electrolytes was made by Benz and Wagner (1961). They measured the e.m.f. of the following cells:

$$Pt, O_{2(g)} \,|\, CaO, CaF_2 \,|\, CaF_2 \,|\, CaSiO_3, SiO_2 \,|\, O_{2(g)}, Pt \qquad (68)$$

$$Pt, O_{2(g)} \,|\, CaO, CaF_2 \,|\, CaF_2 \,|\, Ca_3Si_2O_7, CaSiO_3 \,|\, O_{2(g)}, Pt \qquad (69)$$

$$Pt, O_{2(g)} \,|\, CaO, CaF_2 \,|\, CaF_2 \,|\, Ca_2SiO_4, Ca_3Si_2O_7 \,|\, O_{2(g)}, Pt \qquad (70)$$

On the left-hand electrode of the first cell the reaction is

$$CaO + 2F^-(\leftarrow) = CaF_2 + \tfrac{1}{2}O_{2(g)} + 2e', \qquad (68a)$$

on the right-hand:

$$CaF_2 + SiO_2 + \tfrac{1}{2}O_{2(g)} + 2e'(\leftarrow) = CaSiO_3 + 2F^-(\leftarrow). \qquad (68b)$$

The overall cell process is the formation of silicate

$$CaO + SiO_2 = CaSiO_3; \quad \Delta G_{(68c)} = -2EF. \qquad (68c)$$

TABLE VII.   E.m.f. values of cells which are symmetrical except for the presence of catalysts in various amounts

Phases of system $CaO–SiO_2$	Catalyst mol per mol CaO		Temp., °C	e.m.f., mV
	Left	Right		
CaO	$0.03Cr_2O_3$	—	660	2 ± 0.5
	$0.03K_2Cr_2O_7$	$0.03Cr_2O_3$	850	1.8 ± 0.5
	$0.05PbO$	—	770	0 ± 0.5
$Ca_2SiO_4 + Ca_3Si_2O_7$	$0.03K_2Cr_2O_7$	$0.03Cr_2O_7$	710	2 ± 1
	$0.05PbO$	$0.001PbO$	700	1 ± 1
$Ca_3Si_2O_7 + CaSiO_3$	$0.03K_2Cr_2O_7$	$0.05Cr_2O_3$	710	1 ± 2
$CaSiO_3 + SiO_2$	$0.03Cr_2O_3$	$0.03K_2Cr_2O_7$	730	2 ± 1.5
	$0.01PbO$	$0.05Pb$	720	1 ± 1

Similar equations may be written for the other two cells. In the study of galvanic cells (68)–(70) there were considerable experimental difficulties connected with slow attainment of equilibrium. The equilibrium is more rapidly attained in half-cells of type (68a) by inserting ions of variable valency into the electrodes. For example insertion of Cr_2O_3, $K_2Cr_2O_7$ and PbO (3–5 molar %) considerably accelerates attainment of equilibrium in cells. Indifferent behaviour of these dopents towards electrodes studied may be proved by the data of Table VII.

Benz and Wagner (1961) thought it useful to add CaF_2 powder (10 wt %) to the electrode mixture. The date of the free energy of silicates are given in Table VIII alongside the independent calorimetric data. Applying the technique recommended by Benz and Wagner (1961), Taylor and Schmalzried (1964) measured equilibrium e.m.f. of the cells

$$Pt, O_{2(g)} \mid CaO \mid CaF_2 \mid CaTiO_3, TiO_2 \mid O_{2(g)}, Pt \qquad (71)$$

$$Pt, O_{2(g)} \mid CaO \mid CaF_2 \mid Ca_4Ti_3O_{10}, CaTiO_3 \mid O_{2(g)}, Pt \qquad (72)$$

and calculated the free energy of reactions

$$CaO + TiO_2 = CaTiO_3 \qquad (71a)$$

$$CaO + 3TiO_2 = Ca_4Ti_3O_{10} \qquad (72a)$$

$\Delta G^\circ_{(71a)} = -20 \cdot 0 \pm 1$ (kcal/mol) at 600°C; $\Delta G^\circ_{(72a)} = -15 \cdot 0 \pm 1$ (kcal/mol) at 540°C. Galvanic cell (71) was restudied (Rezukhina et al., 1966) in a wider temperature range (888–972°K) and the data correspond closely with those of independent calorimetric measurements (see Table IX).

TABLE VIII. Integral Molar Free Energies of mixing for the system CaO–SiO_2 at 700°C

Phases	Mol fraction of SiO_2	G^M, kcal	
		E.m.f.†	Calorimetry‡
Ca_2SiO_4	0·333	$-10 \cdot 8 \pm 0 \cdot 1$	$-10 \cdot 5 \pm 0 \cdot 1$
$Ca_3Si_2O_7$	0·40	$-12 \cdot 4 \pm 0 \cdot 15$	
$CaSiO_3$	0·50	$-10 \cdot 6 \pm 0 \cdot 15$	$-10 \cdot 6 \pm 0 \cdot 1$

† Benz and Wagner (1961).

‡ Calculated using data of: Torgeson and Sahama (1948); King (1951); Kelley (1959).

The studies (Rezukhina *et al.*, 1966; Rezukhina and Baginska, 1967; Dneprova *et al.*, 1968) of the following galvanic cells are of interest:

$$Pt, O_{2(g)} \,|\, CaO \,|\, CaF_2 \,|\, CaWO_4, WO_3 \,|\, O_{2(g)}, Pt \qquad (73)$$

$$Pt, O_{2(g)} \,|\, CaO \,|\, CaF_2 \,|\, CaFe_2O_4, Fe_2O_3 \,|\, O_{2(g)}, Pt \qquad (74)$$

$$Pt, O_{2(g)} \,|\, CaO \,|\, CaF_2 \,|\, CaFe_2O_4, Ca_2Fe_2O_5 \,|\, O_{2(g)}, Pt \qquad (75)$$

$$Pt, O_{2(g)} \,|\, CaO \,|\, CaF_2 \,|\, CaO.Nb_2O_5, Nb_2O_5 \,|\, O_{2(g)}, Pt \qquad (76)$$

$$Pt, O_{2(g)} \,|\, CaO \,|\, CaF_2 \,|\, 2CaO.Nb_2O_5, CaO.Nb_2O_5 \,|\, O_{2(g)}, Pt \qquad (77)$$

where the overall cell processes are:

$$CaO + WO_3 = CaWO_4 \qquad (73a)$$

$$CaO + Fe_2O_3 = CaFe_2O_4 \qquad (74a)$$

$$CaO + CaFe_2O_4 = Ca_2Fe_2O_5 \qquad (75a)$$

$$CaO + Nb_2O_5 = CaO.Nb_2O_5 \qquad (76a)$$

$$CaO + CaO.Nb_2O_5 = 2CaO.Nb_2O_5 \qquad (77a)$$

Values of free energy of formation of wolframates, niobates and ferrites from oxides calculated from the equilibrium e.m.f. data are summarized in Table X.

TABLE IX.   The Free Energies of Formation of Calcium Titanate from Oxides

Temperature °K	$-\Delta G°$, kcal	
	E.m.f.	Calorimetry†
888	$19 \cdot 65 \pm 0 \cdot 05$	$20 \cdot 20 \pm 0 \cdot 50$
901	$19 \cdot 65 \pm 0 \cdot 15$	$20 \cdot 21 \pm 0 \cdot 55$
939	$19 \cdot 92 \pm 0 \cdot 10$	$20 \cdot 260 \pm 0 \cdot 55$
948	$19 \cdot 88 \pm 0 \cdot 10$	$20 \cdot 273 \pm 0 \cdot 55$
972	$19 \cdot 74 \pm 0 \cdot 05$	$20 \cdot 30 \pm 0 \cdot 55$

† Calculated using data of Kelley *et al.* (1954).

TABLE X.   Free energies of formation of some binary oxides

Reaction	Constant of equation $\Delta G^\circ = -A + B \times T \times 10^{-3}$, (kcal/mol)		Temperature range, °K
	A	B	
$CaO + WO_3 \rightarrow CaWO_4$	$40\cdot36 \pm 2\cdot50$	$4\cdot0 \pm 2\cdot5$	860–1070
$CaO + Fe_2O_3 \rightarrow CaFe_2O_4$	$7\cdot10 \pm 0\cdot2$	$-1\cdot15 \pm 0\cdot14$	1100–1350
$2CaO + Fe_2O_3 \rightarrow Ca_2Fe_2O_5$	$7\cdot55 \pm 0\cdot5$	$-5\cdot5 \pm 0\cdot4$	1100–1350
$CaO + CaFe_2O_4 \rightarrow Ca_2Fe_2O_5$	$0\cdot44 \pm 0\cdot3$	$-4\cdot34 \pm 0\cdot25$	1100–1350
$CaO + Nb_2O_5 \rightarrow CaO.Nb_2O_5$	$42\cdot0 \pm 1\cdot8$	$5\cdot4 \pm 1\cdot4$	1100–1267
$2CaO + Nb_2O_5 \rightarrow 2CaO.Nb_2O_5$	$50\cdot8 \pm 3\cdot6$	$-5\cdot3 \pm 2\cdot8$	1224–1267
$CaO + CaO.Nb_2O_5 \rightarrow 2CaO.Nb_2O_5$	$8\cdot8 \pm 1\cdot8$	$-10\cdot7 \pm 1\cdot4$	1224–1350

Vetcher and Vetcher (1967b) recommended the use of the galvanic cell

$$Pt, O_{2(g)} | Me_IO, Me_IF_2 | CaF_2 | Me_{II}F_2, Me_{II}O | O_{2(g)}, Pt \qquad (78)$$

to study the exchange reaction of oxides and fluorides:

$$Me_IO + Me_{II}F_2 = Me_IF_2 + Me_{II}O. \qquad (78a)$$

The results were very promising in the study of cell

$$Pt, O_{2(g)} | CaO, CaF_2 | CaF_2 | MgO, MgF_2 | O_{2(g)}, Pt \qquad (78b)$$

Vetcher and Vetcher (1967b) in contrast with other investigators (Benz and Wagner, 1961; Taylor and Schmalzried, 1964) inserted powdered silver metal (1:1 by wt) into an oxide–fluoride mixture instead of Cr_2O_3, $K_2Cr_2O_7$, PbO catalysts. In view of the possible difference of potentials on the two crystal interfaces (Karpatchev and Palguev, 1960) the effect of introduced silver may be attributed to the onset of strong electronic conductivity of the diffusion region on the electrode–electrolyte boundary. When silver was introduced into the electrode mixture the e.m.f. of the cell (78b) became stable. At $T = 900°C$ equilibrium was attained almost immediately, and remained unchanged for 2–3 days. The e.m.f. value of cell (78b) in the interval of 800–950°C is $382 \pm 4\,mV$ (Vetcher and Vetcher, 1967). It means that for the overall cell reaction:

$$CaO + MgF_2 = CaF_2 + MgO \qquad (78c)$$

in this temperature range

$$\Delta S = 0; \quad \Delta G = \Delta H = -2EF = -17.6 \pm 0.2 \text{ (kcal)}.$$

If the change of heat capacity of MgO, CaO, CaF_2 and MgF_2 with temperature is taken into consideration, the value $\Delta H_{298} = -17.7 \pm 0.2$ kcal is calculated. Using the latter value and standard enthalpies of formation of MgO (Gerasimov et al., 1966), CaO (Kubaschewski and Evans, 1955), MgF_2 (Rudzitis et al., 1964), Vetcher and Vetcher (1967b) calculated the enthalpy of formation of CaF_2:

$$\Delta H_{298} = -294.3 \pm 1.2 \text{ (kcal/mol)}.$$

H. Other Halogen-Conductive Solid Electrolytes

Systems of $ThO_2(Y_2O_3)$, $ZrO_2(CaO)$ and CaF_2 type do not exhaust the group of anion-conductive solid electrolytes. Among halides there is an extensive group of crystals for which the transport number of anions approaches 1 (see Table XI). The results of Hood and Morrison (1967), Barsis and Taylor (1966), and Koch and Wagner (1937) make it possible to state that these crystals, similarly to CaF_2, are characterized by Frenkel-type disorder in the anionic sublattice:

$$MeHal_2 = Hal_i' + V_{Hal}{}^{\cdot} + Me_{Me}{}^{\times} \qquad (79)$$

$SrCl_2$ is an exception. The nature of its disorder is doubtful: either that of the anti-Frenkel type with Cl_i' of low mobility or of the Schottky type. The halides enumerated in Table X are no better than CaF_2 when used as solid electrolytes in galvanic cells. What is more, the low melting point of the majority of halides as compared to that of CaF_2 presents an additional difficulty. It may be expected that at the same temperature more fusible halides display greater reactivity to the electrodes, at the same time lowering of temperature results in slower attainment of equilibrium. Thus the use of solid electrolytes other than CaF_2 may be practicable only in studies of compounds that do not include F^--or Ca^{++}-ions and those having an ion common to the electrolyte. Successful studies of the following galvanic cells were done by Kiukkola and Wagner (1957), Benz and Schmalrzied (1961), Taylor and Schmalzried (1964), Egan et al., (1962), Egan (1966):

$$Pb_{(s,1)} \mid PbCl_2(+KCl) \mid PbS, Ag_2S \mid Ag \qquad (80)$$

$$Pt, O_{2(g)} \mid Pb_2SiO_4, Pb_4SiO_6 \mid PbF_2 \mid Pb_2SiO_4, PbSiO_3 \mid O_{2(g)}, Pt \qquad (81)$$

$$Pt, O_{2(g)} \mid PbSiO_3, Pb_2SiO_4 \mid PbF_2 \mid PbSiO_3, SiO_2 \mid O_{2(g)}, Pt \qquad (82)$$

TABLE XI.   Halides which may be used as solid electrolytes

Compound	t_{Hal}	Melting point °C	References
SrF_2		1400	Croatto and Bruno (1948)
BaF_2	$t_{F^-} = 1$	1320	Tubandt et al. (1931)
PbF_2		824	Tubandt and Eggert (1920)
$PbCl_2$		498	Tubandt and Eggert (1920)
$BaCl_2$	$t_{Cl^-} = 1$	960	Tubandt et al. (1931)
$SrCl_2$		873	Hood and Morrison (1967)
$SrCl_2$			Barsis and Taylor (1966)
$PbBr_2$		370	Tubandt and Eggert (1921)
$BaBr_2$	$t_{Br^-} = 1$	847	Tubandt et al. (1931)

$$Pt, O_{2(g)} \mid SrO \mid SrF_2 \mid SrTiO_3, TiO_2 \mid O_{2(g)}, Pt \qquad (83)$$

$$Pt, O_{2(g)} \mid SrO \mid SrF_2 \mid SrTiO_3, Sr_4Ti_3O_{10} \mid O_{2(g)}, Pt \qquad (84)$$

$$Pt, O_{2(g)} \mid MgO \mid MgF_2 \mid MgAl_2O_4, Al_2O_3 \mid O_{2(g)}, Pt \qquad (85)$$

$$Mg, MgCl_2 \mid BaCl_2 \mid AgCl + NaCl \mid Ag \qquad (86)$$

$$Ce, CeCl_3 \mid BaCl_2 \mid AgCl + NaCl \mid Ag \qquad (87)$$

$$Th, ThCl_4 \mid BaCl_2 \mid AgCl + NaCl \mid Ag \qquad (88)$$

$$U, UCl_3 \mid BaCl_2 \mid AgCl + NaCl \mid Ag \qquad (89)$$

$$Co, CoCl_2 \mid BaCl_2 \mid AgCl + NaCl \mid Ag \qquad (90)$$

$$Ni, NiCl_2 \mid BaCl_2 \mid AgCl + NaCl \mid Ag \qquad (91)$$

Free energy values of some reactions calculated from equilibrium e.m.f. data of corresponding cells are summarized in Table XII. In cells (81)–(85) 3 mol % K_2Cr_2O (as a catalyst) and a few per cent of powdered electrolyte were added to the electrodes studied, whereas in cell (80) the solid electrolyte was doped with KCl (0.5% by wt) to increase its conductivity by increasing the anion vacancy concentration. In all cases the cell arrangement did not differ from the normal.

Summing up, one should stress the fact that F^--conductive solid electrolytes are far preferable to oxygen-conductive ones for studying oxide compounds and alloys of metals of high affinity for oxygen. For example, it became possible to obtain free energy of formation of $MgAl_2O_4$ from oxides (Taylor and Schmalzried, 1964) with the help of galvanic cell (85). None of the oxygen-containing electrolytes could be used in this case because oxygen pressure at dissociation of MgO, Al_2O_3 and $MgAl_2O_4$ is much lower than the limiting value of $P_{O_2}(P_{O_2}{}^*)$, at which the onset of considerable electronic conductivity occurs.

III. Cation-Conductive Solid Electrolytes

A. Simple Halides of Silver and Copper as Solid Electrolytes

The group of cation-conductive solid electrolytes includes halides, sulphides, oxides and double salts. Halides of silver and copper are more thoroughly studied. Their transport number of cations approaches 1 (see Table XIII).

TABLE XII.   Free energy values obtained from e.m.f. measurements of cells (80)–(91)

No of cell	Reaction	T, °C	$-\Delta G°$, kcal		References	
			e.m.f.	Comparison data	e.m.f.	Comparison data
1	2	3	4	5	6	7
104	$Pb_{(s,l)} + S_{(l)} = PbS_{(s)}$	250	22·60		Kiukkola and Wagner (1957)	7
		300	22·33			
		327	22·14			
		350	22·05			
		400	21·73			
		427	21·54			
105	$PbO + SiO_2 = PbSiO_3$	640	5·2		Benz and Schmalzried (1961)	
106	$2PbO + SiO_2 = Pb_2SiO_4$	640	7·5			
107	$4PbO + SiO_2 = Pb_4SiO_6$	640	8·5			
107	$SrO + TiO_2 = SrTiO_3$	560	$28·1 \pm 1$		Taylor and Schmalzried (1964)	
108	$SrO + 3SrTiO_3 = Sr_4Ti_3O_{10}$	540	$21·6 \pm 1$			
109	$MgO + Al_2O_3 = MgAl_2O_4$	530	$2·4 \pm 1$			

No of cell	Reaction	T, °C	$-\Delta G°$, kcal		References	
			e.m.f.	Comparison data	e.m.f.	Comparison data
1	2	3	4	5	6	7
110	$Mg + Cl_2 = MgCl_2$	400	127·80	127·30	Egan et al. (1962)	Hamer et al. (1956)
		500	126·18	125·58		
		600	124·54	123·92		
111	$Ce + 1\frac{1}{2}Cl_2 = CeCl_3$	400	221·26	214·2	Egan et al. (1962)	Glassner (1958)
		500	206·58	208·8		
		600	202·02	203·4		
112	$Th + 2Cl_2 = ThCl_4$	400	232·4	234·4	Egan et al. (1962)	Glassner (1958)
		500	224·4	227·2		
		600	216·8	220·0		
113	$U + 1\frac{1}{2}Cl_2 = UCl_3$	400	168·2	178·2	Egan et al. (1962)	Glassner (1958)
		500	163·1	173·1		
		600	158·5	168·0		
114	$Co + Cl_2 = CoCl_2$	400	52·73	55·42	Egan (1966)	Wicks and Block (1963)
		450	51·28	53·82		
115	$Ni + Cl_2 = NiCl_2$	400	48·79	48·42	Egan (1966)	Wicks and Block (1963)
		450	47·32	46·67		

In spite of numerous studies (Teltow, 1949; Kurnick, 1952; Ebert and Teltow, 1955; Schmalzried, 1959; Kröger, 1964) the disorder type in these halides is not known for sure. The majority of investigators agree on Frenkel-type disorder:

$$0 = Ag_i^{\cdot} + V_{Ag}'.\tag{92}$$

At the same time, some authors (Schmalzried, 1959) consider that at high temperatures Frenkel-type disorder of AgBr is replaced by that of the Schottky type:

$$0 = V_{Ag}' + V_{Br}^{\cdot}.\tag{93}$$

The following diagram constructed by Patterson (1970) (Fig. 9) shows the range of the use of halides as solid electrolytes. The domain of $\log P_{x_2}$ and T for each compound where cationic conductivity is observed ($t_{cat} \geqslant 0.99$)

TABLE XIII.   Transport numbers in Silver and Copper Halides

Compound	Melting point, °C	Temperature, °C	t_k^+	t_A^-	t_e^-	References
AgBr$_2$	430	20	1·00	0·00	0·00	
		200–300	1·00	0·00	0·00	Tubandt and
AgCl	455	20	1·00	0·00	0·00	Eggert (1920),
		200–350	1·00	0·00	0·00	Tubbandt and Eggert (1921)
α – AgI	558	150–400	1·00	0·00	0·00	
		18	0·00	0·00	1·00	
		110	0·03	0·00	0·97	
CuCl	430	232	0·50	0·00	0·50	
		300	0·98	0·00	0·02	
		366	1·00	0·00	0·00	Tubandt (1931), Tubandt *et al.* (1927)
		27	0·00	0·00	1·00	
		233	0·14	0·00	0·86	
γ – CuBr	488	299	0·87	0·00	0·13	
		390	1·00	0·00	0·00	
β – CuBr	488	395–445	1·00	0·00	0·00	
		200	0·00	0·00	1·00	
		306	0·32	0·00	0·68	Tubandt *et al.*
γ – CuI	588	358	0·84	0·00	0·16	(1927)
		400	1·00	0·00	0·00	

is shown by the corresponding loop. The low halogenic boundaries of the electrolytic domain for each compound is determined by its thermodynamic stability and may be calculated from the values of free energy of formation (Wicks and Block, 1963). The upper halogenic boundary is determined by the onset of considerable ($t_h \geq 0.01$) hole conductivity which is, evidently, the result of reaction (94) (Wagner, 1956, 1959):

$$\tfrac{1}{2}X_2(\to a\,\mathrm{Ag}X) = X_x{}^\times + V_{\mathrm{Ag}}{}' + h^{\cdot}. \tag{94}$$

The increase of hole conductivity proportionally to the square root of Px_2 (Wagner and Wagner, 1957a; Ilschner, 1958; Raleigh, 1965) agrees with the discussed type of defect formation. It follows from Fig. 9 that a considerable contribution of hole conductivity is observed even at halogen pressure lower than 1 atm. This makes it impossible to use these types of crystals as cation-conductive electrolytes below the given pressure. The high and low temperature boundaries shown in Fig. 9 are determined either by phase transformations, which produce an electron-conductive phase, or by the temperature beyond which quantitative conductivity measurements have not yet been reported. In designating electrolytic domains the data of Raleigh (1965), Ilschner (1958), Wagner and Wagner (1957), Biermann and Jost (1969) and Kurnick (1952) were used. The discrepancies in polarization measurements (Ilschner, 1958) did not allow demarcation of the electrolytic domain of AgCl in Fig. 10. Quantitative description of this domain is impossible until the current–voltage dependence in AgCl is explained.

There were numerous attempts to use silver and cuprous halides as solid electrolytes of e.m.f. cells. For example, Reinhold (1928) studied the galvanic cell

$$C \mid Ag \mid AgCl \mid C, Cl_{2(g)} \tag{95}$$

where the overall cell reaction is:

$$Ag + \tfrac{1}{2}Cl_2 = AgCl. \tag{95a}$$

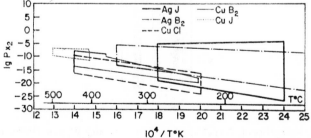

FIG. 9. Electrolytic domains in $\log P_{x_2}$, $1/T$ space for silver and cuprous halides.

The underrated value of free energy of formation of AgCl calculated from e.m.f. data by the equation $\Delta G = -EF$, should be attributed to the onset of hole conductivity of AgCl due to high α_{Cl_2} on the right-hand electrode. Silver- and copper-conductive electrolytes have recently become widely used in thermodynamic and kinetic investigations. Kiukkola and Wagner (1957) measured the e.m.f. of the galvanic cell

$$Pt \mid Ag \mid AgI \mid Ag_2S, S \mid C \qquad (96)$$

$$
\begin{array}{ccc}
2e' & 2Ag^+ & 2e' \\
\leftarrow & \rightarrow & \leftarrow
\end{array}
$$

$$
\begin{array}{c|c}
2Ag = & 2Ag^+ \, (\rightarrow) \\
= 2Ag^+ \, (\rightarrow) & +2e' \, (\leftarrow) \\
+ 2e' \, (\leftarrow) & +S \, (\text{melt}) = \\
& = Ag_2S
\end{array}
$$

to evaluate the free energy of formation of Ag_2S. The set-up of the cell is shown in Fig. 10. In the lower part of the Pyrex tube silver iodide (1) was melted and the graphite electrode (2) and glass tube (3) were immersed. After the AgI had solidified, sulphur was introduced into the glass tube at $T(\text{melt}) > T > 146°\,C$ and the graphite electrode (4) was brought in contact with the AgI. To start the experiment, silver sulphide was formed electro-

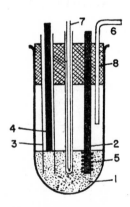

FIG. 10. Cell $Ag|AgI|Ag_2S$, $S(1)|C$, used by Kiukkola and Wagner (1957). (1) AgI, (2) and (4) graphite rods, (3) glass tube, (5) silver wire, (6) argon inlet, (7) thermocouple, (8) rubber stopper.

lytically at the graphite rod (4) by passing a current of 5–20 mA for about half an hour. The study of a similar cell (Kiukkola and Wagner, 1957)

$$Pt \mid Ag \mid AgI \mid Ag_2Se, Se \mid Pt \qquad (97)$$

gave the free energy of formation of Ag_2Se

$$\Delta G^\circ = -13 \cdot 74 - 0 \cdot 0074 \times (T - 500) \text{ kcal/mol (at } T > 500°K).$$

The results of the attempt (Wagner and Wagner, 1957c) to determine the free energy of formation of Cu_2S from elements with the help of cell

$$Pt \mid Cu \mid CuI \mid Cu_2S, S_{(1)} \mid C \qquad (98)$$

were pessimistic, for the Cu_2S crystals interacted with sulphur forming CuS. Besides, CuI contacting with melted sulphur displaued electronic conductivity. To avoid it Wagner and Wagner (1957c) used a more complicated cell:

$$\overbrace{\text{Cu} \mid \text{CuBr} \mid \text{Cu}_2\text{S}}^{A} \quad \text{graphite} \left| \begin{matrix} \text{Cu}_2\text{S} \\ \text{PbS} \end{matrix} \right| \overbrace{\text{PbCl}_2 + 1\% \text{ KCl} \mid \text{Pb}}^{B} \qquad (99)$$

which is in practice a combination of cells A and B, where electrolytes are CuBr and $PbCl_2 + 1\%$ KCl respectively, with the overall cell reaction

$$PbS + 2Cu \rightarrow Cu_2S + Pb. \qquad (99a)$$

The measurements of this cell yielded the stability interval of phases $Pb + PbS + Cu_2S$ and $Cu + PbS + Cu_2S$. E.m.f. $= 0$ corresponds to the phase transition. It follows from the experiment that it occurs at $T = 279°C$, when $\Delta G_{Cu_2S} = \Delta G_{PbS}$. The temperature dependence for this cell is $E = (6 \cdot 5 \pm 1) \times 10^{-3} + 0 \cdot 3 \times 10^{-3} \times (T - 300)$, hence $\Delta G_{300°C} = -277 \pm 46$ cal. Combining the latter with the free energy of formation of PbS ($\Delta G_{300°C} = 22 \cdot 390$ kcal; Kiukkola and Wagner, 1957) one finds that for the reaction

$$2Cu + S = Cu_2S \qquad (98a)$$

$\Delta G^\circ_{300°C} = -22 \cdot 667$ kcal.

Electrolytic properties of silver halides may be varied by doping. For example, Lehovec and Smyth (1962) found that conductivity of AgCl is increased by two orders of magnitude after introduction 0·1 mol % of Ag_2S, evidently due to the reaction

$$Ag_2S(\rightarrow AgCl) = Ag_{Ag}^{x} + Ag_i^{\cdot} + S_{Cl}' \qquad (100)$$

In some cases it was found reasonable to use solid solutions of AgCl–NaCl instead of pure AgCl because of their considerably higher melting temperatures (20% AgCl–80% NaCl–743° C). The thermodynamics of solutions of this type were studied by Panish et al. (1958) by means of measuring the e.m.f. of the cell

$$Ag \mid AgCl - NaCl \mid C,Cl_{2(g)} \qquad (101)$$

A study of the e.m.f. of the cell

$$Br_{2(g)},C \mid AgBr \left| \begin{array}{c} AgBr - \\ -NaBr \end{array} \right| C,Br_{2(g)} \qquad (102)$$

was undertaken (Obrosov and Karpatchev, 1970) with the same purpose. The overall cell process here evidently is

$$AgBr_{(pure)} \rightarrow AgBr (NaCl) \qquad (102a)$$

and

$$\alpha_{AgBr} = \exp\left(-\frac{EF}{RT}\right) \qquad (102b)$$

The studies of nonstoichiometry of cuprous and silver-sulphides by Wagner and Wagner (1957b), Wehefritz (1960) and Wagner (1953) respectively using coulometric technique are of great interest. The following cells are examined:

$$Ag \mid AgI \mid Ag_{2-\delta}S \mid Pt \qquad (103)$$

$$Cu \mid CuBr \mid Cu_{2-\delta}S \mid C \qquad (104)$$

On passing current through the cell (103) from left to right the pellet of Ag_2S receives Ag^+-ions of AgI-crystal and electrons of platinum. As a result the number of moles of Ag in the Ag_2S pellet changes according to the equation

$$\Delta\eta_{Ag} = \frac{It}{F} \qquad (103a)$$

where I is the current, applied for the time t. If the composition of sulphide is $Ag_{2-\delta}S$,

$$\Delta\delta = \frac{It}{\eta_S F} \tag{103b}$$

where η_s is the number of gram atoms of sulphur in the Ag_2S pellet. On measuring the e.m.f. of the open circuit, the silver chemical potential in Ag_2S is found to be

$$\mu_{Ag} - \mu_{Ag}^{\circ} = -EF. \tag{103c}$$

In his experiments Wagner (1953) showed that cubic modification of Ag_2S equilibrated with sulphur or silver has $\delta = 0 \pm 0.00003$ and $\delta = 0 \pm 0.003$ respectively. A considerably wider range of homogenity was observed for cuprous sulphide—from $Cu_{2.000}S$ to $Cu_{1.75}S$ (Wehefritz, 1960). Coulometric titration technique was successfully applied by Kiukkola and Wagner (1957) to study thermodynamics Ag–Te–alloys. The dependence of $\mu_{Ag(in\ Ag-Te)}$ on composition of alloys was established by measuring equilibrium e.m.f. values of the cell

$$Ag \mid AgI \mid (Ag,Te) \tag{104}$$

after every titration step. The mode of dependence helped to ascertain the existence of three phases (α: $Ag_{1.99}Te–Ag_{2.00}Te$; γ: $Ag_{1.88}Te–Ag_{1.91}Te$; ε: $Ag_{1.63}Te–Ag_{1.66}Te$). The free energy of formation of these phases is shown in Table XIV.

Application of the coulometric titration technique in the cells:

$$Cu \mid CuBr \mid Cu_xSe \mid graphite \tag{105}$$

$$Cu \mid CuBr \mid Cu_xTe \mid graphite \tag{106}$$

TABLE XIV.   Free energies of formation of silver tellurides

Phase	$-\Delta G^{\circ}$, kcal	
	250°C	300°C
Ag_2Te	11·4	12·0
$Ag_{1.90}Te$	11·0	11·5
$Ag_{1.64}Te$	9·5	10·0

allowed Lorenz and Wagner (1957) to study the thermodynamics of cuprous selenides and tellurides at 400°C. It was found that selenide Cu_xSe is characterized by wide homogenity, ranging from $Cu_{1.9975 \pm 0.00}Se$ to Cu_xSe where $x < 1.86$. The onset of electronic conductivity in a CuBr electrolyte at small values of μ_{Cn} made thorough evaluation of the composition of Cu_xSe equilibrated with pure selenium impossible. Study of cell (106) demonstrated the existence of three phases in the following ranges of composition: $Cu_2Te–Cu_{1.92}Te$; $Cu_{1.43}Te–Cu_{1.39}Te$; $Cu_{1.32}Te–Cu_{1.30}Te$.

Cation-conductive crystals AgI and CuBr were widely used in electrochemical studies of a number of processes including the following:

(1) evaporation and incorporation of sulphur, selenium and iodine from and into binary crystals containing these components (Rickert, 1961; Ratchford and Rickert, 1962; Mrowec and Rickert, 1962; Birks and Rickert, 1963b).

(2) transportation of Ag^+-ions and electrons across the phase boundary $Ag_2S–Ag$ (Rickert and O'Briain, 1962).

(3) formation of solid NiS on the surface of nickel metal (Mrowec and Rickert, 1961).

(4) transformation of sulphur from one molecular species to another (Birks and Rickert, 1963a).

(5) interaction of copper and sulphur (Donner and Rickert, 1968).
Let us analyse the theoretical basis of these studies. It is known that the e.m.f. of galvanic cell

$$Pt \mid Ag \mid AgI \mid Ag_2S \mid Pt \qquad (107)$$

is related to the chemical potential of Ag in the sample of Ag_2S by the equation

$$\mu_{Ag} - \mu_{Ag}^{\circ} = - EF, \qquad (107a)$$

where μ_{Ag}° is the chemical potential of pure silver metal. In view of the narrow homogenic range of Ag_2S one can write

$$2\mu_{Ag}^{\circ} + \mu_{S}' = 2\mu_{Ag} + \mu_{S} = 2\mu_{Ag}^{x} + \mu_{S}^{\circ}, \qquad (108)$$

where μ_{S}' is the chemical potential of sulphur in Ag_2S equilibrated with silver; μ_{S} is the chemical potential of sulphur in Ag_2S of any composition within the homogenity range; μ_{Ag}^{x} and μ_{S}° are the chemical potentials of silver and sulphur in Ag_2S in equilibrium with sulphur. It follows from eqn (108) that

$$2(\mu_{Ag} - \mu_{Ag}^{\circ}) = \mu_S' - \mu_S \tag{108a}$$

$$2(\mu_{Ag}^{x} - \mu_{Ag}^{\circ}) = \mu_S' - \mu_S^{\circ}. \tag{108b}$$

Taking into consideration that

$$\mu_{Ag}^{x} - \mu_{Ag}^{\circ} = E^*F, \tag{107b}$$

where E^* is the e.m.f. of cell (107) for Ag_2S in equilibrium with sulphur†,

from eqns (107a, 108a, b) one finds

$$\mu_S - \mu_S^{\circ} = 2(E - E^*)F \tag{109}$$

In accordance with $\mu_S = \mu_S^{\circ} + RT \ln a_S$, where a_S- sulphur activity, one may write

$$a_S = \exp \frac{2(E - E^*)F}{RT}. \tag{110}$$

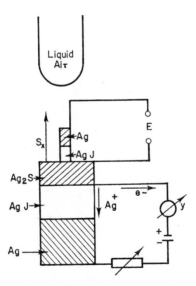

FIG. 11. Arrangement for electrochemical study of vaporization of sulphur from Ag_2S.

† Evidently, the value E^* may be calculated from the standard free energy of formation of Ag_2S from its elements.

The activity of sulphur and partial pressure of sulphur molecules in the gaseous phase may be determined when the Ag_2S in cell (107) is in equilibrium with the gaseous phase and the e.m.f. of the cell is measured. The device shown in Fig. 11 was used (Rickert, 1961) for the study of sulphur evaporation from Ag_2S. It consists as a matter of fact of two cells of type (107), that are placed in vacuum and heated up to 200–400°C. Direct current is continuously passed through one of the cells (the lower one). As a result silver ions are removed from Ag_2S crystal through the AgI membrane while the same quantity of electrons pass through the leads. This results in a decrease of potential of silver in Ag_2S and a corresponding increase of potential of sulphur. This leads to evaporation of sulphur, the rate increasing with μ_s. Evaporating sulphur condenses on the cold finger placed over the sulphide. Eventually a stationary state is achieved, when the quantity of evaporating sulphur is equivalent to the quantity of silver removed from the sulphide when the current passes. The intensity of current in this state is the measure of evaporation rate of sulphur, i.e.

$$J = I/Fq, \tag{111}$$

where J is the rate of evaporation of sulphur, I, the current passing through the cell (107), q the free surface area of the Ag_2S sample. It is obvious that

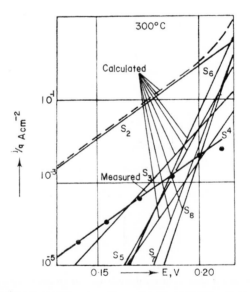

FIG. 12. Sum of maximum possible rates of vaporization of sulphur from an Ag_2S surface, shown as current density as a function of the e.m.f. of cell (107) (dashed line). Straight lines represent the maximum partial currents due to each of the sulphur species.

the chemical potentials of S and Ag in the Ag_2S in the stationary state reach constant values. The potentials may be measured with the help of the ionic contact Ag/AgI (see the upper part of Fig. 11) which also forms a cell of type (107). Activity of sulphur in the gaseous phase is calculated by eqn (110). Current and potential measurements are taken separately and thus the errors determined by polarization effects are minimized. The experimental data (Rickert, 1961) are shown in Fig. 12 in which current density versus e.m.f. of the cell (107) is plotted. In thermodynamic equilibrium sulphur vapour includes molecular species ranging from S_2 to S_8. From the kinetic theory of gases one may calculate the maximum rate of evaporation of a single species as a function of the chemical potential of sulphur in Ag_2S. The summing of the results shown in Fig. 12 proves that the measured evaporation rate is two orders lower than the sum of the maximum rates. On the other hand the dependence of the experimentally measured rate of evaporation (J) on the e.m.f. of the cell shows that $J \sim a_S^2$. Hence, Rickert (1961) concluded that rate of evaporation is limited not by sulphur elimination off the crystal surface but by the formation of molecules of S_2 from adsorbed sulphur atom . The studies of Ag_2Se (Ratchford and Rickert, 1962) and CuI (Mrowec and Rickert, 1962) were made on this analogy. In the study of CuI the galvanic cell

$$Pt \mid Cu \mid CuBr \mid CuI \mid Pt \qquad (112)$$

was used. It appeared that the formation of I_2 molecules in the adsorbed layer limits the rate of evaporation of I from AgI crystals.

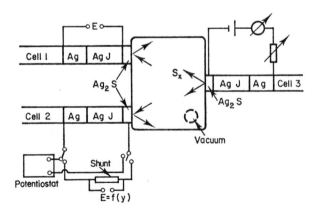

Fig. 13. Principle of the arrangement for electrochemical study of incorporation of sulphur into Ag_2S from the gas phase (Birks and Rickert, 1963)

The design shown in Fig. 13 is intended for electrochemical study of incorporation of sulphur molecules into solid Ag_2S from the gaseous phase (Birks and Rickert, 1963a). Three galvanic cells of type (107) were involved. They were arranged so that the surfaces of the Ag_2S pellets form a part of the inner walls of the vessel directly connected with a high vacuum. The cells have different functions. Cell 3, through which current passes is used to transport sulphur to the evacuated vessel. This sulphur source makes up for the evaporation loss of sulphur from the vessel as well as for the loss resulting from sulphur incorporation into the Ag_2S crystal in cell 2. In the stationary state within the vessel constant sulphur pressure is maintained and α_s is measured by cell 1. The partial pressure phase is calculated by equation

$$P_{S_x} = P^\circ_{S_x} \exp \{2x(E - E^*)F/RT\}, \qquad (113)$$

where $P^\circ_{S_x}$ is the partial pressure of molecules of the type S_x in saturated sulphur vapour. Cell 2 is used to measure the rate of incorporation of sulphur into solid Ag_2S of this cell. The study shows that the stage limiting sulphur incorporation is that where the S–S bond breaks to form adsorbed S atoms (Birks and Rickert, 1963a).

The electrochemical Knudsen cell used by Birks and Rickert (1963b) is of great interest for thermodynamic studies of sulphur vapour. The enthalpy and entropy of transformation of sulphur from one molecular specie to another were determined by the use of this cell (see Table XV).

TABLE XV.   Enthalpies and entropies of reactions of transformation of sulphur (Birks and Rickert, 1963b)

Molecule	Equilibrium	Temperature, °K	H_T°, kcal/mol	S_T°, cal/deg mol
S_2	$2S_{(cond.)} \rightleftarrows S_{2(g)}$	460–670	$28{\cdot}08 \pm 0{\cdot}35$	$32{\cdot}6$
S_3	$2S_3 \rightleftarrows 3S_2$	566–669	$26{\cdot}6 \pm 2{\cdot}0$	$37{\cdot}6 \pm 3{\cdot}0$
S_4	$S_4 \rightleftarrows 2S_2$	615	$28{\cdot}2 \pm 2{\cdot}0$	$36{\cdot}7 \pm 2{\cdot}5$
S_5	$2S_5 \rightleftarrows 5S_2$	565–620	$95{\cdot}5 \pm 4{\cdot}0$	$111{\cdot}4 \pm 8{\cdot}0$
S_6	$\frac{3}{4}S_8 \rightleftarrows S_6$	435–625	$6{\cdot}26 \pm 0{\cdot}33$	$7{\cdot}6 \pm 0{\cdot}8$
S_7	$\frac{7}{8}S_8 \rightleftarrows S_7$	435–625	$5{\cdot}77 \pm 0{\cdot}31$	$7{\cdot}2 \pm 1{\cdot}0$
S_8	$S_8 \rightleftarrows 4S_2$	460–625	$96{\cdot}8 \pm 2{\cdot}2$	$110{\cdot}2 \pm 4{\cdot}2$

The use of galvanic cells of type (107) in the study of silver transport through the phase boundary Ag–Ag$_2$S (Rickert and O'Briain, 1962) gave rise to the following conclusions:

(a) the equilibrium on the phase boundary is independent of the silver transportation rate, (b) the polarization arising from silver transportation is independent of the species (ionic or atomic) of silver transported. The technique of galvanostatic and potentiostatic measurements in cell

$$\text{Pt} \mid \text{Ag} \mid \text{AgI} \mid \text{Ag}_2\text{S NiS} \mid \text{Ni} \mid \text{Pt} \tag{114}$$

allowed the proof of a parabolic law of growth of NiS with high concentration of cationic vacancies (Mrowec and Rickert, 1961).

The use of AgI as a solid electrolyte in the kinetic studies described above is based on the fact that the polarization effect is not observed in this material owing to high mobility of silver ions. On the same grounds AgI was successfully used in a new trend of electrochemical studies (Yuschina and Karpatchev, 1969). This trend is called "chimotronics", the devices—"chimotrons". A special role among chimotrons is played by electrochemically controlled resistances—"memistors". A memistor is a small galvanic cell (see Fig. 14), one electrode of which (the controlling one) is made of metal penetrating into an electrolyte by reason of electrochemical transport; the second (reading electrode) is a thin inert metal film vaporized over a highly resistant base. On both sides of the reading electrode there are leads for measuring resistance values before the information signal and after it. The change in the resistance of the film is proportional to the charge passing through the electrolyte. Designs of this kind may be used as current integrators as well as analogous memory elements since they can contain changeable information. The following conditions are necessary for successful functioning of elements of this type:

(a) direct correspondence between the charge Q passing through the electrolyte and the resistance value R of the memistor, (b) immediate reaction of R to every change of Q, (c) the possibility of continuous cycling of current.

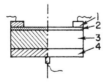

FIG. 14. Memistor design. (1) Ag lead, (2) reading electrode (Ag film resistor), (3) solid electrolyte AgI, (4) controlling electrode.

Because of the high mobility of silver ions in AgI, this substance used as a solid electrolyte of memistors, satisfies the requirements enumerated. Finally, it should be noted that cells of type (107) and

$$Cu \mid CuBr \mid Cu_2S \mid Pt \qquad (115)$$

have been used in the study of kinetics and mechanism of the reduction of silver sulphide by hydrogen (Kobayashi and Wagner, 1957; Schmalzried and Wagner, 1963).

B. Double Salts of Silver and Copper as Solid Electrolytes

The study of state diagrams of systems AgI–MeI (where Me = K^+, $R6^+$, NH_4^+) proved the possibility of forming chemical compounds of the type $MeAg_4I_5$ (Brandley and Greene, 1966, 1967). They are highly conductive to ions. Similar compounds were not found in the analogous series of bromides. The structure of $MeAg_4I_5$ is not yet studied, but it may be supposed that a unit cell of this type is quite large and its symmetry is low (Brandley and Greene, 1967). The temperature dependences of conductivity of poly-crystalline KAg_4I_5 and $RbAg_4I_5$ are expressed as follows:

$$\log \sigma = 3.49 - \frac{680}{T} - \log T \qquad (116)$$

$$\log \sigma = 4.05 - \frac{730}{T} - \log T, \qquad (117)$$

respectively (Brandley and Greene, 1967). The latter shows that at $20° \sigma_{RbAg_4I_5} = 0.124 \pm 0.006\ \text{ohm}^{-1}\,\text{cm}^{-1}$. This is the highest conductivity observed in solids. The transport number measurement in the cell:

$$Ag \mid K_2AgI_3, KAg_4I_5 \mid I_2 \qquad (118)$$

carried out by Tubandt's method, showed that the silver ion is the only charge conductor (Brandley and Greene, 1967). Moving through a rigid lattice formed by J^- and Me^+ ions, the Ag^+ ion meets with the resistance of the lattice; the stronger the bond between Me^+ and I^- the stronger the resistance. The strength of this bond is dependent on the ionic radius of Me^+. This explains the difference in conductivity of KAg_4I_5 and $RbAg_4I_5$. The difference in melting points of these compounds ($T_{\text{melt.}KAg_4I_5} = 253°C$, $T_{\text{melt.}RbAg_4I_5} = 228°C$) is another demonstration of the stronger bond of $K^+–I^-$ as compared to $Rb^+–I^-$. It is interesting to mention that a jump in conductivity is not observed with melting of these compounds. This is an indication that the crystal structure of the compounds is in a very disordered state. From $\sigma(T)$ dependence, the activation energy of conductivity of KAg_4I_5 ($0.13\,eV$) was found (Brandley and Greene, 1967). Nevertheless it does not explain the migration mechanism of Ag^+-ions because the structure of the compound is not known. Moreover Lidiard (1957) doubts whether it is right to conclude a migration mechanism on the basis of $\sigma(T)$ dependence where E is of the order of KT. The use of double salts of the type $MeAg_4I_5$ as possible solid electrolytes is limited by the narrow range of their temperature stability. For example, when T is lower than $38°C$, KAg_4I_5 decomposes according to the reaction

$$2KAg_4I_5 \rightleftarrows 7AgI + K_2AgI_3 \qquad (119)$$

and at $253°C$ it melts incongruently. The range of stability of $RbAg_4I_5$ is tentatively $0–228°C$ (Brandley and Greene, 1967).

CuI, the crystal lattice of which is analogous to that of AgI, also forms the compound KCu_4I_5. The latter is highly ion-conductive owing to high mobility of Cu^{++}-ions. The range of stability of this compound is $257–332°C$ (Brandley and Greene, 1967). The attempt (Takahashi and Yamamoto, 1966) to use silver sulphohalides studied by Reuter and Hardel (1961, 1965a, b, 1966) and by Takahashi and Yamamoto (1965a, b, c, 1966) as solid electrolytes is very tempting. For example, activation energy of ionic conductivity of Ag_3SI is $3.3\,kcal/mol$ (in $AgI–10.0\,kcal/mol$). At room temperature electroconductivity of Ag_3SI is four orders higher than that of AgI which makes $10^{-2}\,ohm^{-1}\,cm^{-1}$. The level of electronic conductivity is very low ($= 10^{-8}\,ohm^{-1}\,cm^{-1}$; Takahashi and Yamamoto, 1966) though at $P_{I_2} = 1$ atm it is evidently considerably higher as a result of the reaction

$$\tfrac{1}{2}I_2 = I_I{}^\times + 3V_{Ag}{}' + V_S{}^{\cdot\cdot} + h^{\cdot}. \qquad (120)$$

This explains the fact that the e.m.f. of the cell

$$Ag_{amalgam} \mid Ag_3SI \mid I_2, C \qquad (121)$$

turned out to be lower than the value calculated from $\Delta G°$ of formation of AgI (at $-75°C$: $E_{term} = 673.5\,mV$, $E_{exp} = 605\,mV$; at $60°C$: $E_{term} = 692\,mV$, $E_{exp} = 681\,mV$) (Takahashi and Yamamoto, 1966). High melting temperature ($700°C$) and low polarization at high current ($1–2\,mA.cm^{-2}$) makes Ag_3SI very useful as electrolyte in galvanic cells working in compensation regime, though the transport number of Ag^+ in Ag_3SI may be lower 1 at high values of μ_{I_2}.†

---

† The use of double silver salts as electrolytes of solid-state fuel cells and memistors may also be recommended (see above).

It is not by chance that only silver and cuprous derivatives are mentioned when speaking of the use of halides as cation-conductive solid electrolytes. Though among alkaline halides there are many crystals with pure cationic conductivity, the possibility of using them as solid electrolytes is very limited owing to their low conductivity. One might also mention a number of materials (e.g. glasses, ceramics) where mobility of alkaline ions is sufficiently high to satisfy the requirements of the e.m.f. technique over a wide temperature range.

C. Sulphides as Solid Electrolytes

The possibility of using sulphides as ion-conductive electrolytes is limited by non-stoichiometry effects which are very typical of these compounds (Fischmeister, 1959; Hirahata, 1951; Kanigarichi et al., 1956). Sulphides of elements of the first and second main subgroups of the periodical table where ionic conductivity prevails are an exception. In fact, the study of pure anhydrous Na_2S by Möbius et al. (1964) showed that charge transport in this crystal is realized only by Na^+-ions ($t_{Na^+} > 0.99$). Sulphur ions in Na_2S, as in other sulphides, move slowly, as might have been expected from the relation of radii:

$$S^{2-} = 1.82 \text{ Å}; Na^+ = 0.98 \text{ Å}; Ca^{2+} = 1.04 \text{ Å}; Y^{3+} = 0.97 \text{ Å}; (\text{Bokiy, 1960}).$$

Worrell et al. (1967) recommended to us CaS doped with Y_2S_3 as a solid electrolyte of e.m.f. cells. It was shown (Worrell et al., 1967; Flahaut et al., 1963) that Y_2S_3 dissolves in the CaS lattice forming cationic vacancies that result in considerable increase of cation mobility. Pure ionic conductivity of CaS (1 wt % Y_2S_3) is found at $T = 700–900°C$ when $P_S < 10^{-6}$ atm. At $P_S > 10^{-6}$ atm appreciable hole conductivity arises. Worrell et al. (1967) refer to studies of the cell

$$Ag, Ag_2S \mid CaS(Y_2S_3) \mid Cu,Cu_2S \tag{122}$$

but do not give results of its measurements.

D. Solid Electrolytes Based on Perowskite-type Structures

Solid electrolyte realizing charge transfer through migration of protons may be useful in creating new solid state fuel cells as well as e.m.f. cells. Danner et al. (1964) suggested the use of compounds of the type ABX_3 ($X = O^{2-}$ and/or F^-) as proton conductors crystallizing in Perowskite-type structures. This structure is realized as a result of cubic dense packing of large cations A^{n+} and anions O^{2-} (F^-). Smaller ions fill in the octahedral holes of this backing. If the sum of positive valencies $m + n = 6$ (where $X = O^{2-}$), one may expect 6 possible combinations corresponding to the change of valency of A from 1 to 6 (see Table XVI).

TABLE XVI. Possible combinations of elements forming Perowskite-type structures (Danner et al., 1964)

Combinations of valencies	A	B
$A^+ B^{5+}$	Li	Ce,
$A^{2+} B^{4+}$	Mg	Ce, Th, U, Zr
$A^{3+} B^{3+}$	Al, Sc, Ga	rare earths, actinides, Y, In, Bi
$A^{4+} B^{2+}$	Sn, Ti, Zr, Ge	Pb, Cd, Ba, Sr, Ca
$A^{5+} B^+$	Nb, Ta	Na, K, Rb, Cs, Tl, NH_4
A^{6+}	Re, W	—

Perowskite-like compounds may strongly deviate from stoichiometry without changing the type of crystalline structure. The composition of the B^{m+} sublattice is most labile. If lanthanum aluminate with a deficit of La in the B^{m+} sublattice, which is expressed by the formula $AlLa_{1-x}(V_{La}{}^x)_x O_3$, is taken and treated with hydrogen, then in accordance with reaction

$$AlLa_{1-x}(V_{La}{}^x)_x O_3 + \tfrac{3}{2}H_2 \rightarrow AlLa_{1-x}H_{3x}O_3 \qquad (123)$$

a structure is formed containing protons in the B^{m+} sublattice. Their high mobility allows conductivity $\sim 10^{-2}$ ohm^{-1} cm^{-1} at 900°C. To attain higher proton mobility (which is extremely important for experiments at low temperatures) vibration frequencies of other ions forming the lattice should be lessened. This lowers the Debay temperature. Two ways are possible here. The first is to substitute ions of lower valency (Ca^{2+}, Ba^{2+}, Sr^{2+}, Na^+ K^+) for part of lanthanum ions. The second is to substitute cations of lower ionization potential and larger ionic radius (Ga^{3+}, Cr^{3+}) for Al^{3+}-cations. In the first case conductivity increases, evidently as a result not only of the increase of proton mobility but also of their concentration in the lattice. The isomorphous structure is preserved even when 30% of the La^{3+} ions are changed for Ca^{2+}, Ba^{2+}, Sr^{2+} ions. But one should bear in mind that even at 10% substitution the missing positive charge is compensated by oxygen vacancies and not by holes (Danner et al., 1965). In non-stoichiometric $AlLa_x O_3$ excess oxygen vacancies arise at $x \leqslant 0.98$. In accordance with this, electrolytes with greater deviation from stoichiometry at oxygen pressure ~ 1 atm display electronic conductivity. The exact value and temperature threshold of onset are not given by Danner et al. (1965).

The charge transfer is non-stoichiometric compounds of the Perowskite-type may be realized not only by the protons but also by other cations of small radius: C^{4+}, B^{3+}, Si^{4+}, Li^+ (Danner et al., 1965). This makes the use of compounds of this type as solid electrolytes in e.m.f. techniques very promising. The possible difficulties of such use of the compounds have not yet described in the literature, so that only general remarks can be made. It is known that Trevoux and Dauge (1965) successfully used C^{4+}-conductive $AlLa_xO_3$ in the fuel cell working on oxidation of carbon, produced by the destruction of hydrocarbons. The scheme of this fuel cell is given in Fig. 15. The working range of the cell is 400–1500°C, optimum temperature 1000°C, $E \cong 600$ mV. To use the Perowskite-type systems in e.m.f. studies one should know transport numbers and the conditions at which electronic conductivity arises. The most important part of the problem is not quite clear. Besides, when speaking of the use of C^{4+}, B^{3+}, Si^{4+} conductive systems in studies of solid phases containing these elements, one should bear in mind that the diffusion rates of carbon, boron and especially of silicon in the majority of compounds are very low. This may prolong the attainment of equilibrium chemical potential. When the diffusion of these elements through a solid electrolyte is faster than that in the material studied increase (or decrease) of activity of these elements within the layer in contact with the solid electrolyte occurs. The result of this is a deviation of the equilibrium e.m.f. value (Kaul et al., 1970).

E. Solid Electrolytes Based on β – Al_2O_3

Ionic conductivity of lamellar structures may be considerably higher than that of ionic crystals of other classes. β-Al_2O_3 is a good example of structures of this kind. The closest approximation to its composition is the formula $Na_2O.ll\ Al_2O_3$. The β-Al_2O_3 structure has now been thoroughly studied (Bragg et al., 1931; Ridgway et al., 1936; Beavers and Ross, 1937). It is lamellar hexagonal lattice with unit cell parameters: $\alpha_0 = 5.58$ Å and $C = 22.45$ Å. Na^+ ions are situated in planes perpendicular to the C-axis as

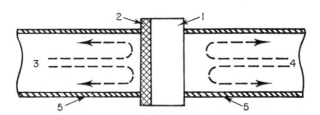

FIG. 15. Fuel cell of Trevoux and Dauge (1965) (1) solid electrolyte $AlLa_xO_3$, (2) carbon layer, (3) hydrocarbon atmosphere, (4) air, (5) stainless steel tubes used as leaders.

far apart as 11·23 Å. Oxygen ions form hexagons in planes above and below the planes containing Na^+ ions. The distance between oxygen-containing planes is 4·76 Å, and the plane containing Na^+ is their plane of symmetry. Oxygen-containing planes are bounded by Na^+ ions and by Al–O–Al bonds (one for each Na^+ ion). The mirror plane is shown in Fig. 16. The large number of vacant interstitials (dotted line in Fig. 16) are typical of this plane. This explains extraordinarily high mobility of singly charged cations in β–Al_2O_3-type structures. The existence of "loose" planes containing Na^+ ions makes it possible to substitute a number of other ions of corresponding radii for Na^+ ions. Ionic exchange in β–Al_2O_3 as well as diffusion in materials thus obtained are most thoroughly studied by Yao and Kummer (1967). It follows from Fig. 17 that crystals in which Na^+ ions are entirely substituted by Ag^+, Tl^+, K^+, and Rb^+ ions may be obtained by treating β–Al_2O_3 with nitrate melts. The direct exchange of Na^+ and Li^+ ions is impossible, though Li–β–Al_2O_3 may be obtained through treatment of Ag^+–β–Al_2O_3 with $LiNO_3$ melt saturated by LiCl at 350°C.

Chlorine ions shift the equilibrium so much that it is possible to get material with the ratio $Ag/Ag + Li < 1·10^{-4}$. Besides, Li^+–β–Al_2O_3 may be obtained as a result of treating NH_4^+–β–Al_2O_3 with $LiNO_3$ melt at 400°C. In this case the equilibrium is shifted in the required direction as a result of decomposition of NH_4NO_3. Indium-substituted β–Al_2O_3 may be obtained by the treatment of Ag^+–β–Al_2O_3 with melted indium metal (350°C, 3 days). Attempts (Yao and Kummer, 1967) to introduce bivalent ions (Pb^{2+}, Sr^{2+}, Ca^{2+}, Zn^{2+}, Fe^{2+}, Mn^2, Cu^{2+}, Hg^{2+}, Sn^{2+}, Cd^{2+}, Ba^{2+}) into β–Al_2O_3 never succeeded in producing material corresponding to the formula $Me\,0·11\,Al_2O_3$. Nevertheless, these attempts should not be considered

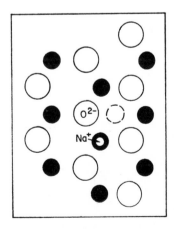

FIG. 16. The arrangements of atoms in the mirror plane of β–Al_2O_3.

exhaustive. Firstly, one cannot be sure that the ionic exchange in each case was thermodynamically advantageous. Correct matching of the exchanging couple "melt–Me^+–β–Al_2O_3" should be based on an evaluation of

$$\Delta G_{\text{exchange}} = \Delta G(Me^{2+}-\beta-Al_2O_3) - \Delta G(Me^+-\beta-Al_2O_3)$$
$$+ \Delta G(Me^+An^-) - \Delta G(Me^{2+}An_2^-). \qquad (124)$$

Secondly, the authors themselves think it possible that in some cases the equilibrium of ionic exchange was not achieved because of diffision diffi-

TABLE XVII. Constants of Equation $D = D_0 \exp\ (-E/RT)$ for the diffusion of Me^+ in $Me^+ - \beta - Al_2\,O_3$ at 200–400°C (Yao and Kummer, 1967)†

Me^+	$D_0 \times 10^4$, $cm^2\ sec^{-1}$	E, kcal/mol
Na^+	2·4	3·81
Ag^+	1·65	4·05
K^+	0·78	5·36
Rb^+	0·34	7·18
Li^+	14·5	8·71

† Diffusion in the direction $\perp$ C-axis.

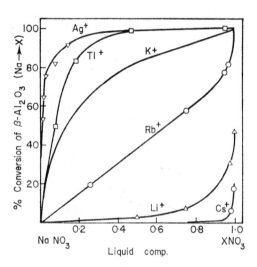

FIG. 17. Equilibria between β–Al_2O_3 and various binary nitrate melts containing $NaNO_3$ and another metal nitrate at 300–350°C (after Yao and Kummer, 1967).

culties (large size of β–Al_2O_3 crystals, low temperature of melts, insufficient experimental time).

The results of Me^+-diffusion measurements in Me^+–β–Al_2O_3 obtained by the tracer technique are shown in Table XVII.

Resistance measurements (Yao and Kummer, 1967) in Na^+–β–Al_2O_3 and Ag^+–β–Al_2O_3 demonstrating the highest diffusion coefficients ($\sim 10^{-7}$ $cm^2\,sec^{-1}$ at 25° C) proved $\sim$ 30 ohm cm at the same temperature. Because of their high conductivity β–Al_2O_3-type materials are already used as solid electrolytes in galvanic elements. The secondary elements described by Hever (1968) are of special interest. Their structure may be expressed schematically as follows:

$$Na_2\,0{\cdot}5(Fe_{0{\cdot}95}\,Ti_{0{\cdot}05}AlO_3)$$

$$\overline{\text{eutectic } \alpha\text{– and } \beta\text{–}Al_2O_3}$$ (125)

$$Na_2\,0{\cdot}5\,(Fe_{0\,95}Ti_{0{\cdot}05}AlO_3)$$

and

$$1{\cdot}3K_2O.0{\cdot}2Na_2\,0{\cdot}1\,O\,(Fe_{1{\cdot}9}Ti_{0{\cdot}1}O_3)$$

$$\overline{1{\cdot}3K_2O.0{\cdot}2\,Li_2\,0{\cdot}1\,OAl_2O_3}$$ (126)

$$1{\cdot}3K_2O.\,0{\cdot}2\,Na_2O.\,1\,O(Fe_{1{\cdot}9}Ti_{0{\cdot}1}O_3)$$

When a voltage is applied, Na^+ cations (125) and K^+ cations (126) move through the electrolyte from one electrode to the other. This is the charging process. The wider the homogeneity range of electrodes used, the greater is the charge accumulated. When the voltage is switched off and the circuit closed there the back movement of cations starts, and continues until the alkaline activities are equal in both electrodes. The optimum temperature for use of these elements is slightly above 300°C. Studies of such cells showed that the diffusion coefficient of $Na^+(K^+)$ ions decreases in proportion to the increase of voltage applied. The mechanism becomes clear if it is accepted that $Na^+(K^+)$ moves through interstitials and vacancies (Hever, 1968). Their occupation increases with the increase in number of moving ions (i.e. at increased voltage). The result is a decrease in diffusion coefficient.

The qualities mentioned (high cationic conductivity, excellent ceramic properties and wide temperature range of stability) make such compounds as β–Al_2O_3 very promising for e.m.f. techniques.

References

Aronson, S. (1964). *In* "Compounds of Interest in Nuclear Reactor Technology", Vol. 10, pp. 247–254. A.I.M.E., Boulder, Colorado.
Aronson, S. (1967). *J. Inorg. nucl. Chem.* **29**, 1611.

Aronson, S. and Auskern, A. (1966). *In* "Thermodynamics", Vol. 1, 165. I.A.E.A., Vienna.
Aronson, S. and Sadofsky, J. (1965). *J. inorg. nucl. Chem.* **27**, 1769.
Barsis, E. and Taylor, A. (1966). *J. chem. Phys.* **45**, 1154.
Beavers, C. A. and Ross, M. A. S. (1937). *Z. Krist.* **97**, 59.
Behl, W. K. and Egan, J. J. (1966). *J. electrochem. Soc.* **113**, 376.
Benz, R. and Schmalzried, H. (1961). *Z. phys. Chem.* **B38**, 295.
Benz, R. and Wagner, C. (1961). *J. phys. Chem.* **65**, 1308.
Biermann, W. and Jost, W. (1969). *Z. phys. Chem.* **25**, 139.
Birks, N. and Rickert, H. (1963a). *Ber. Bunsenges. phys. Chem.* **67**, 501.
Birks, N. and Rickert, H. (1963b). *Ber. Bunsenges. phys. Chem.* **67**, 97.
Bokiy, G. B. (1960). "Kristallochimiya". Moscow State University, Moscow.
Bones, R. J., Markin, T. L. and Wheeler, V. J. (1967). *Proc. Br. Ceram. Soc.* **8**, 51.
Botinck, H. (1958). *Physika* **24**, 639.
Bragg, W. L., Gottfried, C. and West, J. (1931). *Z. Krist.* **77**, 255.
Brandley, I. N. and Greene, P. D. (1966). *Trans. Faraday Soc.* **62**, 2069.
Brandley, I. N. and Greene, P. D. (1967). *Trans. Faraday Soc.* **63**, 424.
Croatto, V. and Bruno, M. (1948). *Gazz. chim. Ital.* **78**, 95.
Danner, G., Forrat, F., Dauge, G. and Billard, M. (1964). Fr. pat. 1.377.296 cl.Hol m.
Danner, G., Forrat, F., Dauge, G. and Billard, M. (1965). Fr.pat. 85.676 cl.HOl m.
Dneprova, V. G., Rezukhina, T. N. and Gerasimov, Y. I. (1968). *Dokl. Akad. Nauk USSR* **178**, 135.
Donner, D. and Rickert, H. (1968). *Z. phys. Chem.* **60**, 11.
Eastman, E. D., Brewer, L., Bromley, L. A., Gilles, P. W. and Lofgren, N. L. (1950). *J. Am. chem. Soc.* **72**, 4019.
Ebert, I. and Teltow, J. (1955). *Ann. Phys.* **15**, 268.
Egan, J. J. (1964). *J. phys. Chem.* **68**, 978.
Egan, J. J. (1966). *In* "Thermodynamics", Vol. 1, p. 157. I.A.E.A., Vienna.
Egan, J. J., McCoy, W. and Bracker, J. (1962). *In* "Thermodynamics of Nuclear Materials", p. 163. I.A.E.A., Vienna.
Eyring, L. and Westrum, E. F. (1953). *J. Am. chem. Soc.* **75**, 4802.
Fischmeister, H. F. (1959). *Acta. chem. Scand.* **13**, 852.
Flahaut, J., Domange, L., Patrie, M., Bostsarron, A. and Guittard, M. (1963). "Advances in Chemistry", **39**, 179.
Gerasimov, Y. I., Krestovnikov, A. N. and Shakhov, A. S. (1966). "Khimicheskaja termodinamika v tsvetnoji metallurgii", Vol. 1. Metallurgizdat, Moscow.
Gingerich, K. A. and Aronson, S. (1966). *J. phys. Chem.* **70**, 2517.
Glassner, A. (1958). ANL Report 575. U.S. Government Printing Office, Washington, D.C.
Hamer, W., Malmberg, M. and Rubin, B. (1956). *J. electrochem. Soc.* **103**, 8.
Heus, R. J. and Egan, J. J. (1966). *Z. phys. Chem.* **49**, 1/2, 38.
Hever, K. (1968). *J. electrochem. Soc.* **115**, 830.
Hinze, J. (1968). Unpublished Research, cited by Patterson (1970).
Hirahata, E. (1951). *J. phys. Soc. Japan* **6**, 428.
Holm, J. L. (1965). *Acta. chem. Scand.* **19**, 1512.
Hood, G. M. and Morrison, I. A. (1967). *J. appl. Phys.* **38**, 4796.
Ilschner, B. (1958). *J. chem. Phys.* **28**, 1109.
Kanigarichi, T., Hihara, T., Tazaki, H. and Hirahata, E. (1956). *J. phys. Soc. Japan* **11**, 606.

Karpatchev, S. V. and Palguev, S. F. (1960). *Trudy Inst. Elektrokhim.* **1**, 79.

Kaul, A. R., Oleinikov, N. N. and Tretyakov, Y. D. (1970). *Electrokhimija* (to be published).

Kelley, K. K. (1959). U.S. Bureau of Mines Bulletin 584 (U.S. Government Printing Office, Washington, D.C.).

Kelley, K. K., Todd, S. S. and King, E. G. (1954). U.S. Bur, of Mines Rept. Invest., 5059.

King, E. G. (1951). *J. Am. chem. Soc.* **73**, 656.

Kiukkola, K. and Wagner, C. (1957). *J. electrochem. Soc.* **104**, 379.

Kobayashi, H. and Wagner, C. (1957). *J. chem. Phys.* **26**, 1609.

Koch, E. and Wagner, C. (1937). *Z. phys. Chem.* **B38**, 295.

Komarov, V. F. and Tretyakov, Y. D. (1970). *Z. phyz. khim.* (to be published).

Kröger, F. A. (1964). "The Chemistry of Imperfect Crystals". North-Holland Publishing Company, Amsterdam.

Kubaschewski, O. and Evans, E. L. (1955). "Metallurgical Thermochemistry". Pergamon Press, London and New York.

Kurnick, S. W. (1952). *J. chem. Phys.* **20**, 218.

Lehovec, K. and Smyth, D. (1962). U.S. Pat. 3.036.144.

Lichter, B. D. and Bredig, M.A. (1965). *J. electrochem. Soc.* **112**, 506.

Lidiard, A. B. (1957). "Handbuch der Physik", **20**, 246.

Lofgren, N. L. and McIver, E. J. (1966). U.K.A.E.A., A.E.R.E. R5169, 17.

Lorenz, G. and Wagner, C. (1957). *J. chem. Phys.* **26**, 1607.

Markin, T. L. and Bones, R. J. (1962). U.K.A.E.A., A.E.R.E. R4178, p. 43.

Matzke, H. and Lindner, R. (1964). *Z. Naturf.* **19a**, 1178.

Möbius, H. H., Witzmann, H. ond Hartung, R. (1964). *Z. phys. Chem.* **227**, 40.

Mollwo, E. (1934). *Nachr. Ges. Wiss. Göttingen* **6**, 79.

Mrowec, S. and Rickert, H. (1961). *Z. phys. Chem.* **28**, 422.

Mrowec, S. and Rickert, H. (1962). *Z. electrochem.* **66**, 14.

Nikitina, G. V. and Romanenko, V. N. (1964). *Izv. Akad. Nauk SSSR* **156**, 6.

Obrosov, V. P. and Karpatchev, S. V. (1970). *Trudy Inst. Electrokhim.* **14**, 119.

Panish, M. B., Blankeship, F. F., Grimes, W. R. and Newton, R. F. (1958). *J. phys. Chem.* **62**, 1325.

Patterson, I. W. (1970). Paper prepared for presentation at the Metallurgical Society of A.I.M.E.

Phillips, W. L. and Hanlon, J. E. (1963). *J. Am. Ceram. Soc.* **46**, 477.

Raleigh, D. O. (1965). *Physics Chem. Solids* **26**, 329.

Ratchford, R. J. and Rickert, H. (1962) *Z. Elektrochem.* **66**, 497.

Reinhold, H. (1928). *Z. anorg. allg. Chem.* **171**, 181.

Reuter, B. and Hardel, K. (1961). *Naturwiss.* **161**, 48.

Reuter, B. and Hardel, K. (1965a). *Z. anorg. Chem.* **340**, 158.

Reuter, B. and Hardel, K. (1965b). *Z. anorg. Chem.* **340**, 168.

Reuter, B. and Hardel, K. (1966). *Ber. Bunsenges.* **70**, 82.

Rezukhina, T. N. and Baginska, Y. (1967). *Elektrokhimiya* **3**, 1146.

Rezukhina, T. N., Levitskii, V. A. and Frenkel, M. J. (1966). *Izv. Akad. Nauk. SSSR, Neorganitcheskie materialy* **2**, 325.

Rickert, H. (1961). *Z. Elektrochem.* **65**, 463.

Rickert, H. and O'Briain, C. D. (1962). *Z. phys. Chem.* **31**, 71.

Ridgway, R., Klein, A. and O'Leary, W. (1936). *Trans. electrochem. Soc.* **70**, 71.

Romanenko, V. N. and Ivanov–Omskij, V. I. (1959). *Dokl. Akad. Nauk. SSSR* **129**, 553.

Rudzitis, E., Feder, H. M. and Hubbard, W. N. (1964). *J. phys. Chem.* **68**, 2978.
Samokhval, V. V., and Vetcher, A. A. (1968). *Zh. Fiz. Khim.* **42**, 644.
Samokhval, V. V. and Vetcher, A. A. (1970). *Dokl. Akad. Nauk. SSSR* **14**, 119.
Samokhval, V. V., Vetcher, A. A. and Panko, E. P. (1969). *In* "Khimitcheskaja svyaz v polyprovodnikakh". p. 195. Nauka i tekhnika, Minsk.
Satow, T. (1967a). *J. nucl. Mat.* **21**, 249.
Satow, T. (1967b). *J. nucl. Mat.* **21**, 255.
Schmalzried, H. (1959). *Z. phys. Chem.* **22**, 3/4, 199.
Schmalzried, H. and Wagner, C. (1963). *Trans. metall. Soc. A.I.M.E.* **227**, 539.
Schnering, H. G. and Bleckmann, P. (1965). *Naturwiss.* **52**, 53.
Short, J. M. and Roy, R. (1964). *J. phys. Chem.* **68**, 3077.
Takahashi, T. and Yamamoto, O. (1965a). *Denki Kagaku* **33**, 346.
Takahashi, T. and Yamamoto, O. (1965b). *Denki Kagaku* **33**, 518.
Takahashi, T. and Yamamoto, O. (1965c). *Denki Kagaku* **33**, 733.
Takahashi, T. and Yamamoto, O. (1966). *Electrochim. Acta.* **11**, 779.
Taylor, R. W. and Schmalzried, H. (1964). *J. phys. Chem.* **68**, 2444.
Teltow, J. (1949). *Ann. Phys.* **5**, 63, 71.
Torgeson, D. R. and Sahama, Th. G. (1948). *J. Am. chem. Soc.* **70**, 2156.
Tretyakov, Y. D. (1968). Vestnik Moskowskogo Gosudarstevennogo Universiteta (Russia) **2**, 88.
Trevoux, P. and Dauge, G. (1965). Fr. pat. 86.382 cl Hol m.
Tubandt, C. (1931). *In* "Landolt-Börnstein, Physikalisch-Chemische Tabellen". Springer, Berlin.
Tubandt, C. and Eggert, S. (1920). *Z. anorg. allgem. Chem.* **110**, 196.
Tubandt, C. and Eggert, S. (1921). *Z. anorg. allgem. Chem.* **115**, 105.
Tubandt, C., Reinhold, H. and Liebold, G. (1931). *Z. anorg. allgem. Chem.* **197**, 225.
Tubandt, C., Rindtorff, and Jost, W. (1927). *Z. anorg. allgem. Chem.* **165**, 195.
Ure, R. W. (1957). *J. chem. Phys.* **26**, 1363.
Vetcher, A. A. and Vetcher, D. V. (1967a). *Zh. fiz. Khim.* **41**, 1288.
Vetcher, A. A., Geiderikh, V. A. and Gerasimov, Y. I. (1965). *Zh. fiz. Khim.* **39**, 2145.
Vetcher, D. V. (1970). Dissertation, Moscow State University, Moscow.
Vetcher, D. V. and Vetcher, A. A. (1967b), *Zh. fiz. Khim.* **41**, 2916.
Wagner, C. (1933). *Z. phys. Chem.* **B21**, 25.
Wagner, C. (1953). *J. chem. Phys.* **21**, 1819.
Wagner, C. (1956). *Z. Elektrochem.* **60**, 4.
Wagner, C. (1959). *Z. Elektrochem.* **63**, 1027.
Wagner, C. (1968). *J. electrochem. Soc.* **115**, 933.
Wagner, J. B. and Wagner, C. (1957a). *J. chem. Phys.* **26**, 1597.
Wagner, J. B. and Wagner, C. (1957b). *J. chem. Phys.* **26**, 1602.
Wagner, J. B. and Wagner, C. (1957c). *J. electrochem. Soc.* **104**, 509.
Wehefritz, V. (1960). *Z. phys. Chem.* **26**, 339.
Weller, P. F. (1966). *Inorg. Chem.* **5**, 736.
Wicks, C. E. and Block, F. E. (1963). "Thermodynamic Properties of 65 Elements —Their Oxides, Halides, Carbides, and Nitrides". U.S. Bureau Mines Bulletin, Government Printing Office, Washington.
Wöhler, L. and Rodewald, G. (1909). *Z. anorg. Chem.* **61**, 54.

Worrell, W. L., Tare, V. B. and Bruni, F. J. (1967). *In* "High Temperature Technology". Proceedings of 3rd International Symposium, I.U.P.A.C., Asilomar, p. 503.
Yao, T. and Kummer, J. T. (1967). *J. inorg. nucl. Chem.* **29**, 2453.
Yushina, L. D. and Karpatchev, S. V. (1969). *Trudy Inst. Elektrokhim.* **13**, 106.
Zintl, E. and Udgard, A. (1939). *Z. anorg. allg. Chem.* **240**, 150.

15. NON-STATIONARY E.M.F. MEASUREMENTS

G. B. Barbi

High Temperature Chemistry Group, Euratom CCR, Ispra, Italy.

I. Introduction

At the beginning of the century some pioneering works (Haber and Tolloczo, 1904; Katayama, 1908) proved the e.m.f. measurements on all-solid galvanic cells to be a useful tool for determining thermodynamic functions. In 1957, the early papers of Kiukkola and Wagner stimulated a general and still increasing revival of interest for this subject.

The first problem to be solved in order to get reliable thermodynamic data from e.m.f. measurements is the choice of a proper intermediate solid electrolyte. This choice depends both on the chemical compatibility of the electrodes and electrolyte materials and the electrical transport properties of the electrolyte; the latter, however, may be strongly affected by the thermodynamic activity of some components of the electrode system.

This problem has been exhaustively studied (Ure, 1957; Steele *et al.*, 1968; Patterson *et al.*, 1967; Bridger *et al.*, 1963) and reviewed (Steele, 1967; Foley, 1969) by many authors since the presence of a pure ionic conductivity is the chief condition to be fulfilled in order to obtain, by means of the basic Nernst relationship, thermodynamic functions by e.m.f. measurements.

In fact, considering the general case of the cell:

$$\text{Pt} \left| \begin{array}{c} \text{Electrode System} \\ \text{at } p_{X_2}{}' \text{ pressure} \end{array} \right| \begin{array}{c} \text{Intermediate} \\ \text{Electrolyte} \end{array} \left| \begin{array}{c} \text{Electrode System} \\ \text{at } p_{X_2}{}'' \text{ pressure} \end{array} \right| \text{Pt} \qquad (I)$$

the fundamental equation (Haber and Tolloczo, 1904; Wagner, 1957; Schmalzried, 1962)

$$E = -\frac{1}{z_{X^-} \cdot F} \int_{\mu_{X_2}{}'}^{\mu_{X_2}{}''} t_{X^-} (\mu_{X_2})\, d\mu_{X_2}$$

(t_{X^-} being the transport number of the X^- ions in the electrolyte, z_{X^-} the valency, μ_{X_2} the chemical potential of X_2 and E the e.m.f. measured from the right to the left) holds only if the non-ionic conductivity induces negligible polarization at the electrode/electrolyte interface. Transport of electrons is always accompanied by a corresponding migration of ions, since the net current flow must be zero in cells operating in open circuit conditions. Because of the relevant polarization processes, the knowledge of the t_{X^-} versus μ_{X_2} function generally is not sufficient to obtain the X_2 chemical potential difference by e.m.f. measurements and this is the reason why the main condition for this problem to be solved is the pure ionic conductivity of the electrolyte.

In searching the electrolytes that may be suitable for our purposes it seems worthwhile to give a criterion for an *a priori* evaluation of their electrical transport properties. The crystals of the utilizable electrolytes have pure ionic bonding, the ionic mobilities being high in ionic crystals, especially those having defective structures; on the other hand, a significant covalency always induces electronic transport characteristics. For instance, the reason why the hypothetical cell:

$$\text{Pt} \left| \text{Me} + \text{Me}_r\text{C} \right| \begin{array}{c} \text{purely C-ion} \\ \text{conducting} \\ \text{electrolyte} \end{array} \left| \text{Me}' + \text{Me}_s'\text{C} \right| \text{Pt} \qquad (II)$$

cannot operate lies in the fact that such an electrolyte has not so far been found, the carbides showing a significant covalency. The same considerations apply to the sulphides.

This represents a remarkable restriction on the employment of solid galvanic cells in the problem of determining the thermodynamic functions. Sometimes, however, this problem may be overcome by taking advantage of the pure ionic conductivity performances of a compound the components of which do not enter the net cell reaction. This is the case of the determination of the free energy of formation of ThC_2. Egan (1964) and Satow (1967) measured the e.m.f. of the cell:

$$Mo \,|\, Th + ThF_4 \,|\, CaF_2 \,|\, ThF_4 + ThC_2 + C \,|\, Mo \qquad (III)$$

where CaF_2 acts as a pure F^--ion carrier. In this case the electrode reactions are:

$$Th + 4F^- \rightleftharpoons ThF_4 + 4\theta$$

$$ThF_4 + 2C + 4\theta \rightleftharpoons ThC_2 + 4F^-$$

the resulting overall process being:

$$Th + 2C \rightleftharpoons ThC_2$$

This point is very important and we shall return to it later, when we describe the behaviour of the three components, three phase electrodes operating in non-stationary conditions.

II. E.m.f. Measurements

The influence of the electrical conductivity properties of the electrolyte, namely a non-negligible electron transport number, is not the only cause inducing erratic behaviour in solid galvanic cells.

Another cause lies in the possible transfer of X_2 *in non-ionic form* from the interface at higher X_2 pressure to the other interface via the gas phase, when the compartments are not separated by a gas-tight diaphragm. Obviously, the same effect holds when the gas phase contains X_2 as impurity, as happens for oxygen-ion-conductive solid cells. We shall now confine our attention to these oxygen cells since they are by far the most important and well-studied, but shall keep in mind the generality of the argument.

The cells

$$Pt, O_2 \,\Big|\, \begin{matrix} \text{pure } O^=\text{-ion conducting} \\ \text{electrolyte} \end{matrix} \,\Big|\, Pt, O_2 \qquad (IV)$$
$$(p_{O_2}') \qquad\qquad\qquad\qquad (p_{O_2}'')$$

and

$$Pt \,\Big|\, \begin{matrix} Me + Me_xO \\ (p_{O_2}') \end{matrix} \,\Big|\, \begin{matrix} \text{pure } O^=\text{-ion conducting} \\ \text{electrolyte} \end{matrix} \,\Big|\, \begin{matrix} Me' + Me_y'O \\ (p_{O_2}'') \end{matrix} \,\Big|\, Pt \qquad (V)$$

although equivalent from the points of view of both thermodynamic and electrical transport properties, may behave differently owing to the different capability of the electrode system to react to the perturbation effects due to oxygen transfer from the gas phase to the interfaces. As a consequence, cells IV and V would develop the same e.m.f., so long as oxygen transport via the gas phase were completely hindered. These conditions have been adopted by Charette and Flengas (1968), their cell having the compartments rigorously separated by a gas-tight diagram of calcia-doped zirconia, having $t_O = 1$ at the pressures of all the systems under their examination.

These conditions may be called "static" since the fluxes of all species i.e. oxygen, oxygen ions, electrons and electron holes are all zero and therefore all the possible electrode processes can be regarded as infinitesimal. This type of cell is schematically represented by Fig. 1a.

On the other hand, we call "stationary" the conditions where, although the electrolyte is still a pure ionic conductor, there is no separation between the compartments of the cell, thus allowing molecular transfer of O_2, but the effects of this transfer via gas phase on the conditions of the electrode/electrolyte interfaces are assumed to be negligible. Such effects, consisting of the chemical reaction of the electrode components with the gas phase, gather at the ternary boundaries between electrode/electrolyte/gas phase and from these points the reaction products tend to diffuse tangentially on to the whole interface (Barbi, 1966). The chemical potentials of the components at the interface could in general no longer remain equal to those at the relevant bulk.

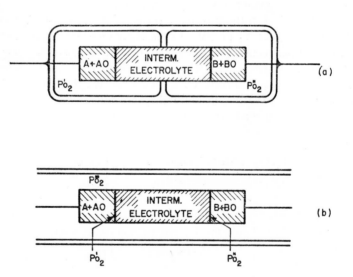

FIG. 1. Working conditions of a solid state electrochemical cell: (a) "static conditions"; (b) "stationary" and "non-stationary" conditions.

For a cell consisting of cylindrical pellets we could consider each electrode/electrolyte interface as resolved into concentric elemental rings behaving as elemental semicells parallel to one another. The overall semicell potential is the integral, normalized over the whole interface, of all the elemental contributions.

Because of the establishment of gradients of chemical potential of the components between bulk and interface, and among the inner and outer regions of the interface itself (Barbi, 1966a), one should expect a deviation of the measured e.m.f., with respect to the value calculated on the basis of the Nernst relationship for the bulk systems. In order to obtain significant values of the thermodynamic functions this deviation must be negligible but for these conditions to be reached it is necessary that the rates of the inter-diffusion processes of the components of the electrodes are high. This fact, however, implies the existence of a lower temperature limit for these cells to be operative, these diffusive processes being thermally activated. This limitation seems to be more restrictive than that pertaining to the electrical resistance of the electrolyte for practically noiseless e.m.f. measurements.

From these points of view, the preparation of the surfaces of the electrodes and the electrolyte plays an important role since a cell assembly of perfectly tallying surfaces reduces the triple boundaries—electrode/electrolyte/gas—to the geometrical boundary. In practice, skilful preparation of perfectly flat and polished surfaces provides the best chance of obtaining significant e.m.f. measurements in stationary conditions.

A problem which still remains to be solved is the evaluation, case for case, of whether or not the perturbating effect of the gas phase may be considered as actually negligible. In practice, a way of determining the magnitude of gas perturbation consists of following the effects of variation of the gas phase conditions on e.m.f. For a cell working at room pressure with an "inert" gas streaming through the assembly, variation in the rate of gas flux generally induces variation in the condition of the oxygen transfer (oxygen impurities are always present in the so-called inert atmosphere) thus resulting in variation of the measured e.m.f. Analogously, variation of the residual pressure in cells working under vacuum could also induce a perturbating effect in the e.m.f. When the stationary e.m.f. values are unaffected by the condition of the gas phase, they can be considered as functions of only the oxygen chemical potential difference of the bulk electrode systems.

III. Galvanic Cells in Non-stationary Conditions

As we saw earlier, the concentration gradients between electrode interfaces and bulk due to the transport of oxygen via the gas phase could make the e.m.f. data useless, since these data represent the situation of the electrode/ electrolyte interface rather than the situation of the bulk.

Furthermore, it often happens that the surface of the metal phase of a metal–metal oxide electrode is covered by an epitaxial oxidation layer. This oxide film represents a steady state condition resulting from the process of oxidation and the process of diffusion of metal atoms through the layer, and in this film an oxygen chemical potential gradient, normal to the metal oxide layer, is present (Wagner, 1952; Hauffe, 1959).

In order to overcome these difficulties, the author devised the "non-stationary" method (Barbi, 1966b), which consists of measuring and recording the e.m.f. after polarization. This technique is independent of the residual oxygen pressure of the gas phase within very broad limits. In particular, the residual oxygen pressure of the gas phase may be kept several orders of magnitude different from those in equilibrium with the electrodes; the only limitation is imposed by the pure ionic transport characteristics of the intermediate solid electrolyte. In the case of the rare earth oxides doped thoria electrolytes, at 1000°C, a residual pressure of less than 10^{-5} atm assures that only negligible electron hole conductivity takes place (Bauerle, 1966).†

The non-stationary technique consists of forcing a constant current to flow through the cell:

$$\text{Pt} \mid \text{Me} + \text{Me}_x\text{O} \mid \text{ThO}_2(+\text{LaO}_{1.5}) \mid \text{Fe} + \text{``FeO''} \mid \text{Pt}$$

in order to reduce the oxide layer covering the metal phase by means of the cathodic reaction:

$$\text{Me}_x\text{O} + 2\theta \rightleftharpoons X\,\text{Me} + \text{O}^=.$$

The oxygen ions transported by the electrolyte oxidize the counter-electrode. The choice of this electrode is critical for a two electrode assembly since it must have a very fast depolarization rate in order also to be utilized as reference electrode. The iron + wüstite system is particularly suitable

† The most widely utilized oxygen solid electrolytes are calcia- (or yttria-) stabilized zirconia and lanthania- (or yttria-) doped thoria. The massive introduction of lower valency cations in the parent Zr^{4+} or Th^{4+} lattices involves a corresponding number of oxygen vacancies to be created in the oxygen sublattice. The concentrations of electrons and electron holes depend on the equilibrium oxygen pressure. It is then possible to define for each electrolyte at any temperature, an oxygen pressure interval where the oxygen ion transport number is greater than a pre-fixed value (for instance $t_{\text{O}=} > 0.99$ which is an empirical, widely-used way of defining the pure oxygen ion conductivity) (Steele and Alcock, 1965). At higher oxygen pressures the electrolyte shows significant p-type electrical conductivity while at equilibrium pressures smaller than the lower extreme of the above-mentioned interval, significant n-type conductivity takes place. At 1000°C the lower limit for zirconia-based electrolytes is about 10^{-15} atm and for lanthania-doped thoria it is remarkably lower. It is felt that at 650°C should not exceed 10^{-50} atm.

from this point of view (Tare and Schmalzried, 1966) and also the function of oxygen chemical potential versus temperature is one of the best studied among all the metal–oxygen systems (Darken and Gurry, 1945; Engell, 1957).

After switching off the current, the e.m.f. versus time decay curve is recorded. These curves, provided a sufficient amount of electric charge is passed, are step-shaped (Fig. 2). The horizontal portion of any intermediate step represents a rest potential associated with the presence of a biphasic metal–oxygen system at both electrode/electrolyte interfaces, whilst the final rest potential coincides with the steady state e.m.f. as measured in stationary conditions. The portions of the curves between two steps represent thermo-dynamic transitions each taking place at least at one interface. So, for metal + metal oxide electrodes, the end of the first horizontal portion of the curve, provided that the residual anodic polarization may be considered negligible, corresponds to the moment at which the activity of the metal at the interface starts to become lower than the activity in the same phase saturated with oxygen. Figure 3 shows the situation of any Me + MeO/ThO$_2$ electrode interface during the entire cycle of a non-stationary e.m.f. measurement. Figure 3a represents the situation before the reduction, Fig. 3b during the passage of the current and Fig. 3c when a partial reduction of the bulk

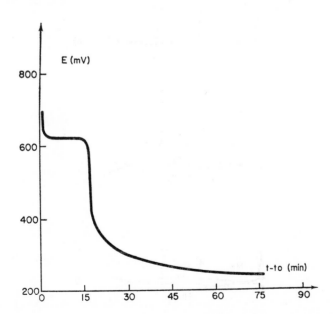

FIG. 2. Typical e.m.f. versus time decay curve after polarization of a solid galvanic cell (Nb + NbO versus Fe + "FeO").

MeO has still been carried out. It is noteworthy that the reduction process also takes place at points on the electrode surface not in contact with the electrolyte; this is because in these interstices the oxygen could become lower than that of decomposition of MeO, since highly negative semicell potentials are set up when sufficiently high polarizarion currents are forced through. A certain time after switching off the current the diffusion of oxygen in the metal phase restores at the Me + MeO/thoria interface the saturation oxygen partial pressure. This situation corresponds to the rest potential portion of the decay curve (Fig. 3d). The successive decay is related to the lowering of the thermodynamic activity of the metal due to the oxide layer reformation at the interface and this decay process lasts as long as the above-mentioned stationary (not representative of the bulk thermodynamic functions) con-

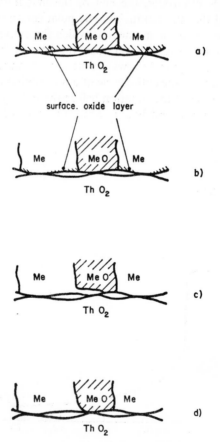

FIG. 3. Schematization of an electrode/electrolyte interface during a non-stationary experiment.

ditions of the interface have been attained. At this last stage of the decay, the corresponding steady state e.m.f. is strongly dependent on many factors, viz. the residual oxygen pressure in the gas phase, its flowing rate (for cell assemblies working at room pressure), the geometry of the cell, the flatness and smoothness of both electrode and electrolyte contact surfaces. From a practical point of view, careful preparation of the surfaces is very important, since, in ideal conditions, i.e. when the ternary contact electrode/electrolyte/ gas is reduced to the geometrical boundary, the influence of the near-the-boundary elemental semicells is reduced to a minimum. These are the conditions providing the best chances of stability of the rest potential even for long periods, because in this ideal case the process of tangential diffusion of the oxidation products from the outermost regions of the electrode surface towards the centre is the sole mechanism determining the successive semicell potential decay. In any case, at least at the beginning of the plateau, the elemental semicell potential contributions due to the outer rings of the interface may be neglected.

The shapes of the e.m.f. decay curves may be different, depending on the characteristics of the electrode system under investigation. For the metal plus metal oxide systems we shall consider here the examples of the $Al + \gamma \cdot Al_2O_3$ and $Nb + NbO$ electrode systems.

1. $Al + \gamma \cdot Al_2O_3$ versus $Fe +$ "FeO" galvanic cell.
The cell:

$$Pt \mid graphite \mid Al + \gamma \cdot Al_2O_3 \mid ThO_2(+LaO_{1.5}) \mid Fe + \text{"FeO"} \mid graphite \mid Pt$$
$$(VI)$$

was the first investigated by Barbi (1965) by means of the non-stationary technique. In this case we are concerned with (1) a very good contact between electrode and electrolyte, (2) a very slow process of diffusion of metal atoms through the surface oxide layer.

The contact surfaces, even if not perfectly flat and smooth at the beginning of the experiments, after a short time become perfectly contiguous owing to the remarkable plasticity of aluminium. As a consequence, after the polarization current has been interrupted, the rate of uptake of oxygen at the interface is determined solely by the rate of tangential diffusion of oxygen from the periphery (in contact with the oxidizing gas phase) towards the centre of the interface.†

† The rate of this process depends on many factors. In particular, the porosity of the intermediate electrolyte facilitates the molecular transport of oxygen from the gas to the interface.

In conclusion we are concerned with two very slow processes: the process of oxidation and that of diffusion of Al from the bulk. The slope of the decay curves is determined by the ratio of the process rates. The e.m.f. value, which in this case must be taken into account to calculate the oxygen partial free energy in equilibrium with Al and $\gamma \cdot Al_2O_3$, is obtained by extrapolation of the curve at $t = t_0$, t_0 being the time at which the current has been switched off (Figs 4 and 5). The extremely low value of the oxygen solubility in aluminium assures that the saturation with oxygen of the metal phase at the interface is accomplished a very short time after t_0.

A general remark concerning the cells working at low temperatures ($T > 700$–$750\,°C$) is the possibility of detecting the phenomena associated with the anodic depolarizarion of the Fe + "FeO" electrode. In particular the Al + $\gamma \cdot Al_2O_3$ versus Fe + "FeO" cell exhibits the first rest potential at about 1680 mV. The difference between the two rest potentials corresponds exactly to the difference between the oxygen chemical potentials in equilibrium with the systems Fe_3O_4 + Fe_2O_3 and Fe + "FeO" (Richardson and Jeffes, 1948). The intermediate rest potential corresponding to the system "FeO" + Fe_3O_4 was never detected during the measurements, but this fact is not surprising since the distance between "FeO" + Fe_3O_4 and Fe + "FeO" oxygen potential levels is only a few millivolts at our working temperatures ($< 660\,°C$) (Barbi, 1964).

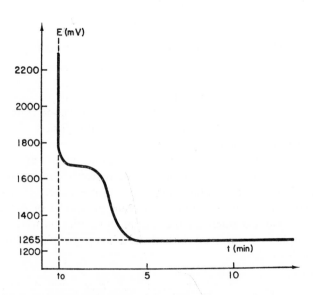

FIG. 4. Decay curve after polarization ($0\cdot3\ mA \times 600\ sec$) of an Al + Al_2O_3 versus Fe + "FeO" solid galvanic chain (cell VI). $T = 930°K$.

A general sketch of the interface during an entire cycle of the experiment is given in Fig. 6. Figure 6a represents the situation before electrolysis, with the aluminium phase covered by an oxidation layer. At the end of the polarization stage it is likely that a proportion of the alumina islands in contact with thoria have also been reduced. As shown in Fig. 6b, once the surface oxide layer has been destroyed, the reduction of the islands starts from the triple boundaries aluminum/alumina/thoria, since the current flow is concentrated at the metal phase regions of the electrode. Figure 6b shows the building up of anodic products at the other interface.

Figure 6c represents the situation at the beginning of the second rest potential, when the system Fe + "FeO" has been restored at the anodic interface and another oxide layer starts to be rebuilt.

2. Nb + NbO *versus* Fe + "FeO" *galvanic cell*

The situation of the cathodic interface of the galvanic cell (Barbi, 1968):

$$Pt \,|\, Nb + NbO \,|\, ThO_2(+LaO_{1.5}) \,|\, Fe + \text{``FeO''} \,|\, Pt \qquad (VII)$$

is very different from the aluminium + alumina cell. Here, the surface of the Nb + NbO electrode is never ideally flat and smooth and it no longer tallies with the thoria surface during the experiment, owing to the much

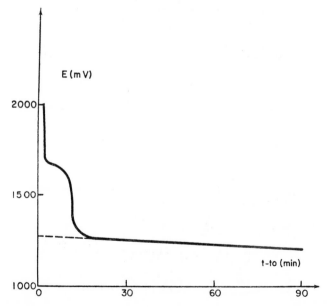

FIG. 5. Decay curve after polarization ($1\ mA \times 600$ sec) of an $Al + Al_2O_3$ versus $Fe + \text{``FeO''}$ solid galvanic chain (cell VI). $T = 900°K$.

lower plasticity of the Nb + NbO dispersion at the working temperature (1000–1300°K) with respect the Al + Al$_2$O$_3$ dispersion. But, on the other hand, the diffusion of metal atoms in the highly defective structure of the oxide layer is considerably faster than in the case of aluminum. At the beginning of the reformation of the oxide layer by the gas phase, the rate-determining step for this process is the transport of oxygen from the gas to the electrode/electrolyte interface. rather than the diffusion of the metal atoms from the bulk. As long as the oxidation process has not com-

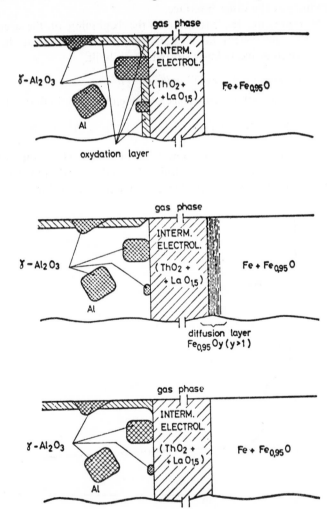

FIG. 6. Schematization of the Al + Al$_2$O$_3$/ThO$_2$ interface (cell VI) during a non-stationary experiment.

pletely destroyed the metal phase, the system remains biphasic at the contact with the electrolyte surface. But once the oxidation of the niobium metal at the interface has been completely accomplished, i.e. when a new oxide layer again completely covers the surface, the e.m.f. falls according to the oxygen potential value of the oxide monophase at the interface. From this time onwards the composition-dependent e.m.f. value is a function of the thickness of the layer which, in turn, is a function of both the diffusion coefficient of the metal and of the residual oxygen pressure in the gas phase.

The imperfectly tallying contact between the electrode and electrolyte surfaces causes the surface oxidation process to take place more uniformly, even if more quickly, than in the case of cell (VII). This is very likely the reason why in the Nb + NbO cell the rest potential value remains quite constant and then falls sharply.

By comparison of Figs 4, 5 and 7 it is possible to observe that the curves of Fig. 7 do not show any rest potential corresponding to a transitory presence of more oxidized biphasic systems at the anodic interface. As we said earlier, this fact is to be ascribed to the remarkably faster anodic depolarization rates for cell (VII) with respect of cell (VI) owing to the higher working temperature.

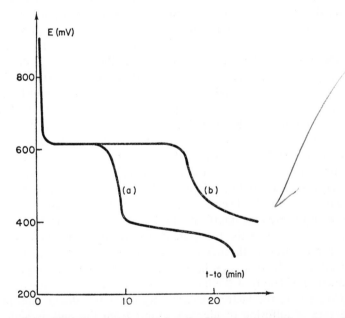

Fig. 7. Influence of the residual pressure on the length of the rest potential portion of a decay curve for the cell VII. $T = 1287°K$, total polarization $0.3 \, mA \times 10 \, sec$: (a) total pressure 1.5×10^{-4} mmHg: (b) total pressure 1×10^{-5} mmHg.

IV. Reliability of the Non-stationary E.m.f. Measurements

In order to evaluate the validity of the non-stationary technique for determining thermodynamic functions, a further step has to be taken. The reliability and usefulness of the e.m.f. data are conditioned by the absence of mixed semicell potentials. This situation may be established by building up more oxygenated phases at the electrode/electrolyte interfaces when an electrode–oxidizing atmosphere is present. Then we shall formulate and discuss a criterion to be utilized in order to exclude the occurrence of mixed potentials in correspondence to any rest potential portion of the decay curves. We shall consider here, as an example, the case of the Nb + NbO/ThO$_2$ semicell of cell (VII.)

In the niobium–oxygen system there exist, besides NbO, more oxygenated phases, namely NbO$_2$ and Nb$_2$O$_5$ and this fact could call into question whether or not the rest potential values obtained by e.m.f. measurements represent the presence of only the Nb + NbO system at the interface. In fact, the contemporary presence of an NbO$_2$ phase would result in a mixed semicell electric potential lying between those corresponding to Nb + NbO and NbO + NbO$_2$ oxygen potential levels.

In order to solve this important question, we may observe that in our case, the experimentally obtained curves exhibit rest potential values which remain extremely constant for long time (Fig. 7). During this time although the oxygen chemical potential remains constant, the overall oxygen content at the interface increases continuously, as the curves will decay after some time. We may therefore conclude that the oxygen coming from the gas phase oxidizes the metal rather than NbO. In other words, NbO$_2$ is not formed on account of the oxygen uptake or, at least, if it were formed it would immediately be reduced to NbO by the metal atoms diffusing from the bulk. In fact, if the continuous uptake of oxygen were to give rise to any stable presence of the NbO$_2$ phase, because of the continuous increment of the ratio of the electrolyte interface in contact with NbO + NbO$_2$ system to that pertaining to the Nb + NbO system, the e.m.f. should decay continuously. The ratio between the rate of diffusion of the metal from the bulk towards the interface and the ratio of the formation of NbO$_2$ determines the slope of the rest potential portion of the curve. Only when the diffusion process of the metal atoms becomes so slow, owing to the increase in thickness of the oxidation layer, that it is not assured a rapid and complete reduction of NbO$_2$ at the interface, does the e.m.f. fall down.

Concerning the factors determining the e.m.f. decay rate, provided that the rate of diffusion of niobium atoms from the bulk is high, the oxygen residual pressure in the gaseous environment does not affect either the rest potential value or its horizontal trend. On the contrary, it shows a strong influence on the duration of the rest potential portion of the decay curve (Fig. 7).

We can therefore formulate this general statement: *the horizontality of the rest potential is the necessary* (although not exclusive) *condition to exclude the presence of mixed potentials* due to more oxidized phases. This condition is both necessary and exclusive if, however, the constancy of the e.m.f. value at different residual oxygen pressures can be demonstrated.

V. Non-stationary Determination on the Thermodynamic Stability of the Intermetallic Phases

Recently the non-stationary technique has also been extended to the ternary systems Me^I + Intermetallic Compound $Me_z^I Me^{II}$ + Oxide of Me^{II}. A three-phase, three-component system defines the values of the chemical potentials at any given temperature.

When Me^{II}, in reacting with oxygen, yields many oxide phases, however, preliminary equilibrium experiments followed by X-ray analyses of the equilibrated system are needed in order to determine which oxide phase is thermodynamically stable under these conditions.

As an example, the thermodynamic properties of $Fe_{7-y}Nb_2$ ($Fe_{7-y}Nb_2$ representing the iron-rich boundary composition of the intermetallic $MgZn_2$-type ε-phase, y varying between zero and 3 according to different authors' investigations) have been determined by measuring the e.m.f. developed by the cell (Barbi, 1969):

$$Pt \,|\, Fe + Fe_{7-y}Nb_2 + NbO_2 \,|\, ThO_2(+LaO_{1.5}) \,|\, Fe + Fe_{0.95}O \,|\, Pt. \quad \text{(VIII)}$$

In this case, preliminary equilibration experiments followed by X-ray diffraction analyses demonstrated that NbO_2 is more stable than NbO under these conditions, the resultant equilibrium niobium activity being lower than that in equilibrium with the $NbO + NbO_2$ biphasic system. In other words, the iron excess behaves as an oxidant of NbO by fixing half of the niobium atoms as intermetallic phase. Thus, the electrode reactions may be written as:

$$Fe_{7-y}Nb_2 + 4\,O^= \rightleftharpoons (7-y)\,Fe + 2NbO_2 + 8\theta$$

and

$$4Fe_{0.95}O + 8\theta \rightleftharpoons 3 \cdot 80Fe + 4\,O^=$$

the resulting net process being:

$$4Fe_{0.95}O + Fe_{7-y}Nb_2 \rightleftharpoons (11 \cdot 80 - y)\,Fe + 2NbO_2.$$

The standard free energy of formation of the iron-rich boundary inter-metallic phase is given by:

$$\Delta G^{\circ}_{f,Fe_{7-y}Nb_2} = 2\Delta G^{\circ}_{f,NbO_2} - 4\Delta G^{\circ}_{f,Fe_{0.95}O} + 8.F.E.$$

Two examples of e.m.f. versus time decay curves after polarization of this cell are shown in Figs 8 and 9. At 1280°K (Fig. 8) the rest potential value of 392 mV corresponds to the presence at the interface of the three phase system. The rest potential is perfectly horizontal, very well defined and it has been used to calculate the free energy of formation of the inter-metallic phase. More generally, however, as a consequence of the multiplicity of processes at the interfaces some decay curves might show secondary rest potentials that cannot be accounted for by the presence of the ternary system at the interface (Barbi, to be published). We saw earlier that the depolarization of the anodic interface may give rise to rest potentials in the decay curves but this is not the only cause of the multiple step trend in some decay curves.

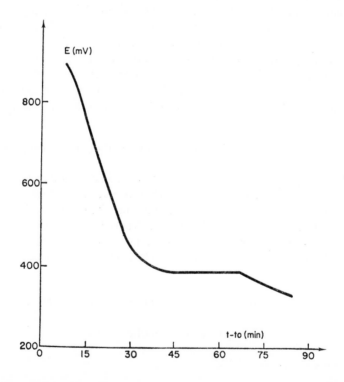

FIG. 8. Decay curve after polarization (1 mA × 180 sec) for cell VIII. $T = 1280°K$.

For instance, in the case of $Fe + Fe_{7-y}Nb_2 + NbO_2$, Fig. 9 shows a first secondary rest potential at about 600 mV that cannot be accounted for by the anodic depolarization process, for two fundamental reasons. Firstly, 1123°K is such a high temperature that 10 min after the interruption of the current the thin anodic Fe_2O_3 has certainly been destroyed, owing to the high diffusion coefficient of iron in iron oxides. Secondly, the difference between the two rest potentials does not correspond to any difference between oxygen potentials of any iron + oxygen biphasic system at this temperature.

On the other hand, the first rest potential of Fig. 9 could easily be explained by assuming a transitory presence at the interface of the system Nb + NbO thus giving rise to a transitory establishment of cell (VII). In fact, the cathodic reduction of NbO_2 does not stop at NbO but yields free niobium metal at the interface. The reoxidation process due to the gas phase oxygen impurities produces niobium monoxide as long as the thermodynamic activity of niobium is higher than that in equilibrium with the $NbO + NbO_2$ system. This means, in practice, that the above-mentioned

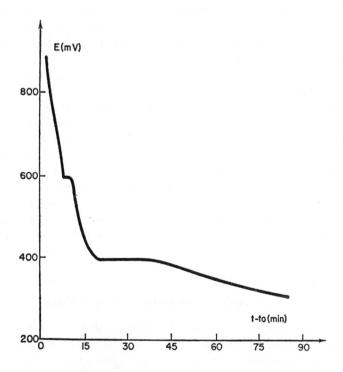

Fig. 9. Decay curve after polarization ($0·1$ mA $\times$ 120 sec) for cell VIII. $T = 1123°K$.

transitory situation lasts as long as the niobium metal phase is present at the interface. When the iron diffusing from the bulk, has consumed all the niobium to yield the intermetallic phase the drop in e.m.f. takes place. Afterwards the e.m.f. stabilizes again at the main rest potential correspond-ing to the three phase system. As the temperature increases, the rate of iron diffusion increases and above a certain temperature the first rest potential vanishes; at $1280°K$ it is no longer detectable (Fig. 8).

From these arguments it follows that it is particularly advisable to work with an electrode containing a strong excess of Me^I. In our case, a strong excess of iron increases the total diffusion flux of this element from the bulk and hastens the process of recombination of niobium to produce the intermetallic phase. Furthermore, it should always be borne in mind that only once the activity of iron reaches a constant value (practically ≈ 1, neglecting the oxygen and Me^{II} solubilities as first order approximation) the activity of oxygen is a single valued function of the activity of Metal II and the evaluation of the stability of the intermetallic phase is possible.

Thus, the decay portion of the curve between the two resting potentials corresponds to the reaction of iron with niobium and to the oxidation of NbO to NbO_2. The portion which follows the second resting potential represents a further oxidation of NbO_2 to more oxygenated phases.

To summarize, the processes occurring at the electrode/electrolyte inter-faces determine the shape of the decay curves. These processes are essentially the rate of diffusion of the excess component from the bulk and the rate of reaction of this component with the cathodic reaction products. The higher the rate of the resulting overall process of formation of the intermetallic phase with respect to the rate of reoxidation at the interface, the more perfectly hori-zontal is the rest potential portion of the decay curves. Otherwise the rest potential is ill-defined and determination of the thermodynamic stability of the intermetallic phase is not possible. At any rate, the general criterion of reliability stated previously for two-phase, two-component thermodynamic systems and concerning the horizontality of the rest potential as the requisite to exclude the occurrence of mixed potentials, can be extended directly to include three-phase, three-components thermo-dynamic systems.

VI. Future Developments—Observations

Because of the lack of consistent thermodynamic data for many inter-metallic compounds, it is expected that the non-stationary technique will be widely applied in the future for determining the free energy of formation of the intermetallic phases. It should be kept in mind, however, that by this method only the thermodynamic data for one-boundary composition may be obtained.

For some applications, however, it is just this knowledge of the activity of the components of the phase at one phase boundary composition that is required. We must mention here, for example, the important technological problem of the structural stability of the phase-dispersed alloys.

It is well known that the particle size distribution of an intermetallic phase, dispersed in a matrix of a metal, and entering as a component, strongly affects the mechanical properties of the dispersion (Mott and Nabarro, 1948; Irmann, 1952). This distribution is altered by thermal treatments (Cochardt, 1957; Komatsu and Grant, 1962) in that the particles tend to grow in size i.e. big particles tend to grow at the expense of smaller ones, which finally disappear (Lenel and Ansell, 1961; Lifshitz and Slyozov, 1961). The kinetic parameters of the transport processes are affected by the concentrations and activities of the components of the dispersed phase. It follows that the knowledge of the activities in equilibrium with the metal matrix gives an approximate indication for forecasting the behaviour of dispersed phase alloys undergoing prolonged thermal treatments.

References

Barbi, G. B. (1964) *J. phys. Chem.* **68**, 2912.
Barbi, G. B. (1966a). *Annu. Chim.* **56**, 992.
Barbi, G. B. (1966b). *Trans. Faraday Soc.* **62**, 1589.
Barbi, G. B. (1968). *Z. Naturf.* **23a**, 800.
Barbi, G. B. (1969). *Z. Naturf.* **24a**, 1580.
Barbi, G. B. (to be published).
Bauerle, J. (1966). *J. chem. Phys.* **45**, 4162.
Bridger, N. J., Denne, W. A., Markin, T. L. and Rand, M. H. (1963). AERE–R4329.
Charette, G. G. and Flengas, S. N. (1968). *J. electrochem. Soc.* **115**, 796.
Cochardt, A. W. (1957). *J. Metals* **9**, 434.
Darken, L. S. and Gurry, R. W. (1945). *J. Am. chem. Soc.* **67**, 1398.
Egan, J. J. (1964). *J. phys. Chem.* **68**, 978.
Engell, H. J. (1957). *Arch. Eisenhüttwes.* **28**, 109.
Foley, R. T. (1969). *J. electrochem, Soc.* **116**, 13C.
Haber, F. and Tolloczko, S. (1904). *Z. Anorg. Chem.* **41**, 407.
Hauffe, K. (1959). "Kinetics of High Temperature Processes", Technol. Press of Mass. Inst. of Techn. 282. p.
Irmann, H. (1952). *Metallurgia* **46**, 125.
Katayama, M. (1908). *Z. phys. Chem.* **61**, 566.
Kiukkola, K. and Wagner, C. (1957). *J. electrochem. Soc.* **104**, 308 and 379.
Komatsu, N. and Grant, N. J. (1962). *Trans. metall. Soc. A.I.M.E.* **224**, 705.
Lenel, F. V. and Ansell, G. S. (1961). Proc. Int. Conf. Powder Metallurgy, Interscience Publ. London. p. 304.
Lifshitz, I. M. and Slyozov, V. V. (1961). *J. Phys. Chem. Solids* **19**, 35.

Mott, N. F. and Nabarro, F. R. N. (1948). Report on Strength of Solids p. 1, Physical Society, London.
Patterson, J. W., Bogren, E. C., and Rapp, R. A. (1967). *J. electrochem. Soc.* **114**, 752.
Richardson, F. D. and Jeffes, J. H. E. (1948). *J. Iron Steel Inst.* **60**, 261.
Satow, T. (1967). *J. nucl. Mater.* **21**, 249.
Schmalzried, H. (1962). *Z. Elektrochem.* **66**, 572.
Steele, B. C. H. (1967). Proc. Symp. Res, Group Inp. College (Nuffield) p. 3. Institute of Mining and Metallurgy, London.
Steele, B. C. H. and Alcock, C. B. (1965) *Trans. metall. Soc. A.I.M.E.* **233**, 1358.
Steele, B. C. H., Powell, B. E. and Moody, P. M. R. (1968). *Proc. Br. Ceram. Soc.* **10**, 87.
Tare, V. B. and Schmalzried, H. (1966). *Trans. metall. Soc. A.I.M.E.* **236**, 444.
Wagner, C. (1952). *J. electrochem. Soc.* **99**, 369.
Wagner, C. (1957). Proc. C.I.T.C.E. VII Meeting p. 361 Butterworths Scientific Publications, London.
Ure, R. W. (1957). *J. Chem. Phys.* **26**, 1363.

16. THERMAL DIFFUSION IN IONIC CRYSTALS

Josette Dupuy

Laboratoire d'Emission Electronique, Faculté des Sciences, Villeurbanne, France.

Introduction

The pioneering work on thermal diffusion of ionic materials, especially ionic crystals, was done between 1929 and 1937 by Reinhold, who dealt with the Soret effect and thermoelectric measurements on AgI–CuI and AgBr–CuBr.

Good theoretical bases were then deduced from non-equilibrium thermodynamics, mainly by de Groot (1951), followed by a whole series of measurements performed by Holtan (1953) concerning non-nomocrystalline ionic salts such as AgI, AgBr, AgCl, $AgNO_3$, $PbCl_2$, $PbBr_2$, PbI_2, CuCl, TlBr.

In 1957 Lidiard and Howard reconsidered the analysis of thermoelectric power of ionic crystals, taking into account the lattice defects, vacancies and interstitials, which create the migration process. Measurements of conductivity and diffusion have been improved over the last 15 years with the use of radioactive tracers, and consequently have stimulated theoretical studies of lattice defects.

The low rate of development of thermal diffusion measurements on ionic crystals must be attributed to the difficulty of interpreting transport mechanisms in non-isothermal conditions. These measurements offer, however, a true theoretical interest for the understanding of matter transport mechanisms and an equally important technological interest (behaviour of materials in nuclear reactor, heat conversion, etc).

We shall try to describe the present state of research on thermal diffusion of ionic crystals. A few results seem to be well established, particularly the phenomenological formulation by non-equilibrium thermodynamic processes and the precise significance of measured quantities.

There are still a number of approximations which render unreliable the experimental results concerning the quantities characterizing thermal diffusion, and hence make very hazardous the agreement which can be deduced by a difficult theoretical interpretation.

The synthetic works where one can find a development of certain aspects of the problems, which are not dealt with here or which place the thermal diffusion in the more general context of matter transport in ionic crystals, are: Lidiard (1957), Allnatt and Jacobs (1961), Howard and Lidiard (1964), Kröger (1964), Adda and Philibert (1966), Allnatt and Chadwick (1967a, b), Manning (1968).

I. Theoretical Aspects

A. Phenomenological Theory of Heat Diffusion

The main emphasis of transport processes through crystalline solids, especially ionic solid conductors, is related to the lattice imperfections: vacancies and interstitial atoms.

The principle of non-equilibrium thermodynamics (de Groot and Mazur, 1962) which exposes in a specialized sense the general equations of transport processes, particularly those under the influence of thermal gradients, must be re-formulated.

This deals essentially with two points:

(1) vacancies are considered as zero-weight components of the system;

(2) constituents are defined by the number of particles in the unit volume instead of by their specific weight. With reference to the now classical discussion given by Alnatt and Jacobs (1961) two important modifications have been added:

(1) restriction on conversation of sites must be substituted for the mass conservation relation (for a region of the crystal without sink and sources of vacancies);

(2) matter flows are referred to their centre of mass in fluids, but for solids we must use the vectorial flows of matter relative to the local crystalline lattice. The reference velocity becomes the mean atomic velocity.

The non-equilibrium thermodynamic theory applied to solids must fulfil the following conditions in order to be satisfactory and complete.

(i) Entropy production must be independent of the reference system (Prigogine's 2nd theorem).

(ii) Onsager's relations hold for the vectorial flows (de Groot and Mazur, 1962, Chapt. 6).

(iii) transformations of fluxes and forces are possible under the same conditions as for scalar flows i.e. with a non-singular matrix which transforms the symmetrical matrix of phenomenological coefficients into another symmetrical one (i.e. Onsager's relations are satisfied).

This results if two restrictions are fulfilled:

$$\left\{ \begin{array}{l} \text{one restriction on the fluxes: } \sum_{k=1}^{n} \mathbf{J}_k' = 0 \text{ which expresses conservation} \\ \text{of sites} \qquad\qquad\qquad\qquad\qquad\qquad\qquad\qquad\qquad\qquad\qquad (1) \\ \text{the other is on the forces: } \sum_{k=1}^{n} N_k' \mathbf{X}_k' = 0 \text{ because we assume that} \\ \text{the system is in mechanical equilibrium.} \end{array} \right.$$

By means of these transformations, we can now use the thermodynamic forces defined by

$$X_k' = - (\nabla \mu_k)_T - e_k \nabla \phi$$
$$X_q' = - \frac{1}{T} \nabla T \qquad\qquad (2)$$

in which $(\nabla \mu_k)_T$ is the part of the gradient of the chemical potential due to the concentration, $\nabla \phi$ is the electric potential gradient due to external fields and e_k is the charge of an atom of species k. The fluxes are the reduced ones:

$$\mathbf{J}_k' = \mathbf{J}_k$$
$$\mathbf{J}_q' = \mathbf{J}_q - \sum_{k=1}^{n} h_k \mathbf{J}_k \text{ where } h_k = \text{partial enthalpy for component } k. \qquad (3)$$

The linear relations between $\mathbf{J}'$ and $\mathbf{X}'$, can be rewritten and considering the three conditions (i), (ii) and (iii), it is possible to introduce the heats of transport by combination of the L_{kq}' coefficients.

$$L_{iq}' = \sum_{k=1}^{n-n_v} L_{ik}' Q_k^{*'}$$

$$J_k' = \sum_{i=1}^{n-n_v} L_{ki}' [X_i' - X_{iv}' + Q_i'^* X_q']$$

(4)

The reciprocal Onsager's relations allow us to write:

$$J_q' = \sum_{i=1}^{n-n_v} Q_i^{*'} J_i + (L_{qq}' - \sum_{i=1}^{n-n_v} L_{iq}' Q_i^{*'}) X_q'$$

When $X_q' = 0$, i.e. in an isothermal system, $Q_i^{*'}$ gives the heat flow resulting from unit flux of component i.

Because of the linear relation between the fluxes of matter, only the heats of transport $(n - n_v)$ are independent. n_v is the number of vacancies and X_{iv}' is the thermodynamic force acting on the vacancies. The sum in expression (4) is over the $(n - n_v)$ which are not vacancies.

The choice of independent flows is related to the diffusion characteristics of the system. Usually, we choose to eliminate the component which does not participate in the process (solvent for fluids). In ionic crystalline systems it depends on the system.

The measurable quantity which can be attained experimentally is a linear combination of the heats of transports of all the individual structure elements. We can obtain experimentally these heats of transport which are characteristics of transports under a temperature gradient by:

(i) *The Soret effect* which is the steady state reached by a system formed from several components under a temperature gradient: a concentration gradient is established, and this balances the thermal forces.

(ii) *A thermoelectric effect* which originates from thermally-induced flows of ions, these flows being different because of the inequality of their mobilities and heats of transport. The thermoelectric effect is derived from the setting up of an electric potential necessary to prevent any current, i.e. to ensure that the contribution of anions and cations to the electric current are equal and opposite.

We shall show, in the two cases which were studied experimentally, the measurement of these values, and their relations with the individual heats of transport.

It is now necessary to specify the theoretical interest of these measurements under thermal forces, i.e. to give a kinetic representation of Q^*.

B. Kinetic Interpretation of Heats of Transport

The thermodynamics of irreversible processes introduce and define the heats of transport as characteristics of thermal diffusion, but do not give any expression deduced from the analysis of their atomic jump mechanism under a temperature gradient.

We can divide the present research into two categories which follow two different main directions: the first category is an extension of the classical concept of rate processes to anisothermal processes; the second looks for the origin of an effective driving force from interactions between diffusing particles and heat-carrying particles, electrons and phonous. The latter were essentially developed for metals.

1. Theories of Isothermal Frequency for a Jump

For a thermally activated process, the probability of the occurrence of a jump to an adjacent vacant site is given by:

$$\omega_0(T) = v \exp\left(-\Delta g_{m/kT}\right)$$

where g_m is the free activation energy and v a vibration frequency.

Allnatt and Chadwick (1967) suppose that in anisothermal systems, the probability of a jump from a plan at T to one at $T + dT$ can be given by:

$$\omega(T, T + \Delta T) = \omega_0(T)[1 + \Delta\omega(T) \cdot \Delta T].$$

The flux of atoms in the direction of increasing temperature between two adjacent atomic planes at a distance dx, the first being at T and the second at $T + \Delta T$, is:

$$\mathbf{J}_1 = \alpha[n_1(T)\, n_v(T + \Delta T)\, \omega(T, T + \Delta T)$$
$$- n_1(T + \Delta T)\, n_v(T)\, \omega(T + \Delta T, T)]$$

n_1 being the concentration of atoms flowing in plane 1.

The first term of the Taylor series for $\mathbf{J}_1$ is

$$\mathbf{J}_1 = -D_1^\circ N\left[\frac{dn_1}{dx} - \frac{n_1}{n_v}\frac{dn_v}{dx} + q_1^* \frac{n_1}{kT^2}\frac{dT}{dx}\right] \tag{5}$$

where: N is the number of sites per unit volume

$$D_1^\circ = \alpha\,\Delta x\,\omega_0 n_v$$

and

$$q_1^* = \Delta h_m - 2kT^2\,\Delta\omega \tag{6}$$

(5) is the general form of Fick's law of diffusion in a temperature gradient for a one-dimensional system.

The first group of kinetic theories try to elucidate the value and significance of $\Delta \omega$.

(a) *The Wirtz* (1943) *model.* The activation energy for a rate process can be divided into three components: Δg_1 supplied by the lattice to initiate the jump, Δg_2 supplied in the atomic plane perpendicular to the line joining the initial and final positions of the jumping atom and Δg_3 to accommodate the atoms in a final position.

In an anisothermal system, Δg_1, Δg_2 and Δg_3 are supplied at three temperatures slightly different from T, $T + \frac{1}{2}\Delta T$ and $T + \Delta T$.

In these conditions, it happens in comparison with (6)

$$q^* = \Delta h_1 - \Delta h_3. \tag{7}$$

Interpretations of enthalpies Δh_1 and Δh_3 differ according to the authors. Some of them identify Δh_1 with migration enthalpy Δh_m and Δh_3 with change in free enthalpy when a vacancy is created.

This identification was critically analysed by Alnatt and Chadwick (1967). This model also predicts $q^* \leqslant \Delta h_m$. According to the preceding authors, it is not obvious that this is a reasonable prediction, and some experimental results are in disagreement with it.

(b) *The Allnatt and Rice* (1960) *Theory.* This theory can be considered as a generalisation of Brinkman's (1954) and Leclaire's (1954) theories. It is intended to calculate the flux, relative to the lattice, of an atom of species i submitted to a Markoffian process in a temperature gradient and a concentration gradient

$$\mathbf{J}_i = n_i \left[\frac{\langle \Delta x_i \rangle}{\Delta T} - \frac{\partial}{\partial x} \frac{\langle \Delta x_i^2 \rangle}{2\Delta T} - \frac{1}{n_i} \frac{\langle (\Delta x_i)^2 \rangle}{2\Delta T} \frac{\partial n_i}{\partial x} \right].$$

For an isothermal random walk:

$$\langle \Delta x_i \rangle = 0$$

$$\langle (\Delta x_i)^2 \rangle = (2/\Gamma)D,$$

where D is the self-diffusion coefficient and Γ the jump frequency.

For an anisothermal walk, these expressions are calculated as functions of the probabilities of finding an atom at the position $x + \Delta x$ at time $t + \Delta t$ if it were at the position x at the time t: assuming that the mean square displacement depends on ∇n and ∇T only at the second order, we can then write for sufficiently small values of ∇n and ∇T

$$\langle (\Delta x_i)^2 \rangle = f[T, n_1 \ldots n_{n-1}].$$

The identification between $\mathbf{J}_i$ and the expression given by the irreversible thermodynamics theory gives:

$$q_1{}^* = \Delta h_m$$

for a vacancy mechanism with a condition of local thermodynamic equilibrium

$$q_i{}^* = \Delta h_m \qquad (8)$$

for an interstitial mechanism.

The basic hypothesises of the random walk model are related first to the mutual independence of the elementary jumps of each particle and second to the local thermodynamic equilibrium postulated for the surroundings of the particle. The latter implies that the energy distribution is approximately uniform and the former implies that the total activation energy which causes the displacement is transferred. The problems arising from this isothermal frequency theory are:

(1) for the "Wirtz model" from the fact that we cannot specify a local temperature (for a site or an atomic plane) in a canonic representation of the system (jump frequency is expressed by Boltzmann's exponential, i.e. it has the characteristic value of an isothermal system);

(2) for the random-walk model from the difficulty of determining the excess of energy in the volume-element surrounding the atom–vacancy pair which leads to the atomic jump. As we know, according to Rice's analysis (1958), an atomic jump takes place because of the in-phase motion of normal modes of high frequency vibrations. In a non-isothermal system, the normal mode distribution influencing the jump frequency of an atom into a vacancy will depend on the temperature gradient in a small but finite volume of element surrounding the atom–vacancy pair. We might then speculate as to the extent of anisotropy of normal mode distribution and the moment of transference between the asymetric phonon distribution and the mobile atoms. This is a very difficult problem which has not yet been solved.

The kinetic interpretation of heats of transport was considered in terms of distribution of the activation energy for diffusion, between the jumping atom and the lattice (Girifalco, 1962; Oriani, 1961).

Recently, Allnatt (1970) has extended the method using the Fokker–Planck kinetic equation to include the case of non-isothermal processes (see also Shimoji and Hoshino, 1967). This theory leads to the following expression of the heat of transport: $q^* = \Delta h_m +$ term involving friction constants characterizing the irreversible aspect of the problem and the anisotropy of temperature.

2. Heats of Transport Interpretation Based on the Analysis of the Driving Forces

The first quantitative estimation is Schottky's (1965) analysis of the perturbation of photon distribution through the existence of a vacancy. This perturbation tends to set up the atom flow. This theory predicts that $\Delta \omega$ in eqn. (6) must be negative and that $q^* = f(1/T^3)$. By analogy with the phenomenon of electromigration, the driving forces must include all causes likely to give a preferential direction of jump to the ions, and consequently express all interaction between heat carriers (electrons and phonons) and lattice imperfections. These forces must be related to the generalized forces of non-equilibrium thermodynamics.

This study was applied generally to metals by analysing phonon diffusion by impurities (Ficks, 1961, 1964) and electron diffusion by impurities (Gerl, 1968; Huntington, 1968). It allows us to account for different values of Q^* in noble and transition metals.

Crolet (1971) did not calculate the driving forces, but tried according to the Manning kinetic theory to analyse the individual atoms jumps. The dissymmetry due to the gradient of temperature is traduced by a effective force related to heats of transport. The resulting contributions of this force come from the vacancy-concentration gradient and from the elementary effects due to the gradient of temperature. These elementary effects are the intrinsic effect justified by the author to be equal to h_m and the phonon diffusion on the jumping atom. This last contribution is estimated as small for the AgBr, Ag, Cu and Au examined. Thus:

$$q^* = h_j{}^m.$$

This brief analysis of the atomic significance of heats of transport gives us an idea of the complexity of the thermotransport problem. The experimental data are not a sufficient basis for the choice between theoretical results because of both their very small number and their scattered nature.

We will analyse the causes of this scattering for each type of experimental study, but before doing this, we wish to point out another difficulty in the interpretation of experimental data arising from vacancy supersaturations.

C. Vacancy Supersaturation

We have seen that a term $(\nabla \mu_k)_T$ appears in generalized thermodynamic force expressions. This term is the gradient of the chemical potential of element k. This element may be a vacancy. In this case μ is not a true Gibbs' chemical potential. However, according to Kröger et al. (1959), this causes no trouble in formulating equilibrium relations deduced from statistical mechanics of the system (Allnatt and Jacobs, 1961).

$$\mu_k = g_k^\circ + kT \log C_k \gamma_k$$

where g_k° is the change in Gibbs free energy when introducing an element k into the perfect crystal at $n_{i=k}$ with P and T fixed

For a Schottky disorder $\mu_+ + \mu_- = 0$

For a Frenkel disorder $\mu_{M^+} = \mu_i + \mu_+$

These relations are deduced from the conditions of electric neutrality, closing and conservation of sites. They express the local thermal equilibrium conditions.

But there exist in the crystal boundaries, dislocations and surfaces which can act as sources or sinks for vacancies and appreciable supersaturation or undersaturation may be obtained (e.g. pore formation).

This may result in a true gradient of concentration with an apparently free enthalpy of formation for defects differing from the thermal equilibrium enthalpy. According to Crolet (1971), a coefficient K characterizing the efficiency of sources and sinks can be defined

$K = 0$ corresponds to total non-efficiency

$$h = Q_s^* - Q_i^* \text{ (for Frenkel disorder)} \neq h_{\text{local equilibrium}}$$

(9)

$K = \infty$ corresponds to total efficiency condition

$$h = h_F = h_{\text{local equilibrium}}$$

II. Experimental Studies and Results

There are two types of experimental study which show the existence of fluxes of matter induced from a gradient of temperature in ionic materials. These are measurement of the thermotransport, and measurement of the thermoelectric effect.

A. Soret Effect in Ionic Materials

1. General Considerations

Between the possible methods of measurement of thermotransport (markers, radioactive tracers, self-diffusion and heterodiffusion), the most convenient for ionic crystals is the steady state established by heterodiffusion. The Soret effect expresses the gradient of concentration set up when an initially homogeneous solution is subjected to a gradient of temperature. A steady state is built up in which the thermal forces X_q are counterbalanced by the forces X_k due to the concentration gradients. The stationary conditions are given: (i) by no current flow

$$\Sigma_k e_k \mathbf{J}_k = 0.$$

(ii) relative to fixed parts of the system far enough from the diffusion zone, for the flows of impurity to be zero everywhere.

For a dilute solution of a radioactive tracer, these conditions give:

$$\frac{\nabla C_k}{C_k} = - \bar{Q}_{eff} \frac{\nabla T}{kT^2}$$

where C_k is the tracer concentration, the gradients are the ones corresponding to steady states at $\bar{Q}_{eff}$ an apparent heat of transport.

The second theoretical stage consists of determining a relation between $\bar{Q}_{eff}$ and heats of transport for species involved in the process.

To achieve this, it is necessary to know:

(i) the lattice defects in the system;

(ii) the nature of the diffusion mechanisms: vacancies, interstitial atoms consisting mainly in anions or cations, or both;

(iii) the nature of interactions between intrinsic defects;

(iv) the nature of the means of incorporation of the impurity into the lattice and the bonds existing between it and the lattice defects.

Usually, the interpretations concerning the three first conditions are different according to the ionic system considered. Referring to the impurity bonds, the electrical conductivity data leads to the same result that a fraction p of the impurities move in the form of strongly bound impurity–vacancy complexes.

Relations $\bar{Q}_{eff} = f(q_i^*)$ will be given for each studied system according to their diffusion characteristics.

We will note briefly the experimental conditions which must be taken into account in order to obtain reliable results. The steady state set up by hetero-diffusion was proved to be the most reliable and precise in ionic crystals (Crolet, 1971). But the diffusion coefficient values for ions in this system need a very long diffusion time to reach the steady state value ($D \simeq 3.05 \times 10^{-11}$ cm^2 s^{-1} at 700° as measured by Bénière (1970) for Sr^{++} in NaCl). The difference between initial and final concentrations decreases approximately like $\exp(- t/\theta)$, according to de Groot (1942) and Tyrrell (1961) where $\theta = l^2/(\pi^2 D^2)$ (l—specimen thickness along the gradient).

A time of 5θ is required to reach a steady state in a sample. It is therefore necessary to assure very good thermal insulation, a very stable and homogeneous gradient of temperature and non-contamination of material. Furthermore, the gradient of temperature must be aligned with the axis of the crystal (according to Allnatt and Chadwick, 1967b) with regard to where the cutting off is done.

Analysis of tracer distribution into the specimen must also take into account edge effects, errors due to sectioning, counting and temperature effects.

The thicknesses generally used are 1 and 2 mm, and the temperature gradients are about $100°C\,cm^{-1}$ and $300°C\,cm^{-1}$.

This gives an idea of the difficulties met when trying to measure the Soret effect and explains why so few measurements in ionic crystals are available despite the interest in these systems.

2. Preliminary Analysis Given by Howard (1957)

This was the first study undertaken to justify the usefulness and reliability of measurements of the Soret effect in ionic crystals. No measurements were available at the time. The equations were derived for divalent cation impurities IX_2 *substituted* in the cation lattice MX. The assumptions are:

(1) the anion diffusion is zero;

(2) the impurity is strongly bound to a vacancy;

(3) the impurity–vacancy system is electrically neutral and the impurity ions move through the crystal only in the form of such complexes;

(4) the unassociated defects are dealt with as if they were components of an ideal solution;

(5) the vacancy may only either change places with an impurity or jump to one of the four nearest adjacent sites; the probabilities for these jumps to occur being respectively ω_2 and ω_1 per unit of time. These probabilities are given by Wirtz formula. When identifying the current density of the impurity ions (calculated from the above assumptions) to the equivalent expression given by the irreversible thermodynamics laws, we find:

$$Q_k{}^* = q_e{}^* - 2q_a{}^*$$

where $q_e{}^* = h_e{}^m - h_e{}^v$ is the heat of transport of an impurity ion jumping into a vacancy, $h_e{}^m$ and $h_e{}^v$ corresponding to the free enthalpy supplied to the ion for jumping and forming a hole, and $q_a = h_a{}^m - h_a{}^v$ is the heat of transport of the host lattice ion adjacent to an impurity ion which jumps into a vacancy.

3. Experimental Results

(a) *Soret effect in* NaCl–SrCl$_2$ *systems.* Allnatt and Chadwick, (1966, 1967b) gave a complete analysis using irreversible thermodynamic methods of the

Soret effect for substitutional cationic tracer impurities into a Schottky disorder or cationic Frenkel disorder.

The general expression found takes into account parameters depending on the theory of defect interaction used for activity coefficients, and mobility of species, i.e. the deviation of the Nernst–Einstein relation of a divalent impurity ion in a univalent crystal. This result comes from the fact that the cross-terms in the equation of irreversible thermodynamics are not eliminated.

A simplification results from a doped crystal in which $n_+ \gg n_-$ or from pure crustals in which the transport number of the anions is small, and the result is given by

$$\bar{Q}_{\text{eff}} = q_I{}^* - 2q_M{}^* + [n(C_I) + h_s)] \frac{l(C_I)}{m(C_I)} \tag{11}$$

where h_s is the enthalpy of formation of a Schottky defect pair, the function l, m, n of the concentration C_i in sites of cation impurities depend on the theory of activity coefficient. I refers to impurity.

With the pair-association theory, we have for Sr^{2+} traces in pure NaCl

$$\left.\begin{array}{c} \bar{Q}_{\text{eff}} = q_I{}^* - 2q_M{}^* - \dfrac{h_s}{2} \\[2mm] \text{for } Sr^{2+} \text{ traces in } SrCl^2 \text{ doped NaCl} \end{array}\right\} \tag{12}$$

$$\bar{Q}_{\text{eff}} = (q_I{}^* - 2q_M{}^*) \left(\frac{1+p}{2} \right) + \chi p$$

where p is the fraction of impurities involved in a complex impurity–cation vacancy at nearest-neighbour separation.

χ the free enthalpy of such an association. The results obtained are given in Table I with

$$Q_k{}^* = q_I{}^* - 2q_M{}^*$$

TABLE I. Thermal diffusion of Sr^{2+} tracer in pure NaCl and in $SrCl_2$ doped NaCl

System	Tracer	$Q_k{}^*$ eV	ΔT °C cm^{-1}
pure NaCl	Sr^{2+}		
NaCl + SrCl^2			
$2\cdot18 \times 10^{-4}$		$-1\cdot56 \pm 0\cdot15$	70
$2\cdot27 \times 10^{-4}$	Sr^{2+}		
$2\cdot55 \times 10^{-4}$			

From this table which shows the agreement between pure and doped crystal values, the authors *can confirm that the anion transport term is not important.*

On the basis of eqn. (6), the discussion requires an approximate evaluation of Δh_I^m and Δh_M^m. Δh_I^m is the activation enthalpy corresponding to the isothermal frequency ω_I for an impurity to jump into an adjacent vacancy. It can be calculated from diffusion measurements. A recent value deduced by Allnatt and Pantelis (1968) is 0.72 eV. Δh_M^m is the activation enthalpy corresponding to the isothermal frequency ω_M of the jumping of a host cation lattice into a vacancy when they are both the closest neighbours of an impurity. It can be estimated from dielectric loss or dielectric relaxation measurements.

The authors adopt 0.74 eV.

Consequently if $\Delta\omega = 0$ $Q_k^*(0) = 1.09 \pm 0.1$ eV and

$$Q_k^* - Q_k^*(0) = -0.45 \pm 0.25 \text{ eV}$$

$$= 2kT^2(2\,\Delta\omega_M - \Delta\omega_I) < 0.$$

Wirtz's theory predicts $q < \Delta h_m$ and hence $\Delta\omega > 0$ which is not compatible with the results.

Schottky's analysis gives $\Delta\omega < 0$ but gives a variation $\Delta\omega \sim T^{-3}$

(b) *The Soret effect in AgCl–CaCl$_2$, AgCl–SrCl$_2$, AgCl–MnCl$_2$ systems and AgBr–CdBr$_2$, AgBr–MnBr$_2$ systems were measured by Crolet (1971).*

It can be assumed that we have, in these systems, a Frenkel disorder, and an anionic perfect lattice with immobile halogen ions.

This point can be used to define equivalences between lattice flows and fluxes relative to either the halogene centre of mass or to the silver centre of mass or to the AgBr or AgCl centre of mass. In the temperature range used (intrinsic range), the association vacancy–impurity is strongly bound.

There are therefore four types of structural elements taking part in the diffusion process, substitutional and interstitial Ag^+, substitutional impurities I, and cationic vacancies. And there is one relation between the substitutional ion fluxes expressing the conservation of sites.

The system of forces and fluxes is made linearly independent by eliminating the vacancies.

Hence

$$\mathbf{J}_j = \Sigma\, L_{jk}[\mathbf{X}_k^+ + Q_k^*\,\mathbf{X}_q], \qquad j = i, s, I$$

where

$$X_k^+ = X_k' - X_{kv}' \begin{cases} X_{kv}' = 0 & k = i \\ X_{kv}' \neq 0 & k = s \\ X_{kv}' \neq 0 & k = I \end{cases} \tag{13}$$

$$X_k^+ = -(\text{grad } \mu_k)_T + (\text{grad } \mu_v)_T - e_k \text{ grad } \phi$$

The electric field current is eliminated by writing the zero current condition

$$\Sigma \, e_j \, \mathbf{J}_j = 0.$$

The steady state condition is that the impurity flow is zero everywhere.

With the strong bond assumption for vacancy–impurity complexes the expression for $\bar{Q}_{\mathrm{eff}}$ becomes:

$$\bar{Q}_{\mathrm{eff}} = q_I{}^* - 2q_s{}^* + h/2 - \chi_a \qquad (14)$$

$q_I{}^*$ is the reduced heat of transport of impurity substitutional ions.

$q_s{}^*$ is the reduced heat of transport of substitutional Ag^+ ions.

According to theoretical conclusions (I.B.2), the author takes for granted that

$$q_I{}^* \simeq h_I^m$$

$$q_s{}^* \simeq h_s^m$$

but h in eqn (14) must vary according to efficiency of sources and sinks

$$h = q_s{}^* - q_I{}^* = h_s^m - h_I^m \quad \text{for total non-efficiency case}$$

$$h = h_F \qquad\qquad\qquad\quad \text{for total efficiency condition.}$$

The results obtained are shown in Table II. The h_I^m values are calculated from diffusion measurements. The h_s^m values are estimated from the ionic thermocurrent and dielectric loss measurements.

The h_i^m value is the mean of the three migration enthalpies corresponding to the three possible interstitialty mechanisms which are calculated from electric conductivity measurements.

$\bar{Q}_{hF}$ is the value calculated using the total efficiency hypothesis

$\bar{Q}_h{}^*$ is the value calculated using the non-efficiency hypothesis

The non-efficiency of sources and sinks of vacancy can be concluded from this result.

At the present time the available results based on considerably detailed experiments dealing with isothermal measurements (conductivity, diffusion, specific heat, thermal expansion) are not in disagreement with the conclusions deduced from non-isothermal measurements. However, there are many problems about the reliability of the values obtained for heats of transport. In particular, because of the possible influence of local thermal non-equilibrium which make these values ambiguous, comparison with the theroetical predictions may be called in question.

TABLE II. Thermal diffusion of Sr^{2+}, Cd^{2+}, Mn^{2+} tracer in pure AgCl and pure AgBr

System	Tracer	$\bar{Q}_{mes}$ eV	$\bar{Q}_{hF}^*$	$\bar{Q}_h^*$	Notice
AgCl	Sr^{2+}	-1 ± 0.1	0.2	-0.4	2 different values
			-0.2	-0.8	available for h_s^m
	Cd^{2+}	-0.62 ± 0.06	0.15	-0.42	3 different values
			0.20	-0.33	available for χ
			0.10	-0.52	
			0.33	-0.24	2 different values
			0.38	-0.15	available for h_s^m
			0.28	-0.34	
	Mn^{2+}	-0.56 ± 0.05	0.33	-0.24	
AgBr	Cd^{2+}	0 ± 0.1	0.27	-0.16	2 values available
			0.32	-0.11	for χ
	Mn^{2+}	0 ± 0.1	0.37	-0.06	

B. Thermoelectric Effect

1. Theoretical Expression

The thermoelectric effect in an ionic crystal is related to the thermally induced flow of ions which move with different mobilities and different heats of transport. If we maintain no current, an electrical potential will be set up to balance the effect of the temperature gradient. The electrical potential gradient is proportional to the temperature gradient: the ratio $\nabla\phi/\nabla T$ is the homogeneous thermoelectric power $\theta_{homogene}$. In the case of an ionic lattice with Schottky defects, four different types of structural elements can participate in the diffusion process: M^+, X^-, $\square^+$, $\square^-$. There are four flows J_k' of which only two are independent, owing to conservation of sites in each sublattice (eqn (1))

$$J_M^{+\prime} + J_+' = J_X^{-\prime} + J_-' = 0.$$

The reduced thermodynamic forces given by eqn (2) are further reduced to two independent ones because of the mechanical equilibrium condition (eqn 1).
We can write

$$X_k^+ = X_k' - X_{kv}' \quad \text{with} \quad k = M^+, X^-, kv = +, -.$$

Only two heats of transport are then independent. They are the only observable ones:

$$Q_k^+ = Q_k'^* - Q_{kv}'^* \quad \text{with} \quad k = M^+, X^-.$$

It is easiest, according to the statistical description for the crystal, to speak in terms of lattice defects.

If we note

$$Q_1^{+} = -Q_M^{+} \quad \text{and} \quad Q_2^{+} = -Q_X^{+}$$

we can then write that the cationic vacancy flow is

$$\mathbf{J}_1 = L_{11} | X_1^{+} + Q_1^{+} X_q | + L_{12} | X_2^{+} + Q_2^{+} X_q |$$

and the anionic vacancy flow is

$$\mathbf{J}_2 = L_{21} | X_1^{+} + Q_1^{+} X_q | + L_{22} | X_2^{+} + Q_2^{+} X_q |.$$

The L_{ii} coefficients are related to the diffusion coefficients $(n_i/kT)D_i$ in which n_i is the number of vacancies in the sublattice i per unit volume. The L_{ij} coefficients are related to the departure from the Nernst–Einstein relation. These are generally small and can be neglected. This assumes small interactions between ions and vacancies and little pairing between themselves

$$\mathbf{J}_1 = D_1 \left| -\nabla n_1 + \frac{en_1}{kT}\nabla\phi - \frac{Q_1^{+} n_1}{kT^2}\nabla T \right|$$

$$\mathbf{J}_2 = D_2 \left| -\nabla n_2 - \frac{en_2}{kT}\nabla\phi - \frac{Q_2^{+} n_2}{kT^2}\nabla T \right| \tag{15}$$

The relation expressing the zero electric current condition is:

$$\mathbf{J}_1 - \mathbf{J}_2 = 0.$$

Hence, we can derive

$$\frac{\nabla\phi}{\nabla T} = \frac{kT}{e}\frac{D_1(\nabla n_1/\nabla T) - D_2(\nabla n_2/\nabla T)}{D_1 n_1 + D_2 n_2} + \frac{1}{eT}\frac{D_1 n_1 Q_1^{+} - D_2 n_2 Q_2^{+}}{D_1 n_1 + D_2 n_2} \tag{16}$$

(i) For a pure ionic crystal, if local thermal equilibrium may be assumed:

$$n_1 = n_2$$

$$\frac{kT}{n_1}\frac{\operatorname{grad} n_1}{\operatorname{grad} T} = \frac{kT}{n_2}\frac{\operatorname{grad} n_2}{\operatorname{grad} T} = \frac{h}{2T}. \tag{17}$$

If we note $\phi = D_1/D_2$

$$eT\frac{\nabla\phi}{\nabla T} = eT\,\theta_{\text{homog}} = \frac{\phi(Q_1^{+} + \frac{1}{2}h) - (Q_2^{+} + \frac{1}{2}h)}{\phi + 1}. \tag{18}$$

This can be written with transport numbers, according to Allnatt and Jacobs (1961)

$$eT\theta_{\text{homog}} = t_+ |Q_1{}^+ + \tfrac{1}{2}h| - t_- |Q_2{}^+ + \tfrac{1}{2}h| \tag{19}$$

and can be modified according to Jacobs and Knight (1970) by writing

$$Q_1{}^+ + Q_2{}^+ = -Q_{MX}{}^+$$

which gives

$$eT\theta_{\text{homog}} = t_+(h - Q_{MX}{}^+) - Q_2{}^+ - \frac{kT^2}{n_1} \frac{\text{grad } n_1}{\text{grad } T}. \tag{20}$$

(ii) For ionic crystals doped with a divalent cation impurity, the θ_{homog} expression can be calculated from (16) if we can make some assumptions on the nature of current carriers and the nature of bonds between vacancies and impurities. The most simple case occurs when: (a) the conductivity is assured mainly by only one type of carrier, and (b) the pairing association theory may be applied to vacancy–divalent cation impurity. In this case, we have

$$\xi = \frac{\text{concentration of free cationic vacancies in doped crystal}}{\text{concentration of free cationic vacancies in pure crystal}}$$

Then:

$$Te\,\theta_{\text{homog}} = \frac{\xi^2(Q_1{}^+) - \phi(Q_2{}^+ + h)}{\xi^2 + \phi} - \frac{kT^2}{en_1} \frac{\text{grad } n_1}{\text{grad } T} \tag{21}$$

Problems arise from the fact that the electrical potential gradient induced by the temperature gradient is measured by electrodes, and the experimental values include, in addition to θ_{homog}, a term which is derived from the temperature dependence of the contact potential at the electrode–ionic crystal interface. The uncertainty about the estimation of this additional term makes the thermoelectric measurement analysis as questionable as the Soret effect analysis. First, we will analyse the theoretical approach and then examine the approximation made for each experimental study.

2. Problems Concerning the Heterogeneous Thermopower

We shall suppose in this discussion that thermodynamic equilibrium is established at the crystal–electrode interface by using an electrode material having an element common with the ionic crystal (metal or halogen—M or X^2). Lawson (1962) attempted to deny this equilibrium, but this objection does not seem likely to hold because the Nernst equation is verified when different partial pressures are used in a gaseous electrode (*cf.* Section B.4, also Howard and Lidiard, 1964, p. 190). The problem arises from the necessity of calculating the chemical potential of a structural element belonging to an ionic crystal. The thermodynamic equilibrium (for example at an M/MX interface) corresponding to a minimum Gibbs' function is given by the equality of the electrochemical potentials μ in both phases

$$\bar{\mu}_M + (M) = \bar{\mu}_M + (MX)$$

$$\mu_M + (M) - e\phi_M(M) = \mu_M + (MX) - e\phi(MX)$$

where μ is the chemical potential and ϕ the electrical potential

$$\mu_{M^+}(MX) - \mu_{M^+}(M) = e\,|\,\phi(M) - \phi(MX)| \tag{22}$$

$\mu_{M^+}(M)$ must be identified with the chemical potential of the metal and calculated by $\int C_p/T\ dT$, $\mu_{M^+}(MX)$ must take into account the influence of the defects on the equilibrium properties of the crystal.

This has been pointed in particular by Grimley and Mott (1947), Howard and Lidiard (1957a, b) and Allnatt and Jacobs (1961).

The calculation of $\mu_{M^+}(MX)$ results from two equilibrium conditions:

one related to the interface: $\bar{\mu}_{M^+} = \bar{\mu}_{M^+}{}^s$

$$\tag{23}$$

the second related to the crystal: $\bar{\mu}_{M^+}{}^s = \bar{\mu}_M{}^B - \mu_+{}^B$

s corresponds to the crystal surface, B corresponds to the crystal bulk.

From eqns (22) and (23)

$$e(\phi_M - \phi_{MX}) = \mu_{M^+}{}^B - \mu_+{}^B - \mu_M(M)$$

$$e\phi_{\text{heter}} = e\,\frac{d\phi(M/MX)}{dT} = S_M(M) - S_{M^+}{}^B + S_+{}^B \tag{24}$$

If equilibrium is set up with the halogen phase, the equations obtained have the same form because we never assume that the metal or halogen is incorporated with the solid phase. An adsorbed phase could possibly be formed in which the halogen would have very specific properties. It is necessary to verify that such a process cannot take place, i.e. the interface equilibrium is governed by a Nernst relation in function with the halogen partial pressure. This has been done for oxides but not yet for alkaline halogenures (see however Jacobs and Knight, 1970).

We write

$$e\phi_{\text{heter}}' = W + S_+{}^B \tag{25}$$

where W is a term which is called "Wagner's approximation" because this author tried to evaluate it by identifying

$$\mu_{M^+}{}^B = \tfrac{1}{2}\mu_{MX}. \tag{26}$$

This identification is based on Einstein's model for ionic solids in which $+$ and $-$ ions have similar symmetries and characteristics, and for which the bond forces are purely Coulombian. We refer the reader to Lidiard (1957), Allnatt and Jacobs (1961) and Howard and Lidiard (1964) where the calculation of chemical potentials of structural elements of a pure or doped crystal is analysed from hypotheses on the nature of the defects and their mutual influences.

(i) For a *pure crystal*, the quasi-chemical reaction of formation of a pair of Schottky defects obeys the relation

$$\mu_+ + \mu_- = 0$$

with $\mu_+ = \mu_+^\circ + kT \log \dfrac{n+}{N}$ (with ideal solution hypothesis for defects)

$$\mu_- = \mu_-^\circ + kT \log \frac{n-}{N}$$

Hence we have

$$S_+ + S_- = 0$$

$$S_+ = S_+^\circ - k \log C_+ - \frac{kT}{C_+} \frac{\operatorname{grad} C_+}{\operatorname{grad} T}$$

$$S_- = S_-^\circ - k \log C_- - \frac{kT}{C_-} \frac{\operatorname{grad} C_-}{\operatorname{grad} T}$$

In these relations superscript 0 denotes quantities related to the crystal without defect. C_r is the concentration of vacancies in the corresponding sublattice.

Entropy variations S_+ and S_- are the sum of a configurational contribution and of another contribution which has an essentially vibrational origin in alkali halide compounds.

For the local equilibrium conditions we have $C_+ = C_-$, hence

$$\frac{kT}{C_+} \frac{\operatorname{grad} C_+}{\operatorname{grad} T} = \frac{kT}{C_-} \frac{\operatorname{grad} C_-}{\operatorname{grad} T} = \frac{h}{2T}.$$

In the case of an electrode which is reversible in relation to M, then:

$$e\theta_{\text{heter}} = S_M(M) - \tfrac{1}{2}S_{MX} - S_+^{\circ} - k\log C_+ - \frac{kT}{C_+}\frac{\text{grad } C_+}{\text{grad } T}. \qquad (27)$$

(ii) Let us consider a crystal *doped with divalent ions.* according to the hypothesis that (a) a kind of vacancy is created essentially by adjunction of divalent ions, and the concentration of the thermally activated vacancy is very small compared to that of the former; (b) N_k complexes are formed by one impurity and one cation vacancy which become closer and closer to each other, and hence leave N_j free impurities. Equilibrium conditions imply that

$$\mu_+ + \mu_j = \mu_k.$$

The chemical potential μ_+, and consequently s_+ is modified by a supplementary term $k\log\zeta$ which takes account the change in concentration of cation vacancies due to doping.

We have

$$e\,\theta_{\text{heter}} = S_M(M) - \tfrac{1}{2}S_{MX} + S_+^{\circ} - k\log C_+ - \frac{kT}{C_+}\frac{\text{grad } C_+}{\text{grad } T} - k\log\zeta \qquad (28)$$

We shall see through various experimental studies how the terms of eqns (27) and (28) have been evaluated. These terms govern the information that we can obtain about the heats of transport.

Note: Usually, we do not try to evaluate the term grad C_+/grad T because it is eliminated with the corresponding term in θ_{homogene} (20) and (21).

3. Experimental Results on Alkali Halides

(a) *General characteristics*

(i) The experimental problems met in this kind of study are less important than the ones met in the Soret effect analysis.

We must ensure: purity of the material, a homogeneous distribution of doping elements, a stable and small temperature gradient and a measurement as precise as possible of the electrical potential gradient corresponding to the temperature gradient (local potential measurement).

If a halogen electrode is used, it is necessary to know very precisely the partial gas pressure at the contact.

(ii) The results always have certain identical characteristics: the measured thermoelectric power is always negative with reference to the usual conventions, i.e. the cold electrode is positive, its absolute value decreases with temperature, more or less according to the disorder in the sublattice which contains the most mobile ions; the thermoelectric power is highly influenced by impurity concentrations, at least in the small temperature range.

(b) *Results relative to silver halides*

The thermoelectric properties of these compounds have been measured by Holtan (1953), Patrick and Lawson (1954), Christy et al. (1959, 1961), Weininger (1964), Ruch and Dupuy (1965, 1966), Kvist et al. (1966) and can be seen in Figs 1, 2 and 3 respectively, for AgCl, AgBr and AgI.

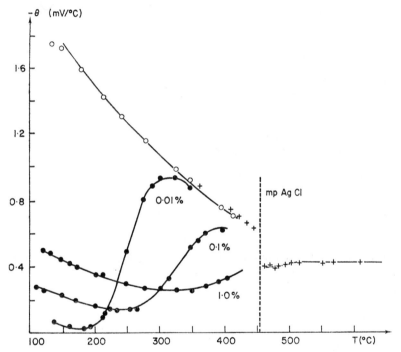

FIG. 1. Thermoelectric power of pure AgCl (○ Christy, 1961), (+ Dupuy and Ruch, 1966) and AgCl containing CdCl$_2$ (● Christy, 1961) as a function of temperature.

The most complete analysis has been done in function with the temperature and the concentration of doping CdX$_2$ ($X = $ Cl and Br) by Christy (1961). He shows that: AgX contains atomic imperfections Ag$_i^+$ and cation vacancy as a result of Frenkel disorder; concentration isotherms confirm the existence or the non-existence of vacancy–impurity associations according to the possible relations between ξ, impurity concentration C_i and vacancy concentration in the pure salt, C_0.

The discussion is about the analysis of the variations of $\Delta\theta(\xi) = \theta(\xi) - \theta$ (pure salt: $\xi = 1$) where θ is the thermoelectric power of an ionic compound which shows mainly Frenkel disorder and for which $\theta_{homogene}$ and $\theta_{heterogene}$

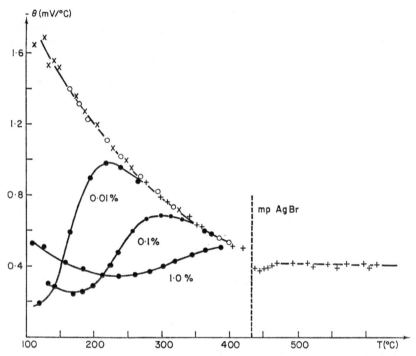

FIG. 2. Thermoelectric power of pure AgBr ($\times$, $\bigcirc$ Christy and Lawson, 1954), ($+$ Ruch and Dupuy, 1965), and AgBr containing $CdBr^2$ ($\bullet$ Christy, 1959) as a function of temperature.

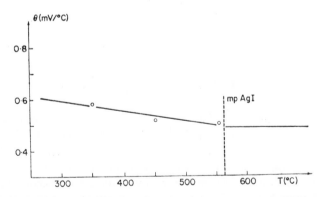

FIG. 3. Thermoelectric power of pure AgI as a function of temperature according to Kvist (1964).

are given by expressions such as (21) and (28); the quantities in these expressions corresponding to the structural elements, Ag_i^+ and cation vacancy.

$$eT\theta(\xi) = \frac{\xi^2(-Q_M^+ + h_F) - \phi Q_i^+}{\xi^2 + \phi} - kT \log \xi - T|S_i^B - S_i^\circ| + TS_M +$$

$$+ kT \log C_+ \qquad (29)$$

$$eT \Delta\theta(\xi) = \frac{\phi(\xi^2 - 1)(-Q_M^+ + Q_i^+ + h_F)}{(\xi^2 + \phi)(\phi + 1)} - kT \log \xi. \qquad (30)$$

The variations of $\Delta\theta(\xi)$ are independent of the hypotheses concerning evaluation of S_{iB} and S_i°.
If we note

$$Q = -Q_M^+ + Q_i^+ + h_F \qquad (31)$$

we have

$$eT \Delta\theta(\xi) = \frac{\phi(\xi^2 - 1)Q}{(\xi^2 + \phi)(\phi + 1)} - kT \log \xi \qquad (32a)$$

$$eT\theta(1) = \frac{Q}{1 + \phi} - Q_i^+ + kT \log C_+ - T(S_i^B - S_i^\circ) + TS_M \qquad (32b1)$$

or

$$eT\theta(1) = \frac{Q}{1 + 1/\phi} - Q_M^+ + kT \log C_+ + T(S_i^B - S_i^\circ) + TS_M \qquad (32b2)$$

$$\text{with} \quad S_i^\circ + S_v^\circ = 0.$$

(i) There is *good agreement* (as shown in Fig. 4) between the *experimental values of* $\Delta\theta(\xi)$ and the *values calculated from eqn* (32a), where ξ and ϕ are deduced from conductivity measurements.

(ii) In the small temperature range $\xi \gg 1$, the variation of θ is the form of $eT\theta = AkT \log C_i + B$ where
$A = 1$ if there are no vacancy–impurity associations
$A = \frac{1}{2}$ if there is any association, thermoelectric measurements agree with the *non-existence of associations*.

(iii) Without any supplementary hypothesis about the interface, no study of transport can be carried out except through the analysis of the values of Q given by eqn (31).

722   Josette Dupuy

Variations of $Te\Delta\theta(\xi)$ combined with the relation $\xi = f(C_i)$ allow us to calculate ϕ, C_0 and Q.

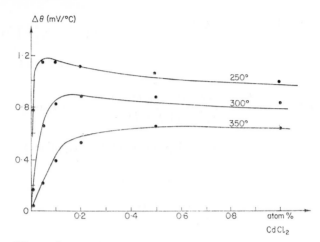

Fig. 4. The difference between the thermoelectric powers of doped and pure AgCl (Christy, 1961) (—theoretical points).

The *results obtained for ϕ and C_0 are in agreement with those obtained from conductivity measurements.* The latter measurements also give a value for the formation energy of a Frenkel defect pair: h_F. We can then calculate $|-Q_M{}^+ + Q_i{}^+|$ whose variations with respect to temperature are given in Fig. 5.

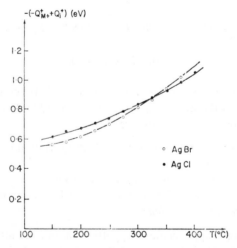

Fig. 5. Sum of the heats of transport for silver ions and interstitials in AgBr and AgCl (Christy, 1961).

The sum of the heats of transport of the structural elements involved in the thermal diffusion mechanism is negative, high, and depends very much on the temperature.

This result diasgrees with the heat of transport theories which predict the identification of Q^+ with h_m and its independence of the temperature.

(iv) $Q_i^+ + T(S_i^B - S_i^\circ)$ and $Q_M^+ - T \mid S_i^B - S_v^\circ \mid$ may be estimated from (32b) by introducing the measured experimental values of Q, ϕ and C_0 which were determined before.

We can obtain the heat of transport values if we know how to evaluate the partial entropies of the structure elements. S_i^B is calculated by Wagner's approximation. S_i° is deduced from the calculation of the change in the vibration modes when an interstitial ion is introduced in the lattice. In the case of a solid whose nearest neighbours' characteristics are modified the most simple approximation, based on Einstein's model, gives:

$$S_v^\circ \simeq 2k$$

and

$$S_i^\circ < 0.$$

But $S_i^\circ + S_v^\circ = S$, a value which can be reached experimentally and is of the order of $12.5\,k$ and $13.9\,k$ respectively for AgBr and AgCl according to Christy (1961).

These values imply that each contribution S_i° and S_v° is high and positive, and consequently cannot be simply evaluated from the nearest neighbour single harmonic freqeuncy approximation.

Remarks

1. The specific heats given for AgCl and AgBr are very debatable, especially for AgBr. But their analysis leads on the one hand to S_{AgX}, which means that it is involved in the computation of Wagner's approximation of $S_{Ag^+}{}^B$, and on the other hand to h and S (free enthalpy and entropy of creation of the Frenkel's defects) through the characteristics of the line $\log T^2 \, \Delta C_p = f(1/T)$.

In a recent experimental study, Leonardi (1969) specified the values relative to AgBr and compared them to the values deduced from conductivity and thermal expansion measurements.

The various values which were measured are given in Table III. S_i°; S_v° and $S_{Ag^+}{}^B$ could not be evaluated farther because of the more complex nature of the bonds in silver halides.

Individual heats of transport cannot be given; the conclusions deduced from the values of Q are the least reliable: high values, very much dependent on the temperature (Fig. 5) and very different from the sum of the corresponding activation and migration energies (Table III).

TABLE III. Enthalpy h and entropy s of formation of Frenkel defects pairs; heats of transport measured and calculated.

compound	h(eV)	s/k	h_i^+ (e)	h_v^- (eV)	$-Q_M^+ + Q_i^+$ (eV)	$Q_M^+ - T\,S_v^o$ (eV)	Q_M^+ (eV)	Q_i^+
AgBr after Christy (1961) data for C_p	1·29	12·5	0·15	0·30	200°C $-0·6$ 350°C $-0·95$	0·33 0·40	0·57 0·71	$-0·03$ $-0·24$
after Leonardi (1969) data for C_p	0·91	5·7				200°C 0·32 350°C 0·31	0·56 0·62	$-0·04$ $-0·33$
AgCl			0·15	0·33	150°C $-0·68$ 400°C $-1·05$	0·34 0·44		

Q^{M+} is evaluated with $S_v^o \simeq 6k$

2. An identical result for the order of magnitude is found in melted silver halides (after Leonardi, 1969; Connan *et al.*, 1969; Poillerat, 1970). The computation of heats of transport through a *justified* evaluation of $S_{Ag^+}(AgX)$ gives results which are much higher than those of alkali halides.

3. We must note that the variations of the thermoelectric power of AgI in the intrinsic range which are given in Fig. 3 are much less influenced by the temperature.

This is the same for thallium halides, as we will see in *c* below, and for melted alkali halides and silver halides.

It is regrettable that a study as complete as these for AgBr, AgCl and TlCl has never been done for AgI

It is possible that the disorder existing in the sublattice of the Ag^+ ions in the intrinsic range as assumed by Ubbelonde (1957) might be related to the non-independence of the heats of transport from the temperature and that it allows a more complete analysis analogous to that done for TlCl.

(c) *Results concerning thallium halogenures*

These measurements were performed by Christy and Dobbs (1967). They deal with TlCl, TlCl + $PbCl_2$, TlBr. The experimental technique used takes into consideration the very high oxidability of *thallium* when it is *used as an electrode constituent*.

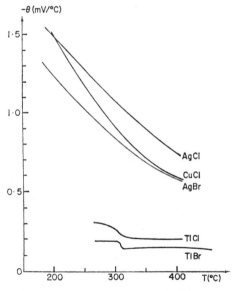

Fig. 6. The thermoelectric power of pure thallium halides compared to pure silver halides and cuprous halides (Christy and Dobbs, 1967).

The results are analysed in the same way as those for silver salts, taking into account the fact that the *lattice defects* in these crystals are of the *Schottky-type*.

The experimental values are given in Fig. 6.

The values of ξ for $\xi \gg 1$, i.e. in the small temperature range, are not known. A *non-association between* Pb^{++} *and the next nearest neighbour cation vacancy* is assumed by the author to analyse the variations of $\Delta\theta(\xi)$ isothermal curves. It matches wery well with the theoretical curves calculated by choosing C_+^0, h, S/k and Q, and this leads to the following conclusions:

1. there is good agreement (Fig. 7) with the theoretical expression (32a) applied to the Schottky disorder.

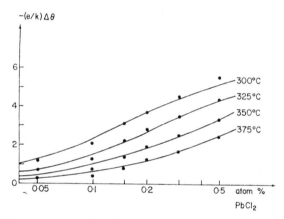

Fig. 7. The difference between the thermoelectric powers of $TlCl - PbCl_2$ and pure $TlCl$ (Christy and Dobbs, 1967).

2. we can calculate $Q_M^+ + Q_X^+$.

The values used and obtained are given in Table IV.

TABLE IV.   Enthalpy h and entropy S of defect formation and heats of transport according Christy's and Dobbs' (1967) evaluation

TlCl	$h(eV)$	S/k	$Q(eV)$	$Q_M^+ + Q_X^+(eV)$
non-association assumption	1·34	10·3	−0·11	1·45 independent of T
association assumption	1·5	14		2

3. The *non-independence of Q* on the temperature allows the author to assume identical behaviour by the individual heats of transport. It is possible to analyse the values of the thermoelectric power of the pure salt. accounting for the value of Q measured from $\Delta\theta(\xi)$ and for $S_{T1+}{}^{B}$, estimated by Wagner's approximation. We then deduce—$Q_{M^+} + TS_+^o$ and $Q_X{}^+ + TS_-^o$ whose linear variations with temperature give the results shown in Table V.

TABLE V.   Individual heats of transport estimated

TlCl	$Q_M{}^+$(eV)	$h_+{}^m$(eV)	$Q_X{}^+$(eV)	$h_-{}^m$(eV)	S_+^o meV °C^{-1}	S^o meV °C^{-1}
non association assumption	0·94	0·5	0·51	0·2	0·32	0·57

The individual heats of transport values can be given for thallium halides because their sum does not vary with the temperature.

The *order of magnitude* obtained is *in disagreement with the theoretical predictions.*

(d) *Results concerning alkali halides*

The alkali halides studied were in the intrinsic range of *Schottky-type conductors.*

The + and − carriers have very different mobility characteristics

for NaCl: $t_+ \simeq 1$

for KCl   the anion's mobility increases with the temperature but

and NaBr: remains smaller than that of the cation vacancy

for KBr: the anion's mobility may become equal to that of the cation vacancy

CsBr: $t_- > t_+$

The conductivities given contribute together to the conclusion of a *complete association between divalent ions and nearest neighbour cation vacancies in a doped crystal.*

The thermoelectric measurements need the use of a halogen electrode. Two kinds have been used: either by adopting a particular geometry of the halogen support—usually carbon—and by assuming as impervious the electrode–ionic crystal contact (e.g. Leonardi, 1969); or by adopting a contact between neutral metal, ionic crystal, and halogen atmosphere at a well-defined pressure (Hoshino and Shimoji, 1970; Jacobs and Knight, 1970).

In both cases, it is necessary to verify both Nernst's law and that the partial pressure which acts at the contact corresponds to the pressure of equilibrium. This can be done by taking, for example, a partial pressure equal to 1 and verifying that the e.m.f. of the formation cell corresponds to the calculated thermodynamic value.

Verification is also needed to check that the created centres (in this case centres F) perturb neither the homogeneous mechanisms not the interface reaction. For $P_{Br^2} = 70$ mmHg, Hoshino and Shimoji (1970) find an excess of bromine in the optical spectrum of NaBr and KBr. According to these authors, the contribution of the holes to the total current can be considered as negligible.

We have plotted in Fig. 8 all the experimental results measured by various authors for KCl, NaBr and KBr.

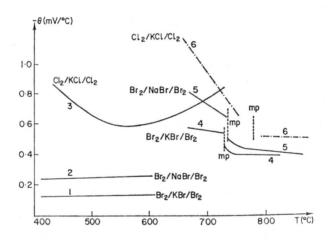

FIG. 8. The thermoelectric power of pure alkali halides measured with halogen electrode. (1, 2-Hoshino and Shimoji, 1970; 3-Jacobs and Knight, 1970;4, 5-Leonardi, 1969; 6-Bert and Dupuy, 1970).

The results obtained with the same products doped with divalent ions are generally higher in absolute value than those relative to the corresponding pure salt. They usually show a more regular variation than the ones relative to doped silver halogenures. But in this study, only one concentration of doping elements was used.

(i) *Jacobs and Knight* (1970) analyse their results obtained for KCl, KCl–SrCl$_2$ on the basis of eqn (20) for $\theta_{homogeneous}$ and of the relation equivalent to (27) for $\theta_{heterogene}$ in the case of an electrode reaction.

$$Cl^-(S) + \tfrac{1}{2}H_2(g) = HCl\,(g) + e^-(Pt)$$

$$Te\,\theta = t_+(h - Q_{MX}^+) + Q_X^+ + T(-S_{HCl}{}^{(g)} + \tfrac{1}{2}S_{H_2}{}^{(g)} + S_{Cl^-}{}^B) - TS_-^\circ$$
$$+ kT\log C_-. \quad (33)$$

The values of θ corresponding to two concentrations of cation vacancies differ only by t_+ and c_-. Their difference allows us to calculate

$$Q_{MX}^+ = 1\cdot99 \qquad 0\cdot14\,eV = Q_M{}^+ + Q_X{}^+.$$

The value of $Q_X{}^+$ is obtained in the same way as by Christy (1961), by evaluating $S_{Cl}{}^B$ from Wagner's approximation and S_-° from $S = 0\cdot16 \times 10^{-3}\,eV\,°C^{-1}$ with Einstein's approximation for a change in the frequency of the nearest ionic neighbour of the vacancy.

The resulting values, which are independent of the temperature, for the individual heats of transport are given in Table VI.

TABLE VI. Heats of transport and activation energy for KCl according to Jacobs and Knight (1970)

K	$Q_{MX}{}^+$ (eV)	$Q_M{}^+$ (eV)	$Q_X{}^-$ (eV)	$\Delta h_+{}^m$	$\Delta h_-{}^m$	ref. Δh
KCl	1·99	0·72	1·27	0·68	0·95	Aschner (1966)
						Fuller (1961)
				0·79	0·85	Bénière (1970)

(ii) *Hoshino and Shimoji* (1970 measured the thermoelectric power of KBr, KBr + BaBr$_2$ and NaBr, NaBr + BaBr$_2$. They used Wagner's approximation for the calculation of $\theta_{heterogene}$ and neglected terms due to the defects They also used the limit form of $Te\,\theta_{homogene}$ for $\zeta \gg 1$ for the computation of $\theta_{homogene}$

$$Te\,\theta_{homog}(\xi) = Q_+{}^+ + \frac{p\chi}{1+p}$$

where p = association degree

χ = free enthalpy of association

The results of this analysis are given in Table VII.

TABLE VII.   Heats of transport of a cation and anion for KBr and NaBr crystals according Hoshino and Shimoji (1970)

	Q_M^+ (eV)		h_m^+ (eV)	Q_X^+ (eV)	h_m^- (eV)
	$p = 1$ association	$p < 1$ dissociation			
KBr	$-0\cdot5 \pm 0\cdot2$	$-0\cdot3 \pm 0\cdot3$	$0\cdot83$	$1\cdot1$	$1\cdot04$
NaBr	$-0\cdot5 \pm 0\cdot1$	$0\cdot4 \pm 0\cdot1$	$0\cdot80$	$3\cdot1$	$1\cdot17$

(iii) *Leonardi* (1969) measured the thermoelectric power of the series LiBr, NaBr, KBr and CsBr. The performances of the halogene electrode, used with a pressure equal to atmospheric pressure, were established by verifying, first, the relation: $dE/dT = \theta_M - \theta_{Br^2}$ for AgBr where θ_M and θ_{Br^2} are the thermoelectric powers of AgBr, measured respectively with silver electrodes and bromine electrodes, and dE/dT is the temperature coefficient of the formation cell. Leonardi secondly verified the agreement between the measured electromotive force of the formation cell and its value computed by the thermochemical method.

This verification was applied to molten AgBr, but was unfortunately not done in the case of the solid. These results, which are given in Fig. 9 for a temperature range close to the melting point, differ from the previous measurements. Wagner's approximation was used to calculate $\theta_{heterogene}$ and it was assumed that Q_M^+ and Q_X^+ were independent of the temperature for computing $\theta_{homogene}$, whose variations are due to the change in transport numbers with the temperature.

In spite of the number of thermoelectric power measurements on ionic crystals performed during the past few years, the values for heats of transport are not yet well established, as can be seen from Tables VI, VII and VIII.

TABLE VIII.   Heats of transport of a cation and anion for NaBr, KBr, CsBr Crystals according to Leonardi (1969)

	Q_M^+ (eV)	Q_X^+ (eV)	h_m^+ (eV)	h_m^- (eV)	t^+	References
NaBr	$1\cdot40$	$3\cdot43$	$0\cdot62$	$1\cdot17$	$0\cdot844(973°K)$	Schamp and Katz (1954)
KBr	$1\cdot72$	$2\cdot09$	$0\cdot68$	$0\cdot87$	$0\cdot579(973°K)$	Rolfe (1964)
CrBr	$1\cdot36$	$0\cdot83$	$0\cdot58$	$0\cdot27$	$0\cdot561(873°K)$	Lynch (1960)

The reason for this is the large number of approximations made in calculations dealing with the interface contribution. Just as for the Soret effect results, a number of results agree with isothermal transport results, but precise values for heats of transport are not yet available. However, it seems that there is poor agreement with the theories which predict a heat of transport smaller than or equal to the activation energy in the isothermal diffusion process. Moreover, a problem arises from the local thermodynamic equilibrium of the defects. It has always been assumed that this has been verified but it can be questionable according to experimental conditions. Complementary measurements could be made with thermopiles of the type $M'/MX/M'$. However, these lead to a number of problems which will be discussed later.

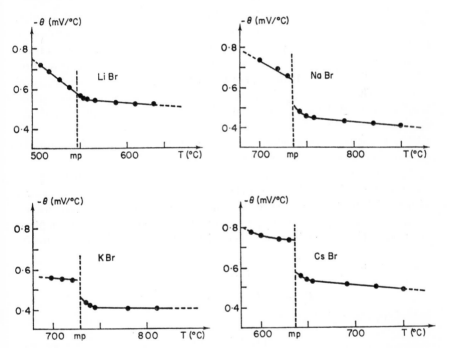

FIG. 9. The thermoelectric power of pure alkali bromides as a function of temperature (Leonardi, 1969).

4. Experimental Results Concerning Metallic Oxides Thermopiles

As far as we know, no study has yet been made of the Soret effect in metallic oxides. On the contrary, however, thermoelectric power measurements (or Seebeck effect measurements) have currently been used, together with conductivity measurements to determine the nature of charge carriers, their origin and the mechanisms which create the conduction process. Indeed, the data vary according to the conduction mechanism, i.e. purely ionic or mixed form intrinsic or extrinsic origin.

We shall give the results obtained up to now concerning the compounds for which conduction is purely ionic. We shall limit ourselves to give the main directions of interpretation of thermoelectric measurements of the semi-conductor type in order to emphasise the common points in the interpretation of thermal diffusion data.

The parameters, variations of which have been used in the experimental study, are those which influence the conduction mechanism: temperature, purity, atmosphere, crystallization state.

(a) *metallic oxides having ionic conduction*

We find common points with the previous studies relative to alkali halides. They concern the interface reaction and the estimation of its contribution to the total measured experimental value. However, the analysis of the homogeneous component has not been oriented toward the interpretation of thermal diffusion mechanisms, nor toward the part played by the structural elements. The study of the influence of the atmosphere on metallic oxide conduction needed the development of experimental techniques for determining the conductivity and measuring the polarization (Robert, 1957; Kleitz, 1970). The thermoelectric power benefited from it, at least for the electrochemical aspect, i.e. the interface reaction.

If purely ionic conduction is assured by O^{--}, the electrode reaction for the system metallic oxide–oxygen electrode will give a variable contribution with partial pressures of oxygen. We suppose that into the contact between electrode (phase 1) and oxide (phase 2) there is production or consumption of oxygen. The electrochemical potential for O^{--} into two phases must be the same.

It results in $\bar{\mu}^1_{O--} = \bar{\mu}^2_{O--}$

$$\mu^1_{O--} - 2F\phi_1 = \mu^2_{O--} - 2F\phi_2.$$

Indeed the electrode reaction is

$$\tfrac{1}{2}O_2 + 2e \rightleftharpoons O^{--}$$

and

$$\mu^1_{O--} = \tfrac{1}{2}\mu^1_{O_2} + 2\mu_e{}^1.$$

It results in

$$\tfrac{1}{2}\mu_{O_2}{}^1 + 2\mu_e{}^1 - 2F\phi_1 = \mu^2_{O--} - 2F\phi_2$$

$$F(\phi_1 - \phi_2) = \tfrac{1}{4}\mu^1_{O_2} + \mu_e{}^1 - \tfrac{1}{2}\mu^2_{O--}$$

$$F\theta_{\text{heter}} = -\tfrac{1}{4}S_{O_2}(p) + \tfrac{1}{2}S^1_{O--} - S_e.$$

The chemical potential μ_{O_2} is of the form $\mu(p_0) + RT \log (p_0/p)$, it gives:

$$S_{O_2}(p) = S_{O_2}(p_0) + R \log \frac{p}{p_0}$$

$$F\theta_{\text{heter}} = - \tfrac{1}{4}S_{O_2}(p_0) - \tfrac{1}{4}R \log \frac{p}{p_0} + \tfrac{1}{2}S_{O^{--}}^1 - S_e$$

If the electrode is reversible with respect to the oxygen ion, the thermo-electric power will vary linearly with $\log \rho$. This was verified experimentally by Fischer (1967) for zirconia stabilized by chalk or ytterbium oxide (Fig. 10).

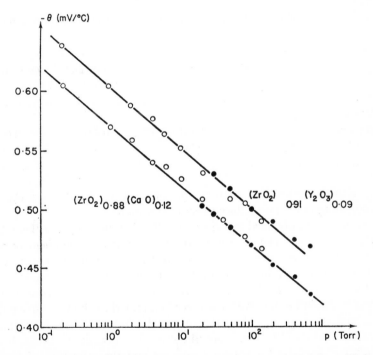

FIG. 10. The thermoelectric power of $(ZrO_2)_{0.88}$ $(CaO)_{0.12}$ and $(ZrO_2)_{0.91}$ $(Y_2O_3)_{0.09}$ as a function of oxygen pressure (Fischer, 1967).

The theoretical slope of $\theta = f (\log p)$ is equal to $0.0216\,\text{mV}\,^\circ\text{C}^{-1}$. The experimental values are respectively $0.0219\,\text{mV}\,^\circ\text{C}^{-1}$ for $(ZrO_2)_{0.88}$ $(CaO)_{0.12}$ and $0.0220\,\text{mV}\,^\circ\text{C}^{-1}$ for $(ZrO_2)_{0.91}$ $(Y_2O_3)_{0.09}$.

The homogeneous thermoelectric power is given according to the authors, by the relation

$$TF\theta_{homog} = \tfrac{1}{2}Q_O^{+}{}_{--}$$

assuming that the ions O^{--} transport the current.

This expression is analogous with the global expressions obtained by Holtan (1953) without considering the structural elements.

The evaluation of $Q_O^{+}{}_{--}$ from experimental data meet the same difficulty as alkali halides: the evaluation of the molar partial entropy of O^{--} in the oxide.

Table IX shows the estimations of Fischer, obtained by using Wagner's approximation (modified by Pitzer to account for the mass difference).

TABLE IX.   Heat of transport of oxygen ion evaluated from zirconia stabilized thermopiles

	$T \, °K$	$\theta \, mV \, deg^{-1}$	t_-	$S_{O^{--}} \, mV.mol^{-1} \, deg^{-1}$	$Q_O^{+}{}_{-} \, (eV)$
$(ZeO_2)_{0.88} \, (CaO)_{0.12}$					
$159 \, torr = p_A$	1073	-0.460	1	0.421	<0.01
$(ZrO_2)_{0.91} \, (Y_2O_3)_{0.09}$	—	-0.492	1	0.357	-0.06

$S_{O^{--}}$ is the transported entropy of the ion $O^{--} = S_{O^{--}} + Q_O^{+}{}_{-}/T$ corresponding to the non-reduced heat of transport.

Its value can be obtained experimentally but does not have any theoretical significance directly related to transport mechanisms in a gradient of temperature. The orders of magnitude obtained for the heats of transport are small. The minus sign implies—according to the authors—the confirmation that there is diffusion of negative carriers by a vacancy mechanism. This interpretation, in keeping with the preceding remark, is however debatable.

The analysis of $\theta_{homogene}$ along the lines of Howard and Lidiard (1957), and Allnatt and Jacobs (1961) was carried out by Grunberg et al. (1969a, b) for zirconia stabilized by CaO, taking into account the migration of oxygen vacancies which have a positive charge and of Ca^{--} impurities which are substituted on zirconia sites. The latter ones migrate as neutral tight bound complexes: imputiry–cation vacancy.

$$\beta = \frac{\text{complex concentration}}{\text{Ca concentration}} = \text{association degree.}$$

The expression given for the thermoelectric power is:

$$FT\theta = -\left(Q_v - \frac{\beta}{1+\beta}H_p'\right) - \left(TS_{--}^{\circ} - kT\log C_{--} + H_p'\frac{\beta}{1+\beta}\right)$$
$$\text{I}$$

$$+ (\tfrac{1}{2}TS^{\circ}(O_2) - kT\log p_{02}) - S_{0--}^{\circ}$$
$$\text{II}\qquad\qquad\qquad\text{III}$$

The first bracket is the $TF\theta_{\text{homogene}}$ expression corresponding to a conduction due mainly to anion vacancy. The other terms correspond to the interface reaction in terms of structural elements, including especially the mobile ones i.e. vacancies $\square_{O--}$

$$\tfrac{1}{2}O_2 + 2e^- + \square_{O^{2-}} \rightleftharpoons O_0.$$

I corresponds to the enthalpy of the anionic vacancy S_{O--} i.e. of the defect involved in the interface reaction, II corresponds to the electrode (neglecting the term due to electrons), III is the chemical potential of the oxygen ion in the bulk of the crystal.

This last expression was not used to interpret experimental data. It would certainly be possible, by analogy with silver halides and thallium halides, with different concentrations and maintaining a purely ionic conduction, to obtain some data concerning the nature of carriers and thermal diffusion characteristics, independently of interface terms.

Let us note an application that Guillou et al. (1969a, b) deduced from their measurements: the electrothermal oxygen pump which bases the enriching of a gas with oxygen on the transport of oxygen in stabilized zirconia in a temperature gradient.

This process could be rendered more worthwhile by a proper choice of compounds whose conduction would be mainly ionic for one type of charge carrier with high heat of transport.

(b) *Some aspects of the Seebeck effect in metallic oxides with non-ionic conduction*

Atmosphere and temperature are two important parameters which influence the thermoelectric results relative to metallic oxides with mixed conduction.

These results are generally parallel to those for conductivity, and are a useful tool for determining the charge carriers' nature and origin. They may be ionic, and may result from the extrinsic process (doping), or from a lattice disorder which favours a certain concentration of the giving or the accepting type; or they may be due to electronic transitions.

Usually, the conductivity characteristics (polarization effect, order of magnitude of the mobility, activation energy) are used to choose between these four formation processes.

On the contrary, however, the mechanism which makes the conduction process take place is not as clear as in the case of classical semiconductors.

There exist two theories: one by Morin (1959) based on a band scheme which predicts an order of magnitude for the thermoelectric power given by

$$\theta = \frac{k}{e}\left| A + \frac{E_F}{kT} \right| \simeq \frac{k}{e}\left(A + \log\frac{N_v}{p} \right).$$

N being the effective state density of the valence band and p the density of carriers. A is a term related to kinetic energy whose form depends on the conduction mechanism.

The second, which is the "hopping model" or "polaron", developed by Heikes and Johnston (1957) and Heikes et al. (1963) assumes that the carrier is trapped in its site by its own distortion effect. Its motion is possible only through a thermally activated mechanism which can take place only if the distortions around the initial and final sites are equivalent.

This model was compared with the lattice defects' diffusion mechanism and used especially for interpreting the thermoelectric power measurements, on VO_{2+x} by de Coninck and Devreese (1969) and on FeO by Emi and Shimoji (1968).

In the case of FeO the *charge carrier is a positive hole* created by the dissolution reaction of oxygen in FeO

$$\tfrac{1}{2}(O_2)(g) = O^{--} + \square_{2+} + 2 \oplus$$

$\square_{2+}$ being a Fe^{2+} vacancy.

The slope U_s of the straight line $\theta = f(1/T)$ is given by $U_s = Q_\oplus{}^x + \Delta H_\oplus^\circ$.

$Q_\oplus{}^*$ is the reduced heat of transport of the hole, and $\Delta H_\oplus^\circ$ its heat of formation.

The apparent activation energy U_σ for the conductivity is equal to

$$U_\sigma = \Delta H_\oplus^\circ + \Delta H_\oplus{}^x$$

where $\Delta H_\oplus{}^*$ is the activation enthalpy.

For FeO: $\Delta H_\oplus{}^x \simeq Q_\oplus{}^x$ according to Rice and Allnatt's theory (Emi and Shimoji, 1968).

For UO_{2+x} $U_\sigma > U_s$, hence $Q_\oplus{}^x < \Delta H_\oplus{}^x$, according to the Wirtz theory (de Coninck and Devreese, 1969).

There is, however, a certain ambiguity about the interpretation of thermo-electrical results because we must consider the interface. Usually, the electrode influence is not taken into account for these compounds. In the opposite case, there results a difference in the analysis which is well illustrated by the interpretation of the measurements done on FeO by Tannhauser (1967) and by the interpretation of these same results by Emi and Shimoji (1968). The diffusion mechanism assumed by the two authors is a "hopping" mechanism through positive holes. But, the change in sign of θ with temperature is explained in two different ways: the first author explained it by a change of sign of the carriers and the second one by a variation of the heat of transport and consequently by a variation of the activation enthalpy for jumping into the hole which might be counterbalanced by the interface contribution.

III. Information Concerning Heats of Transport by Thermopiles *M'/MX/M'*

A certain number of authors determined the thermoelectric power of ionic crystals with metal electrodes of various natures.

The results are characterized by high dispersion and instability. The first question which a electrochemist asks himself concerns the definition of the heterogeneous contact M'/MX. Does an equilibrium exist or not? If one does exist, then by which mechanism and by which entity is it governed?

A.   Nature of the Contact

Several assumptions have been made about the origin of the heterogeneous contribution of θ.

1. *There exists a thermodynamic equilibrium between the two phases* (i.e. ionic and metallic phases) which is assured by

(i) *the ions*: According to Howard (1958), there is formation of atoms of the metallic compound at the electrode surface by transfer of electrons from metal. The resulting heterogeneous contribution to the thermoelectric power is equivalent to that obtained with reversible electrodes.

(ii) *the electrons*: According to Allnatt and Jacobs (1961), an electron transfer can take place between the crystal and the metal; in MX, the electrons split up between the conduction band and the centres F formed from anion vacancies. There is consequently a non-stoichiometry in the crystal arising from the excess of electrons. However, the calculation of the chemical potential of the electrons gives—according to Jacobs and Maycock (1966)—a value for the number of electrons arising from non-stoichiometry that is much too high compared to the experimental results.

TABLE X.   Summary references for $M'/MX/M'$ measurement.

Material	Electrode		References
Cd/I_2, Pb/I_2, Th/I_2 Cd/Br_2, Pb/Br_2, Th/Br_2, Cd/Cl_2, Pb/Cl_2, Th/Cl_2, AgI, $NaNO_3$, KNO_3	Polycrystalline	Pt	Thiele (1925)
NaCl, MCl	Polycrystalline	Pt, Pt-Rh-Au.	Nikitinskaya–Morin (1955)
LIF		Ag	Curien–Mihailovic (1959)
KCl	Monocrystalline	Pt	Allnatt and Jacobs (1962)
NaCl	Polycrystalline	Pt	Christy–Hsuen–Muller (1963)
Li_2SO_4	Polycrystalline	Pt, Zu	Kvist, Lunden (1964)
KCl	Monocrystalline	Pt	Jacobs Maycock (1966)
NaCl, NaI Ka, KI CsCl	Polycrystalline	W	Greenberg (1966)
NaCl	Monocrystalline	Pt	Allnatt Chadwick (1967)
NaBr	—	Pt, C	Shimoji Hoshino (1967)
KBr	—	Pt, C	
IK	—	C Pt	
AgBr	Polycrystalline	Au	J. Dupuy (1968)
AgCl		C W	J. Bert J. Dupuy (1970)
KCl	Monocrystalline	Pb_f C	J. Bert J. Dupuy (1970)

Allnatt and Chadwick (1967a) tried to generalize this model by assuming non-stoichiometry due to a lack of electrons rather than to an excess. There would then be formation of a centre V_k instead of a centre F, with the possibility of formation of a centre $(V_k^x)_2$. The degree of non-stoichiometry calculated by the authors is high but not unreasonable. The main objection which can be raised about this mechanism concerns the existence of a centre V_k^x which is not well established in the pure crystal.

2. All the previous assumptions deal with considerations which are concerned more with the crystal than the interface. They were disproved by Jacobs and Maycock (1966) who proposed the hypothesis *that no charge crosses the interface: the electrode is assumed to be ideally polarized with respect to the crystal.*

The analysis of the heterogeneous contribution of the thermoelectric power, based on this hypothesis, was made by Allnatt and Jacobs (1968) by considering that the space charge diffused in the crystal is of the Gouy–Chapman type. It corresponds to a surface charge due to the difference in energies of formation for cation vacancies and anion vacancies.

The authors developed Gibbs' adsorption equation applied to charged species, taking into account the conditions of equality between cation and anion sites and between opposite charges on the two sides of the interface. Owing to the equilibrium between the surface and the bulk of the crystal, the surface concentration for both types of vacancy is a Boltzmann's distribution. We can compute a charge density, the excess in surface charge concentration and the potential difference between crystal and metal. We then deduce the expression of the heterogeneous thermoelectric power which is given by

$$e\theta_{\text{heter}} = -\frac{e}{e^{y/2}+1}\frac{h}{T} + \frac{\partial\mu}{\partial C_-}\frac{\partial C_-}{\partial T} \qquad (34)$$

where $y = e[(\phi_S - \phi_B)/kT]$, $\phi_S - \phi_B$ = potential difference between the surface and the bulk of the crystal h is the energy of formation of a pair of Schottky defects.

The difficulty is in the evaluation of $\phi_S - \phi_B$

For a pure crystal, the thermoelectric power results from the form of equations (20) and (34)

$$Te\,\theta = -Q_M{}^+ - t_-\,|\,h - Q_{MX}{}^+\,| + \frac{h}{1 + e^{y/2}}\cdot$$

For a doped crystal in the conditions $C_I \simeq C_+$

$$Te\,\theta = -Q_M{}^+ - \frac{\xi\chi}{1 + \xi}$$

where ζ and χ have their usual meaning $\xi = C_k/C$ and χ is the enthalpy of the association vacancy–impurity.

Remark

The general equations have been obtained by accounting for all the structure elements in the surface zone: M^+, X^- $\square^+$, $\square^-$ and complexes, but not accounting for a gradient of concentration, in the diffused zone, for the complexes.

B. Experimental Results

1. *Influence of Experimental Conditions.* This influence was studied by Dupuy (1968) on the results of thermoelectric power measurements for solid and melted silver halides made with platinum electrodes. The very important influence of the *atmosphere*, of the *purity of the material* (both crystalline and metallic material) and of the *nature of the electrode material* (W, C, Pt, Au) were particularly emphasised.

2. The two kinds of hypothesis imply different experimental results. Indeed, if *equilibrium* is ensured by an *element of the ionic compound*, we must expect a certain analogy between the results of the thermopiles $M'/MX/M'$ and $M/MX/M$ or $X^2/MX/X^2$.

This has not been verified in detail for alkali halides but the results obtained by Bert and Dupuy (1970) are very different, depending on whether the thermoelectric power of AgCl is measured with electrodes made of silver of chlorine or wrapped carbon. If *equilibrium* is ensured by an *electron transfer*, the results must be (the metallic thermoelectric effects apart) almost independent of the choice of M'.

Now, the order of magnitude and the variation with the temperature are different for thermoelectric measurements for NaBr, KBr, NaCl, KCl with C, Pt, Rh and W electrodes, as can be seen in Fig. 11.

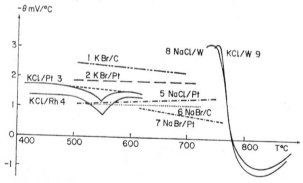

Fig. 11. The thermoelectric power of alkali alides measured with various "inert" electrodes (1, 2, 6, 7–Hoshino and Shimoji, 1967, 1968; 3, 4 Jacobs and Maycock, 1966; 5 Allnatt and Chadwick, 1967; 8 Greenberg, 1966).

If *the interface behaves like an ideal conductor*, it must fulfil certain requirements:

(i) in particular, there is no transfer of charge between metal and salt

(ii) the structure of the double layer being of the Gouy–Chapman type must not comprise any specific interactions and in particular interactions related to adsorption phenomena.

The more precise surface studies made by physical measurements like LEED, AUGE electrons, field ionic emission etc, indicate a high density of ionic surface states and the complexity of the surface interactions in halogenures of alkaline metals with a metallic substrate. It is then necessary to specify whether the interface M'/MX is influenced only by the properties of the space charge of the crystal since this is due to the defects which have the smallest formation energy, there must be very important differences between crystals of the NaCl type and crystals of the CsCl type, because the metal slightly influences the interface properties.

On the other hand, if more specific bond forces are to be expected, the impurities may considerably change the nature and density of surface states.

Usually, the conductivity measurements are done in conjunction with thermoelectric ones. They point out particularly the nature and characteristics of the lattice defects of the material studied. It would be desirable to associate polarization measurements with the previous ones; this might allow us to justify the choice of the systems as well as the interpretation of the data.

In polarization measurements, the current–potential characteristics reflect the polarization state of the interface, and hence justify the choice of the metal. The variations of capacity with the potential allow us to specify the nature of the interface polarization and its change with the temperature. Hence these measures should be a valuable tool for interpreting the thermoelectric power measurements on the basis of Jacob's model.

The values of the heats of transport for KCl, computed from thermopile measurements ($Cl^2/KCl/Cl^2$ and $Pt/KCl/Pt$) are given in Table XI. The interpretation of the measurements relative to the first thermopile is that given in Section II.B.4, and the interpretation of the results relative to the second one is based on the model of the electrode ideally polarized with respect to KCl.

With this model, the value obtained for Q^+ varies with the value chosen for $\phi_S - \phi_B$

$$\phi_S - \phi_B = -(g_+ + kT \log n_+{}^B) + (\mu_{M^+}{}^B - \mu_{M^+}{}^S).$$

The first contribution is Lehovec's (1953); but it neglects the surface states and accounts only for the gradient of vacancies. The second one accounts for the surface properties and is very difficult to evaluate. Grimley and Mott (1947) evaluate it at 0·5 V, which would give a total value equal to zero.

Allnatt and Jacobs (1968) give a value of 0·22 V which is consistent with the values obtained for θ in the small temperature range when a precipitation of impurities occurs.

TABLE XI.   Heats of transport evaluation for KCl crystal from different thermopile devices.

Electrode material		Q_M^+ (eV)	Q_X^+ (eV)	h_m^+ (eV)	h_m^- (eV)
Pt	$(\phi_S - \phi_B)_{11} = 0\cdot5$	1·13			
	$y = 0$		0·4		
	$(\phi_S - \phi_B)_{11} = (0\cdot22)_V$	0·97		0·68	0·95
Cl²		0·72	1·27		

This table shows that the disparity of the values obtained does not allow us to conclude either that the theories used for interpreting the results relative to thermopiles with reversible electrodes are inadequate, or that there is disagreement with the capicator model.

It is not the intention of this chapter to conclude that thermal diffusion is unable to give us an idea of the properties of lattice defects in ionic crystals, but the discrepancy between results obtained for the same system by different experimental approaches must be eliminated in some other way. In particular the supersaturation problem must be elucidated. Available experimental values for heats of transport will then be able to support a difficult theoretical investigation.

Acknowledgments

I would like to thank Drs. A. R. Allnatt, R. W. Christy, R. Fischer, P. W. M. Jacobs, M. Shimoji, P. Lawson and A. Kvist, who have permitted me to include their results, Prof Deporte for his comments on metallic oxides, and J. L. Crolet for his criticism of the manuscript. Lastly I wish to acknowledge the assistance of Mrs. Piquet, Mrs. Villard and Mr. Bert in the preparation of this paper.

List of Symbols

$\square_+$, $\square_-$: cationic and anionic vacancies

n_r: number of species r per unit volume

c_+, c_-: site fractions of cation and anion vacancies.

i, $+$ or 1, $-$ or 2,: interstitial cation, cation vacancy, anion vacancy.

k: complexes

j: free impurities

ζ: $\dfrac{\text{concentration of free cationic vacancies in doped crystal}}{\text{concentration of free catonic vacancies in pure crystal}}$

$c_+{}^0$: site fraction of cation vacancies in pure crystal

h: free enthalpy of formation of the defect pair

χ: free energy of association

μ_r, S_r: chemical potential and partial entropy of r

$\bar{\mu}_r$: electrochemical potential of r

ϕ: internal potential

μ_r°, S_r°: chemical potential and entropy change on introducing species of r into a crystal without defect

p: partial pressure

$Q_k{}^*$: reduced heat of transport of species k

$'$: the quantities referred to lattice

$+$: design linear independent quantities;

$\qquad X_k{}^+ =$ thermodynamic force

$\qquad Q_k{}^+ =$ heat of transport of k, only observable

θ_{heter}, θ_{homog}: heterogeneous and homogeneous part of the thermoelectric power

t_r: transport number of r

Δh_m, or $h_k{}^m$: heat of activation for motion of a species k

References

Allnatt, A. R. (1970). Europhys. Conf. "Atomic Trans, in Solids and liquids" Marstrand Sweden (June).

Allnatt, A. R. and Chadwick, A. V. (1966). *Trans. Faraday Soc.* **62**, 1726–35.

Allnatt, A. R. and Chadwick, A. V. (1967a). *J. chem. Phys.* **47**, 2372–78.

Allnatt, A. R. and Chadwick, A. V. (1967b). *Trans. Faraday Soc.* **63**, 1929–42.

Allnatt, A. R. and Chadwick, A. V. (1967c). *Chem. Rev.* 681–705.

Allnatt, A. R. and Jacobs, P. W. M. (1961). *Proc. R. Soc. Lond.* **260A**, 350–369.

Allnatt, A. R. and Jacobs, P. W. M. (1962). *Proc. R. Soc. Lond.* **267A**, 31–44.

Allnatt, A. R. and Jacobs, P. W. M. (1968). *Can. J. chem.* **46**, 1635–41.

Allnatt, A. R. and Pantelis, P. (1968). *Trans. Faraday Soc.* **548**, 2101–2105.

Adda, Y. and Philibert, J. (1966). La diffusion dans les solides BSTN.

Allnatt, A. R. and Rice, S. A. (1960). *J. chem. Phys.* **33**, 573–78.

Aschner, J. F. (1954). Thesis Univ. Illinois.

Bénière, F. (1970). Thesis Paris–Orsay Serie A 683.

Bert, J. (1970) Thesis Special. Univ. Lyon.

Bert, J. and Dupuy, J. (1970). Europhys. Conf. Marstrand Sweden.

Brinkman, J. (1954). *Phys. Rev.* **93**, 345.

Christy, R. W. (1961). *J. chem. Phys.* **34**, 1148–55.

Christy, R. W. (1962). Techn. Rapport. 3, VS Naval Res.

Christy, R. W. and Dobbs, H. S. (1967) *J. chem. Phys.* **46**, 722.

Christy, R. W., Fukushima, E. and Li, H. T. (1952). *J. chem. Phys.* **30**, 136–8.

Christy, R. W., Hsuen, Y. W. and Muller, R. C. (1963). *J. chem. Phys.* **38**, 1647–51.

Coninck, de R. and Devreese, J. (1969). *Phys. Stat. Solodi* **32**, 823–829.

Connan, R., Dupuy, J., Leonardi, J. and Poillerat, G. (1969). *J. chem. Phys.* p64–69.

Crolet, J. L. (1971). Thèse Orsay.

Curien, H. and Mihailovic, Z. (1959). CR. Ac. Sc. **248**, 2982–3.

Dupuy, J. (1968). Euchem Conf. Pavie (Italia) Mai.

Dupuy, J. and Ruch, J. (1966). CR Acad, Sc. **262**, 1512–5.

Emi, T. and Shimoji, M. (1968). *Act. Metall.* **16**, 1093–1100.

Evans, B. E., Swanson, L. W. and Bell, A. E. (1968). *Surf. Sci.* **11**, 1–18.

Ficks, V, B. (1961). *Sov. Phys. Sol. State* **3**, 724.

Ficks, V. B. (1964). *Sov. Phys. Sol. State* **5**, 2549.

Fischer, W. (1967). *Z. natursfors.* **22a**, 1575–81.

Fuller, R. G. (1966). *Phys. Rev.* **142**, 524.

Gerl, M. (1968). Rapport C.E.A. R3487.

Girifalco, L. A. (1962). *Phys. Rev.* **128**, 2630–8.

Girifalco, L. A. (1963). *J. chem. Phys.* **39**, 2377.

Greenberg, J. (1966). *J. electroch. Soc.* **113**, 937–40.

Grimley. T. B. and Mott, N. F. (1947). *Trans. Faraday Soc.* **11**, 3a.

de Groot, S. R. (1942). *Physica* **9**, 699.

de Groot, S. R. (1947). *J. phys. Rod.* **87**, 193–200.

de Groot, S. R. (1951). "Thermodynamics of irreversible Processes". North Holland Publish. Co., Amsterdam.

de Groot, S. R. and Mazur, P. (1962). "Non Equilibrium Thermodynamics". North Holland Publish. Co., Amsterdam.

Grunberg, M., Guillou, M. and Godard, G. (1969a). Rapport E.D.F. 350012.

Guillou, M., Lecante, A. and Grunberg, M. (1969b). Reunion SFT "Conversion de l'energie par les piles à combustibles et la MHD (24 Janvier).

Heikes, R. R. and Johnston, W. D. (1957). *J. chem. Phys.* **26**, 582–87.

Heikes, R. R., Maradudin, A. and Miller, R. C. (1963). *Ann. Phys.* **8**, 733–46.

Holtan, H. (1953). Thesis Utrecht.

Hoshino, H. and Shimoji, M. (1967). *J. phys. Chem. Sol.* **28**, 1169–1175.

Hoshino, H. and Shimoji, M. (1968). *J. phys. Chem. Sol.* **29**, 1431–41.

Hoshino, H. and Shimoji, M. (1969). *J. phys. Chem. Sol.* **30**, 1603–07.

Hoshino, H. and Shimoji, M. (1970). *J. phys. Chem. Sol.* **31**, 1553–64.

Howard, R. E. (1957). *J. chem. Phys.* **27**, 1337–81.

Howard, R. E. (1958). Ph .Thesis Oxford Univ.

Howard, R. E. and Lidiard, A. B. (1957a). *Phil. Mag.* **2**, 1462–67.

Howard, R. E. and Lidiard, A. B. (1957b). *Dis. Faraday Soc.* **23**, 113–21.

Howard, R. E. and Lidiard, A. B. (1964). *Reports on Progress in Physics* **27**, 61.

Huntington, H. B. (1968). *J. phys. Chem. Solid.* **29**, 1641–1651.

Jacobs, P. W. M. and Knight, P. C. (1970). *Trans. Faraday Soc.* **569**, 1227–1235.

Jacobs, P. W. M. and Maycock, J. W. (1966), *Trans. Aime* **236**, 165.

Kleitz., M. (1970). Thèse univeristé de Grenoble.

Kröger, F. A. (1964). "Chemistry of Imperfect Crystals". North Holland Publish. Co. Amsterdam.

Kröger, F. A., Stieles, F. and Vink, H. J. (1959). Philips Res, Report. **14**, 557–601.

Kvist, A. (1964). *Z. fur Natursf* **19a**, 1159–LL60.

Kvist, A. and Lunden, A. (1964). *Z. fur Natursf* **19a**, 1058–64.

Kvist, A., Randsalu, A. and Svenson, I. (1966). *Z. fur Natursf* **21a**, 284.

Lawson, A. W. (1962). *J. appl. Phys.* **31**, 466–73.

LeClaire, A. D. (1954). *Phys. Rev.* **93**, 344.

Lehovec, K. (1953). *J. chem. Phys.* **21**, 1123–28.

Leonardi, J. (1969). Thèse d'Etat, Srtasbourg.

Lidiard, A. B. (1957). *Handbuch Phys.* **20**, 246–349.

Manning, J. R. (1968). "Diffusion kinetics for atoms in crystals". D. Van Nostrand Cie. Inc.

Morin, F. J. (1959). *Phys. Rev. Letters* **3**, 34.

Oriani, R. A. (1961). *J. chem. Phys.* **34**, 1773–1777.

Patrick, L. and Lawson, A. W. (1954), *J. chem. Phys.* **22**, 1492–95.

Poillerat, G. (1970). Thèse d'Ingénieur Docteur Strasbourg.

Reinhold, H. (1957). See Allnatt, A. R. and Chadwick, A. V.

Rice, S. A. (1958). *Phys. Rev.* **112**, 804–11.

Robert, G. (1957). Thèse d'Etat, Grenoble.

Rolfe, J. (1964). *Canad. J. phys.* **42**, 2195.

Ruch, J. and Dupuy, J. (1965) Cr. **261**, 957–960.

Schamp, H. W. and Katz, E. (1954). *Phys. Rev.* **94**, 828.

Shimoji, M. and Hoshino, H. (1967). *J. phys. Chem. Sol.* **28**, 1155–67.

Schottky, G. (1965). *Phys. Stat. Sol.* **8**, 357.

Tannhauser, D. S. (1962). *J. phys. Chem. Sol.* **23**, 25–34.

Thiele, J. (1925). *Phys. Zeits* **8**, 321–29.

Tyrrell, H. J. V. (1961). "Diffusion and heat flow in liquids". Butterworth Co. Ltd., London.

Ubbelohde, A. R. (1957). *Quem. Rev.* **11**, 246,

Weininger, J. L. (1964). *J. electrochem. Soc.* July 769–74.

Wirtz, K. (1943). *Physikalis Zeitz* **44**, 221–231.

Wirtz, K. (1944). *Zeitsf für Phys.* **124**, 482–500.

Wirtz, K. and Hiby, J. W. (1943). *Physik Zeit.* **44**, 369–382.

17. THERMODYNAMIC EQUILIBRIUM DIAGRAMS

W. L. Worrell

School of Metallurgy and Materials Science, University of Pennsylvania, Philadelphia, Pennsylvania, U.S.A.

and

J. Hladik

Laboratorie de Physique Générale, Faculté des Sciences de l'Universite de Paris, France.

I. Introduction

Knowledge of the high-temperature thermodynamic properties of electrolytes is essential for the prediction and understanding of their chemical stability in various environments and in the presence of various electrodes. One of the easiest ways to picture and to summarize such thermodynamic data is to construct a Pourbaix–Ellingham or a Pourbaix diagram for the particular electrolyte. It is the purpose of this chapter to review briefly the construction of such diagrams and to discuss their applications in investigations using solid and liquid electrolytes.

II. Pourbaix–Ellingham Diagrams

A. Thermodynamic Stability

At constant temperature and pressure, the thermodynamic stability of any solid compound is indicated by the magnitude of $\Delta G°$, the standard free energy. The standard free energy of formation of any oxide, for example, is related to the oxygen pressure in equilibrium with the pure metal and its oxide by:

$$\Delta G° = \pm RT \ln p_{O_2}. \tag{1}$$

Thus a method frequently used to compare the thermodynamic stability of various oxides over an extended temperature range is to plot $\pm RT \ln p_{O_2}$ (or $\Delta G°$) as a function of temperature (Ellingham, 1944; Elliott and Gleiser, 1960; Richardson and Jeffes, 1948). Similarly, in a complex system, one component of which is oxygen, Pourbaix and Rorive-Boute (1948) have expressed reaction equilibria in terms of two variables: $- RT \ln p_{O_2}$ and T.

At the end of a recent paper by Guillou et al. (1968), Pourbaix has discussed the application of such diagrams in research using solid-state electrolytes. These diagrams previously have been designated Pourbaix–Ellingham diagrams (Worrell and Chipman, 1964). Such diagrams have been used to picture the thermodynamically stable regions in various metal–carbon–oxygen ternary systems and to estimate the minimum temperature necessary to obtain a metallic phase or a specified carbide in high-temperature carbothermic reduction processes (Worrell and Chipman, 1964; Worrell, 1965).

With two independent components such as metal and oxygen the phase rule requires that the degrees of freedom are four minus the number of phases. Four phases, therefore, specify a point; three phases determine a line, and two phases define an area in phase space. Noting the presence of a vapour phase, three condensed phases will meet at a point, two condensed phases will specify a line, and one condensed phase will define a region in a two-dimensional Pourbaix–Ellingham diagram.

B. Zirconia and Thoria Electrolytes

At this point, it is helpful to consider specific oxides. Zirconia (ZrO_2) and thoria (ThO_2) are chosen as examples because of their many applications as solid-oxide electrolytes. It should be emphasized that the thermodynamic calculations summarized below, strictly apply only for the pure oxides. Solid-oxide electrolytes are solid solutions of either zirconia or thoria with either calcia (CaO) or yittria (Y_2O_3). However, the effect of the small decrease in the thermodynamic stability of ThO_2 or ZrO_2 in solid solution is negligible when compared with the large magnitude of their standard free energy. Thus the Pourbaix–Ellingham diagrams which are calculated for the pure oxides are also applicable for solid solutions such as ZrO_2 (CaO), ZrO_2 (Y_2O_3), ThO_2 (CaO) and ThO_2 (Y_2O_3).

The standard free energy of formation equations for ZrO_2 and ThO_2 between 300 and 2000°K (Coughlin, 1954) are:

$$Zr(s) + O_2(g) \rightarrow ZrO_2(s)$$

$$\Delta G_f^\circ = -259{,}700 + 43{\cdot}6T \tag{2}$$

$$Th(s) + O_2(g) = ThO_2(s)$$

$$\Delta G_f^\circ = -292{,}500 + 45{\cdot}8T. \tag{3}$$

Using these equations, Pourbaix–Ellingham diagrams were constructed for ZrO_2 and ThO_2 and are shown in Figs 1 and 2, respectively.

The melting point of zirconium is 2130°K, while that of thorium is 2023°K (Kubaschewski et al., 1967).

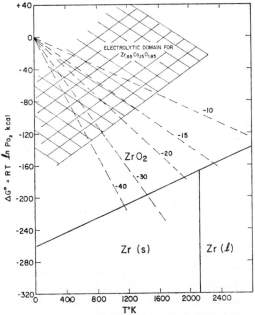

FIG. 1. Pourbaix–Ellingham diagram for ZrO_2.

The dotted lines shown in Figs 1 and 2 are those calculated for the indicated equilibrium oxygen pressures of 10^{-10}, 10^{-15}, 10^{-20}, 10^{-30} and 10^{-40} atm. These lines are very useful in estimating the chemical stability of zirconia-based and thoria-based electrolytes when they are in contact with highly reducing electrodes. For example, if the equilibrium oxygen pressure of a metal–metal oxide electrode is 10^{-30} atm at 1500°K, Figs 1 and 2 indicate that this electrode would reduce ZrO_2 to zirconium but that ThO_2 would be stable. Although a thoria-based electrolyte is thermodynamically stable under these conditions, it shows appreciable n-type electronic conductivity.

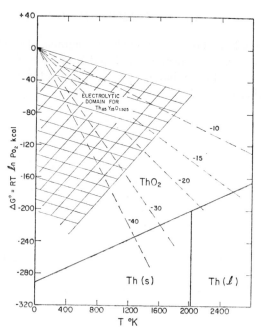

FIG. 2. Pourbaix–Ellingham diagram for ThO_2.

C. Electrolytic Domains

It should be emphasized that the regions of application for most solid-state electrolytes are dictated not only by their thermodynamic stability but also by their degree of ionic conductivity. Patterson (1971a, b) has recently discussed and summarized the high-temperature conductivity data for a number of solid-state electrolytes. He has defined an electrolytic domain as the region in which the ionic contribution to the total conductivity is 99% or greater. The electrolytic domain for $Zr_{0.85}$ $Ca_{0.15}$ $O_{1.85}$ and for $Th_{0.85}$ $Y_{0.15}$ $O_{1.925}$ as given by Patterson (1971a, b) has been superimposed as cross-hatched areas on Figs 1 and 2 respectively. The electrolytic domains illustrated in Figs 1 and 2 are strictly applicable only for the indicated composition of the solid-oxide electrolyte. In general, the electrolytic domain will shift with composition because the ionic conductivity changes with composition or vacancy concentration.

Any successful application of the $Zr_{0.85} Ca_{0.15} O_{1.85}$ or the $Th_{0.85} Y_{0.15} O_{1.925}$ electrolyte must be in the electrolytic domain region illustrated in Figs 1 and 2. Consider our previous example of a metal–metal oxide electrode which has an equilibrium oxygen pressure of 10^{-30} atm at 1500°K. Figures 1 and 2 indicate that the lower limit of the electrolytic domain region is $10^{-12.4}$ and 10^{-16} atm for $Zr_{0.85} Ca_{0.15} O_{1.85}$ and $Th_{0.85} Y_{0.15} O_{1.925}$ respectively. Thus either electrolyte would exhibit n-type electronic conductivity at 1500°K when p_{O_2} is 10^{-30} atm.

Finally, it is of interest to compare the width of the electrolytic domain for these two solid-oxide electrolytes at a particular temperature. For $Zr_{0.85} Ca_{0.15} O_{1.85}$, Fig. 1 indicates that at 1200°K the electrolytic domain extends from an equilibrium pressure of 10^{+8} atm to $10^{-19.3}$ atm. For $Th_{0.85} Y_{0.15} O_{1.925}$, Fig. 2 indicates that at 1200°C the electrolytic domain extends from 10^{-6} atm to $10^{-26.3}$ atm. Thus the $Zr_{0.85} Ca_{0.15} O_{1.85}$ electrolyte can be used successfully in a region of p_{O_2} between 10^8 and 10^{-6} atm where the thoria-based electrolyte would exhibit p-type electronic conductivity. On the other hand, the $Th_{0.85} Y_{0.15} O_{1.925}$ electrolyte is applicable between $10^{-19.3}$ and $10^{-26.3}$ atm p_{O_2} where the zirconia-based electrolyte would exhibit n-type electronic conductivity.

III. Pourbaix Diagrams

A. Redox Potentials

1. Potential Scale

Chemical and electrochemical reactions in many solvents and molten salts have been reviewed by Charlot and Tremillon (1963). They showed that the electrochemical concepts used in aqueous solution may be transposed to inorganic solvents and molten electrolytes.

Some of these concepts have also been used for solid electrolytes but no Pourbaix diagrams have yet been used. Therefore, it seems interesting to construct theoretical diagrams of this kind for solid electrolytes (Hladik, 1970). Before beginning this work we shall thoroughly review the necessary principles.

The oxido-reduction reactions are electron exchange reactions between a reductant of system 1 and the oxidant of system 2:

$$Red_1 \rightleftarrows Ox_1 + n_1 e$$

$$\underline{Ox_2 + n_2 e \rightleftarrows Red_2}$$

$$n_1 Ox_2 + n_2 Red_1 \rightleftarrows n_1 Red_2 + n_2 Ox_1.$$

In an electrolyte where a free, solvated electron does not exist, a redox system:

$$Ox + ne \rightleftarrows Red$$

is characterized by a redox potential E, defined by the Nernst formula:

$$E = E^\circ + \frac{RT}{nF} \log_e \frac{[Ox]}{[Red]}.$$

E° is the standard potential characteristic of a given system, and is measured by taking an arbitrary origin. A potential scale is obtained by plotting the E° values of each redox system on a linear diagram.

In addition the equilibrium constant of the reaction may be calculated. For example, consider the two following systems:

$$a Ox_1 + ne \rightleftarrows a\,Red_1 \qquad E_1$$

$$\underline{b\,Red_2 \rightleftarrows bOx_2 + ne \qquad E_2}$$

$$a Ox_1 + b\,Red_2 \rightleftarrows a\,Red_1 + b\,Ox_2.$$

The equilibrium constant:

$$K = \frac{[Red_1]^a\,[Ox_2]^b}{[Ox_1]^a\,[Red_2]^b}$$

is then related to the redox potential by:

$$\log K = (E_1 - E_2)\frac{nF}{RT}$$

and to the free energy ΔG of the reaction, by $RT \log K = -\Delta G$

2. Potential Scale Limits

Consider the case of NaCl electrolyte which can be reduced.

$$Na^+ + e \rightarrow Na \qquad E^\circ(25^\circ C) = 3\cdot 98\,V$$

and oxidized:

$$Cl^- - e \rightarrow \tfrac{1}{2} Cl_2 \qquad E^\circ(25^\circ C) = 0\,V.$$

The decomposition potential of NaCl is $3\cdot 98$ V. Therefore, the potential domain in which one can perform oxido-reduction reactions is limited in the NaCl solvent.

Generally, each solvent has a certain decomposition potential range corresponding to the reduction and oxidation of the solvent itself.

3. Equilibrium Diagrams

The method for calculating the equilibrium diagrams has been described by Pourbaix (1963). The equilibrium potentials concern the redox potentials and the metal dissolution potentials, as well as the gas electrode potentials. These various aspects are only three special cases of the same electrode potential which is the equilibrium potential of an electrochemical reaction.

(a) Aqueous solutions.

In the case of electrochemical reactions in aqueous solutions which may be written:

$$a\text{Ox} + c\text{H}_2\text{O} + ne^- \rightleftarrows b\,\text{Red} + m\text{H}^+ \tag{1}$$

the equilibrium condition is (at 25°C)

$$E = E° - \frac{0\cdot 0591}{n} mpH + \frac{0\cdot 0591}{n} \log \frac{[\text{Ox}]^a}{[\text{Red}]^b}.$$

Considering the potential and the *pH* as independent variables, and the logarithmic function as a parameter, we can then draw the potential $-pH$ equilibrium diagram for the electrochemical reaction (1).

(b) Ionized molten salts.

The equilibrium diagrams have been extended to the various solvents and recently, to the molten salts (Charlot and Tremillon, 1963).

For example, redox systems have been studied in LiCl–KCl with BaO variable concentrations (Delarue, 1960; Molina, 1960; Leroy, 1962). Consider, for example, the case of the vanadium oxygenized compounds in molten alkali halides. Their redox properties vary with the O^{2-} ion concentrations in the molten solution. For the V^{IV}/V^V system, let us consider the reaction:

$$\text{VO}_2{}^+ + e \rightleftarrows \text{VO}^{2+} + O^{2-}. \tag{2}$$

The equilibrium potential is given by

$$E = E° + \frac{RT}{F} \log_e \frac{[\text{VO}_2{}^+]}{[\text{VO}^{2+}][O^{2-}]}.$$

We may write, by definition:

$$pO^{2-} = -\log O^{2-}$$

therefore, the preceding equation becomes:

$$E = E^\circ + \frac{RT}{F}pO^{2-} + \frac{RT}{F}\log_e \frac{[VO_2{}^+]}{[VO^{2+}]}.$$

Considering the potential and the pO^{2-} as independent variables and the logarithmic function as a parameter, we can then draw the potential $-pO^{2-}$ equilibrium diagram for the electrochemical reaction (2).

B. Theoretical Decomposition Potentials

The decomposition potentials of solid electrolytes can be calculated from thermodynamic data. We have the following equation between the e.m.f. of the corresponding chemical cell and the free energy change ΔG

$$\Delta G = -nFE \tag{3}$$

where E is the decomposition potential.

Although experimental measurements of decomposition voltages usually do not agree with theoretical values, for several reasons (including the chemical interaction of the electrodes with the electrolyte, the production of electrodes in a non-standard state, deviations from isothermal conditions and electrode polarization) it is nevertheless important to know the theoretical values for the systems under study. These values aid in elucidating mechanisms and in interpretations of experimental observations on solid electrolytes.

The method of evaluating the standard reversible e.m.f. of cells composed of a solid electrolyte and two electrodes, one of which is reversible to the anion and the other to the cation of the solid electrolyte, follows well-known thermodynamic relations. The e.m.f. of such cells, for example:

Metal/metallic chloride/chlorine gas; is related to the free energy change for the chemical reaction.

Metal + chlorine gas → metallic chloride.

According to convention, this free energy change is the free energy of formation ΔG°, of the chloride when the chloride metal and gas are in their standard states. It is related to the heat of formation, ΔH°, by the equation:

$$\Delta G^\circ = \Delta H^\circ - T\,\Delta S^\circ \tag{4}$$

where $\Delta S°$ is the difference in the entropy of the products and the reactants of the chemical reaction. Tables of thermodynamic data give $\Delta G°$ and/or $\Delta H°$ for 298·16°K.

The following equation:

$$\frac{d(\Delta G°/T)}{dT} = \frac{-\Delta H°}{T^2} \tag{5}$$

is used to obtain $\Delta G°$ at temperatures above 298·16°K.

However, $\Delta H°$ is a function of temperature and depends on the difference between the heat capacities of the products and reactants, as expressed by:

$$\Delta H° = \Delta H_0° + \int_{298·16}^{T_2} \Delta C_p \, dT \tag{6}$$

where $\Delta H_0°$ is evaluated from the heat of formation at 298·16°K as shown below, and ΔC_p represents the sum of the heat capacities of the products of a reaction minus the sum of the heat capacities of the reactants. It is usually expressed as a function of temperature by an empirical equation:

$$\Delta C_p = \Delta a + (\Delta b)T + \frac{\Delta c}{T^2} \tag{7}$$

where a, b and c have known numerical values. Substitution of eqn. (7) in eqn. (6) and integration gives:

$$\Delta H_0° = \Delta H° - (\Delta a)T - \frac{(\Delta b)T^2}{2} + \frac{\Delta c}{T}. \tag{8}$$

Substitution of eqn. (8) in eqn. (5) and integration yields:

$$\Delta G° = \Delta H_0° - \Delta aT \ln T - \frac{(\Delta b)T^2}{2} - \frac{\Delta c}{2T} + AT \tag{9}$$

which gives $\Delta G°$ as a function of temperature. If $\Delta G°$ and $\Delta H_0°$ are known at some temperature, usually 298·16°K, the integration constant A may be evaluated.

Hamer et al. (1956) have calculated the theoretical electromotive forces for cells containing a single solid or molten chloride electrolyte. Reference to Table I shows that the relative position of the elements is nearly the same whether the electrolyte phase is solid or molten. The values obtained for the solid electrolyte (providing the solid phase is anhydrous) and those for aqueous systems differ in the free energy associated with the process of dilution of the saturated solution of the electrolyte to a solution of unit mean activity.

In another paper, some similar series are given for solid and molte
fluorides, bromides and iodides (Hamer et al., 1965). These are given, i
series arrangement, in Tables II, III, and IV. Theoretical electromotive force
for cells using an oxide electrolyte have also been calculated by Hamer (196!
and are listed in Table V.

A heavy vertical line in the tables indicates that the values to the left c
the line are for solid electrolytes while those to the right are for molte
electrolytes. When a sublimation (S), vaporization (V), or decomposition (D
temperature is given in parenthesis, the e.m.f. directly above it is for tha
temperature rather than for the one heading the column.

C. Equilibrium Diagrams in Solid Carbonates

Equilibrium diagrams of solid electrolytes may be extended by using thermc
dynamic data. In extending this treatment to solid carbonates it has bee
necessary to find a function analogous to pH, which will express the acidit
of the system. We can use, for example with molten salts, the function:

$$pCO_2 = -\log_{10} [CO_2]. \tag{1(}$$

At first, it is desirable to establish a potential scale in solid carbonates
The oxidation of the carbonate anion may be used as reference potential
we may write:

$$CO_3^{2-} - 2e \rightarrow CO_2 + \tfrac{1}{2} O_2.$$

If the CO_3^{2-} activity is close to unity, we have:

$$E_{ref} = E^\circ - \frac{2\cdot3}{2F} RT \log_{10} [CO_2] [O_2]^{1/2}. \tag{11}$$

The redox potential will depend on the pressures of carbon dioxide an
oxygen in the gas phase. A reference scale of potential can be obtained b
choosing $E_{ref} = 0$ for the mixture of CO_2 and O_2 at 1 atm total pressure
with $CO_2 : O_2 = 2$. Substituting into the eqn. (11), with the appropriat
values, one has at 25° C:

$$E_{ref} = 0\cdot0123 - 0\cdot0296 \, p \, CO_2 + 0\cdot0148 \log [O_2]. \tag{12}$$

Figure 3 is the diagrammatic representation of eqn. (12) where the depend
ence of E_{ref} on pCO_2 is plotted. Each line corresponds to a different partia
pressure of oxygen, the values of $\log [O_2]$ being plotted on the right side.

There is a lower limit of electroactivity in the electrolyte. This may com
about by the decomposition of the solid electrolyte:

$$M_2CO_3 \rightarrow 2M + CO_2 + \tfrac{1}{2} O_2.$$

The appearance of alkali metal at unit activity will hold the redox potential at:

$$E_{M/M^+} = E^{\circ}_{M/M^+} + 2 \cdot 3 \frac{RT}{F} \log_{10} [M^+].$$

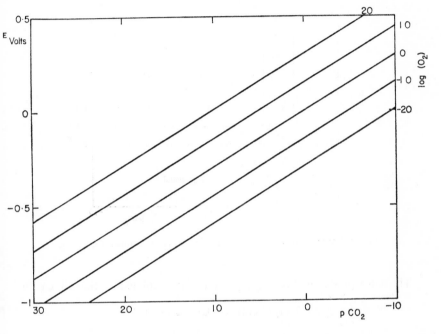

FIG. 3. Oxidation of the carbonate anion. Potential scale, E versus pCO_2.

If, for example, the solid is a pure sodium carbonate, the lower limit of electroactivity is a horizontal line at the standard potential of sodium. For the reaction:

$$Na_2 CO_3 \text{ (solid)} \rightarrow 2Na \text{ (solid)} + CO_2(\tfrac{2}{3} \text{ atm}) + \tfrac{1}{2} O_2(\tfrac{1}{3} \text{ atm})$$

the free energy change is:

$$\Delta G^{\circ} = 156 \cdot 14 \text{ kcal.}$$

Hence:

$$E^{\circ}_{Na/Na^+} = - 3 \cdot 38 \text{ V.}$$

The E/pCO_2 grid provides a suitable framework for the electrochemical behaviour of the different metals. Figure 4 is the silver diagram and its construction exemplifies the general treatment for other temperatures and metals. This diagram is divided into three regions in which only one silver phase may be present at unit activity.

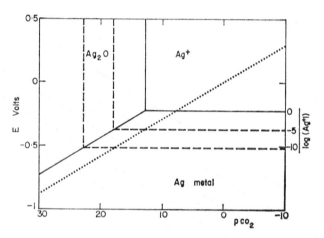

FIG. 4. Pourbaix diagram for silver in carbonates. Temperature, 25°C.

The redox potential of the Ag/Ag$^+$ system is calculated for the cell reaction:

$$Ag_2CO_3 \text{ (solid)} \rightarrow 2Ag + CO_2 + \tfrac{1}{2}O_2$$

the free energy change being:

$$\Delta G° = 10\cdot22 \text{ kcal.}$$

Hence, the standard potential of silver is:

$$E°_{Ag/Ag^+} = -0\cdot221 \text{ V}$$

and the Ag/Ag$^+$ electrode potential, relative to reference scale, is:

$$E_{Ag/Ag^+} = -0\cdot221 + 0\cdot0296 \log [Ag^+]. \tag{13}$$

Equation (13) generates a series of isoactivity lines perpendicular to the E-axis.

In the Ag$^+$ domain, Ag$^+$ ions may be present at activities ranging from zero to unity.

The boundary between the Ag^+ and Ag_2O areas is calculated from the decomposition reaction:

$$Ag_2CO_3(\text{solid}) \rightarrow Ag_2O(\text{solid}) + CO_2$$

whose the free energy change is:

$$\Delta G^\circ = 17\cdot 63 \text{ kcal.}$$

The equilibrium constant is:

$$K = \frac{[Ag_2O]\,[CO_2]}{[Ag_2\,CO_3]} = 10^{-12\cdot 8}$$

and therefore:

$$pCO_2 = 12\cdot 8 - \log(Ag^+) \tag{14}$$

The $\log(Ag^+)$ scale for the Ag_2O domain is obtained by plotting eqn. (14) which generates a series of lines parallel to the E-axis. The limit between the Ag_2O and Ag domain is therefore the intersection of the lines given by eqn. (13) and (14).

The E/pCO_2 diagrams do not provide a complete description of the electrochemical behaviour of metal and carbonate systems and the thermodynamic approach sheds no light on the structural and kinetic features which are so often important. The merit of this treatment lies in the fact that a better knowledge of some electrochemical phenomena can be gained.

References

Charlot, G. and Tremillon, B. (1963). "Les réactions chimiques dans les solvants et les sels fondus". Gauthier–Villars, Paris.
Coughlin, J. P. (1954). *U. S. Bureau of Mines Bull.* 542.
Delarue, G. (1960). Thesis, Paris.
Ellingham, H. J. T. (1944). *J. Soc. chem. Ind., Lond.* **63**, 125.
Elliott, J. F. and Gleiser, M. (1960). "Thermochemistry for Steelmaking", Vol. 1 Addison-Wesley, Reading, Mass.
Guillou, M., Millet, J. and Palous, S. (1968). *Electrochim. Acta* **13**, 1411.
Hamer, W. J. (1965). *J. electroanal. Chem.* **10**, 140.
Hamer, W. J., Malmberg, M. S. and Rubin, B. (1956). *J. electrochem. Soc.* **103**, 8.
Hamer, W. J., Malmberg, M. S. and Rubin, B. (1965). *J. electrochem. Soc.* **112**, 750.
Hladik, J. (1970). *C. R. Acad. Sci.* **270**, 1771.
Kubaschewski, O., Evans, E. LL. and Alcock, C. B. (1967). "Metallurgical Thermochemistry" 4th Ed. Pergamon Press, London.
Leroy, M. (1962). *Bull. Soc. chem. Fr.* 968.

References continued on page 798.

TABLE I.   Standard electromotive forces for single solid or molten metallic chlorides*

Order 25°C	Metal ion	E°						
		25°C (aqueous) volts	25°C (solid) volts	Uncert-ainty (25°C) volts	100°C volts	200°C volts	300°C volts	350°C volts
1	Ra^{+2}	4·28	4·366	±0·07	4·272	4·189	4·108	4·068
2	K^{+1}	4·285	4·232	±0·01	4·158	4·056	3·954	3·904
3	Sm^{+2}	<4·52	4·206	±0·15	4·147	4·071	3·998	3·962
4	Ba^{+2}	4·26	4·202	±0·01	4·139	4·056	3·975	3·935
5	Cs^{+1}	4·283	4·189	±0·01	4·109	4·002	3·896	3·843
6	Rb^{+1}	4·285	4·177	±0·04	4·101	3·998	3·897	3·846
7	Sr^{+2}	4·25	4·048	±0·01	3·987	3·909	3·832	3·794
8	Li^{+1}	4·405	4·011	±0·09	3·955	3·881	3·800	3·761
9	Na^{+1}	4·074	3·980	±0·004	3·910	3·810	3·712	3·663
10	Ca^{+2}	4·230	3·888	±0·02	3·830	3·754	3·680	3·643
11	La^{+3}	3·88	3·565	±0·03	3·504	3·426	3·350	3·313
12	Ce^{+3}	3·84	3·517	±0·03	3·456	3·378	3·303	3·266
13	Pr^{+3}	3·83	3·481	±0·01	3·420	3·342	3·266	3·229
14	Nd^{+3}	3·80	3·430	±0·01	3·369	3·291	3·215	3·177
15	Pm^{+3}	3·78	3·410	±0·03	3·353	3·279	3·208	3·173
16	Sm^{+3}	3·77	3·380	±0·10	3·322	3·249	3·178	3·143
17	Eu^{+3}	3·77	3·340	±0·10	3·283	3·210	3·139	3·104
18	Gd^{+3}	3·76	3·317	±0·01	3·260	3·187	3·116	3·081
19	Tb^{+3}	3·75	3·261	±0·03	3·204	3·130	3·059	3·025
20	Dy^{+3}	3·71	3·206	±0·03	3·149	3·075	3·004	2·970
21	Y^{+3}	3·72	3·163	±0·04	3·106	3·032	2·961	2·926
22	Ho^{+3}	3·68	3·134	±0·01	3·077	3·003	2·932	2·897
23	Er^{+3}	3·66	3·119	±0·01	3·062	2·989	2·918	2·883
24	Tm^{+3}	3·64	3·086	±0·01	3·029	2·956	2·884	2·850
25	Yb^{+3}	3·63	3·075	±0·10	3·017	2·944	2·873	2·838
26	Mg^{+2}	3·73	3·070	±0·004	3·006	2·922	2·840	2·800
27	Lu^{+3}	3·61	3·049	±0·01	2·988	2·909	2·834	2·796
28	Sc^{+3}	3·44	2·946	±0·01	2·885	2·807	2·731	2·694
29	Zr^{+2}	—	2·905	±0·43	2·847	2·772	2·699	2·664
30	U^{+3}	3·16	2·846	—	2·788	2·713	2·639	2·602
31	Th^{+4}	3·26	2·840	±0·05	2·779	2·699	2·622	2·584
32	Zr^{3}	—	2·790	±0·29	2·736	2·668	2·603	2·571

$E°$							
400°C	450°C	500°C	550°C	600°C	800°C	1000°C	1500°C
volts	volts	volts	volts	volts	volts	volts	volts
4·029	3·990	3·952	3·913	3·876	3·723	3·569	13·098
3·854	3·805	3·755	3·707	3·658	13·441	3·155	2·598
3·926	3·891	3·856	3·822	3·787	13·661	3·559	3·317
3·888	3·848	3·808	3·768	3·728	3·568	13·412	3·079
3·791	3·739	3·692	3·645	3·599	13·362	3·078	2·667 (1300V)
3·795	3·745	3·695	3·645	3·595	13·314	3·001	2·428 (1381V)
3·757	3·720	3·684	3·648	3·612	3·469	13·333	2·177
3·722	3·684	3·646	3·608	3·571	13·457	3·352	3·122 (1382V)
3·615	3·566	3·519	3·471	3·424	3·240	13·019	2·366 (1465V)
3·607	3·570	3·534	3·498	3·462	13·323	3·208	2·926
3·277	3·241	3·205	3·170	3·134	2·997	12·876	2·607
3·229	3·193	3·157	3·121	3·086	2·945	12·821	2·540
3·192	2·156	3·120	3·085	3·049	2·911	12·795	2·523
3·140	3·103	3·067	3·030	2·994	12·856	2·736	2·455
3·139	3·105	3·072	3·038	3·006	12·884	2·784	2·554
3·109	3·075	3·041	3·008	2·975	12·861	2·763	2·712 (1107D)
3·070	3·036	3·002	2·969	2·936	12·828	2·815 (>827D)	—
3·047	3·013	2·979	2·946	2·913	12·807	2·709	2·483
2·990	2·956	2·923	2·890	12·858	12·858	2·754	2·657
2·935	2·901	2·868	2·835	2·802	12·690	12·690	2·599
2·892	2·858	2·824	2·791	2·758	12·643	2·548	2·329
2·863	2·829	2·796	2·762	2·729	12·610	2·511	2·283
2·849	2·815	2·781	2·748	2·715	12·589	2·488	2·257 (1497V)
2·815	2·781	2·748	2·715	2·682	2·553	12·447	2·221 (1487V)
2·804	2·770	2·736	2·703	2·670	2·542	12·434	2·449 (1027D)
2·760	2·720	2·680	2·641	2·602	12·460	2·346	1·974 (1418V)
2·760	2·723	2·687	2·652	2·616	2·478	12·356	2·108 (1477V)
2·657	2·621	2·585	2·549	2·514	2·375	12·264 (967V)	—
2·629	2·594	2·560	2·526	2·508 (577D)	—	—	—
2·566	2·530	2·494	2·458	2·423	2·280	12·162	1·886
2·546	2·509	2·472	2·435	2·399	12·264	2·208 (921V)	—
2·540	2·500	2·492 (477D)	—	—	—	—	—

TABLE I.—*Continued*

Order 25°C	Metal ion	$E°$						
		25°C (aqueous) volts	25°C (solid) volts	Uncertainty (25°C) volts	100°C volts	200°C volts	300°C volts	350°C volts
33	Hf^{+4}	3·06	2·537	±0·32	2·481	2·409	2·340	2·328 (317S)
34	U^{+4}	2·86	2·493	—	2·436	2·362	2·289	2·252
35	Be^{+2}	3·21	2·435	±0·11	2·382	2·315	2·252	2·222
36	Mn^{+2}	2·54	2·287	±0·02	2·235	2·166	2·098	2·065
37	Zr^{+4}	2·89	2·266	±0·22	2·209	2·135	2·063	2·041 (331S)
38	Ti^{+2}	2·99	2·255	±0·22	2·202	2·134	2·069	2·037
39	$(Al^{+3})_2$	3·02	2·200	±0·01	2·150	2·097 (180S)	—	—
40	Ti^{+3}	ca 2·57	2·154	±0·14	2·097	2·024	1·954	1·919
41	V^{+2}	ca 2·54	2·103	±0·43	2·044	1·970	1·898	1·863
42	Tl^{+1}	1·696	1·916	±0·02	1·865	1·798	1·732	1·696
43	Zn^{+2}	2·123	1·914	±0·01	1·854	1·776	11·706	1·680
44	Cr^{+2}	2·27	1·846	±0·06	1·795	1·729	1·664	1·632
45	$(Si^{+3})_2$	—	1·778	±0·07	1·736	1·713 (145V)	—	—
46	Cd^{+2}	1·763	1·775	±0·01	1·715	1·637	1·560	1·521
47	Ti^{+4}	—	1·748	±0·11	1·700	1·678 (136V)	—	—
48	V^{+3}	ca 2·23	1·735	±0·43	1·674	1·596	1·576 (227D)	—
49	Ga^{+2}	ca 1·81	1·713	±0·22	1·659	l(200D)	—	—
50	In^{+1}	ca 1·61	1·709	±0·22	1·654	1·581	11·520	1·492
51	Cr^{+3}	2·10	1·706	±0·07	1·646	1·567	1·489	1·451
52	In^{+2}	ca 1·66	1·657	±0·06	1·601	1·528	11·464	1·435
53	In^{+3}	1·702	1·641	±0·04	1·588	1·519	1·451	1·417
54	Pb^{+2}	1·486	1·627	±0·01	1·569	1·493	1·420	1·383
55	Ga^{+3}	1·89	1·619	±0·01	11·572	1·528 (200V)	—	—
56	Sn^{+2}	1·496	1·607	±0·02	1·566	1·490	11·428	1·400

$E°$							
400°C	450°C	500°C	550°C	600°C	800°C	1000°C	1500°C
volts	volts	volts	volts	volts	volts	volts	volts
—	—	—	—	—	—	—	—
2·217	2·181	2·146	2·111	12·078	1·974	1·953 (827V)	—
2·192	12·167	2·144	2·122 (547V)	—	—	—	—
2·032	1·999	1·967	1·935	1·902	11·807	1·725	1·649 (1190V)
—	—	—	—	—	—	—	—
2·006	1·975	1·945	1·914	1·885D	—	—	—
—	—	—	—	—	—	—	—
1·886	1·852	1·836 (475D)	—	—	—	—	—
1·828	1·794	1·761	1·727	1·695	1·566	1·441	11·269 (1377V)
1·660	11·629	1·606	1·583	1·561	1·473	1·470 (806V)	—
1·655	1·629	1·603	1·577	1·552	1·476 (756V)	—	—
1·600	1·568	1·537	1·505	1·474	1·352	11·262	1·137 (1302V)
—	—	—	—	—	—	—	—
1·481	1·442	1·403	1·364	11·331	1·193	1·002 (980V)	—
—	—	—	—	—	—	—	—
—	—	—	—	—	—	—	—
—	—	—	—	—	—	—	—
1·465	1·439	1·414	1·389	1·364	1·360 (609V)	—	—
1·412	1·374	1·336	1·299	1·261	1·113	1·006 (947S)	—
1·407	1·380	1·361 (485V)	—	—	—	—	—
1·384	1·352	1·321 (498S)	—	—	—	—	—
1·345	1·307	11·271	1·243	1·215	1·112	1·039 (954V)	—
—	—	—	—	—	—	—	—
1·373	1·346	1·320	1·295	1·270	1·259 (623V)	—	—

TABLE I.—Continued

Order 25°C	Metal ion	25°C (aqueous) volts	25°C (solid) volts	Uncertainty (25°C) volts	100°C volts	200°C volts	300°C volts	350°C volts
						$E°$		
57	Fe^{+2}	1·80	1·565	±0·02	1·516	1·541	1·388	1·357
58	Si^{+4}	—	1·484	±0·005	1·466 (57V)	—	—	—
59	Co^{+2}	1·637	1·464	±0·02	1·408	1·337	1·269	1·236
60	Ni^{+2}	1·610	1·412	±0·04	1·355	1·282	1·210	1·174
61	V^{+4}	—	1·323	±0·32	1·281	1·253 (152V)	—	—
62	Cu^{+1}	0·839	1·232	±0·02	1·191	1·140	1·093	1·071
63	Sn^{+4}	1·35	1·228	±0·01	1·184	1·176 (113V)	—	—
64	Ge^{+4}	—	1·225	±0·16	1·190 (83V)	—	—	—
65	Fe^{+3}	1·396	1·197	±0·06	1·147	1·084	11·023	1·016 (319V)
66	Ag^{+1}	0·560	1·137	±0·01	1·093	1·037	0·984	0·959
67	Sb^{+3}	—	1·122	—	11·077	1·028	1·019 (221V)	—
68	Bi^{+3}	—	1·102	—	1·051	0·986	10·926	0·896
69	$(Hg^{+1})_2$	0·571	1·092	±0·01	1·022	0·930	0·840	0·800
70	As^{+3}	—	1·019	—	0·986	0·973 (130V)	—	—
71	Hg^{+2}	0·506	0·952	±0·02	0·892	0·814	10·743	0·741 (304V)
72	Cu^{+2}	1·023	0·846	±0·11	0·791	0·721	0·654	0·621
73	Ir^{+1}	—	0·794	±0·13	0·751	0·699	0·650	0·626
74	Pd^{+2}	0·373	0·768	±0·06	0·714	0·646	0·581	0·549
75	Mo^{+2}	—	0·759	±0·06	0·711	0·651	0·593	0·566
76	Sb^{+5}	—	0·742	—	0·701	0·665 (172V)	—	—
77	Ir^{+2}	<0·26	0·733	±0·06	0·685	0·625	0·567	0·539
78	Mo^{+3}	1·56	0·723	±0·07	0·669	0·602	0·536	0·519 (327D)
79	Ir^{+3}	ca 0·21	0·665	±0·04	0·610	0·539	0·472	0·439
80	Mo^{+4}	—	0·650	±0·05	0·600	0·582 (127D)	—	—

$E°$							
400°C	450°C	500°C	550°C	600°C	800°C	1000°C	1500°C
volts	volts	volts	volts	volts	volts	volts	volts
1·327	1·297	1·267	1·237	1·207	ll·118	1·050	1·041 (1026V)
—	—	—	—	—	—	—	—
1·203	1·171	1·140	1·109	1·079	l0·977	0·900	0·881 (1050V)
1·139	1·104	1·070	1·036	1·003	0·875n	0·763 (987V)	—
—	—	—	—	—	—	—	—
1·050	ll·035	1·024	1·013	1·003	0·970	0·943	0·862
—	—	—	—	—	—	—	—
—	—	—	—	—	—	—	—
—	—	—	—	—	—	—	—
0·935	0·911	l0·896	0·883	0·870	0·826	0·784	0·665
—	—	—	—	—	—	—	—
0·867	0·844 (441V)	—	—	—	—	—	—
0·730	0·663	0·597	0·540 (543D)	—	—	—	—
—	—	—	—	—	—	—	—
—	—	—	—	—	—	—	—
0·589	0·558	0·528	0·506 (537D)	—	—	—	—
0·603	0·580	0·558	0·536	0·515	0·433 (799D)	—	—
0·518	0·487	0·457	0·427	0·397	l0·331 (737D)	—	—
0·538	0·511	0·485	0·471 (527D)	—	—	—	—
—	—	—	—	—	—	—	—
0·512	0·485	0·458	0·432	0·406	0·320 (771D)	—	—
—	—	—	—	—	—	—	—
0·406	0·374	0·342	0·311	0·280	0·180 (765D)	—	—
—	—	—	—	—	—	—	—

TABLE I.—*Continued*

Order 25°C	Metal ion	$E°$						
		25°C (aqueous) volts	25°C (solid) volts	Uncert-ainty (25°C) volts	100°C volts	200°C volts	300°C volts	350°C volts
81	W^{+2}	—	0·629	±0·11	0·581	0·521	0·463	0·435
82	Mo^{+5}	—	0·597	±0·02	0·550	l0·491	0·463 (268V)	—
83	Rh^{+3}	ca 0·56	0·593	±0·04	0·539	0·471	0·406	0·374
84	Rh^{+2}	ca 0·76	0·575	±0·06	0·524	0·460	0·398	0·368
85	Pt^{+1}	—	0·572	±0·13	0·525	0·465	0·407	0·380
86	Pt^{+2}	ca 0·16	0·564	±0·06	0·513	0·449	0·387	0·357
87	W^{+4}	—	0·564	±0·03	0·513	0·449	0·432 (227D)	—
88	Ru^{+3}	—	0·546	±0·04	0·494	0·428	0·364	0·333
89	W^{+5}	—	0·538	±0·03	0·491	0·431	l0·392 (276V)	—
90	Rh^{+1}	ca 0·76	0·520	±0·13	0·478	0·425	0·376	0·352
91	W^{+6}	—	0·513	±0·03	0·464	0·401	l0·343	0·325 (337V)
92	Pt^{+3}	—	0·484	±0·07	0·426	0·351	0·279	0·243
93	Pt^{+4}	—	0·465	±0·03	0·412	0·344	0·278	0·261 (327D)
94	Au^{+3}	−0·140	0·211	±0·03	0·162	0·101	l0·043	0·021
95	Au^{+1}	ca −0·320	0·182	±0·04	0·138	0·082	0·037 (287D)	—
96	C^{+4}	—	0·178	±0·005	0·147 (77V)	—	—	—

S = sublimation; V = vaporization; D = decomposition.

* Values listed at the sublimation, vaporization, or decomposition temperatures obviously apply to cells with the solid or molten electrolyte, that is, slightly below the sublimation, vaporization, or decomposition temperatures.

$E°$							
400°C	450°C	500°C	550°C	600°C	800°C	1000°C	1500°C
volts	volts	volts	volts	volts	volts	volts	volts
0·408	0·393 (427D)	—	—	—	—	—	—
—	—	—	—	—	—	—	—
0·343	0·312	0·281	0·251	0·221	0·104	0·020 (948D)	—
0·339	0·310	0·281	0·253	0·225	10·142	0·093 (958D)	
0·353	0·326	0·300	0·274	0·257 (583D)	—	—	—
0·328	0·299	0·270	0·242	0·225 (581D)	—	—	—
—	—	—	—	—	—	—	—
0·303	0·273	0·243	0·214	0·185	0·170 (627D)	—	—
—	—	—	—	—	—	—	—
0·328	0·305	0·282	0·260	0·238	0·153	0·084 (965D)	—
—	—	—	—	—	—	—	—
0·209	0·185 (435D)	—	—	—	—	—	—
—	—	—	—	—	—	—	—
−0·001	—	—	—	—	—	—	—
—	—	—	—	—	—	—	—
—	—	—	—	—	—	—	—

TABLE II.   Standard electromotive forces for single, solid, or molten metallic fluorides*

Order 25°C	Metal ion	$E°$						
		25°C (aqueous) volts	25°C (solid) volts	Uncertainty (25°C) volts	100°C volts	200°C volts	300°C volts	350°C volts
1	Eu^{+2}	6·266	6·234	±0·15	6·167	6·080	5·996	5·955
2	Li^{+1}	5·911	6·054	±0·09	5·981	5·881	5·775	5·722
3	Ca^{+2}	5·736	6·021	±0·01	5·953	5·864	5·776	5·732
4	Sm^{+2}	<6·03	6·017	±0·15	5·950	5·863	5·779	5·738
5	Sr^{+2}	5·756	6·012	±0·002	5·943	5·854	5·768	5·726
6	Ba^{+2}	5·766	5·952	±0·02	5·885	5·797	5·712	5·670
7	Ra^{+2}	5·786	5·941	±0·07	5·871	5·781	5·694	5·651
8	La^{+3}	5·386	5·811	±0·10	5·743	5·656	5·571	5·530
9	Ce^{+3}	5·346	5·739	±0·10	5·671	5·584	5·499	5·457
10	Pr^{+3}	5·336	5·710	±0·10	5·645	5·563	5·483	5·444
11	Nd^{+3}	5·306	5·652	±0·10	5·584	5·496	5·411	5·369
12	Pm^{+3}	5·286	5·637	±0·10	5·573	5·490	5·410	5·370
13	Na^{+1}	5·580	5·607	±0·01	5·534	5·429	5·325	5·273
14	Sm^{+3}	5·276	5·594	±0·10	5·530	5·447	5·367	5·328
15	Gd^{+3}	5·266	5·580	±0·10	5·515	5·433	5·352	5·313
16	K^{+1}	5·791	5·525	±0·02	5·446	5·338	5·230	5·176
17	Tb^{+3}	5·256	5·522	±0·10	5·457	5·375	5·295	5·255
18	Dy^{+3}	5·216	5·493	±0·10	5·429	5·346	5·266	5·226
19	Y^{+3}	5·236	5·478	±0·10	5·414	5·331	5·251	5·212
20	Ho^{+3}	5·186	5·449	±0·10	5·385	5·302	5·222	5·183
21	Mg^{+2}	5·236	5·438	±0·02	5·369	5·278	5·188	5·144
22	Rb^{+1}	5·791	5·421	±0·04	5·348	5·250	5·153	5·105
23	Lu^{+3}	5·116	5·407	±0·10	5·342	5·259	5·179	5·140
24	Er^{+3}	5·166	5·406	±0·10	5·342	5·259	5·179	5·140
25	Eu^{+3}	5·276	5·392	±0·10	5·327	5·245	5·164	5·125
26	Tm^{+3}	5·146	5·392	±0·10	5·327	5·245	5·164	5·125
27	Cs^{+1}	5·789	5·217	±0·01	5·140	5·037	4·935	4·884
28	Yb^{+3}	5·136	5·175	±0·10	5·111	5·028	4·948	4·908
29	Sc^{+3}	5·946	5·059	±0·10	4·999	4·921	4·845	4·809
30	Th^{+4}	4·766	4·925	±0·11	4·864	4·786	4·710	4·673
31	Zr^{+3}	—	4·813	±0·36	4·753	4·676	4·601	4·565
32	Be^{+2}	4·716	4·705	±0·11	4·652	4·586	4·524	4·404
33	Zr^{+4}	4·396	4·586	±0·33	4·527	4·452	4·380	4·345
34	U^{+4}	4·366	4·564	—	4·505	4·430	4·357	4·322
35	Hf^{+4}	4·566	4·477	±0·43	4·419	4·344	4·271	4·236
36	Ti^{+3}	ca 4·076	4·329	±0·22	4·274	4·204	4·137	4·104
37	$(Al^{+3})_2$	4·526	4·250	±0·07	4·188	4·107	4·026	3·986
38	V^{+3}	4·086	3·896	±0·43	3·841	3·771	3·704	3·671
39	Mn^{+2}	4·046	3·881	±0·11	3·828	3·760	3·695	3·663
40	Ti^{+4}	—	3·794	±0·22	3·741	3·673	3·619 (284V)	
41	Cr^{+2}	3·771	3·708	±0·07	3·655	3·587	3·523	3·492
42	V^{+2}	ca 4·046	3·664	±0·43	3·606	3·531	3·460	3·425
43	Cr^{+3}	3·606	3·601	±0·07	3·543	3·471	3·401	3·366
44	Zn^{+2}	3·629	3·591	±0·07	3·535	3·465	3·397	3·365

			$E°$				
400°C	450°C	500°C	550°C	600°C	800°C	1000°C	1500°C
volts	volts	volts	volts	volts	volts	volts	volts
5·914	5·874	5·834	5·795	5·756	5·602	5·453 I	5·101
5·669	5·617	5·564	5·512	5·461	5·256 I	5·071	4·495
5·689	5·646	5·603	5·560	5·517	5·350	5·182 I	4·785
5·698	5·657	5·617	5·578	5·539	5·385	5·236	4·884
5·684	5·643	5·602	5·562	5·522	5·364	5·203 I	4·768
5·629	5·588	5·547	5·507	5·468	5·310	5·154 I	4·803
5·608	5·566	5·524	5·483	5·442	5·277	5·111	4·566
5·489	5·448	5·408	5·368	5·329	5·174	5·020	4·648
5·416	5·376	5·335	5·295	5·255	5·097	4·938	4·555
5·405	5·367	5·329	5·291	5·254	5·109	4·965	4·621
5·327	5·286	5·245	5·204	5·164	5·004	4·843	4·458
5·332	5·294	5·256	5·218	5·181	5·035	4·894	4·560
5·221	5·170	5·119	5·068	5·017	4·818	4·529	3·781
5·289	5·250	5·213	5·175	5·138	4·992	4·850	4·517
5·274	5·236	5·198	5·161	5·123	4·977	4·836	4·504
5·123	5·070	5·017	4·965	4·913	4·674	4·355	3·630
5·217	5·178	5·140	5·103	5·065	4·920	4·778	4·447
5·188	5·149	5·111	5·074	5·036	4·891	4·749	4·419
5·173	5·135	5·097	5·059	5·022	4·876	4·735	4·407
5·144	5·106	5·068	5·030	4·993	4·847	4·706	4·376
5·100	5·056	1·013	4·969	4·926	4·746	4·567	3·994
5·057	5·009	4·961	4·913	4·865	4·583	4·274 I	3·670
5·100	5·062	5·025	4·987	4·950	4·804	4·662	4·336
5·101	5·063	5·025	4·987	4·950	4·804	4·662	4·333
5·086	5·048	5·010	4·973	4·935	4·790	4·648	4·316
5·086	5·048	5·010	4·973	4·935	4·789	4·648	4·320
4·833	4·783	4·733	4·682	4·632	4·367	4·055 I	3·677
4·870	4·831	4·793	4·756	4·719	4·573	4·431	4·104
4·772	4·736	4·701	4·666	4·631	4·495	4·363	4·076
4·637	4·600	4·565	4·529	4·494	4·355	4·220	3·962
4·529	4·493	4·458	4·424	4·289	4·255	4·123	3·825
4·464	4·435	4·407	4·379	4·352	4·247	4·073 I	4·058
4·310	4·276	4·242	4·208	4·175	4·045	3·964 (930S)	
4·237	4·252	4·217	4·183	4·149	4·015	3·881 I	3·626
4·202	4·168	4·134	4·100	4·067	3·939	3·860 (927S)	
4·072	4·040	4·009	3·978	3·947	3·828	3·712	3·499
3·946	3·906	3·867	3·828	3·789	3·629	3·471	3·275
3·639	3·608	3·577	3·546	3·516	3·398	3·284	3·087
3·632	3·601	3·570	3·540	3·510	3·391	3·289	3·044
3·461	3·430	3·400	3·371	3·341	3·227	3·115	2·883
3·391	3·358	3·325	3·292	3·260	3·133	3·011	2·754
3·333	3·300	3·267	3·234	3·202	3·076	2·954 I	2·954
3·332	3·299	3·265	3·231	3·198	3·068	2·912	2·439

TABLE II.—*Continued*

Order 25°C	Metal ion	$E°$						
		25°C (aqueous) volts	25°C (solid) volts	Uncertainty (25°C) volts	100°C volts	200°C volts	300°C volts	350°C volts
45	Ga^{+3}	3·496	3·440	±0·14	3·376	3·292	3·211	3·171
46	Fe^{+2}	3·306	3·417	±0·04	3·362	3·291	3·223	3·191
47	Cd^{+2}	3·269	3·356	±0·02	3·303	3·234	3·169	3·135
48	In^{+3}	3·208	3·353	±0·04	3·289	3·206	3·123	3·083
49	V^{+4}	—	3·307	±0·43	3·253	3·186	3·121	3·104 (327D)
50	Co^{+2}	3·143	3·231	±0·02	3·177	3·110	3·045	3·014
51	Ni^{+2}	3·116	3·226	±0·02	3·169	3·096	3·026	2·992
52	Pb^{+2}	2·992	3·221	±0·04	3·155	3·082	3·011	2·975
53	Mn^{+3}	2·583	3·209	±0·07	3·152	3·079	3·009	2·975
54	Fe^{+3}	2·902	3·166	±0·19	3·109	3·036	2·966	2·932
55	Sb^{+3}	—	2·946	—	2·898	2·839	2·782 |	2·758
56	Bi^{+3}	—	2·891	—	2·834	2·761	2·690	2·652
57	V^{+5}	—	2·706	± 0·17	2·657 |	2·652 (111V)		
58	Tl^{+1}	3·202	2·602	±0·22	2·549	2·481	2·416 |	2·385
59	Cu^{+2}	2·529	2·526	±0·09	2·471	2·400	2·333	2·301
60	Co^{+3}	3·288	2·472	±0·14	2·415	2·342	2·273	2·239
61	Ti^{+3}	2·145	2·306	±0·14	2·250	2·180	2·113	2·078
62	Pd^{+2}	1·879	2·212	±0·22	2·158	2·091	2·026	1·995
63	Rh^{+3}	2·066	2·081	±0·29	2·025	1·952	1·882	1·849
64	Ag^{+1}	2·067	1·917	±0·04	1·871	1·815	1·762	1·736
65	Hg^{+2}	2·012	1·800	±0·22	1·735	1·653	1·574	1·535
66	Pd^{+3}	—	1·518	±0·29	1·457	1·380	1·360 (227D)	
67	Au^{+3}	1·366	1·243	±0·43	1·194	1·131	1·071	1·042
68	Au^{+2}	<1·381	1·019	±0·43	0·966	0·899	0·834	0·803
69	Au^{+1}	1·186	0·607	±0·17	0·565	0·513	0·464	0·440

S = sublimation; V = vaporization; D = decomposition.

* Values listed at the sublimation, vaporization, or decomposition temperatures obviously apply to cells with the solid or molten electrolyte, that is, slightly below the sublimation, vaporization, or decomposition temperature.

			$E°$				
400°C	450°C	500°C	550°C	600°C	800°C	1000°C	1500°C
volts	volts	volts	volts	volts	volts	volts	volts
3·132	3·093	3·055	3·017	2·980	2·923 (952S)		
3·158	3·126	3·094	3·062	3·030	2·905	2·780	2·529
3·101	3·068	3·035	3·002	2·969	2·826	2·605	2·109
3·043	3·003	2·964	2·925	2·887	2·736	2·589 I	2·447
2·983	2·952	2·922	2·892	2·862	2·746	2·631	3·391
2·958	2·924	2·890	2·857	2·825	2·697	2·573	2·338
2·938	2·901	2·865	2·829	2·793	2·654	2·525 I	2·350
2·941	2·908	2·875	2·842	2·810	2·682	2·557 I	2·383
2·899	2·865	2·832	2·800	2·767	2·640	2·513 I	2·354
2·746 (376S)							
2·615	2·578	2·542	2·506	2·470 I	2·381 (1027V)		
2·359	2·333	2·308	2·284	2·260	2·234 (655V)		
2·269	2·237	2·206	2·176	2·145	2·028 I	1·922	1·693
2·205	2·172	2·139	2·107	2·075	1·949 I	1·826	1·670 (1327V)
2·044	2·011	1·978	1·945 I	1·918	1·815	1·718 (927V)	
1·964	1·934	1·904	1·874	1·845	1·741	1·660 I	1·574
1·815	1·782	1·750	1·717	1·684	1·554	1·424 I	1·099 (1427S)
1·711 I	1·690	1·674	1·660	1·646	1·597	1·551	1·509 (1147V)
1·476	1·414	1·353	1·293	1·233 I	1·180 (647D)		
1·014	0·986	0·958	0·931	0·904 I	0·809	0·734	0·555 (1197V)
0·772	0·742	0·712	0·683	0·654 I	0·538	0·421	0·144 (1252V)
0·415	0·391	0·367	0·342 I	0·318	0·220	0·123	−0·121 (1202V)

TABLE III.   Standard electromotive forces for single, solid, or molten metallic bromide*

Order 25°C	Metal ion	$E°$						
		25°C (aqueous) volts	25°C (solid) volts	Uncertainty (25°C) volts	100°C volts	200°C volts	300°C volts	350°C volts
1	Cs^{+1}	4·004	3·988	±0·01	3·918	3·824	3·730	3·683
2	Ra^{+2}	4·001	3·957	±0·07	3·890	3·803	3·718	3·676
3	K^{+1}	4·006	3·946	±0·01	3·874	3·775	3·677	3·628
4	Rb^{+1}	4·006	3·936	±0·04	3·866	3·773	3·681	3·635
5	Ba^{+2}	3·981	3·814	±0·01	3·750	3·669	3·590	3·552
6	Sm^{+2}	<4·24	3·708	±0·15	3·649	3·572	3·498	3·462
7	Sr^{+2}	3·971	3·621	±0·01	3·559	3·479	3·402	3·363
8	Na^{+1}	3·795	3·613	±0·02	3·544	3·445	3·348	3·299
9	Li^{+1}	4·126	3·569	±0·09	3·514	3·441	3·363	3·324
10	Ca^{+2}	3·951	3·416	±0·02	3·357	3·280	3·206	3·170
11	La^{+3}	3·601	3·122	±0·04	3·061	2·983	2·907	2·870
12	Ce^{+3}	3·561	3·050	±0·04	2·989	2·911	2·835	2·798
13	Pr^{+3}	3·551	3·011	±0·04	2·951	2·874	2·799	2·763
14	Nd^{+3}	3·521	2·986	±0·04	2·928	2·852	2·778	2·742
15	Pm^{+3}	3·501	2·934	±0·04	2·877	2·803	2·732	2·697
16	Sm^{+3}	3·491	2·891	±0·10	2·833	2·760	2·689	2·653
17	Gd^{+3}	3·481	2·862	±0·04	2·805	2·731	2·660	2·625
18	Tb^{+3}	3·471	2·819	±0·04	2·761	2·688	2·616	2·581
19	Dy^{+3}	3·431	2·775	±0·04	2·714	2·636	2·560	2·522
20	Y^{+3}	3·451	2·761	±0·04	2·700	2·621	2·545	2·508
21	Ho^{+3}	3·401	2·746	±0·04	2·686	2·607	2·531	2·493
22	Er^{+3}	3·381	2·717	±0·04	2·657	2·578	2·502	2·464
23	Eu^{+3}	3·491	2·689	±0·10	2·631	2·558	2·486	2·451
24	Tm^{+3}	3·361	2·689	±0·04	2·628	2·549	2·473	2·435
25	Lu^{+3}	3·331	2·645	±0·04	2·584	2·506	2·429	2·392
26	Mg^{+2}	3·451	2·606	±0·01	2·548	2·474	2·403	2·368
27	Zr^{+2}	—	2·528	±0·43	2·471	2·397	2·326	2·292
28	Sc^{+3}	3·161	2·506	±0·01	2·446	2·367	2·291	2·253

$E°$							
400°C	450°C	500°C	550°C	600°C	800°C	1000°C	1500°C
volts	volts	volts	volts	volts	volts	volts	volts
3·637	3·590	3·544	3·498	3·452 I	3·204	2·903	2·468 (1300V)
3·635	3·594	3·553	3·513	3·473	3·310 I	3·154	2·667
3·580	3·532	3·485	3·438	3·391 I	3·187	2·904	2·385 (1383V)
3·589	3·543	3·497	3·452	3·407 I	3·151	2·852	2·348 (1352V)
3·514	3·476	3·438	3·401	3·364	3·216 I	3·084	2·791
3·426	3·391	3·355	3·321	3·286 I	3·165	3·062	2·822
3·326	3·288	3·251	3·214	3·178 I	3·053	2·930	2·595
3·251	3·203	3·156	3·108	3·061	2·889	2·667	2·107 (1392V)
3·287	3·249	3·212	3·176 I	3·147	3·040	2·940	2·798 (1310V)
3·134	3·099	3·064	3·028	2·994 I	2·861	2·740	2·447
2·833	2·797	2·761	2·725	2·690 I	2·553	2·438	2·167
2·761	2·724	2·688	2·652	2·616 I	2·482	2·363	2·085
2·727	2·691	2·655	2·620	2·585 I	2·462	2·354	2·092
2·706	2·670	2·634	2·599	2·564 I	2·441	2·327	2·060
2·662	2·628	2·594	2·561	2·527 I	2·413	2·314	2·086
2·619	2·585	2·551	2·517	2·484 I	2·372	2·328 (887D)	
2·590	2·556	2·522	2·488	2·455 I	2·329	2·227	1·998 (1487V)
2·547	2·512	2·478	2·445	2·412	2·282 I	2·177	1·950 (1487V)
2·485	2·448	2·412	2·376	2·341	2·201 I	2·079	1·829 (1477V)
2·471	2·434	2·398	2·362	2·326	2·187 I	2·062	1·815 (1467V)
2·456	2·420	2·383	2·347	2·312	2·172 I	2·047	1·085 (1467V)
2·427	2·391	2·354	2·283	2·283	2·143 I	2·013	1·774 (1457V)
2·417	2·382	2·348	2·315	2·282 I	2·164	2·120 (887D)	
2·398	2·362	2·325	2·289	2·254	2·114 I	1·984	1·755 (1437V)
2·355	2·318	2·282	2·246	2·211	2·071 I	1·949	1·726 (1407V)
2·333	2·299	2·265	2·232	2·199 I	2·078	1·980	1·824 (1227V)
2·257	2·224	2·190	2·157	2·124 I	1·992	1·860	1·710 (1227V)
2·216	2·180	2·143	2·107	1·072	1·932 I	1·846 (929S)	

TABLE III.—*Continued*

Order 25°C	Metal ion	E°						
		25°C (aqueous) volts	25°C (solid) volts	Uncertainty (25°C) volts	100°C volts	200°C volts	300°C volts	350°C volts
29	Zr^{+3}	—	2·433	±0·29	2·373	2·296	2·222	2·186
30	Yb^{+3}	3·351	2·428	±0·10	2·368	2·289	2·213	2·175
31	U^{+3}	2·881	2·397	—	2·342	2·272	2·205	2·171
32	Th^{+4}	2·981	2·378	±0·11	2·318	2·241	2·164	2·127
33	Hf^{+4}	2·781	2·201	±0·32	2·142	2·066	1·993	1·977 (322S)
34	U^{+4}	2·581	2·060	—	2·003	1·929	1·858	1·823
35	V^{+2}	ca 2.261	2·016	±0·43	1·958	1·883	1·810	1·775
36	Zr^{+4}	2·611	2·006	±0·22	1·948	1·873	1·801	1·766
37	Ti^{+2}	2·711	1·995	±0·22	1·940	1·869	1·801	1·768
38	Mn^{+2}	2·261	1·908	±0·04	1·854	1·786	1·719	1·687
39	Be^{+2}	2·931	1·869	±0·11	1·818	1·755	1·695	1·666
40	Ti^{+3}	ca 2.281	1·843	±0·14	1·788	1·717	1·649	1·615
41	$(Al^{+3})_2$	2·741	1·761	±0·03 |	1·707	1·649	1·617 (257V)	
42	Tl^{+1}	1·417	1·738	±0·02	1·686	1·619	1·553	1·517
43	Ti^{+4}	—	1·731	±0·11 |	1·710	1·687	1·680 (230V)	
44	In^{+1}	ca 1·331	1·717	±0·22	1·659	1·582 |	1·156	1·487
45	Zn^{+2}	1·844	1·624	±0·02	1·567	1·494	1·424	1·390 |
46	Cr^{+2}	1·985	1·539	±0·11	1·486	1·418	1·353	1·321
47	Cd^{+2}	1·484	1·537	±0·02	1·475	1·395	1·318	1·278
48	Ge^{+4}	—	1·409	±0·16 |	1·364	1·316 (189V)		
49	Pb^{+2}	1·207	1·366	±0·02	1·309	1·234	1·160	1·122
50	$(Si^{+3})_2$	—	1·344	±0·14 |	1·298	1·245	1·213 (240V)	
51	Sn^{+3}	1·217	1·331	±0·02	1·280	1·215 |	1·153	1·122
52	In^{+3}	1·423	1·321	±0·04	1·264	1·189	1·114	1·078

$E°$							
400°C	450°C	500°C	550°C	600°C	800°C	1000°C	1500°C
volts	volts	volts	volts	volts	volts	volts	volts
2·150	2·115	2·080	2·044	2·010	1·875 I	1·857 (827 dp)	
2·138	2·102	2·065	2·029	1·994	1·854 I	1·726 (ca 1000D)	
2·138	2·106	2·073	2·041	2·009 I	1·889	1·793	1·569
2·090	2·053	2·017	1·981	1·945 I	1·820	1·788 (857V)	
1·789	1·755	1·721 I	1·691	1·663	1·573 (ca 766V)		
1·740	1·706	1·672	1·639	1·605	1·476 I	1·374	1·269 (1227V)
1·761 (357S)							
1·735	1·703	10671	1·639	1·608 I	1·511	1·424	1·331 (1227V)
1·654	1·623	1·591	1·560	1·529 I	1·422	1·334	1·322 (1027V)
1·638	1·610 I	1·584	1·581 (527V)				
1·582	1·550	1·518	1·486	1·454	1·331 (927 dp)		
1·480	1·444 I	1·416	1·391	1·367	1·275	1·267 (819V)	
1·458	1·429	1·401	1·374	1·347	1·318 (662V)		
1·357	1·329	1·301	1·272	1·245	1·192 (702V)		
1·289	1·258	1·227	1·197	1·167	1·049 I	0·954	0·900 (1127V)
1·238	1·198	1·159	1·120 I	1·085	0·944	0·876 (863V)	
1·089	1·059 I	1·031	1·003	0·976	0·876	0·823 (914V)	
1·093	1·063	1·036	1·008	0·981	0·960 (639V)		
1·062 (371S)							

TABLE III.—*Continued*

Order 25°C	Metal ion	E° 25°C (aqueous) volts	25°C (solid) volts	Uncertainty (25°C) volts	100°C volts	200°C volts	300°C volts	350°C volts
53	Ga^{+3}	1·611	1·297	±0·01	1·244	1·244 I (314V)	1·183	
54	Fe^{+2}	1·521	1·247	±0·01	1·194	1·127	1·062	1·030
55	V^{+4}	—	1·225	±0·43 I	1·183	1·131	1·108 (247V)	
56	Co^{+2}	1·358	1·149	±0·02	1·097	1·030	0·965	0·934
57	Ni^{+2}	1·331	1·106	±0·04	1·049	0·977	0·907	0·872
58	Cu^{+1}	0·560	1·049	±0·03	1·000	0·939	0·881	0·853
59	Ag^{+1}	0·282	1·011	±0·01	0·966	0·911	0·860	0·836
60	Si^{+4}	—	0·999	±0·01 I	0·953	0·922 (153V)		
61	Sn^{+4}	1·074	0·987	±0·01 I	0·939	0·880	0·877 (207V)	
62	$(Hg^{+1})_2$	0·292	0·942	±0·01	0·871	0·779	0·690	0·646
63	Fe^{+3}	1·117	0·888	±0·06	0·836	0·769 I	0·716	0·629
64	Sb^{+3}	—	0·831	— I	0·776	0·719	0·672 (288V)	
65	Bi^{+3}	—	0·824	—	0·774	0·710 I	0·657	0·630
66	B^{+3}	—	0·773	±0·06 I	0·742 (91V)			
67	Hg^{+2}	0·227	0·737	±0·01	0·663	0·566 I	0·483	0·469 (319V)
68	Cu^{+2}	0·744	0·662	±0·02	0·609	0·541	0·476	0·459 (327D)
69	As^{+3}	—	0·636	— I	0·597	0·551	0·542 (221V)	
70	Mo^{+2}	—	0·585	±0·15	0·538	0·477	0·419	0·391
71	Rh^{+1}	0·481	0·564	±0·12	0·521	0·468	0·418	0·394
72	Mo^{+3}	1·281	0·520	±0·07	0·467	0·399	0·333	0·301
73	Rh^{+3}	0·281	0·506	±0·04	0·452	0·384	0·318	0·286
74	Pd^{+2}	0·094	0·492	±0·06	0·441	0·376	0·314	0·284
75	Ir^{+1}	—	0·477	±0·12	0·435	0·382	0·332	0·308
76	Mo^{+4}	—	0·434	±0·04	0·381	0·314	0·249 I	0·221 (347V)

$E°$							
400°C volts	450°C volts	500°C volts	550°C volts	600°C volts	800°C volts	1000°C volts	1500°C volts
0·999	0·968	0·937	0·907	0·876 I	0·780	0·730 (927V)	
0·902	0·872	0·841	0·811	0·781 I	0·684	0·635 (927V)	
0·838	0·804	0·771	0·738	0·705	0·576 I	0·527 (877V)	
0·826	0·799 I	0·774	0·754	0·736	0·767	0·605	0·494 (1318V)
0·814 I	0·795	0·781	0·767	0·754	0·706	0·659	0·520
0·582	0·573 (407D)						
0·669	0·647	0·625	0·604	0·583	0·571 (627V)		
0·604	0·578	0·573 (461V)					
0·363	0·336	0·309	0·283	0·257	0·198 (727D)		
0·370	0·347	0·324	0·301	0·279	0·224 (727D)		
0·270	0·253 (427D)						
0·254	0·223	0·192	0·176 (527D)				
0·254	0·225	0·196	0·167	0·139 I	0·061	0·028 (927V)	
0·285	0·262	0·240	0·217	0·196	0·184 (627D)		

TABLE III.—*Continued*

Order 25°C	Metal ion	25°C (aqueous) volts	25°C (solid) . volts	Uncertainty (25°C) volts	100°C volts	200°C volts	300°C volts	350°C volts
77	Pt^{+4}	—	0·390	±0·04	0·337	0·268	0·203	0·185 (347D)
78	Ir^{+2}	−0·019	0·369	±0·07	0·315	0·247	0·182	0·150
79	W^{+2}	—	0·369	±0·15	0·321	0·260	0·202	0·174
80	In^{+3}	−0·069	0·361	±0·07	0·308	0·240	0·174	0·142
81	W^{+4}	—	0·325	±0·05	0·272	0·203	0·138	0·120 (327S)
82	Os^{+2}	0·231	0·325	±011	0·277	0·217	0·159	0·131
83	W^{+5}	—	0·321	±0·13	0·271	0·207 ∣	0·149	0·134 (333V)
84	Ru^{+3}	—	0·289	±0·04	0·236	0·168	0·102	0·085 (327D)
85	Pt^{+3}	—	0·289	±0·10	0·236	0·167	0·101	0·069
86	Pt^{+2}	ca −0·119	0·282	±0·07	0·228	0·160	0·094	0·063
87	W^{+6}	—	0·275	—	0·267 (37D)			
88	Pt^{+1}	—	0·260	±0·13	(dp)			
89	Au^{+1}	−0·599	0·169	±0·04	0·125	0·069	0·063 (212D)	
90	Au^{+3}	−0·419	0·153	±0·07	0·106	0·053 (187D)		

S = sublimation; V = vaporization; D = decomposition; dp = disproportionates.
* Values listed at the sublimation, vaporization, or decomposition temperatures obviously apply to cells with the solid or molten electrolyte, that is, slightly below the sublimation, vaporization, or decomposition temperature.

				$E°$				
400°C volts	450°C volts	500°C volts	550°C volts	600°C volts	800°C volts	1000°C volts	1500°C volts
0·119	0·088	0·071 (477D)					
0·146	0·126 (437D)						
0·110	0·094 (427D)						
0·103	0·076	0·049	0·023	−0·003	I −0·094	−0·162	−0·202 (1127V)
0·037	0·035 (405D)						
0·031	0·025 (410D)						

TABLE IV. Standard electromotive forces for single, solid, or molten iodides*

Order 25°C	Metal ion	$E°$						
		25°C (aqueous) volts	25°C (solid) volts	Uncertainty (25°C) volts	100°C volts	200°C volts	300°C volts	350°C volts
1	Cs^{+1}	3·559	3·557	±0·01	3·491	3·397	3·303	3·257
2	Rb^{+1}	3·561	3·474	±0·04	3·406	3·315	3·224	3·179
3	K^{+1}	3·561	3·441	±0·01	3·369	3·269	3·171	3·122
4	Ra^{+2}	3·556	3·350	±0·07	3·283	3·195	3·110	3·068
5	Eu^{+2}	4·036	3·231	±0·15	3·172	3·095	3·021	2·985
6	Ba^{+2}	3·536	3·187	±0·01	3·124	3·042	2·963	2·924
7	Sm^{+2}	<3·80	3·122	±0·15	3·063	2·987	2·913	2·876
8	Na^{+1}	3·350	3·035	±0·01	2·967	2·871	2·776	2·729
9	Sr^{+2}	3·526	3·014	±0·01	2·953	2·874	2·797	2·759
10	Li^{+1}	3·681	2·914	±0·09	2·860	2·788	2·712	2·675
11	Ca^{+2}	3·506	2·845	±0·01	2·784	2·705	2·628	2·591
12	La^{+3}	3·156	2·486	±0·01	2·423	2·341	2·261	2·222
13	Ce^{+3}	3·116	2·457	±0·01	2·398	2·321	2·246	2·209
14	Pr^{+3}	3·106	2·414	±0·01	2·352	2·272	2·194	2·156
15	Nd^{+3}	3·076	2·371	±0·01	2·309	2·229	2·151	2·113
16	Pm^{+3}	3·056	2·327	±0·02	2·266	2·188	2·111	2·074
17	Sm^{+3}	3·406	2·284	±0·10	2·220	2·138	2·058	2·019
18	Gd^{+3}	3·036	2·201	±0·01	2·138	2·056	1·977	1·938
19	Tb^{+3}	3·026	2·183	±0·03	2·118	2·035	1·953	1·913
20	Dy^{+3}	2·986	2·154	±0·01	2·090	2·007	1·927	1·887
21	Y^{+3}	3·006	2·139	±0·03	2·077	1·995	1·916	1·878
22	Ho^{+3}	2·956	2·110	±0·01	2·046	1·962	1·881	1·841
23	Er^{+3}	2·936	2·096	±0·01	2·034	1·954	1·876	1·837
24	Tm^{+3}	2·916	2·067	±0·01	2·006	1·926	1·849	1·812
25	Eu^{+3}	3·046	2·038	±0·10	1·974	1·890	1·809	1·769
26	Zr^{+2}	—	2·038	±0·43	1·980	1·905	1·833	1·798

$E°$							
400°C	450°C	500°C	550°C	600°C	800°C	1000°C	1500°C
volts	volts	volts	volts	volts	volts	volts	volts
3·211	3·164	3·119	3·073	3·028 │	2·791	2·500	2·080 (1280V)
3·134	3·089	3·045	3·000	2·956 │	2·705	2·406	1·967 (1304V)
3·074	3·026	2·979	2·931	2·884 │	2·686	2·387	1·923 (1324V)
3·027	2·986	2·945	2·905	2·865 │	2·713	2·578	2·111
2·948	2·913	2·878 │	2·846	2·819	2·712	2·613	2·384
2·886	2·848	2·811	2·773	2·737 │	2·598	2·479	2·205
2·840	2·805	2·769 │	2·738	2·710	2·604	2·504	2·269
2·682	2·636	2·589	2·544	2·498 │	2·353	2·135	1·703 (1304V)
2·722	2·685	2·648 │	2·617	2·589	2·480	2·370	2·063
2·639 │	2·603	2·572	2·541	2·511	2·398	2·292	2·208 (1171V)
2·554	2·517	2·480	2·444	2·407 │	2·272	2·150	2·015 (1227V)
2·184	2·146	2·108	2·070	2·033 │	1·891	1·770	1·535 (1402V)
2·173	2·137	2·101	2·066	2·031 │	1·897	1·779	1·560 (1397V)
2·118	2·081	2·044	2·007	1·971 │	1·836	1·720	1·507 (1377V)
2·075	2·037	2·000	1·962	1·925 │	1·782	1·657	1·440 (1367V)
2·037	2·000	1·964	1·928	1·892 │	1·752	1·639	1·443 (1367V)
1·980	1·942	1·904	1·866	1·829	1·682 (820D)		
1·900	1·861	1·824	1·787	1·749	1·603 │	1·470	1·282 (1337V)
1·874	1·835	1·796	1·758	1·719	1·570 │	1·430	1·240 (1327V)
1·848	1·810	1·771	1·733	1·696	1·548 │	1·409	1·228 (1317V)
1·839	1·801	1·764	1·727	1·690	1·545	1·404 │	1·237 (1307V)
1·801	1·762	1·723	1·685	1·647	1·497	1·351	1·174 (1297V)
1·800	1·762	1·725	1·688	1·652	1·509	1·370 │	1·212 (1277V)
1·774	1·737	1·701	1·664	1·628	1·487	1·350 │	1·207 (1257V)
1·729	1·691	1·651	1·613	1·575	1·425	1·369 (877D)	
1·763 │	1·733	1·709	1·685	1·661	1·571	1·484	1·473 (1027V)

TABLE IV.—*Continued*

Order 25°C	Metal ion	$E°$ 25°C (aqueous) volts	25°C (solid) volts	Uncertainty (25°C) volts	100°C volts	200°C volts	300°C volts	350°C volts
27	Lu^{+3}	2·886	1·995	±0·01	1·932	1·851	1·772	1·733
28	Mg^{+2}	3·006	1·951	±0·01	1·893	1·819	1·747	1·712
29	Zr^{+3}	—	1·922	±0·29	1·861	1·781	1·704	1·683 (327 dp)
30	Sc^{+3}	2·176	1·908	±0·01	1·847	1·768	1·692	1·655
31	Yb^{+3}	2·906	1·807	±0·10	1·742	1·659	1·577	1·538
32	U^{+3}	2·436	1·767	—	1·714	1·647	1·581	1·549
33	Hf^{+4}	2·336	1·648	±0·32	1·586	1·507	1·430	1·392
34	Th^{+4}	2·536	1·496	±0·11	1·435	1·356	1·279	1·241
35	Zr^{+4}	2·166	1·496	±0·22	1·438	1·363	1·290	1·255
36	U^{+4}	2·136	1·467	—	1·410	1·336	1·265	1·230
37	V^{+2}	ca 1·816	1·453	±0·43	1·399	1·331	1·266	1·235
38	Ti^{+2}	2·266	1·409	±0·22	1·351	1·277	1·205	1·169
39	Mn^{+2}	1·816	1·388	±0·04	1·333	1·263	1·196	1·163
40	Tl^{+1}	0·972	1·388	±0·02	1·334	1·264	1·196	1·159
41	Cr^{+2}	1·541	1·279	±0·15	1·226	1·157	1·091	1·059
42	Ti^{+3}	ca 1·846	1·250	±0·14	1·194	1·122	1·052	1·020 (347 dp)
43	Ti^{+4}	—	1·212	±0·11	1·159	1·094	1·037	1·010
44	Be^{+2}	2·486	1·203	±0·11	1·150	1·084	1·020	0·990
45	Zn^{+2}	1·399	1·185	±0·01	1·130	1·060	0·993	0·960
46	$(Al^{+3})_2$	2·296	1·184	±0·01	1·129	1·057	0·999	0·971
47	In^{+1}	ca 0·886	1·171	±0·22	1·107	1·020	0·931	0·387
48	Cd^{+2}	1·039	1·141	±0·02	1·086	1·016	0·949	0·914
49	Ge^{+2}	0·636	1·084	±0·33	1·031	0·962	0·905	0·879
50	Ge^{+4}	—	1·019	±0·16	0·963	0·896	0·838	0·810

E°

400°C	450°C	500°C	550°C	600°C	800°C	1000°C	1500°C
volts	volts	volts	volts	volts	volts	volts	volts
1·695	1·657	1·619	1·582	1·545	1·400	1·259 |	1·137 (1207V)
1·676	1·643	1·609	1·575	1·542 |	1·423	1·353 (927V)	
1·617	1·581	1·544	1·508	1·473	1·333	1·257 (909S)	
1·498	1·459	1·420	1·382	1·344	1·194	1·048	1·029 (1027D)
1·517 1·355	1·486 1·335 (427S)	1·455	1·424	1·393	1·270	1·146	0·837
1·203	1·166	1·129	1·093 |	1·060	0·939	0·918 (837V)	
1·220 |	1·199 (431S)						
1·195	1·160	1·126	1·092	1·058	0·923	0·788	0·383
1·203	1·172	1·142	1·112	1·083 |	0·970	0·915 (927V)	
1·135	1·100	1·066	1·033	0·999 |	0·894	0·799	0·787 (1027V)
1·130	1·098	1·066	1·034	1·002 |	0·903	0·890 (827V)	
1·122 |	1·087	1·059	1·032	1·005	0·902	0·890 (823V)	
1·027	0·995	0·964	0·933	0·903 |	0·784	0·771 (827V)	
0·995 (377V)							
0·960	0·930 |	0·910 (487V)					
0·928 |	0·895	0·868	0·841	0·815	0·750 (727V)		
0·952 (386V)							
0·853	0·819	0·785	0·752	0·720	0·650 (715V)		
0·882	0·860	0·839	0·819	0·799	0·711 (796V)		
0·854	0·829	0·805	0·781	0·758	0·724 (677V)		
0·795 (377V)							

TABLE IV.—*Continued*

Order 25°C	Metal ion	E°						
		25°C (aqueous) volts	25°C (solid) volts	Uncertainty (25°C) volts	100°C volts	200°C volts	300°C volts	350°C volts
51	Pb^{+2}	0·762	1·008	±0·01	0·944	0·870	0·799	0·763
52	In^{+3}	0·978	0·911	±0·04	0·860	0·793 ∣	0·738	0·713
53	Ga^{+3}	1·166	0·853	±0·01	0·797	0·725 ∣	0·665	0·638 (349V)
54	Sn^{+2}	0·772	0·849	±0·07	0·794	0·725	0·652 ∣	0·612
55	Cu^{+1}	0·115	0·821	±0·04	0·771	0·707	0·645	0·615
56	Ag^{+1}	−0·163	0·788	±0·01	0·743	0·696	0·659	0·641
57	Fe^{+2}	1·076	0·759	±0·04	0·706	0·639	0·574	0·542
58	$(Hg^{+1})_2$	−0·153	0·677	±0·02	0·609	0·521	0·444 (290D)	
59	Co^{+2}	0·913	0·646	±0·02	0·596	0·531	0·469	0·439
60	Hg^{+2}	−0·218	0·607	±0·01	0·542	0·459 ∣	0·387	0·358
61	B^{+3}	—	0·549	±0·07 ∣	0·510	0·464	0·460 (210V)	
62	Ni^{+2}	0·886	0·542	±0·02	0·487	0·415	0·346	0·312
63	Si^{+4}	—	0·452	±0·01	0·399 ∣	0·335	0·282 (288V)	
64	Sb^{+3}	—	0·452	—	0·403 ∣	0·341	0·288	0·263
65	Bi^{+3}	—	0·441	—	0·386	0·315	0·244	0·207
66	Mo^{+2}	—	0·369	±0·15	0·315	0·247	0·182	0·150
67	Rh^{+3}	−0·164	0·361	±0·14	0·308	0·239	0·174	0·156 (327D)
68	Pd^{+2}	−0·351	0·325	±0·17	0·272	0·203	0·133	0·106
69	Rh^{+1}	0·036	0·325	±0·13	0·277	0·217	0·159	0·132
70	As^{+3}	—	0·318	—	0·268 ∣	0·209	0·157	0·132
71	Ir^{+1}	—	0·304	±0·12	0·261	0·208	0·159	0·135
72	Mo^{+4}	—	0·304	±0·04	0·253	0·187	0·120	0·086
73	Ir^{+2}	−0·464	0·217	±0·07	dp			
74	Pt^{+3}	—	0·217	±0·10	0·160	0·086	0·032 (277D)	
75	Pt^{+4}	—	0·206	±0·04	0·150	0·078	0·024 (277D)	

$E°$							
400°C	450°C	500°C	550°C	600°C	800°C	1000°C	1500°C
volts	volts	volts	volts	volts	volts	volts	volts
0·726 ǀ	0·696	0·670	0·644	0·620	0·532	0·504 (872V)	
0·687	0·663	0·639 (500V)					
0·589	0·550	0·520	0·491	0·462	0·396 (714V)		
0·586	0·557	0·528	0·500 ǀ	0·474	0·396	0·324	0·244 (1207V)
0·624	0·607	0·590	0·574 ǀ	0·563	0·528	0·496	0·401
0·511	0·480	0·499	0·418 ǀ	0·390	0·306	0·295 (827V)	
0·409	0·379	0·350 ǀ	0·326	0·306	0·229	0·219 (827V)	
0·355 (354V)							
0·278	0·245	0·212	0·179	0·147	0·053 (747S)		
0·238	0·225 (427V)						
0·170	0·164 (408D)						
0·119	0·088	0·058	0·027	−0·004	−0·127	−0·249	−0·555
0·074 ǀ	0·046	0·021	−0·002	−0·026	−0·082 (827V)		
0·104 ǀ	0·090 (427D)						
0·108	0·101 (414V)						
0·111	0·088	0·065	0·054 (527D)				
0·053 ǀ	0·038 (422V)						

TABLE IV.—*Continued*

Order 25°C	Metal ion	$E°$						
		25°C (aqueous) volts	25°C (solid) volts	Uncertainty (25°C) volts	100°C volts	200°C volts	300°C volts	350°C volts
76	Pt^{+2}	−0·564	0·195	±0·07	0·142	0·073	0·008	−0·010 (327D)
77	Au^{+1}	−1·044	0·194	±0·04	0·165	0·138 (177D)		
78	Ir^{+3}	−0·514	0·152	±0·07	0·060	−0·059	−0·176	−0·233
79	W^{+2}	—	0·130	±0·15	0·77	0·009	−0·009 (227V)	
80	Rh^{+2}	0·036	0·108	±0·11	dp			
81	Pt^{+1}	—	0·108	±0·13	0·061	0·000	−0·058 (300 dp)	
82	W^{+4}	—	0·108	±0·11	0·055	0·029 (137D)		

S = sublimation; V = vaporization; D = decomposition; dp = disproportionates.
* Values listed at the sublimation, vaporization, or decomposition temperatures obviously apply to cells with the solid or molten electrolyte, that is, slightly below the sublimation, vaporization, of decomposition temperature.

$E°$							
400°C	450°C	500°C	550°C	600°C	800°C	1000°C	1500°C
volts	volts	volts	volts	volts	volts	volts	volts
−0·290	−0·321 (427D)						

TABLE V.   Standard E.M.F. for single, solid, or molten oxides

Metal	25°C (aqueous) volts	25°C (solid) volts	Uncertainty 25°C volts	100°C volts	200°C volts	300°C volts	350°C volts	400°C volts	450°C volts	500°C volts
Zr^{2+}	—	5·287G	±0·053	5·298	5·199	5·100	5·051	5·002	4·953	4·904
Ca^{2+}	3·418h	3·131	±0·002	3·090	3·037	2·984	2·931	2·931	2·905	2·881
La^{3+}	3·298h	3·085L	±0·13	3·027	2·979	2·933	2·909	2·886	2·863	2·840
Ac^{3+}	3·091i	3·079G	±0·031	3·048	3·005	2·964	2·943	2·922	2·902	2·882
Pr^{3+}	3·237h	3·058L	±0·06	3·022	2·975	2·930	2·906	2·884	2·861	2·838
Nd^{3+}	3·242h	3·040L	±0·06	3·009	2·964	2·921	2·899	2·878	2·858	2·836
Th^{4+}	2·880h	3·018L	±0·001	2·982	2·933	2·883	2·858	2·833	2·808	2·784
Be^{2+}	3·021i	3·014	±0·065	2·975	2·922	2·869	2·842	2·815	2·789	2·764
Th^{2+}	—	3·003G	±0·030	2·970	2·924	2·878	2·855	2·832	2·809	2·786
Ce^{3+}	3·275h	2·999G	±0·030	2·963	2·915	2·868	2·844	2·820	2·796	2·772
Ba^{1+}	—	2·970G	±0·030	2·931	2·865	2·802	2·771	2·731	2·700	2·669
Sm^{3+}	3·233h	2·963G	±0·030	2·928	2·881	2·833	2·809	2·786	2·762	2·738
Mg^{2+}	3·091h	2·952	±0·001	2·910	2·854	2·798	2·770	2·742	2·714	2·686
Li^{1+}	3·363h	2·903	±0·002	2·857	2·533	2·406	2·341	2·277	2·212	2·147
Am^{3+}	3·107h	2·902G	±0·029	2·868	2·824	2·779	2·757	2·735	2·714	2·692
Sr^{2+}	3·276h	2·901	±0·002	2·861	2·809	2·758	2·733	2·708	2·683	2·659
Y^{3+}	3·120h	2·894G	±0·029	2·859	2·813	2·767	2·744	2·721	2·699	2·667
Sc^{3+}	3·013h	2·826G	±0·028	2·792	2·744	2·697	2·673	2·650	2·626	1·602
Hi^{4+}	2·901oh	2·797L	±0·067	2·757	2·705	2·654	2·628	2·603	2·577	2·552
U^{4+}	2·582h	2·786	±0·12	2·793	2·704	2·659	2·636	2·613	2·591	2·568
Ba^{2+}	2·309h	2·738	±0·002	2·701	2·651	2·693	2·579	2·555	2·532	2·508
Al^{3+}	2·701h	2·723	±0·009	2·683	2·629	2·575	2·549	2·522	2·494	2·468
U^{2+}	—	2·667G	±0·027	2·632	2·588	2·544	2·522	2·500	2·478	2·456
Pu^{3+}	2·821h	2·666G	±0·027	2·633	2·590	2·548	2·527	2·507	2·486	2·466
Ra^{2+}	2·548o	2·548L	±0·12	2·507	2·458	2·409	2·395	2·361	2·338	2·314
Ti^{2+}	2·537o	2·537G	±0·025	2·497	2·447	2·397	2·372	2·347	2·323	2·299
Pu^{4+}	2·457h	2·537G	±0·025	2·500	2·457	2·413	2·392	2·371	2·349	2·328
Pa^{4+}	—	2·526G	±0·025	2·487	2·439	2·391	2·367	2·343	2·319	2·295
Np^{4+}	2·529h	2·526G	±0·025	2·494	2·448	2·404	2·381	3·359	2·338	2·316
Am^{4+}	2·533h	2·504L	±0·038	2·468	2·424	2·379	2·357	2·334	2·312	2·291
Ti^{3+}	2·501o	2·501	±0·026	2·463	2·413	2·366	2·342	2·318	2·295	2·272
Pu^{2+}	—	2·385G	±0·024	2·355	2·320	2·285	2·268	2·252	2·236	2·219
Pr^{4+}	2·385o	2·385L	±0·038	2·347	2·299	2·250	2·225	2·202	u	
Ce^{4+}	2·374o	2·374	±0·3	2·341	2·298	2·256	2·234	2·213	2·182	2·171
Ti^{4+}	2·091ho	2·219	±0·084	2·177	2·132	2·087	2·065	2·043	2·021	2·000
Si^{2+}	—	2·125G	±0·021	2·089	2·039	1·991	1·966	1·941	1·916	1·891
Si^{4+}	2·101i	2·083	±0·056	2·045	1·999	1·954	1·931	1·908	1·886	1·863

$E°$										
550°C	600°C	800°C	1000°C	1500°C	1750°C	2000°C	2250°C	2500°C	2750°C	3000°C
volts	volts	volts	volts	volts	volts	volts	volts	volts	volts	volts
4·856	4·807	4·614	4·381	3·947	3·715	3·389	3·254	nd		
2·855	2·830	2·728	2·626	2·354	2·114	1·882	1·657	1·439	1·227l	nd
2·817	2·794	2·703	2·550	2·317	2·204	2·901	1·982	nd		
2·862	2·843	2·766	2·687	2·503	2·422l	2·350	nd			
2·815	2·793	2·702	2·608	2·370	2·254l	2·139	2·027	1·917	1·808	1·701
2·816	2·796	2·717	2·642	2·334	2·214	2·095	1·979l	1·970	u	
2·759	2·734	2·636	2·539	2·296	2·173	2·049	1·925	1·799	1·672	1·542
2·737	2·712	2·607	2·505	2·309	2·130	1·828	1·670l	nd		
2·764	2·741	2·649	2·557	2·325	2·207l	2·100	2·002	1·905	1·810	1·742 (2977V)
2·748	2·724	2·627	2·526	2·280l	2·165	2·066	1·969	1·873	1·778	1·685
2·638	2·608l	2·518 (767 V)								
2·714	2·691	2·596	2·507	2·260	2·135	2·021	1·917l	1·813	1·709	1·607
2·659	2·631	2·508	2·366	1·905	1·636	1·370	1·107	0·847	0·588l	0·332
2·081	2·016	1·752	1·489	1·689l	1·291	0·933	0·578	0·470 (2327 V)		
2·670	2·648	2·563	2·477	2·262	2·161l	2·069	nd			
2·634	2·610	2·513	2·409	2·105	1·871	1·642	1·416l	nd		
2·654	2·632	2·545	2·459	2·250	2·141	2·036	1·934	nd		
2·579	2·555	2·461	2·367	2·127	2·002	1·878	1·755l			
2·528	2·502	2·405	2·308	2·074	1·960	1·848	1·738	1·631	1·525l	nd, u
2·546	2·523	2·434	2·342	2·110	1·995	1·881	u			
2·485	2·461	2·361	2·224	2·021	1·860l	1·673	1·497	1·323	1·167 (2727 V)	
2·441	2·414	2·301	2·188	1·909	1·772	1·637l	nd			
2·434	2·411	2·313	2·213	1·963	1·831	1·699	1·568	nd		
2·446	2·426	2·336	2·253	2·061l	1·986	1·917	1·850	1·785	1·722	1·663 (2977 V)
2·291	2·267	2·173	2·077	1·788	1·335	1·335	1·112	0·893	u	
2·274	2·250	2·155	2·062	1·838	1·731 (1737D)	nd				
2·308	2·287	2·200	2·114	1·906	1·805	1·707l..0·620		1·543	1·466	1·390
2·272	2·248	2·152	2·056	1·815	1·693	1·570	1·447l	nd		
2·294	2·273	2·115	2·018	1·750	1·623	1·498	1·375l	1·337	nd	
2·268	2·247	2·162	2·076	1·863	1·762	1·663	u			
2·249	2·226	2·136	2·046	1·828	1·725	1·614l	1·509	1·418	1·329	1·241
2·203	2·187	2·115	2·043l	1·931	1·878	1·826	1·815 (2052 V)			
2·150	2·129	2·943	1·054	1·734	1·626	1·519	nd			
1·978	1·955	1·869	1·776	1·557	1·449l	1·340	1·247	1·155	1·064	1·001 (2927 D)
1·866	1·841	1·740	1·638	1·368	1·203	1·039	0·875l	nd		
1·841	1·822	1·734	1·647	1·412l	1·288	1·166 (1977D)				

TABLE V.—*Continued*

Metal	$E°$ 25°C (aqueous) volts	25°C (solid) volts	Uncertainty 25°C volts	100°C volts	200°C volts	300°C volts	350°C volts	400°C volts	450°C volts	500°C volts
Np^{5+}	2·025oh	2·077G	±0·021	2·041	1·996	1·952	1·930	1·908	1·886	1·864
U^{6+}	2·045o	2·045	±0·07	2·011	1·967	1·921	1·899	1·877	1·855	1·832
B^{3+}	2·045o	2·045	±0·003	2·011	1·965	1·920	1·897	1·875 l	1·852	1·839
Ta^{5+}	2·041o	2·041	±0·06	2·005	1·957	1·909	1·886	1·862	1·838	1·815
Pa^{5+}	—	2·025G	±0·020	1·988	1·940	1·891	1·867	1·843	1·819	1·795
V^{2+}	—	2·016G	±0·033	1·992	1·944	1·898	1·875	1·852	1·830	1·807
Nb^{2+}	—	1·973G	±0·02	1·936	1·893	1·850	1·829	1·808	1·787	1·766
Nb^{4+}	—	1·964L	±0·05	1·932	1·886	1·840	1·817	1·796	1·773	1·750
V^{3+}	1·959o	1·959	±0·045	1·925	1·881	1·837	1·815	1·795	1·773	1·753
Na^{1+}	2·678h	1·951	±0·002	1·904	1·830	1·757	1·721	1·686	1·651	1·616
Mn^{2+}	1·951h	1·882	±0·002	1·854	1·816	1·779	1·760	1·742	1·723	1·705
Nb^{5+}	—	1·873L	±0·036	1·838	1·791	1·746	1·722	1·700	1·677	1·655
Cr^{3+}	1·883h	1·806	±0·005	1·669	1·598	1·527	1·491	1·455	1·419	1·383
V^{4+}	—	1·724	±0·011	1·690	1·646	1·604	1·583	1·562	1·541	1·521
Ga^{3+}	1·647h	1·714L	±0·008	1·673	1·613	1·553	1·523	1·493	1·462	1·433
K^{1+}	2·651h	1·652L	±0·018	1·577	1·496	1·415	1·376	1·335	1·296	1·255
Zn^{2+}	1·646h	1·649	±0·001	1·610	1·558	1·507	1·482	1·456	1·431	1·403
Ga^{1+}	1·630o	1·630L	±0·009	1·582	1·518	1·453	1·420	1·389	1·357	1·325
Ca^{4+}	1·550o	1·550L	±0·013	1·516	1·473	1·440 (275 D)				
Mn^{3+}	1·387h	1·534L	±0·010	1·500	1·456	1·412	1·390	1·368	1·346	1·324
Rb^{1+}	2·548h	1·507L	±0·006	1·455	1·381	1·306	1·270	1·232	1·294	1·157
Sr^{4+}	1·507h	1·607L	±0·007	1·469	1·418	1·410 (215 D)				
Mg^{4+}	1·497o	1·497G	±0·015	1·511 (88 D)						
V^{5+}	1·492o	1·492	±0·012	1·457	1·412	1·368	1·346	1·323	1·302	1·280
Ba^{4+}	1·472o	1·472L	±0·010	1·435	1·387	1·337	1·319	1·288 l	1·264	1·243
Li^{2+}	1·463o	1·463L	±0·004	1·453	1·380 (197 D)					
In^{3+}	1·401h	1·449L	±0·002	1·411	1·358	1·304	1·277	1·250	1·223	1·196
Ge^{2+}	1·431o	1·432L	±0·25	1·389	1·331	1·276	1·276	1·221	1·194	1·166
Cs^{1+}	2·452h	1·422L	±0·04	1·350	1·250	1·150	1·101	1·051	1·001 l	0·953
P^{5+}	1·887i	1·420G	±0·014	1·384	1·333	u				

				$E°$						
550°C	600°C	800°C	1000°C	1500°C	1750°C	2000°C	2250°C	2500°C	2750°C	3000°C
volts	volts	volts	volts	volts	volts	volts	volts	volts	volts	volts
1·842	1·820	1·809 (627 D)								
1·810	1·789	1·766 (652 D)	u							
1·820	1·801	1·727	1·655	1·476	1·387	1·297	1·207 (2247 V)	u		
1·792	1·768	1·674	1·583	1·356	1·246l	1·142	nd			
1·771	1·747	1·650	1·554	1·314	1·191l	1·081	nd			
1·785	1·763	1·675	1·590	1·382	1·283	1·174l	1·097	1·025	0·954	0·883
1·746	1·725	1·646	1·569	1·390	1·307	1·228	1·153l	nd, u		
1·746	1·725	1·646	1·569	1·390	1·307	1·228	1·153l	nd, u		
1·729	1·707	1·621	1·536	1·329	1·229	1·129l	1·047	0·971	0·896	0·824
1·732	1·712	1·632	1·553	1·366	1·271l	1·184	1·110	1·037	0·965	0·895
1·582	1·549	1·422l	1·256	0·743 (1275 S)						
1·686	1·668	1·591	1·515	1·305	1·190l	1·103	nd, u			
1·632	1·610	1·522	1·439l	1·245	1·169	1·096	1·023	0·950	0·878	0·827 (2927 V)
1·346	1·310	1·165	1·019	0·651	0·465l	u				
1·499	1·479	1·400	1·321	1·131l	1·053	0·978	0·907	0·837	0·770	0·703
1·403	1·374	1·257	1·145	0·881l	0·709	0·559	0·459	0·380	0·324 (2627 V)	
1·215	1·176	1·071	0·805	0·347	0·131	−0·051	nd, u			
1·376	1·348	1·239	u							
1·294	1·262l	1·201 (727 V)								
1·302	1·281	1·197	1·111	0·892 (1347D)						
1·119	1·082l									
1·259	1·238l	1·164	1·054	0·931	0·852	0·773	0·757 (2·52 V)	u		
1·223	1·203	1·119	1·104 (837 D)							
1·169	1·142	1·035	0·926	0·660l	0·530	0·429	nd			
1·140	1·112	1·052 (710 S)								
0·909	0·866	0·830 (642 V)								

TABLE V.—*Continued*

	$E°$									
Metal	25°C (aqueous) volts	25°C (solid) volts	Uncertainty 25°C volts	100°C volts	200°C volts	300°C volts	350°C volts	400°C volts	450°C volts	500°C volts
Cr^{4+}	—	1·388G	±0·014	1·356	1·310	1·265	1·242	1·220	1·208 (427 D)	
Ge^{4+}	1·401i	2·377L	±0·13	1·334	1·278	1·222	1·195	1·167	1·139	1·111
W^{4+}	—	1·349L	±0·007	1·315	1·271	1·228	1·206	1·185	1·164	1·144
Sn^{4+}	1·237h	1·346	±0·003	1·307	1·254	1·200	1·173	1·145	1·118	1·090
Sn^{2+}	1·321h	1·333	±0·003	1·303	1·258	1·168	1·135	u		
W^{6+}	1·321	±0·01	±0·001	1·286	1·242	1·199	1·178	1·157	1·136	1·115
Fe^{3+}	1·170h	1·280	±0·001	1·245	1·199	1·154	1·132	1·110	1·088	1·066
Mo^{4+}	1·272o	1·272L	±0·008	1·235	1·191	1·148	1·127	1·106	1·084	1·064
Fe^{3+}	1·277h	1·266	±0·002	1·237	1·200	1·165	1·147	1·129	1·112	1·095
In^{2+}	—	1·258G	±0·013	1·220	1·167	1·113	1·086	1·059	1·032	1·005
H^{1+}	1·229h	1·2229	±0·002	1·169 (100 V)						
Mn^{4+}	1·201o	1·208	±0·001	1·172	1·124	1·077	1·053	1·030	1·006	0·983
Cd^{2+}	1·209h	1·166	±0·030	1·127	1·078	1·028	1·004	0·976	0·949	0·922
Ni^{2+}	1·119h	1·121	±0·007	1·084	1·037	0·990	0·967	0·943	0·920	0·897
Na^{2+}	1·114o	1·114L	±0·047	1·075	1·018	0·962	0·933	0·905	0·876	0·849
Co^{2+}	1·134h	1·106	±0·002	1·069	1·024	0·980	0·957	0·953	0·912	0·898
K^{2+}	1·085o	1·085L	±0·026	1·043	0·984	0·927	0·897	0·869	0·840I	0·813
Sb^{3+}	1·061i	1·077	±0·009	1·041	0·995	0·949	0·938	0·903	0·880	0·858
Te^{2+}	—	1·062G	±0·011	1·027	0·977	0·928	0·904	0·879	0·854	0·825
Sb^{4+}	—	1·019	±0·001	0·982	0·934	0·887	0·863	0·840	0·816	0·793
As^{3+}	1·081i	0·995	±0·005	0·959	0·913	0·871I	0·857	0·841	0·826	0·824 (457 V)
Tc^{4+}	—	0·985G	±0·010	0·948	0·901	0·854	0·830	0·807	0·783	0·760
Pb^{2+} red	0·952h	0·981	±0·001	0·944	0·893	0·844	0·820	0·794	0·768I	0·742
Re^{4+}	0·977o	0·977L	±0·006	0·943	0·897	0·852	0·829	0·806	0·785	0·763
Bi^{2+}	—	0·943L	±0·011	0·908	0·859	0·809	0·780	0·751	0·720	0·691
Re^{6+}	0·946i	0·918G	±0·092	0·887	0·846	0·812	u			
Rb^{2+}	0·906o	0·906L	±0·02	0·860	0·799	0·738	0·708	0·678	0·647	0·617
Sb^{5+}	0·801i	0·869	±0·020	0·834	0·783	0·733	0·708	0·694 (380 D)		

$E°$										
550°C	600°C	800°C	1000°C	1500°C	1750°C	2000°C	1150°C	2500°C	2750°C	3000°C
volts	volts	volts	volts	volts	volts	volts	volts	volts	volts	volts
1·083	1·056	0·947	0·826l	0·574	0·449	0·326	0·206	0·157	u	
1·122	1·102	1·022	0·946l	0·781	0·715	0·689				
						(1852D)				
1·063	1·036	0·930	0·824	0·566l	0·439	0·330	0·227	0·064	−0·097	−0·210
										(2927 V)
1·094	1·094	1·073	0·991	0·991	0·724	0·651	0·629			
						(1827 V)				
1·045	1·024	0·939	0·855	0·645	nd, u					
1·043	1·021	0·939	0·859	0·662	0·568l	0·497				
						(1977D)				
1·077	1·060	0·982	0·920	0·767	0·695	0·622	0·549	0·498		
								(2427V)		
0·979	0·952	0·846	0·742l	0·516	0·418					
					(1727V)					
0·960	0·937	0·828	0·806	u						
			(847 D)							
0·895	0·867	0·742	0·538	0·036	−0·022					
					(1559S)					
0·875	0·852	0·763	0·677	0·476	0·345l	0·253	nd			
0·820	0·792	u								
0·876	0·854	0·765	0·676	0·448	0·316l	0·223	0·131	0·041	−0·004	
									(2627V)	
0·789	0·766	0·662	0·497	0·099	0·078					
					(1527 V)					
0·836	0·814l	0·731	0·654	0·496						
				(1425 V)						
0·795	0·764l	0·650	0·559	0·261	0·259					
					(1502 V)					
0·770	0·747	0·643	u							
0·737	0·714	0·622	0·532	0·314	0·211	0·110l	0·021	−0·060	−0·139	−0·218
0·717	0·692	0·595	0·505	0·318						
				(1472 V)						
0·741	0·719	0·634	0·550l	0·377	0·306	0·236	0·171	0·106	0·044	−0·012
										(2977 V)
0·662	0·632	0·514l	0·405	0·153	0·080					
0·588l	0·560	0·416	0·241	−0·183	u					

TABLE V.—*Continued*

Metal	25°C (aqueous) volts	25°C (solid) volts	Uncertainty 25°C volts	100°C volts	200°C volts	300°C volts	350°C volts	400°C volts	450°C volts	500°C volts
Cr^{6+}	0·867o	0·867G	±0·009	0·832 I	0·791	0·757	0·741	0·725	0·709	0·693
Bi^{3+}	0·858o	0·858	±0·002	0·823	0·777	0·731	0·705	0·679	0·654	0·628
Cs^{2+}	0·848o	0·848	±0·04	0·811	0·753	0·701	0·675	0·649	0·623	0·598
As^{4+}	—	0·810G	±0·008	0·775	0·726	0·678	0·654	0·630	0·606	0·581
Tc^{6+}	—	0·802G	±0·008	0·767	0·723	0·679	0·657	0·636	0·615	0·593
As^{5+}	1·097i	0·801	±0·001	0·765	0·716	0·668	0·644	0·620	0·596	0·572
Re^{7+}	0·985i	0·788G	±0·079	0·755	0·712 I	0·671	0·655	0·650 (363 V)		
Cu^{1+}	0·759o	0·758	±0·010	0·729	0·690	0·652	0·634	0·614	0·596	0·578
K^{3+}	0·723o	0·723L	±0·09	0·689	0·642	0·596	0·573	0·550 I	0·529	0·510
Re^{3+}	—	0·720G	±0·072	0·691 I	0·658 (187 V)					
Tl^{1+}	0·744h	0·718	±0·007	0·666	0·598	0·537 I	0·510	0·482	0·455	0·428
In^{1+}	—	0·716G	±0·007	0·669	0·618	0·559 I	0·533	0·512	0·491	0·470
Te^{4+}	0·971i	0·700	±0·001	0·517	0·471	0·425	0·401	0·379	0·358	0·332
Tc^{7+}	—	0·691G	±0·007	0·520	0·488 I	0·455	0·451 (311 V)			
Rb^{3+}	0·668o	0·668L	±0·05	0·630	0·578	0·526	0·499	0·473	0·447 I	0·423
Cu^{2+}	0·620h	0·659	±0·004	0·626	0·578	0·531	0·508	0·485	0·462	0·439
Cs^{3+}	0·622o	0·622L	±0·040	0·587	0·541	0·497	0·475	0·453	0·432	0·411
Pb^{4+}	0·567o	0·567	±0·002	0·530	0·482	0·438 (290 D)				
P^{4+}	—	0·553G	±0·006 I	0·513	0·478 (180 S)					
K^{4+}	0·540o	0·540L	±0·513	0·513	0·479	0·445	0·429	u		
Rb^{4+}	0·513o	0·513L	±0·06	0·481	0·437	0·394	0·372	0·351 I	0·324	0·313
Na^{4+}	0·504o	0·504L	±0·06	0·481	0·449	0·417	0·401	0·385	0·369	0·354
Cs^{4+}	0·501o	0·501L	±0·16	0·467	0·422	0·378	0·355	0·334 I	0·322	0·304
Po^{4+}	0·901i	0·499	±0·012	0·463	0·416	0·368	0·343	0·317	0·292	0·267
Se^{4+}	0·767i	0·450L	±0·002	0·417	0·369	0·320	0·305 (330 S)			
Ru^{4+}	0·441o	0·441L	±0·011	0·346	0·358	0·311	0·287	0·263	0·240	0·216
Rh^{1+}	0·414o	0·414L	±0·013	0·393	0·363	0·335	0·321	0·307	0·293	0·280
Os^{4+}	0·383o	0·383	±0·01	0·506	0·461	0·414	0·391	0·368	0·345	0·322
Os^{8+}	0·383o	0·383	±0·01 I	0·359	0·350 (130 V)					
Rh^{3+}	0·361o	0·361L	±0·008	0·324	0·277	0·229	0·206	0·182	0·160	0·136

$E°$										
550°C	600°C	800°C	1000°C	1500°C	1750°C	2000°C	2250°C	2500°C	2750°C	3000°C
volts	volts	volts	volts	volts	volts	volts	volts	volts	volts	volts
0·677	0·662	0·619 (727 V)								
0·603	0·579	0·4791	nd							
0·5731	0·549	0·528 (650 D)								
0·557	0·533	0·4161	0·311	0·080	nd					
0·572	0·551	0·467	nd							
0·549	0·525	0·421	0·407 (827 D)							
0·560	0·541	0·469	0·4001	0·229	0·212	0·202 (1800D)				
0·490	0·471	0·482 (702 V)								
0·459 (537 V)										
0·307	0·2821	0·718	0·084	−0·176	−0·309	nd, u				
0·400	0·378	0·260	0·124	u						
0·416	0·393	0·303	0·215	0·078	u (1336D)					
0·393	0·376	0·285	0·175	−0·091	−0·219	nd, u				
0·285	0·267	0·169	nd							
0·3381	0·326	0·283	0·220	0·209 (1027 V)						
0·287	0·274	0·173	0·072	−0·170	−0·287	−0·401	nd			
0·2411	0·220	0·135	0·050	−0·251	−0·397	−0·539	nd			
0·193	0·169	0·076	−0·015	−0·074						
0·267	0·254	0·203	0·155	0·126	nd, u					
0·300	0·277	0·255 (650 D)								
0·112	0·090	−0·908	−0·138	−0·197 (1115D)						

TABLE V.—*Continued*

Metal	25°C (aqueous) volts	25°C (solid) volts	Uncertainty 25°C volts	100°C volts	200°C volts	300°C volts	350°C volts	400°C volts	450°C volts	500°C volts
Rh^{2+}	0·347o	0·347L	±0·013	0·314	0·274	0·234	0·215	0·196	0·176	0·158
K^{6+}	0·316o	0·318G	±0·003 I	0·286	0·253	0·221	0·206	0·191	0·178 (442 V)	
Pd^{2+}	0·332h	0·312L	±0·008	0·269	0·215	0·160	0·134	0·107	0·079	0·053
Ir^{3+}	0·304o	0·304L	±0·05	0·274	0·235	0·197	0·178	0·159	0·141	0·123
Ir^{4+}	0·304o	0·304L	±0·04	0·270	0·231	0·193	0·173	0·155	0·136	0·117
Hg^{2+}	0·303o,l	0·303l	±0·001	0·263	0·208	0·153	0·126	0·075	0·024	0·026 (500 D)
Ru^{8+}	0·287o	0·287G	±0·029 I	0·285	u					
Pt^{2+}	0·251h	0·239G	±0·024	0·198	0·153	0·109	0·088	0·066	0·045	0·024
Pt^{4+}	0·226i	0·206G	±0·021	0·177	0·131	0·086	0·064	0·041 I	0·019	0·007 (447 D)
Ag^{1+}	0·057o	0·057	±0·001	−0·097 (187 D)						
Ag^{2+}	−0·028o	−0·028L	±0·05	−0·039 (100 D)						
Au^{3+}	−0·228h	−0·282	±0·14	−0·355	−0·415 (160 D)					

S = sublimation; V = vaporization; D = decomposition.
h = hydroxide.
oh = $HfO(OH)_2$; $NpO_2(OH)$.
o = oxide.
ho = TiO_2H_2O.
G = A. GLASSNER—*The Thermodynamic Properties of the oxides, fluorides and chlorides to 2500°K* Argonne national laboratory. Report ANL-5750.
L = W. M. LATIMER—*The oxidation states of the elements and their potential in aqueous solutions.* Prentice-Hall. New York, 1952.

					$E°$					
500°C	600°C	800°C	1000°C	1500°C	1750°C	2000°C	2250°C	2500°C	2750°C	3000°C
volts	volts	volts	volts	volts	volts	volts	volts	volts	volts	volts
0·139	0·120	0·049	−0·019	−0·060 (1121D)						
0·027	0·001	−0·101	−0·138 (877 D)	u						
0·105	0·086	0·014	−0·053l	−0·204	−0·272	−0·336 (1977 V)				
0·099	0·081	0·011	−0·057	−0·089 (1100D)						

nd, u

Molina, G. (1960). Thesis, Paris.

Patterson, J. W. (1971a). Chap. 5. *In* "The Physics of Electronic Ceramics" (Eds. L. L. Hench and D. B. Dove), Marcel Dekker Inc. New York.

Patterson, J. W. (1971b). *J. electrochem. Soc.* **118,** 1033.

Pourbaix, M. (1963). "Atlas d'équilibres électrochemiquis", Gauthier–Villar, Paris.

Pourbaix, M. J. and Rorive-Boute, C. M. (1948). *Discuss. Faraday Soc.* **4,** 139.

Richardson, F. D. and Jeffes, J. H. E. (1948).   *J. Iron Steel Inst.* **160,** 261.

Worrell, W. L. (1965). *Trans. metall. Soc. A.I.M.E.* **233,** 1173.

Worrell, W. L. (1965). *Can. met. Quart.* **4,** 87.

Worrell, W. L. and Chipman, J. (1964). *J. phys. Chem.* **68,** 860.

18. COMPARISON BETWEEN ELECTRODE POTENTIALS IN SOLID AND LIQUID ELECTROLYTES

J. Hladik

Université de Dakar, Sénégal, West Africa

I. Universal Reference Electrode

A. The Hydrogen Electrode

The hydrogen electrode was universally adopted as the comparison electrode of the potentials of various electrochemical systems. This electrode is the best one. It has been used for a long time and is capable of a high degree of reproducibility. Consequently it is the best choice when establishing potential scales in all kinds of electrolytes, be they non-aqueous, molten or solid. However, this choice should not be considered as ultimate and an absolute potential scale is preferable. Thus, in thermodynamics, Kelvin's temperature scale has replaced the Celsius empirical scale.

The choice of a particular reference electrode, no matter what it be, will permit the comparison of potential scales in various solvents. Actually, the reference electrodes in each kind of solvent are very diverse and the choice of a unique reference electrode is deemed necessary.

B. Comparison Between Electrode Potentials

The use of an aqueous solution to define the hydrogen electrode, in principle, limits its use to aqueous electrolytes. Now electrochemistry has progressively extended itself to non-aqueous solvents, molten salts and solid electrolytes.

It is, therefore, necessary to compare the potential scales in the various solvents. Several means have been proposed by Strehlow (1952) and Koepp et al. (1960).

(a) *"Mixed"* cells. When two identical electrodes, for example silver, are plunged into two Ag^+ ionic solutions in two different solvents, an e.m.f. is established between the electrodes which is the sum of a junction potential and of:

$$\Delta E = \frac{RT}{F} \text{Log}_e \frac{[Ag^+]_{s_1}}{[Ag^+]_{s_2}}$$

$[Ag^+]_{s_1}$ and $[Ag^+]_{s_2}$ being the Ag^+ activities in each solvent respectively.

(b) *Calculation of solvation energies.* The difference of the standard potential of a system, chosen then as a reference, for two difference solvents is determined by calculation. For example, for the Rb^+/Rb system, the difference of the Rb^+ solvation energies in each of the two solvents is calculated by using the Born equation (Pleskov, 1947; Strehlow, 1952; Koepp et al., 1960).

II. Comparison Between Potentials in Aqueous and Molten Media

The standard potentials of numerous electrochemical systems having been determined in molten salts, one then looked for means to compare their potential scales in the various electrolytes. It is ascertained that the normal potential of a metal depends chiefly on the nature of the anion of a solvent and only slightly on the cation (formation of the complexes with the ions of the solvent). (Klemm and Biltz, 1926; Drossbach, 1956). In fact, generally, the cations of the solvent fasten on to the anions and the anions of the solvent on to the cations.

The determination of the standard potentials in molten salts and the comparison of the potential scales, have posed the problem of choosing one origin for the potentials. Various reference electrodes have been used in molten salts. One reference electrode, slightly sensitive to the solvent's nature has been proposed (Biltz and Klemm, 1926; Cambi and Devoto, 1927; Heymann et al., 1943). This electrode is made up of sodium or a mixture of tin–sodium, isolated from the solution by a sheet of glass which contains some Na^+ ions, and is susceptible to being used in every solvent containing some Na^+ ions.

A direct comparison between the potential of an electrode in an aqueous electrolyte and that of a second electrode in a molten electrolyte can also be made by using a "mixed" cell. The idea of such a comparison was first put into effect by Abraham (1963). The aqueous solution and the molten salt ($TlNO_3$) contained a common solute, silver nitrate. The electrolytic junction between the two compartments was made by using an asbestos cord impregnated with organic materials. The results allowed a definition of the standard potential for the silver electrode in water using a reference electrode in molten salts.

The problem of the junction between aqueous and molten media has again been undertaken by Abraham and Hechler (1968). The junction is simplified by putting the aqueous and molten media directly in contact in a glass capillary. The authors have measured the e.m.f. of the following cell:

$$Ag/AgNO_3, TlNO_3//AgNO_3, TlNO_3/Ag$$
aqueous solution   molten eutectic

This simplified technique is used when studying the differences between potentials of silver electrodes plunged respectively into an aqueous electrolyte and a $CsNO_3 - TlNO_3-Cd(NO_3)_2$ molten mixture. (Abraham and Hechler, 1969). The authors studied these "mixed" cells varying the $AgNO_3$ concentration both in aqueous and molten electrolytes.

A series of measurements of potentials was made by Boxall (1970) in molten alkali salts. The electrode potentials in molten nitrites, nitrates, chlorides, sulphates and carbonates are compared with the hydrogen electrode potential, at 25°C.

III. Comparison Between Potentials in Aqueous and Solid Media

A. Experimental Technique

The comparison technique of potentials, in two different kinds of solvents, by means of a "mixed" cell, can very easily be extended to solid electrolytes.

Such a comparison was made at the time of the study of a "mixed" cell formed by aqueous and solid media (Hladik, 1970). The electrode in the aqueous solution is a saturated HgCl/Hg electrode; the solid electrolyte is the $CuCl_2$–LiCl–KCl mixture. The junction between aqueous and solid electrolytes was made with the help of a sintered glass impregnated with the aqueous solution. The mixed cell can be written as:

$$Cu/CuCl_2 + LiCl + KCl/sintered/\quad KCl \quad /HgCl, Hg$$
solid   glass   saturated
                         solution

The measurements are obtained at 25°C.

The capillary technique, used in molten salts by Abraham and Hechler (1968), can very easily be applied to solid electrolytes. Figure 1 depicts the mixed cell thus obtained. The solid electrolyte is contained in a glass tube to which is sealed a capillary tube. The solid electrolyte is melted which permits the filling up of the capillary. The tube is then chilled to the desired temperature. The level of the solid in the capillary reaches its upper extremity. The glass tube is then immersed in the liquid electrolyte which is in contact with the solid electrolyte.

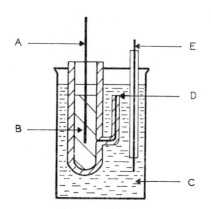

Fig. 1. "Mixed" cell-A and E, silver electrodes; B, solid electrolyte; C, aqueous solution; D, capillary.

B. Silver Electrode in Nitrates (after Hechler, 1970)

1. Experimental Cell

The apparatus depicted in Fig. 1 is used to study the "mixed" cell:

$$Ag/AgNO_3, TlNO_3//AgNO_3, TlNO_3/Ag$$

molten or	aqueous
solid electrolyte	solution

The $AgNO_3$–$TlNO_3$ mixture is composed of:

$$0.533 \text{ in } AgNO_3$$
$$0.467 \text{ in } TlNO_3$$

The melting point of this binary is 86°C. The aqueous solution contains the same nitrates and the solute concentrations are 0.1m/1 in $AgNO_3$, 0.0875 m/1 in $TlNO_3$.

When the mixture is chilled, the electrolyte contained in the capillary can be cracked; this is due to the contraction of the mixture when it is cooled.

Then the electrical contact between the electrode in the aqueous solution and the one in the solid electrolyte is interrupted. It is then necessary to melt the mixture again and then rechill it slowly.

More often, the glass tube is broken after a certain length of time following solidification. This is due to the slow transformation of the metastable phase into the stable phase. The capillary (length: 4 cm) prevents water diffusion towards the silver electrode embedded in the solid electrolyte.

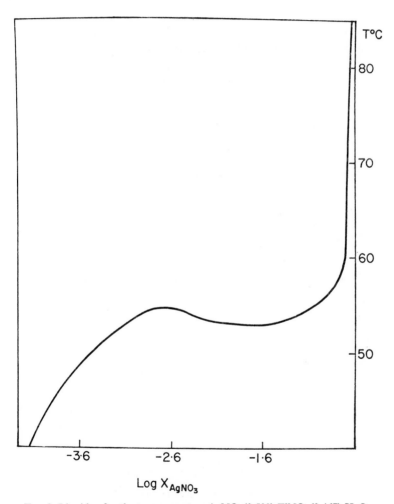

FIG. 2. Liquidus for the ternary system $AgNO_3(0{\cdot}533)$ $TlNO_3(0{\cdot}467)$ H_2O.

The internal resistance of the solid-state cell is about 10^7 to 10^8 Ohms. Stable and reproducible values can be obtained for the e.m.f., the uncertainty is less than ± 0.5 mV for the experiments performed between 100°C and 50°C; below this temperature the reproducibility is only ± 1.5 mV.

2. Ternary System $AgNO_3$–$TINO_3$–H_2O

The liquidus curve for the $AgNO_3$–$TINO_3$–H_2O mixtures has been determined (Fig. 2), using the nitrate proportions, $AgNO_3 : TINO_3 = 0.533 : 0.467$.

The ternary mixture is capable of an important superfusion and, therefore, ultrasonics are used to activate the crystallization. The temperatures in the beginning of crystallization are thus determined with an uncertainty less than 1°, for a given composition of the ternary system.

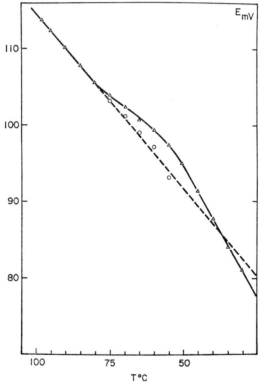

FIG. 3. E.m.f. of the "mixed" cell

Ag/Ag NO₃, TlNO₃//AgNO₃, TlNO₃/Ag
molten or solid aqueous solution

The liquidus is depicted in Fig. 2. This curve has a maximum for a percentage of 62·1 (weight of the salts).

3. E.m.f. of the "Mixed" Cell

The experimental results for the e.m.f. of the mixed cell are shown by the graph in Fig. 3.

These e.m.f.'s are the sum of a junction potential and of the potential difference between the silver electrodes in the two electrolytes. The latter value can be calculated by evaluation of the junction potential.

The extreme concentrations of the gradient, in the upper extremity of the capillary, are, on the one hand, the concentration of the aqueous solution, and, on the other hand, the concentration corresponding to the solubility of $AgNO_3$ and $TlNO_3$. Using an approximation taken from the liquidus curve, one can evaluate this solubility and with the use of the classical equation, the junction potential varies from about 30 mV to 40 mV in the temperature range 50 to 100°C (Table I). Therefore, the potential difference of the silver electrode in the binary $AgNO_3$–$TlNO_3$ and its standard potential in water, in the temperature range 100°C and 50°C, is $-65·7 \pm 5$ mV.

TABLE I.   "Mixed" cell of silver nitrate. E.m.f and junction potential

t	E	$\cdot f$	$-\left[\dfrac{RT}{F}\log a\right]$	$-\left[E - \dfrac{RT}{F}\log a\right]$	$-E_J$	$U_{AgTl} - U_0$
98	113·7	0·607	218·2	104·5	37·7	−66·8
95	112·3	0·609	216·3	104·0	37·4	−66·6
90	110·0	0·612	213·2	103·3	36·9	−66·4
85	107·7	0·616	210·1	102·4	36·2	−66·2
80	105·5	0·619	207·0	101·5	35·3	−66·2
75	103·9	0·622	203·9	100·0	34·5	−65·5
70	102·3	0·625	200·8	98·5	33·6	−64·9
65	100·9	0·628	197·8	98·1	32·6	−65·5
60	99·3	0·631	194·7	95·4	31·7	−63·7
55	97·3	0·634	191·7	95·1	29·7	−66·1
50	94·9	0·636	188·6	93·7	—	—
45	91·2	0·639	185·6	94·4	—	—
40	85·9	0·641	182·6	96·7	—	—
35	82·6	0·644	179·5	96·9	—	—
30	79·3	0·646	176·5	97·2	—	—
25	75·1	0·648	173·5	98·4	—	—

References

Abraham, M. (1963). Thesis Strasbourg, France.

Abraham, M. and Hechler, J. J. (1968). *Electrochimica Acta.* **13,** 1681.

Abraham, M. and Hechler, J. J. (1969). *Electrochimica Acta.* **14,** 725.

Biltz, W. and Klemm, W. (1926). *Z. anorg. allgem. Chem.* **152,** 267.

Boxall, L. G. (1970). Thesis. University of Saskatchewan, Saskatoon (Canada).

Cambi, L. and Devoto, G. (1927). *Gazz. Chim. Ital.* **57,** 836.

Drossbach, P. (1956). *Z. Electrochem.* **60,** 387.

Hechler, J. J. (1970). Unpublished results.

Heymann, E., Martin, R. J. L. and Mulcahy, M. F. R. (1943). **47,** 473.

Hladik, J. (1970). *Comptes Rendus Ac. Sc.* **270,** 1771.

Klemm, W. and Biltz, W. (1926). *Z. anorg. allgem. chem.* **152,** 225.

Koepp, H. M., Wendt, H. and Strehlow, H. (1960). *Z. Elektrochem.* **64,** 483.

Pleskov, W. A. (1947). *Uspekhi khim.* **16,** 254.

Strehlow, H. (1952). *Z. Electrochem.* **56,** 827.

D. ELECTRODE PROCESSES

19. SOLID STATE CELLS, IMPEDANCE AND POLARISATION

K. O. Hever†

School of Chemistry, City of Leicester Polytechnic, Leicester, England

I. Introduction

A. Uses of Solid State Cells

The current blossoming of interest in solid state electrochemical cells, signified by the inclusion in the Electrochemical Society Fall meeting of a symposium on solid electrolytes and solid state batteries, has gone past the stage of using such cells primarily as analytical tools, e.g. for the determination of the thermodynamic quantities of solid state reactions. Solid state cells have become of interest both from a theoretical view point and the possibility of constructing devices either totally novel or not easily duplicated in the field of electronics. Some examples of this trend are logic elements based on semi-conducting glasses (Ovshinsky, 1968) or on silver iodide electrolyte (Mrgudich, 1965), all solid state batteries (Gutmann *et al.*, 1967,

† Deceased.

809

1968), solid electrolyte batteries (Lehovec and Broder, 1954) and fuel cells ("Advances in Chemistry," 1965), thermoelectrochemical cells (Weininger, 1964) and electrochemical integrators (Raleigh, 1967). The possibility of constructing discrete components (Breyer and Bauer, 1966), dubbed "faradaics", such as constant phase shift elements, an all solid state capacitor (Hever, 1968) with unusually high energy density and even an electrochemical transistor (Bardeen and Letaw, 1954) and an electrochemical oscillator (Tamamushi, 1966) have been described. The possibilities for the construction of novel devices seem, in this field, to be unlimited but electrolytics suffer in comparison to electronics in two major ways.

(1). The speed of ion movement in electrolytes is characteristically slow compared to electronics in semi-conductors or compared to the speed with which gases can be mixed and reacted; this implies that electrolytic devices will be limited as components to DC or low frequency AC applications and limited as to power sources to situations where high energy density and lower power density (e.g. space applications) are required.

(2). The chemical components of aqueous electrolyte batteries are seldom thermodynamically stable, thus aqueous batteries of e.m.f. greater than the decomposition voltage of water, 1·23 volts, are in commercial use despite the unavoidable presence of corrosion. For many applications 1% or so of self discharge per week is not unacceptable (e.g. the automobile battery).

It is in situations where the side effects of corrosion—low reliability and durability, limited range of operating conditions, the production of gases—are unacceptable, that solid state devices offer real advantages; thus no battery for remote location use will be ignored on the grounds of power cost if commensurate increases in reliability and durability are offered. Mechanical and electrical stability, operation over a wide temperature range and in a gravity-free environment, ease of construction and, very often, cheapness of raw materials are the principal advantages that solid state electrochemical cells offer.

From a theoretical viewpoint the absence of convection simplifies the situation and diffusion layers of several millimetres building up on the time scale of minutes can be observed. The extent of an excess or deficit in the concentration of mobile ion in the diffusion layer is obviously governed by the crystal structure of the solid ion conductor, in particular the amount of nonstoichiometry that the crystal can tolerate without change of phase. A coulometric experiment on a solid can thus provide information on the crystal structure and the range of stoichiometry over which the phase is stable. As in aqueous electrochemistry the current passing through an electrode measures the rate of the electrode process under a known driving force and polarisation methods can therefore be used as an analytical technique to study the kinetics of solid state reactions or gas–solid reactions if the atmosphere is not inert. The use of solid state cells in thermodynamic studies has been reviewed by Wagner ("Advances in Electrochemistry and Electrochemical Engineering", 1966; Kiukkola and Wagner, 1957) and a review concentrating on kinetic studies has been given by D. O. Raleigh ("Progress in Solid State Chemistry", 1967). To illustrate the possible techniques one example is given here.

In the solid state cell

$$Ag \mid AgI \mid \alpha\text{-}Ag_2S \mid Pt \qquad \qquad (I)$$

above 177°C the electrolyte is the purely cationic conductive (Jost and Weiss, 1954) AgI while Ag_2S is a mixed Ag^+ and electron conductor (Wagner, 1933a, b; Rahlts, 1935). The cell potential therefore fixes the activity of metallic silver in Ag_2S relative to that of pure silver at the same temperature. Wagner (1953) using a potentiostatic technique measured the charge passed on changing the cell potential, i.e. the activity of Ag in Ag_2S, and established the range of stoichiometry of $\alpha\text{-}Ag_2S$ as Ag_xS where $x = 2 \cdot 000 - 2 \cdot 002$ at 200°C.

For a 1 mm thick sample of Ag_2S Wagner found the equilibration time to be ~ 1 sec., at 220°C, which leads to an estimate of the diffusion coefficient $D = x^2/t \sim 10^{-2} \, cm^2 \, sec^{-1}$ for silver metal in Ag_2S. In a similar manner Raleigh (1967) investigated the concentration of silver in silver–gold alloy as a function of activity by the use of the cell

$$Ag \mid AgBr \mid Au \qquad \qquad (II)$$

and estimated the diffusion coefficient of silver in gold from the polarization decay transient. Placing the right hand electrode of cell (I) in an atmosphere of sulphur allowed Reinhold (1934) to obtain the free energy of formation of $\alpha\text{-}Ag_2S$, $-10 \cdot 15 \, kcal \, mole^{-1}$ at 150°C, from the cell potential. The kinetics of heterogeneous reduction of Ag_2S by H_2 was investigated by Kobayashi and Wagner (1957) using a H_2 atmosphere for the right hand electrode; they concluded, that since the rate of reduction as measured by the cell current was independent of the activity of silver metal and hence independent of the activity of sulphur, the rate determining step must be

$$S^{2-} + H_2 = H_2S + 2\,electrons.$$

The rate of reduction of Ag_2S by vacuum (Rickert, 1961) was found to be second order in sulphur activity indicating as rate determining step

$$2S \,(absorbed) = S_2 \,(gas).$$

Replacing the right hand electrode platinum contact of cell (I) by a reactive metal, e.g. Ni, allowed Mrowec and Rickert (1963) to investigate the kinetics of the solid state reaction

$$Ni + \alpha\text{-}Ag_2S = NiS + 2Ag$$

and they concluded that at 400–450°C the rate determining step was the diffusion of Ni^{2+} in NiS and the rate of polarization decay was used to estimate the diffusion coefficient of nickel ion vacancies. From an experimental viewpoint the failure of cells to move to another voltage level on charge/discharge indicates the production or removal of a phase and this fact can be used to delineate phase diagrams. In thermodynamic studies it is important to establish the absence of self discharge of the cell, which would imply electronic conductivity in the electrolyte; polarisation measurements can help establish which reaction the cell e.m.f. corresponds to (Barbi, 1968). For a more detailed appreciation of the role of polarisation measurements as a technique the review and original articles are recommended.

B. Types of Polarisation

As in the case of aqueous cells at room temperature, the polarisation to be expected in solid state cells can be divided into ohmic, diffusion (or "concentration") charge transfer (or "activation") and crystallisation types. The specific conductance of a "good" ionically conducting solid at room temperature of about 10^{-3} mho cm^{-1} is poor in comparison to most aqueous electrolytes, 1–10^{-1} mho cm^{-1}, and ohmic polarisation tends to dominate the performance of solid state cells. To overcome this defect solid state cells are usually investigated at temperatures above room temperature, 100–1000°C, and the result is to depress the importance of charge transfer and crystallisation overvoltage due to their high activation energy, > 20 kcal mole^{-1}, compared to diffusion and ohmic effects of 4–10 kcal mole^{-1}. Thus with but few exceptions the bulk of experimental work on polarisation can be accounted for in terms of these two types of effects only. The separation of the two types in any particular experiment has proved more difficult since it is usually experimentally impossible to detect diffusion layers by the use of a reference electrode placed close to the working electrode. The separation of the two effects is normally inferred from polarisation decay experiments (current-time under controlled voltage or voltage-time under controlled current). In particular, ohmic effects should always be linear, current proportional to voltage in steady state experiments and response proportional to displacement from equilibrium in experiments on transients; a single relaxation time is also to be expected. Diffusion effects should be linear only when the diffusion equation can be linearised, i.e. when exp $(eV/kT) - 1$ can be put as (eV/kT) i.e. at polarisations of a few milli-volts only. The difficulty lies in the fact that simple Debye relaxation is seldom observed even in cases clearly under ohmic control and that capacitances are observed which are much greater than can be accounted for either on the basis of linear effects, i.e. geometric capacitance between the electrolyte and electrodes considered as infinitely conductive or non-linear effects, i.e. the capacitance due to the presence of space-charge.

This chapter discusses some solid state ion conductors and non-metallic electron conductors from which solid state cells might be constructed; the results of polarisation measurements are then discussed. It is suggested that the cause of unusually large capacitances lies in a slight electronic conductivity of ionically conducting samples and the effect of this on polarisation measurements calculated from the theory of the transmission line using dielectric loss as an example. The results suggest that the failure of the Debye theory does not necessarily imply the failure of a two parameter model.

C. Solid Ion Conductors

Crystalline solids owe their ionic conductivity to the presence of disorder in the crystal structure. The disorder may take the form of ions appearing in interstitial positions (Frenkel defects), or vacant lattice sites (Schottky defects). Solid ion conductors fall, loosely, into two groups: those in which the lattice defects must be produced either thermally or by the introduction of foreign ions into the lattice, called "valency-controlled doping", and those in which a considerable number of defects occur naturally in the solid due to its structure. The first group is characterised by a low conductivity and a high activation energy for conduction. Since the experimentally determined activation energy is the sum of the energies for migration of the defect and its creation, the concentration of defects at low doping levels, $\leqslant 1\%$, is often equal to the concentration of the doping ion and thus the mobility of the defect ion can be calculated. Knowledge of the nature, number and mobility of defects makes this group of great interest from a theoretical viewpoint. A large number of conductors of this type have been listed ("American Institute of Physics Handbook", 1963) and a typical example is KCl. In this crystal, above 400°C, both anions and cations participate to a comparable extent (Hantlemann and Wagner, 1950) in the conduction process and therefore a Schottky disorder model involving an equal number of cation and anion vacancies to maintain electrical neutrality is indicated. Substitutional doping of K^+ by a univalent ion, e.g. Na^+ should not, therefore, alter the conductivity markedly and this has been found to be the case for K^+ doped into NaCl (Etzel and Maurer, 1950). Valency controlled doping by $PbCl_2$ (Lehfeldt, 1933) or $SrCl_2$ (Kiukkola and Wagner, 1957) must lead to an equivalent number of cation vacancies or, much less likely, an equivalent number of anion interstitials. Doping KCl with $SrCl_2$ causes a marked increase in the conductivity by an amount proportional to the concentration of Sr^{2+} and the crystal becomes a pure cationic conductor; equating the concentration of cation vacancies of the concentration of Sr^{2+} allows the mobility of vacancies to be estimated: the activation energy for mobility, E_a, was found to be 18 kcal mole^{-1}. Knowing also the conductivity of undoped KCl the concentration of defects in the pure crystal can be calculated and hence the activation energy for the creation of defects, $E_a = 48$ kcal mole^{-1}.

Above 500°C most of the alkali halides conduct by both cation and anion migration whereas at lower temperatures the conduction process is dominated by the ion of lower activation energy. Thus the sodium, potassium and rubidium halides become cation conductors while the caesium halides and thallium chloride become anion conductors. The lithium halides are cation conductors over the whole temperature range. Lead diiodide also conducts by virtue of Schottky defects. Other conductors in this group are the anionic Frenkel defect conductors (Koch and Wagner, 1937); $PbCl_2$ either pure or doped with KCl and the fluorite structure conductors, CaF_2 and ZrO_2 which may be doped with CaO or Y_2O_3. The fluorite lattice MX_2 consists of metal ions M each coordinated to eight X atoms situated at the corners of a cube; these cubes are joined by corners to form empty cubes as secondary sites for additional anions in cubic close packing up to one additional anion per formula unit, i.e. up to MX_3. Alternatively, substitution of M by a metal of lower valency may lead to anion vacancies (Wadsley, 1955). Electrolytes based on ZrO_2 are of considerable interest since their ability to transport O^{2-} leads to uses in the study of the thermodynamics of oxides (Kiukkola and Wagner, 1957), in high temperature fuel cells and as oxygen sensors. The high temperature fluorite form of ZrO_2 is stabilised by the presence of CaO, Y_2O_3 or La_2O_3; X-ray and density measurements indicate that the doping cation substitutes for Zr (Bogren *et al.*, 1967), charge compensation is achieved, at least up to 15% of doping, by the appropriate number of anion vacancies which are the mobile specie. Pure CaF_2 is an anionic conductor (Ure, 1957) where F^- vacancies and F^- interstitials are created thermally, their mobilities have been determined by measurements on crystals doped with NaF, where vacancies dominate, and YF_3, where interstitials dominate.

Ionic conductors in the second group are characterised by a high conductivity, $> 10^{-1}$ mho cm^{-1} at a suitable temperature, a low activation energy for conduction, < 10 kcal mole^{-1}, and a relative insensitivity of conductance to doping agents or impurities. The vacancies necessary for conduction exist as a natural consequence of the lattice structure and an activation energy is necessary only for migration into the vacancy; it is ion conductors in this group, e.g. Li_2SO_4, AgI, β-alumina, which offer the greater possibility for the construction of electrochemical devices.

β-Li_2SO_4 which has a monoclinic unit cell undergoes a transition (Bielen et al., 1962) at 585°C to the cubic α-Li_2SO_4 which has the spinel structure; the lithium ions occupy the octahedral holes of the spinel lattice and because of their small size are free to migrate through the crystal, the conductivity (Kvist and Lunden, 1965) of Li_2SO_4 at 600°C is 0.3 mho cm^{-1}. Doping with small amounts of Na^+ decreases the conductivity while K^+ and Rb^+ increase it indicating that it is the size of the doping ions which is the controlling factor (Kvist, 1966a). Evidence that the conductivity is a cooperative motion among Li^+ ions has been given from isotope measurements (Kvist, 1966b). Thus Li_2SO_4 is a conductor in which the size of the mobile ion allows the holes between O^{2-} ions rather than O^{2-} vacancies to provide a conduction path. Another example where the cations are distributed over a large number of sites is the cationic Frenkel defect conductor α-AgI. The cubic zinc blende γ-AgI changes to the hexagonal zinc oxide β-AgI at 136°C in which the Ag^+ and I^- ions have mutual tetrahedral coordination. At 146°C β-AgI undergoes a phase change to α-AgI in which the I^- body centred cubic sublattice provides a framework for the Ag^+ ions which distribute themselves randomly on forty-two sites per two silver ions and provide a purely cationic conductor of 1.31 mho cm^{-1} at the transition point ("Ionic Crystals, Lattice Defects and Nonstoichiometry", 1968).

The sodium ion conductor beta–alumina has received considerable interest due to its use in the sodium–sulphur battery; it has the empirical formula $Na_2O \cdot 11\,Al_2O_3$ and has a hexagonal layer structure consisting of Al_9O_{16} spinel blocks separated by low density planes containing $NaAl_2O$ groups; these planes are 11.23 Å apart and ion conduction occurs only within these planes which are perpendicular to the axis of the unit cell. The self diffusion of Na^+, Ag^+, K^+, Rb^+ and Li^+ in single crystals of beta–alumina has been measured by the radioactive tracer technique and a strong dependence of diffusion coefficient on ion size was observed (Kummer and Yao, 1967). Beta–alumina may be classified (Schlenk, 1951) as an occlusion compound in which a stable oxide, in this case the deficient spinel compound γ-Al_2O_3, crystallises around alkali ions which are occluded with anion vacancies. Occlusion compounds have been extensively reviewed by Wadsley and these compounds result when simple binary oxides are crystallised in the presence of ions too large to substitute for the metal or to find suitable interstitial sites. The central characteristic of such compounds is their wide range of nonstoichiometry. The host oxide may arrange itself so as to form cages, tunnels or layers which contain sites available for occlusion (Wadsley,

1955). From the viewpoint of ionic conductivity cage structures such as the silicates of the ultramarine family, the clathrate compounds of beta–quinol (Palin and Powell, 1949), and the heteropoly anions (Dawson, 1953; Schoemaker et al., 1954) of tungsten and molybdenum are not of interest since the voids are invariably too large to incorporate unhydrated ions. Tunnel occlusion compounds are theoretically less suitable as ionic conductors in polycrystalline form than layer structures due to the greater probability of the conducting tunnel being blocked at the intercrystalline boundary, however the large number of such compounds that are known (Wadsley, 1955) provides an interesting field in which to look for ionic conductivity. Examples in this group are the tunnel alumino–silicates such as the minerals eucryptite (Winkler, 1948), beryl (Bragg and West, 1926) and milarite (Ito et al., 1952) and the low alkali bronzes. The tungsten bronzes MWO_3 are perovskite lattices based on WO_3 octahedra. When the amount of alkali is lowered to M_xWO_3 where $x \leqslant 0.33$ the WO_3 octahedra rearrange to form 4 and 5 (Magnéli, 1949) or 6 (Magnéli, 1953) membered rings enclosing tunnels which contain one alkali metal ion for every two tungsten octahedra running lengthwise down a tunnel. Ionic conductivity might be expected in a tunnel bronze in which the alkali ion is a loose fit as, for example, in the form of K_xWO_3 based on six membered rings (Magnéli, 1953). Similar compounds are formed by MnO_2 as host oxide in the mineral hollandite (Byström and Byström, 1950) $BaMn_8O_{16}$ in which the Ba^{2+} alternate with vacancies down tunnels of square cross section; isomorphous with hollandite is the mineral priderite (Norrish, 1951) KTi_8O_{16} based on TiO_3 octahedra. More complex tunnel structures are exhibited by the vanadium bronzes (Ozerov, 1954) $M_xV_2O_5$ where $x \leqslant 0.33$ and isomorphous crystals (Wadsley, 1955; Flood et al., 1953) exist for $M =$ Na, Ag, Li. Some evidence of ion conductivity has been obtained for the vanadium bronzes. An interesting point arising from these compounds is the method of charge compensation for the occluded alkali ions which in minerals occurs by the replacement of cations by ions of lower valency, e.g. Al^{3+} for Si^{4+} in the aluminosilicates and when prepared in the laboratory by the most used method, hydrolysis of a mixture of the fluorides, by reduction of the necessary number of metal ions, e.g. Ti^{4+} to Ti^{3+}. In KTi_8O_{16} the presence of the lower valency state is clearly indicated by the colour (black) and the presence of a high electronic conductivity (Reid and Sienko, 1967). From a practical viewpoint a very poor ionic conductor could be used to form a highly conducting solid by ion exchange at an elevated temperature with an ion of suitable size, provided that the exchanged material is stable at the temperature used. Finally, doubly charged ions seem to show a much lower mobility than singly charged ions in solids, probably due to the greater coulombic interaction with the lattice with increasing charge (Kummer and Yao, 1967).

D. Non-metallic Electron Conductors

The principles underlying the preparation and properties of non metallic semiconductors have been extensively investigated by Verwey and co-workers ("Conference on Semiconducting Materials", 1951; Kröger and Verwey, 1951; Haayman et al., 1950). Of the metal halides, oxides and sulphides the oxides have received most attention. Unlike typical metallic semiconductors, doping levels in oxides are very high, typically 10^{-1} compared to 10^{-7} while mobilities are very low, typically $10 \, cm^2 \, V^{-1} \, s^{-1}$ compared to $10,000 \, cm^2 \, V^{-1} \, s^{-1}$. Verwey concludes that a general condition for appreciable electronic conductivity in substances of this type is the presence of ions of the same element with different valency at crystallographically equivalent lattice points. Thus the semiconductor Fe_3O_4 of specific resistance $\rho \sim 10^{-2}$ ohm cm is beleived (Haayman et al., 1947) to contain, at $< 120°K$, ordered Fe^{2+} and Fe^{3+} alternating in a sequence of octahedral holes in the spinel structure, i.e.

$$Fe^{3+}(Fe^{2+}, Fe^{3+}) \, O_4.$$

The transition at 120°K involves the disordering of the Fe^{2+} Fe^{3+} Fe^{2+} sequence and a sharp decrease in specific resistence results (Haayman and Verwey, 1941). Fe_3O_4 is unusual in that it is an undoped stoichiometric semiconductor, a more common type involves nonstoichiometry and a metal in two valency states, e.g. 'FeO' which contains an excess of oxygen, some Fe^{3+} ions and cation vacancies and can be written

$$(Fe^{2+}_{1-3x} \, Fe^{3+}_{2x} \, Hole_x) \, O$$

where $x \leqslant 0.1$ and FeO is a p-type semiconductor. The oxygen deficient oxide ZnO, an n-type semiconductor, involves Zn metal interstitials. The nonstoichiometric semiconductors, in general, show only slight deviations from stoichiometry and a low conductivity, their thermal stability and reproducibility of preparation are poor and their applications therefore limited. An exception here are the tungsten and vanadium bronzes already mentioned which show a wide range of nonstoichiometry and high conductivity, e.g. when $NaWO_3$ is heated with WO_3 the perovskite structure is preserved and a bronze Na_xWO_3 produced:

$$(Na_x \, hole_{1-x})(W_x^{5+} \, W_{5-x}^{6+}) \, O_3.$$

Similarly silver, lithium and sodium vanadium bronzes are readily made from the vanadate and V_2O_5 and show a very high n-type semiconductivity (Sienko and Sohn, 1966). On crystallising bronzes oxygen is given off to the atmosphere.

Of greater interest are the semiconductors produced by controlled valency doping in which stoichiometry is maintained but the need for electrical neutrality requires the production of unusual valency states; it is the over-riding energetic importance of the particular close-packed oxide lattice together with the greater entropy of a one phase system compared to a two phase system which makes doping possible. Thus pure NiO of $\rho \sim 10^8$ ohm cm can take up considerable quantities of Li_2O at 1200°C; the Li^+ enters substitutionally into Ni^{2+} sites and charge balance is achieved by the production of an equal number of Ni^{3+} ions; the necessary oxygen is taken from the atmosphere. The specific resistance of the p-type material

$$(Li_x Ni_x^{3+} Ni_{1-2x}^{2+}) O$$

where $x = 0.1$ is about 1 ohm cm. An n-type material of similar conductivity can be made by the production of Fe^{2+} in Fe_2O_3 by sintering Fe_2O_3 and TiO_2 powders at 1300°C. It is important that the doping ion itself should not change valency, thus in Fe_2O_3, Ti Zr Ta Nb Sn act successfully but V Cr Mn do not. Doping levels are usually limited to a few per cent, thus although $FeTiO_3$ has the same structure as Fe_2O_3 the conductance no longer drops beyond 5% doping indicating that the extent of charge compensation by Fe^{2+} is limited. Verwey has established that the doping ion should change the valency of the electron conducting ion by ± 1 thus KCr_3O_8, i.e. $K Cr^{3+} Cr_2^{6+} O_8$ is an insulator and has pointed out that the doping ion and electron conducting ion need not hold similar crystallographic positions as they do not in the perovskite conductor

$$(Sr_{1-x} La_x)(Ti_{1-x}^{4+} Ti_x^{3+}) O_3.$$

If foreign ions are present in the crystallographic positions active for conduction they act as scattering centres, e.g. $CoFe_2O_4$ with a Co deficit is n-type and a Co excess is p-type but to the extent of $\zeta = 10^3$–10^4 ohm cm only. Finally it is not always possible to predict the effect of added ions thus Fe^{4+} can be produced in $LaFeO_3$ by doping with Sr^{2+} but the incorporation of Zn^{2+} Ni^{2+} Co^{2+} or Mg^{2+} into the ferrite magnetic material $BaO.6Fe_2O_3$ produces n-type materials of expanded unit cell and derived from the compounds $BaO.6Fe_2O_3.Fe_3O_4$ and $BaO.6Fe_2O_3.2Fe_3O_4$. It is obvious that electronic conduction in oxides is greatly affected by both the amount and nature of impurities and often by the nature of materials in contact with the solid, e.g. the atmosphere and the electrodes. Ways of establishing the presence or absence of electronic conductivity in the presence of ionic conductivity are considered next.

E. Mixed Ion and Electron Conductors

Due to the much greater mobility of electrons or electron holes compared to ions the extent of semiconduction in a solid often depends on small deviations from stoichiometry which can be caused by impurities or the atmosphere enclosing the sample, thus a dependence of total conductivity on the atmosphere is a powerful indication of the presence of semiconduction. Many oxides and sulphides of the transition elements are naturally non-stoichiometric and show mixed ion and electron conduction, e.g. FeS, Ag_2S, FeO, Cu_2O, ZnO. The cuprous halides show mixed conduction between 200–300°C; at low temperatures they are semiconductors and at high temperatures cation conductors ("Ionic Crystals, Lattice Defects and Non-stoichiometry", 1968). At sufficiently high temperatures the ZrO_2 type anion conductors show mixed conduction, e.g. at 870°C the conductivity (Kiukkola and Wagner, 1957) of 0·85 ZrO_2 0·15 CaO was found to be independent of the partial pressure of oxygen (P) but that of 0·85 ThO_2 0·15 CaO varied markedly. The pressure dependence of semiconductivity indicates the nature of the nonstoichiometry and differentiates between p- and n-types. In the case of NiO at high temperatures the conductivity (Mitoff, 1961) increases with oxygen partial pressure as $\sigma \propto P^{1/6}$, indicating p-type conductivity via the equilibrium

$$4Ni^{2+} + O_2 = 4Ni^{3+} + 2O^{2-} + 2 \text{ cation vacancies.}$$

Thus the equilibrium constant is

$$K = \frac{[Ni^{3+}]^4 [\text{cation vacancy}]^2}{P}$$

and putting [cation vacancy] $= \frac{1}{2}[Ni^{3+}]$ (required for electrical neutrality)

$$K = \frac{[Ni^{3+}]^6}{4P} .$$

The temperature must be high enough to establish the equilibrium and to dissociate carrier–vacant lattice site pairs, e.g. σ for Cu_2O, expected to depend on P as $\propto P^{1/8}$, varies experimentally as $P^{1/7}$. Both p- and n-type PbS can be prepared under a sulphur atmosphere; the conductivity (Bloem, 1956) at exact stoichiometry, which occurs at a sulphur pressure of 0·2 atmosphere at 950°C, is at a pronounced minimum.

In cases where the pressure dependence of conductivity is impracticable other methods may be used to detect semiconductivity. A simple qualitative test to differentiate p- and n-types is the Seebeck effect where the hot junction is positive for n-type conductors and negative for p-type. The Soret effect for ion conductors is negligibly small in comparison.

The Hall effect provides both the concentration of carriers and their sign but it is a difficult technique on poorly conducting samples since the effect is proportional to current.

Deviations from Faraday's Laws are an unambiguous test for semi-conductivity but again are difficult on poorly conducting samples of solids due to the large quantity of electricity needed to deposit measurable amounts of products and the possibility of diffusion and reaction of the products.

Wagner and Wagner (1957) have used a polarisation method to detect semiconductivity in cuprous halides by the use of a cell with one electrode blocking

$$Cu \mid CuX \mid Graphite.$$

On polarising the cell at less than the decomposition potential of CuX, with the graphite electrode positive, a transient current flows while a steady state distribution of copper ions is set up. When a steady state has been reached the current is carried exclusively by electron holes. Assuming that the entire polarising voltage (E) appears across the CuX graphite boundary then it will provide an increase in the electron hole concentration at the right hand side of the CuX sample by a factor of $\exp(EF/RT)$ over the concentration of electron holes in CuX coexisting with metallic copper at the left hand side of the CuX sample. Assuming that the holes diffuse by virtue of the concentration gradient and migration is negligible, then the current density will be given by

$$I = \frac{RT}{LF} \sigma^\circ_+ [\exp(EF/RT) - 1]$$

where L is the thickness of the CuX sample, F is the Faraday constant and σ°_+ is the electron hole conductivity of CuX coexisting with metallic copper. A linear relationship between $\log I$ and E for $EF/RT \gg 1$ was found experimentally and σ°_+ values obtained from the intercept; the slope was close to the theoretical value of $F/2\cdot30\,RT$. Thus electronic conductivity was investigated in samples where cationic conductivity was a factor 10^5 larger. The Wagner polarisation technique has been extended by Patterson et al., (Bogren et al., 1967) to electrolytes exhibiting comparable amounts of electron and electron hole conduction in the presence of a high level of ionic conductivity. The current–voltage characteristic for the one electrode blocking cell is

$$I = \frac{RT}{FL} \left\{ \sigma^\circ_- \left[1 - \exp\left(-\frac{EF}{kT} \right) \right] + \sigma^\circ_+ [^- \exp(EF/RT) - 1] \right\}$$

and therefore, for $EF/RT \gg 1$

$$I = \frac{RT}{FL} [\sigma_-^\circ + \sigma_+^\circ \exp (EF/RT)]$$

and plot of I versus $\exp (EF/RT)$ should give a straight line of intercept $(RT/FL) \sigma_-^\circ$ and slope $(RT/FL) \sigma_+^\circ$. This method has been successfully applied to $Zr_{0.85}$, $Ca_{0.15}$, $O_{1.85}$ and σ_{ionic}, σ_+°, σ_-°, determined for various electrode systems, temperatures and oxygen pressure values. The method has also been applied to Ilschner's data on AgBr in equilibrium with silver.

Due to the strong dependence of electronic conductivity on nonstoichiometry the thermodynamic state of the sample should be well defined. In the examples above, this is achieved by equilibrating the sample with metal, e.g. Cu in the case of CuX, and the activity of halogen therefore has its lowest possible value. When inert electrodes are used, e.g. Pt, Au, then the state of the sample can be fixed by control of the atmosphere, e.g. in the case of halides by the halogen partial pressure of the atmosphere. Even in the case of typical ionic conductors such as the silver halides, electron hole conduction becomes appreciable at high halogen partial pressures; thus the cell

$$Ag \mid AgBr \mid C, Br_2(g)$$

used by Reinhold (1928) exhibited an e.m.f. 3·8 mv lower than predicted from the free energy of formation of AgBr due to self discharge caused mainly by electron hole construction in AgBr promoted by Br_2 gas. Highly refractory materials, e.g. the semiconducting oxides, are usually investigated in a polycrystalline sintered state and equilibrium with the atmosphere is often sluggish, causing a difference in composition between the surface and the bulk of the crystallites. The presence of intercrystallite barrier layers causes a marked frequency dependence ("Conference on Semiconducting Materials", 1951) of capacitance and resistance, which is considered in the next section.

II. Linear Polarisation Effects

A. Dielectric Relaxation

The most simple idealised experiment to measure the ionic conductivity of a solid would be to place against the two ends of the sample a highly ionically conducting solution, preferably a solution of the ion which is mobile in the solid, and make a DC circuit using reversible electrodes dipped into the solutions. In practice this experiment is made very difficult either by the fact that the sample is not 100% dense and the solution leaks through it or, if the sample is sufficiently dense, because of surface reactions on the solid which makes ion transfer at the solution/solid boundary much slower than in the bulk solid (Hever, unpublished experiments). Consequently conductivity measurements are invariably made using AC methods and solid electrodes either pressed against or evaporated on to the sample. The disadvantage of AC methods is that they measure the total ion flow: the "true" conductivity, i.e. the ohmic portion that corresponds to the idealised experiment above plus displacement (transient) currents which show up as a time dependent DC conductivity or a frequency dependent AC conductivity. The displacement currents can arise in two ways: (a) The rotation of dipoles and the release of ions from impurity centres or crystal faults or any process which involves the movement of ions over a distance of lattice dimensions. On a macroscopic scale the potential will be linear with distance and Ohm's Law and the Superposition Principle will be obeyed; the "true" conductivity is either that at $t = 0$ for the DC case or that obtained by extrapolation to zero frequency in the AC case; the dispersion of the conductivity can be described by reference to a simple equivalent circuit consisting of frequency-independent circuit elements. (b) The growth or decay of space charges arising when ions which are free to move throughout the bulk of the sample arrive at the electrodes and either cannot be discharged (perfectly blocking electrode) or can be discharged only at a finite rate (partially blocking electrode), neglecting the effect of a finite rate of generation/recombination

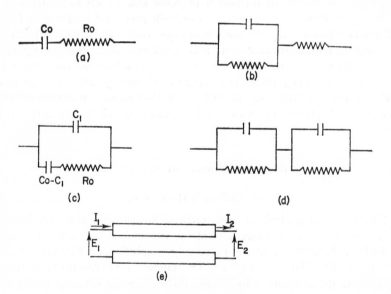

Fig. 1. Equivalent circuits.

which only appears experimentally in studies on semiconductors then a space charge equilibrium occurs when the migration of charges due to the field is balanced by the diffusion of charges due to their distribution. The potential will be non-linear with distance, and Ohm's Law and the Super-position Principle will fail; the true conductivity is that at $t = 0$ for the DC case or that obtained by extrapolation to infinite frequency in the AC case. The dispersion of the conductivity is usually described in terms of an equivalent circuit containing frequency dependent elements and the problem of non-linearity is solved by using linear approximations to non-linear equations which are valid for small applied potentials,

$$V \ll \frac{kT}{e}, \text{ i.e. } V \ll 25 \text{ mv.}$$

The case (a) above is the dielectric relaxation (DR) effect and is discussed next; case (b) has been termed the space–charge polarisation (SCP) effect and is discussed in a subsequent section.

The most simple equivalent circuit for the DR effect is the series RC circuit of Fig. 1 (a). The response to a step voltage at time $t = 0$ should be a conductivity (σ) decreasing exponentially as

$$\sigma(t) = \sigma(0) \exp(-t/\tau)$$

where τ, the relaxation time, is given by $\tau = RC$. Sutter and Nowick (1963) investigated NaCl single crystals in the cell

$$\text{Ag} \mid \text{NaCl} \mid \text{Ag}$$

in the temperature range 50–$200°$ C and found that the $\sigma(t)$ versus t curve independent of the applied voltage step over the range 2–200 V, unequivocally indicating DR polarisation. An ohmic "tail" was observed on the polarisation curves indicating that the electrodes were not totally blocking and the time dependence of conductivity could be expressed empirically, except for very short or very long times, by an inverse power relation

$$\sigma(t) - \sigma(\infty) \propto \frac{1}{t^n} \quad n = 0\cdot3 \leftrightarrow 0\cdot5$$

and $\sigma(t)$ versus $\log t$ was strongly concave towards the origin. As pointed out by Sutter and Nowick the shape of the conductivity decay curve could not be expressed as a single exponential or even as a superposition of exponentials with a narrow range of time constants; in fact an average time constant is meaningless for an inverse power shape and an extrapolation to $t = 0$ impossible. The inverse power function has been used by several workers (Bell and Manning, 1940; Brennecke, 1940) and seems to be of general applicability, but has little theoretical justification since it predicts an infinite current at $t = 0$ and an infinite charge passed at infinite time but it does not predict a peak in the dielectric loss curve. Sutter and Nowick also found that the ohmic tail was replaced by a second decay curve describable in terms of a single relaxation time when an air gap was introduced between the sample and either electrode by a mica spacer, indicating that a more appropriate equivalent circuit contains a leakage resistance and parallel air gap capacitances in series with the sample resistance, Fig. 1(b). Electrolysis over sixteen hours, equivalent to the removal of 20 planes of sodium, did not affect subsequent conductivity measurements. Thus Sutter and Nowick present impressive evidence for the DR effect in the case of "pure" NaCl crystals under their particular experimental conditions. Also their samples are highly insulative, $\sigma \sim 6 \times 10^{-14} \, \text{ohm}^{-1} \, \text{cm}^{-1}$, and thus the greater part of the applied voltage will appear across the sample and the DR effect, which is a bulk effect, can be expected. AC polarisation measurements on more conductive ($\sim 10^{-4} \, \text{ohm}^{-1} \, \text{cm}^{-1}$) single crystal KCl samples, "pure" or doped with $SrCl_2$, have been made by Jacobs and Maycock (1963) using the cell

$$Pt \mid KCl \mid Pt$$

in the temperature range 587–690°C. These authors found non ohmic behaviour under high DC voltages and noted that the differential AC and DC conductances were identical at DC voltages greater than four volts. The differential AC capacity decreased rapidly with DC voltage, becoming equal to the geometric capacitance between the electrodes: results that are consistent with polarisation caused by finite rate of discharge, i.e. an SCP effect. At low DC voltages Ohm's Law was accurately obeyed and AC capacitance measurements were found to be independent of the applied AC

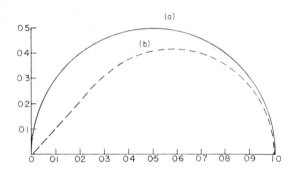

FIG. 2. Normalised Cole–Cole plots for (a) Debye Theory; (b) Transmission line.

voltage, to depend on angular velocity (w) according to $C \propto w^{-2}$, to depend on the length of the sample (L) according to $C \propto L^{-2}$ and increase with temperature in the manner of an activated process. These results are consistent with the Debye theory ("Polar Molecules", 1929; "Theory of Dielectrics", 1950) of the DR effect which assumes blocking electrodes and fields which obey an exponential decay function. Linearity is assumed by the use of the Superposition Principle. Employing the equivalent circuit of Fig. 1(a) the complex impedance is

$$\mathbf{Z} = R_0 + \frac{1}{iwC_0} = \frac{iwR_0C_0 + 1}{iwC_0}.$$

The impedance of a complex capacitance $\mathbf{C}$ is

$$\mathbf{Z} = \frac{1}{iw\mathbf{C}}$$

and by comparison

$$\mathbf{C} = \frac{C_0}{1 + iwR_0C_0}$$

the real part of which is a frequency dependent capacitance C

$$C = \frac{C_0}{1 + w^2\tau^2}$$

where τ the relaxation time is defined by $\tau = R_0C_0$; under the conditions $w\tau \gg 1$ this equation reduces to

$$C = \frac{C_0}{w^2\tau^2}$$

and gives the observed frequency dependence. $\tau = R_0C_0$ is independent of the area of the electrode but proportional to the sample length, and hence C should vary as L^{-2} while the temperature dependence of C should be that of σ^2 for simple Debye theory. The physical interpretation of C_0 is that of the geometric capacitance of the system with the sample considered infinitely conductive. It is important to point out that Jacobs and Maycock explained their results on the basis of Macdonald's linearised SCP theory, for the case of both electrodes blocking, since this theory involves more than the three assumptions of Debye theory and since the Jacobs and Maycock results seem to be truly linear, the use of Macdonald's equations seems to be unduly restrictive. The earlier paper on KCl by Allnatt and Jacobs (1961) also interprets the effects on an SCP theory and in these measurements the frequency slope of polarisation capacity was 1·62 while that of polarisation conductance was ~1·2, compared to the theoretical values of 1·5 on SCP. The polarisation capacity was again found to be independent of applied voltage; a further difficulty involved the much higher (a factor of 10^3 or 10^4) experimental capacity over that predicted .It seems that KCl with "inert" electrodes may well fall between the conditions necessary for clear-cut DR or SCP effects.

B. Frequency Dependent Dielectric Relaxation

In the preceding section the frequency dependence of polarisation capacity was derived using the equivalent circuit of Fig. 1(a). The use of bridge networks where the sample is electrically balanced against a parallel R–C circuit also gives a measure of the excess resistance of the sample above the idealised experiment value and this excess series resistance is usually expressed as a negative polarisation parallel conductance. Although experimentally less accurate than polarisation capacity measurements, and seldom attempted, the frequency dependence is derived here for completeness.

From Fig. 1(a)

$$\mathbf{Z} = \frac{iw\tau + 1}{iwC_0} \frac{1}{\mathbf{Z}} = \frac{iwC_0}{1 + iw\tau}$$

and $\Delta(1/R)$ is the real part of $1/\mathbf{Z} - 1/R_0$, i.e.

$$\Delta(1/R) = -\frac{1}{R_0(1 + w^2\tau^2)} \, ;$$

for $w\tau \gg 1$ the polarisation conductance varies as w^{-2}.

The circuit of Fig. 1(a) is insufficient to account quantitatively for DR effects. A minor objection is the neglect of the inter-electrode capacity (C_1) which must dominate high-frequency impedance measurements; this can easily be included by use of the circuit of Fig. 1(c) and the results normalised to show the frequency dependence by the Cole–Cole (1941) plot of the quadrature components of the complex normalised capacity:

$$\frac{\mathbf{C} - \mathbf{C}_1}{\mathbf{C}_0 - \mathbf{C}_1} = \frac{1}{1 + iw\tau}$$

whose components are

$$\frac{1}{1 + w^2\tau^2} \text{ and } \frac{w\tau}{1 + w^2\tau^2}.$$

The Cole–Cole plot of Debye behaviour is thus the semicircle of Fig. 2. The frequency dependence of the components is symmetrical on a $\log(w\tau)$ plot about $w\tau = 1$, Fig. 3, as can be seen by defining a variable z as $z = \ln(w\tau)$ when the components are

$$\frac{1}{1 + e^{2z}} = \frac{e^{-z}}{e^z + e^{-z}} \text{ and } \frac{1}{e^z + e^{-z}}.$$

A major objection to the simple Debye theory for solids is that the predicted curves of Fig. 3 and Fig. 4 are not observed (Cole, 1938). The experimental curves of Fig. 4 though symmetrical are less sensitive to frequency, i.e. the Cole–Cole semicircle appears as a segment. Kauzmann (1942) has discussed this failure and compared various attempts to account for it in terms of a distribution of relaxation times, i.e. the assumption of a frequency-dependent equivalent circuit. An example is the Fuoss–Kirkwood (1941) distribution function which has been employed (Kummer *et al.*, 1969) to account for the dielectric loss curves of substituted β-alumina. The distribution function leads to formulae of the Debye type with $w\tau$ replaced by $(w\tau)^\lambda$ where λ is an empirical parameter of each material. Distribution functions are of doubtful use except for curve-fitting as it is not yet possible to predict parameter values; also a superposition of relaxation times cannot account for an inverse power decay curve, although this may be more a criticism of the inverse power function, or for the experimentally observed skewed Cole–Cole plot in which marked deviations from a semicircle occur in the high frequency range. Ullman (1968) has pointed out that a non-exponential decay function is to be expected when the dipole is subject to environmental changes, e.g. free volume, on the time scale of decay. From a mechanistic viewpoint the appearance of non-Debye DR curves opens the possibility that the sample is exhibiting SCP polarisation since these are shown in the final section to be a characteristic feature of SCP.

III. Non-linear Polarisation Effects

A. Intraphase Diffusion

The preceding section discussed the dielectric relaxation effect which may be expected to occur in highly resistive samples when the bulk of the applied e.m.f. will lie across the sample. In highly conducting samples the electrodes cannot be assumed to be reversible even in cases where reversibility might be expected, e.g. AgI | Ag (Mrgudich, 1960). The consequence of a finite rate of discharge is to cause a change in the density of charge carriers at the ends of the sample (SCP effect). The blocking action increases the resistance above the infinite frequency value and introduces a parallel capacitance. Both excess resistance and parallel capacitance are frequency and temperature dependent and often dependent on the thermal and mechanical history of the sample. The theory of the SCP effect has been developed by many authors using a microscopic approach, i.e. assuming the nature, number and mobility of the charge carriers, the mobilities are assumed to be independent of concentration and the Poisson–Boltzmann equation reduced to one dimension (corresponding experimentally to thin samples of large electrode area) and linearised by the assumption of small applied voltages, in some cases the degree of dissociation and the effect of finite recombination time of charge carriers are taken into account. The microscopic approach suffers from the mathematical complexity of the final equations, from some artificiality in defining partially blocking electrodes (Chang and Jaffé, 1952) and from the conceptual difficulty that it is often not clear whether the initial assumptions are required or are unduly restrictive (Macdonald, 1970). The microscopic approach also suffers in predicting parallel capacitances often far smaller than those observed (Allnatt and Jacobs, 1961; Lawson, 1962; Raleigh, 1966). To avoid these difficulties a macroscopic approach based on equivalent circuits will be used here; the more important literature

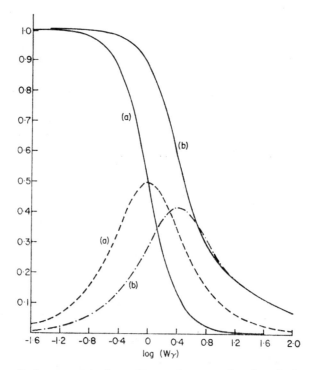

FIG. 3. Normalised components of complex capacitance as a function of reduced frequency (a) Debye Theory; (b) Transmission line.

references on SCP theory are given in the appendix. The equivalent circuit for the SCP must represent the fact that polarisation will occur only at the interfaces for high frequencies and that the AC polarisation waves will penetrate further into the bulk as frequency decreases, i.e. the transmission line of Fig. 1(e). Although transmission line theory is well known by electrical engineers it is less well known by chemists and the main results are derived here. Assuming the transmission line to have a complex series impedance per unit length z and a complex shunt admittance y then Ohm's Law yields

$$\frac{dE}{dx} = -zI \text{ and } \frac{dI}{dx} = -yE \text{ therefore } \frac{d^2E}{dx^2} = yzE.$$

Let $E = f(x)e^{iwt}$, then $f''(x) = a^2f(x)$ where $a = (yz)^{\frac{1}{2}}$. The general solution is

$$f(x) = [A_1 \cosh (ax) + A_2 \sinh (ax)] e^{iwt}$$

which gives

$$E = [A_1 \cosh (ax) + A_2 \sinh (ax)] e^{iwt}$$

$$I = -\frac{a}{z}[A_1 \sinh (ax) + A_2 \cosh (ax)] e^{iwt}.$$

The boundary condition at $x = 0$ gives $A_1 = E_1$ and $A_2 = -I_1 z/a$. Therefore

$$I = \left[\frac{-E_1}{b} \sinh (ax) + I_1 \cosh (ax) \right] e^{iwt}$$

where $b = (z/y)^{\frac{1}{2}}$.

Assuming the boundary condition $I_2 = 0$ the complex impedance Z is

$$\frac{E_1}{I_1} = Z = \frac{\cosh (ah)}{\sinh (ah)} \cdot b.$$

For a series impedance and shunt capacitance line $z = R'$ and $y = iwC'$ giving $a = (iw\tau)^{\frac{1}{2}}$ where $\tau = R'C'$ has the dimensions of time length^{-2}. For $(w)^{\frac{1}{2}} L > 2$ then

$$Z = b = \left(\frac{R'}{iwC'} \right)^{\frac{1}{2}} = \left(\frac{R'}{2wC'} \right)^{\frac{1}{2}} (1 - i).$$

Including an infinite frequency series resistance R_0 the line can therefore be represented as a series RC circuit of values

$$R_s = R_0 + \left(\frac{R'}{2wC'}\right)^{\frac{1}{2}}$$

$$\frac{1}{wC_s} = \left(\frac{R'}{2wC'}\right)^{\frac{1}{2}}$$

where the polarisation excess series resistance and capacitance depend on frequency as $w^{-\frac{1}{2}}$. Almost all experimental investigations to date have been under conditions where the polarisation resistance is very small compared to R_0 and parallel RC bridge circuits have been used. The series-parallel conversion formulae used are

$$\frac{1}{R_p} = \frac{1}{R_s}\frac{(w\tau_s)^2}{1 + (w\tau_s)^2}$$

$$C_p = \frac{C_s}{1 + (w\tau_s)^2}$$

where $\tau_s = R_s C_s$.

For polarisation series resistance small compared to R_0 and in the high frequency limit, the parallel excess conductance and parallel polarisation capacitance are

$$\Delta\left(\frac{1}{R_p}\right) = \frac{1}{R_p} - \frac{1}{R_0} = -\frac{1}{R_0}\left(\frac{R'}{2wC'}\right)^{\frac{1}{2}}$$

and

$$C_p = \frac{1}{w^2 R_0{}^2 C_s} = \frac{1}{wR_0{}^2}\left(\frac{R'}{2wC'}\right)^{\frac{1}{2}}$$

i.e. the polarisation parallel capacity depends on frequency as $w^{-3/2}$.

Exact (Macdonald, 1955) or approximate (Allnatt and Jacobs, 1961; Jackson and Young; Friauf, 1954) $w^{-3/2}$ behaviour has been observed in solids with blocking or partly blocking electrodes. Friauf (1954) investigated the parallel polarisation capacity in the cell

$$Au\,|\,AgBr\,|\,Au$$

in the temperature range 210–270°C using $< 0\cdot1$ V rms polarisation. C_p was found to have $w^{-3/2}$ frequency dependence, to increase with temperature according to an activation energy of 20–30 kcal mole^{-1}, in fair agreement with the activation energy of R_0 for single crystals of 17·8 kcal mole^{-1}. The predicted dependence on sample length, $C_p \propto L^{-2}$, was not observed but has been observed in other cases. No low frequency effects, i.e. saturation of capacity and rapid rise of resistance were observed and no experimental indication of effects caused by a finite formation–recombination rate obtained. The $w^{-3/2}$ behaviour of C_p is also obtained by Macdonald (1955) in photoconducting KBr crystals but the SCP theory may not be applicable since C_p was independent of AC voltage to within 1% up to 300 V rms. Furthermore the Cole–Cole plot of $(1/R_p w)$ versus C_p was distorted in the low frequency region, which could not be accounted for on the basis of a distribution of relaxation times or on the basis of the Cole–Cole plot of a material showing the SCP effect which distorts the semicircle in the high frequency region. Furthermore C_p was independent of L. Macdonald attributes these effects to a charge free layer at the electrodes which would be in series with the SCP capacity. Low frequency saturation of capacity was observed.

B. Interphase Diffusion Effects

The SCP theory is not limited to cases where space charge exists but may also be applied to a concentration gradient of neutral species; in interphase diffusion the quantity of matter transferred may be very large and lead to immense capacitances. Apart from the experiments on Ag_2S quoted earlier, two cells involving interphase solid solid diffusion effects have been reported. Raleigh (1966) has investigated the capacitance of the silver bromide metal interface by use of the cell

$$C, Br_2 \mid AgBr \mid Pt \text{ or } Au$$

at 246 and 293°C using the potential step method and measuring the current transient. The starting potential was held in the range 0·4–0·7 V, left electrode positive, i.e. less than the decomposition potential of AgBr. Raleigh found a typical double layer charging current of up to 1 mA peak decaying on the time scale of milliseconds superimposed on the steady-state hole current of $\sim 20\mu$ amp. The transient was approximately independent of starting voltage, direction of the voltage step and temperature, and proportional to the size of the step and the area of the AgBr/metal interface. The polarisation effect therefore seems to be a double layer capacitance effect of the Helmholtz type of value 200–300μF cm^{-2}. Capacitances of this size cannot be accounted for on the basis of space charge effects but the stored charge is of the order consistent with one lattice layer of anions or cations. Raleigh also noted effects consistent with plating out of silver and its slow diffusion (on the time scale of minutes) into the metal which were investigated (Raleigh, 1967) using the cell

$$Ag \mid AgBr \mid Au$$

at 400°C. The application of a polarising e.m.f. to the cell, silver electrode negative, defined the silver activity at the AgBr/Au interface which was maintained by the cell current as diffusion into the bulk proceeded. The current density therefore declined as $t^{-\frac{1}{2}}$ in accord with the results of diffusion theory

$$i = F \Delta c \left(\frac{D}{\pi t}\right)^{\frac{1}{2}}$$

where Δc is the concentration increment conditioned by the voltage step increment. The $t^{-\frac{1}{2}}$ dependence failed at times less than 10 s where double layer charging occurred and at times greater than 100 s where the hole current predominated. Diffusion coefficient (D) values were in accord with those determined by sectioning techniques.

In an identical method D values were obtained (Hever, 1968) for sodium and potassium metal at 300°C and 500°C in ceramic materials based on γ-Fe_2O_3 isomorphous with β-alumina and whose single crystals showed the 2-dimensional diffusion property of the alkali ions and which were made to function as electrodes by valency controlled doping of the polycrystalline samples with titanium dioxide. The two identical electrodes were separated by an ionically conducting but electronically insulating β-alumina electrolyte and diffusion of alkali metal under the voltage jump driving force occurred, as with Raleigh's work, by internal diffusion of ions and external flow of electrons. The chemical phase change was observed on charge–discharge of the cell which behaved for DC applied voltages less than 0·5 V as a capacitor ~ 20 F. This immense capacitance arises from the fact that the whole electrode volume is operating as the capacitance rather than one plane as in Raleigh's work and each electrode is behaving as a parallel plate condenser of plate separation ~ 2 Å corresponding to the Na^+–Fe^{2+} distance and repeat distance 11 Å, corresponding to the distance between alkali ion conducting planes. Obviously the transfer of 20 coulombs per volt between electrodes of volume 0·1 cm^3 is possible only because of the high non-stoichiometry of the crystal structure. The AC impedance of the cells was measured over the range 0·001 cycles per minute to several cycles per second and followed theoretical expressions developed from the solution of the diffusion equation over the whole frequency range of interest including low frequency effects. These two experiments on interphase diffusion can therefore be explained both qualitatively and quantitatively on the basis of SCP theory since diffusion and transmission line theories are equivalent.

C. The Magnitude of Equivalent Circuit Elements

The microscopic theory gives directly the value of observed circuit elements in terms of the fundamental quantities: densities of charge carriers, mobility, valency, while that of equivalent circuits needs further assumptions concerning the physical basis of the effect before fundamental quantities can be evaluated. An example is the transmission line theory of the SCP effect where the equation

$$\frac{\partial^2 i}{\partial x^2} = R'C'\frac{\partial i}{\partial t}$$

is solved and the physical process of diffusion where $\partial C/\partial t = (D\partial^2 c/\partial x^2)$ is solved. In either case a physical model is necessary to make the electrical circuit equal to physical process transformation. It is the calculation of space-charge capacitance on the basis of a physical model which has proved most difficult, the experimental values being unacceptably high. The assumption of a Boltzmann distribution of charge at high defect concentrations is not reasonable (Grimley and Mott, 1947; Grimley, 1950) but the presence of very large experimental capacitances caused by ions crossing the interface in order to equilibrate the activity of the mobile specie indicates that a space charge may not be responsible for the effect. The fact that polarisation measurements on solids are usually undertaken at elevated temperatures where even "inert" electrodes such as platinum may chemically react with the sample (Allnatt and Jacobs, 1961) supports this idea. Furthermore all electrodes are oxidising or reducing with respect to the sample and tend to induce p- or n-type conductivity in it; the number of electrons transferred to the bulk necessary for slight electronic conductivity may be far greater than can be accomodated in a space charge layer a few angstrom units thick and thus lead to high experimental capacitances.

IV. Conclusion

The DR and SCP effects account separately for some of the experimental work to date but other experimental results fall between the two, e.g. $w^{-3/2}$-C_p dependence with Ohm's Law obeyed. Even in cases believed to be pure DR effect the Cole–Cole plot may not be a semicircle and a distribution of relaxation times insufficient to account for the discrepancy. The presence or absence of linearity over a wide voltage range is clearly the primary difference between the effects but is often given only a cursory examination. Where possible reference electrodes can provide information (Briain and Rickert, 1962) as to the high resistance region of the cell where the bulk of the voltage drop will occur. The possibility of activation polarisation analogous to that observed in aqueous systems must be considered but only one case has been reported to date, that of Karpachev *et al.* (1961) who established linear Tafel plots, i.e. ln *i* versus polarisation plots for the cell

$$Pt\ CO_2, CO \mid CaO \cdot ZrO_2 \mid Pt, O_2$$

presumably due to the rate determining step

$$CO_2 + 2\ electrons = CO + O^{2-}$$

where O^{2-} forms part of the stabilised zirconia lattice. Other examples of activation polarisation may be expected and in this connection K. J. Vetter notes that the frequency dependence of the diffusion impedence will be disturbed by nonuniform electrode surfaces since the active areas on the electrode will be subject to linear diffusion at low frequencies when the penetration of the diffusion wave is much greater than the separation of active centres but subject to spherical diffusion at much higher frequencies.

As a comparison between SCP and DR effects the author has calculated the normalised quadrature components of dielectric loss for the transmission line. In this case the full impedance expression of the line for any frequency is

$$Z = \frac{1}{C'}\sqrt{\frac{2}{w}}\ \tau(F_1 - iF_2)$$

where $\tau = R'C'$ and F_1 and F_2 have the values

$$2F_1 = \frac{\sinh w' - \sin w'}{\cosh w' - \cos w'}$$

$$2F_2 = \frac{\sinh w' + \sin w'}{\cosh w' - \cos w'}$$

and w', a reduced angular velocity, is $w' = (2w\tau)^{1/2}$.
This value of the impedance corresponds to a complex capacitance given by

$$C = \frac{C_1}{(w'\tau)^{\frac{1}{2}}(iF_1 + F_2)}$$

which has real and imaginary parts

$$\frac{F_2}{w'(F_1{}^2 + F_2{}^2)} \quad \text{and} \quad \frac{F_1}{w'(F_1{}^2 + F_2{}^2)}.$$

These functions are plotted versus log w in Fig. 3 and against each other (Cole–Cole presentation) in Fig. 2. It is interesting that a deviation from Debye theory occurs in the high frequency region, i.e. for $w\tau \leqslant 4$ when

$F_1 = F_2 = F_1^2 + F_2^2 = \frac{1}{2}$. The normalised imaginary component is then $1/(2w\tau)^{\frac{1}{2}}$ compared to the Debye expression $1/(w\tau)$, i.e. on the basis of the Kirkwood empirical expression $\lambda = \frac{1}{2}$.

The low frequency approximation, $w\tau \leqslant 2$, gives $F_1 = w'/6$ and $F_1^2 + F_2^2 = 1/(w')^2$ giving a normalised imaginary component of $w\tau/3$ compared to the Debye value of $w\tau$ and an empirical value $\lambda = 1$. The Cole–Cole plot has a maximum < 0.5 and is a skewed segment. In view of the similarity of this Cole–Cole plot based on the transmission line, i.e. on SCP theory, to experimental curves usually ascribed to the DR effect, it appears that the frequency dependence of equivalent circuit elements may not be sufficient to unambiguously differentiate between the two theories. The linear nature of the DR effect and non linear nature of the SCP effect would still seem to be unambiguous. In this connection an interesting equivalent circuit is that of Fig. 1(d) which describes the decrease of dielectric constant and resistivity of a sintered polycrystalline ferromagnetic ceramic material with frequency ("Conference on Semiconducting Materials", 1951). This appears to be a DR effect where macroscopic linearity occurs but the polarisation appears across thin layers of poor conductor separated by large regions of fairly well conducting material. The insulating materials probably lie at grain boundaries. This type of behaviour is reported to be characteristic of ceramic semiconductors and low frequency dielectric constants as high as 10^6 observed.

V. Appendix

Following the experimental discovery of polarisation resistance and polarisation capacity by Wien (1896) and others (Warburg, 1899) a theory to explain these effects based on diffusion alone was given by Warburg (1899; 1901). The presence of a Helmholtz double layer was assumed together with the Warburg diffusion impedance by Krüger (1903). A suggestion that space charge effects should be included by Jaffé ("The Physics of Crystals") was followed by the space charge polarisation theory of Jaffé (1933) which was extended by Chang and Jaffé (1952) to cases of partial blocking and without the assumption of a constant thickness diffusion layer. Similar but independent theories of a less approximate nature were given by Macdonald (1953) and by Friauf (1954) and extended by Beaumont and Jacobs (1967). A diffuse layer containing unsymmetrical valence types at equilibrium was considered by Grahame (1953) and the Macdonald–Friauf AC theory generalised for unsymmetrical valence types by Baker and Buckle (1968) and by Macdonald (1970). A one blocking electrode model was considered by Ferry (1948) and extended to the two blocking electrode case by Buck (1969).

References

Advances in Chemistry (1965). Series 47. "Fuel Cell Systems". American Chemical Society, Washington.

"Advances in Electrochemistry and Electrochemical Engineering" (1966). **4**, Interscience, New York.

Allnatt, A. R. and Jacobs, P. W. M. (1961). *J. Phys. Chem. Solids.* **19**, 281.

American Institute of Physics Handbook (1963). Section 9f. McGraw–Hill.

Baker, C. G. J. and Buckle, E. R. (1968). *Trans. Faraday Soc.* **64**, 469.

Barbi, G. B. (1968). *Z. Naturforsch. A.* **23**, (6), 800.

Bardeen, J. and Letaw, H. (1954), *J. App. Phys.* **25**, 600.

Beaumont, J. H. and Jacobs, P. W. M. (1967). *J. Phys. Chem. Solids.* **28**, 657.

Bell, M. E. and Manning, M. F. (1940). *Rev. Mod. Physics.* **12**, 215.

Bielen, H., Eysel, W., Hahn, Th. and Weber, F. (1962). *Chem. Erde* **22**, 175.

Bloem, J. (1956). *Philips Res. Reports.* **11**, 273.

Bogren, E. C., Patterson, J. W. and Rapp, R. A. (1967). *J. Electrochem. Soc.* **114**, 752.

Bragg, W. L. and West, J. (1926). *Proc. Roy. Soc.* (London). **A111**, 691.

Brennecke, C. G. (1940). *J. App. Phys.* **11**, 202.

Breyer, B. and Bauer, H. H. (1966). *J. Electroanal. Chem.* **11**, 65.

Briain, C. O. and Rickert, H. (1962). *Z. Physik. Chem. N. F.* **31**, 71.

Buck, R. P. (1969). *J. Electroanal. Chem.* **23**, 219.

Byström, A. and Byström, A. M. (1950). *Acta. Cryst.* **3**, 146.

Chang, H. and Jaffé, G. (1952). *J. Chem. Phys.* **20**, 1071.

Cole, K. and Cole, R. (1941). *J. Chem. Phys.* **9**, 341.

Cole, R. (1938). *J. Chem. Phys.* **6**, 385.

Conference on Semiconducting Materials (1951). Butterworths, 151.

Dawson, B. (1953). *Acta. Cryst.* **6**, 113.

Etzel, H. W. and Maurer, R. J. (1950). *J. Chem. Phys.* **18**, 1003.

Ferry, J. D. (1948). *J. Chem. Phys.* **16**, 737.

Flood, H., Krog, Th. and Sörum, H. (1953). *Tidsskr. Kjemi, Berguesen Met.* **5**.

Friauf, R. J. (1954). *J. Chem. Phys.* **22**, 1329.

Fuoss, R. M. and Kirkwood, J. G. (1941). *J. Am. Chem. Soc.* **63**, 385.

Grahame, D. C. (1953). *J. Chem. Phys.* **21**, 1054.

Grimley, T. B. (1950). *Proc. Roy. Soc.* (London) **A201**, 40.

Grimley, T. B. and Mott, N. F. (1947). *Disc. Far. Soc.* **1**, 3.

Gutmann, F., Hermann, A. M. and Rembaum, A. (1967). *J. Electrochem. Soc.* **114**, 323.

Gutmann, F., Hermann, A. M. and Rembaum, A. (1968). *J. Electrochem. Soc.* 359.

Haayman, P. W. and Verwey, E. J. W. (1941). *Physica.* **8**, 979.

Haayman, P. W., Romeyn, F. C. and Verwey, E. J. W. (1947). *J. Chem. Phys.* **15**, 181.

Haayman, P. W., Romeyn, F. C. and Verwey, E. J. W. (1950). *Philips Res. Rep.* **5**, 173.

Hantelmann, P. and Wagner, C. (1950). *J. Chem. Phys.* **18**, 72.

Hever, K. O. (1967). Unpublished experiments.

Hever, K. O. (1968). *J. Electrochem. Soc.* **115**, 831.

"Ionic Crystals, Lattice Defects and Nonstoichiometry" (1968). Greenwood, N. N. Butterworths, London.

Ito, T., Morimoto, N. and Sadanga, R. (1952). *Acta. Cryst.* **5**, 209.

Jackson, B. J. H. and Young, D. A. (1967). *Trans. Far. Soc.* **63**, 2246.

Jacobs, P. W. M. and Maycock, J. N. (1963). *J. Chem. Phys.* **39**, 757.
Jaffé, G. (1933). *Ann Physik.* **16**, 217, 249.
Jost, W. and Weiss, K. (1954). *Z. physik. Chem.* (Frankfurt) **2**, 112.
Karpachev, S. P., Neuimin, A. D. and Palguev, S. F. (1961). *Dokl. Akad. Nauk. SSSR.* **141**, 402.
Kauzmann, W. (1942). *Rev. Mod. Physics.* **14**, 12.
Kiukkola, K. and Wagner, C. (1957). *J. Electrochem. Soc.* **104**, 308, 379.
Kobayashi, H. and Wagner, C. (1957). *J. Chem. Phys.* **26**, 1609.
Koch, E. and Wagner, C. (1937). *Z. physik. Chem.* **B38**, 295.
Kröger, F. A. and Verwey, E. J. W. (1951). *Philips Tech. Rev.* **13**, (4) 90.
Krüger, F. (1903). *Z. phys. Chem.* **45**, 1.
Kummer, J. T. and Yao, Y. F. (1967). *J. inorg. nucl. Chem.* **29**, 2453.
Kummer, J. T., Radzilowski, R. H. and Yao, Y. F. (1969). *J. App. Physics.* **40**, (12) 4716.
Kvist, A. (1966a). *Z. Naturforsch.* **a21** (8) 1221.
Kvist, A. (1966b). *Z. Naturforsch.* **a21** (4) 487.
Kvist, A. and Lunden, A. (1965). *Z. Naturforsch.* **20a**, 235.
Lawson, A. W. (1962). *J. Appl. Physics.* **33**, 466.
Lehfeldt, W. (1933). *Zeits f. Physik.* **85**, 717.
Lehovec, K. and Broder, T. (1954). *J. Electrochem. Soc.* **101**, 208.
Macdonald, J. R. (1955). *J. Chem. Phys.* **23**, 275.
Macdonald, J. R. (1953). *Phys. Rev.* **912**, 4.
Macdonald, J. R. (1970). *Trans. Far. Soc.* **66**, 943.
Magnéli, A. (1949). *Arkiv. Kemi.* **1**, 223.
Magnéli, A. (1953). *Acta. Chem. Scand.* **7**, 315.
Miller, C. W. (1932). *Phys. Rev.* **22**, 622.
Mitoff, S. P. (1961). *J. Chem. Phys.* **35**, 882.
Mrgudich, J. N. (1960). *J. Electrochem. Soc.* **107**, 475.
Mrgudich, J. N. (1965). U. S. Patent 3,170,817.
Mrowec, S. and Rickert, H. (1963). *Z. physik. Chem. N. F.* **36**, 329.
Norrish, K. (1951). *Minerolog. Mag.* **29**, 496.
Nowick, A. S. and Sutter, P. H. (1963). *J. Appl. Phys.* **34**, 734.
Ovshinsky, S. R. (1968). U.S. Patent 3,395,445.
Ozerov, R. P. (1954). *Dokl. Akad. Nauk. SSSR.* **99**, 93.
Palin, D. E. and Powell, H. M. (1949). *J. Chem. Soc.* 816.
Polar Molecules (1929). P. Debye, The Chemical Catalogue Company, New York.
"Physics of Crystals" (1928). Joffé, A. McGraw-Hill, New York.
Progress in Solid State Chemistry (1967). **3**, Pergamon.
Rahlts, P. (1935). *Z. physik. Chem.* **B31**, 157.
Raleigh, D. O. (1966). *J. Phys. Chem.* **70**, 689.
Raleigh, D. O. (1967). *J. Electrochem. Soc.* **114**, 493.
Reid, A. F. and Sienko, M. J. (1967). *Inorg. Chem.* **6**, 321.
Reinhold, H. (1928). *Z. Anorg. Allgem. Chem.* **171**, 181.
Reinhold, H. (1934). *Z. Electrochem.* **40**, 361.
Rickert, H. (1961). *Z. Electrochem.* **65**, 463.
Schlenk, W. (1951). *Fortschr. Chem. Forsch.* **2**, 92.
Schoemaker, D. P., Pauling, L. and Waugh, J. L. T. (1954). *Acta. Cryst.* **7**, 438.
Tamamushi, R. (1966). *J. Electroanal. Chem.* **11**, 65.
"Theory of Dielectrics" (1950). H. Fröhlich, University Press, Oxford.
Ullman, R. (1968). *J. Chem. Phys.* **49**, (2) 831.

Ure, R. W. Jr. (1957). *J. Chem. Phys.* **26,** 1363.
Vetter, K. J. (1967). Electrochemical Kinetics, Academic Press, London.
Wadsley, A. D. (1955). *Rev. of Pure and Applied Chem.* **5,** (3) 165.
Wagner, C. (1933a). *Z. physik. Chem.* **B21,** 42.
Wagner, C. (1933b). *Z. physik. Chem.* **B23,** 469.
Wagner, C. (1953). *J. Chem. Phys.* **21,** 1819.
Wagner, C. and Wagner, J. B. (1957). *J. Chem. Phys.* **26,** 1597.
Warburg, E. (1899). *Wied. Ann.* **67,** 493.
Warburg, E. (1901). *Drud. Ann.* **6,** 125.
Wein, M. (1896). *Wied. Ann.* **58,** 37.
Weininger, J. L. (1964). *J. Electrochem. Soc.* **111,** 769.
Winkler, H. G. F. (1948). *Acta. Cryst.* **1,** 27.

20. REFERENCE ELECTRODES

J. Hladik

Université de Dakar, Sénégal, West Africa

I. Reference Electrodes Properties

A. Reversible and Irreversible Electrodes

If numerous reference electrodes have been used by the electrochemists for experiments with solid electrolytes, it seems that only a few of them have been occupied by the characteristics of the reference electrodes. Steele (1970) noted "there is, for example, very little quantitative information regarding the current densities that can be tolerated by the various reference electrodes which have been employed." Due to this lack of experimentation, we shall describe mainly some typical arrangements where reference electrodes have been used.

In many studies the authors used comparison electrodes. They are either an unpolarizable counter-electrode or a third electrode which plays the role of a pseudo-reference electrode. The simplicity of these arrangements make them very useful, but even under such conditions, the results are not very reliable; this kind of electrode will also be described.

Since electrodes are always reversible systems, we shall now examine the distinction between reversible and irreversible processes.

The distinction between these two processes has been discussed by Ives and Janz (1961) in their book "Reference Electrodes". However, the use of the current-potential curves of each component of a redox system, seems the simplest way to make this distinction.

Suppose that we perform an oxidation such as:

$$\text{Red} - ne \rightarrow \text{Ox}$$

where Red and Ox designate the two components of the redox-system. If the anode potential increases, we can measure the increasing of the electrolytic current. The magnitude of the current being proportional to the number of electrons exchanged during the unit of time, the curves $i = f(E)$ give a measure of the rate at which the reaction occurs. The current-potential curves allows, therefore, the prediction of the rate of the electrochemical reaction for each value of the electrode potential.

For each compoent, Red and Ox, of the system, we can separately register the curves $i = f(E)$. These curves can occupy various relative positions on the diagram i versus E.

In the case of Fig. 1 we can see that there does not exist any potential value for which the two reactions occur with an appreciable rate. In the case of Fig. 2 we can see that for a potential like E_1, the oxidation of Red occurs with a finite rate i_1, while the reduction of Ox occurs with a rate i_2. The equilibrium potential corresponds to a total current equal to zero, $i_1 + i_2 = 0$.

In Fig. 1, an equilibrium potential exists but the curve does not allow an exact determination of the position of the point E_{eq}. In the case of Fig. 2, it is possible to determine the position of the equilibrium potential E_{eq}. The

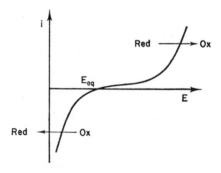

FIG. 1. Current-potential curve. Irreversible electrochemical process.

curve of the total current, $i = i_1 + i_2$, can be drawn, versus E, and corresponds to the simultaneous presence of Red and Ox compound. The electrode, plunging into the electrolyte, takes precisely the equilibrium potential E_{eq}. At this potential the composition of the electrolyte remains stable, the oxidation rate being equal to the reduction rate.

A redox system whose curves $i = f(E)$ correspond to the case of Fig. 2 is called a reversible system (a "fast" system, according to some authors). The other case corresponds to an irreversible system (a "slow" system).

For a reversible system, the oxidation of the reducing agent is easily obtained if the potential E is slightly higher than E_{eq}. One can, therefore, perform an oxidation or a reduction with a finite rate in conditions very near to the reversibility.

B. Reference Electrodes Characteristics

The result of the preceding considerations is that the reversible systems will be used for reference electrodes, their equilibrium potential being well defined and stable.

It is indispensable that no other irreversible reaction appreciably perturbs the reversible electrochemical process which determines the reference potential. If it is disturbed one will obtain a "mixed" potential which is not wholly reversible (Charlot *et al.*, 1959).

There are other criteria which reference electrodes must satisfy. They are:

Electrochemical: e.m.f. well defined and stable in time; weak polarizability; easy reproducibility with well defined compounds; possible utilisation in a given temperatures range; e.m.f. independent of the solid electrolyte composition in contact with the reference.

Technological: easy construction and manipulation; well defined, stable junction potential; contamination risks by foreign substances eliminated.

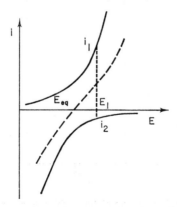

FIG. 2. Current-potential curve. Reversible electrochemical process.

Many of these criteria will be more or less well satisfied according to the nature of the materials used, their degree of purity, their stability and other characteristics proper to a given substance.

The electrode polarizability should be weak, i.e. the electrode rapidly returns to its equilibrium potential after a certain current has passed through it. This is theoretically the case for reversible systems but concentration polarization can occur for certain reference electrodes. Polarizability is studied by applying a fixed potential to the reference electrode, and, after removing the imposed potential, the depolarization curve $E = f(t)$ is registered. For example, the polarization curve of the reference electrode platinum/platinum chloride in solid LiCl–KCl is shown on Fig. 3 (Hladik, 1966).

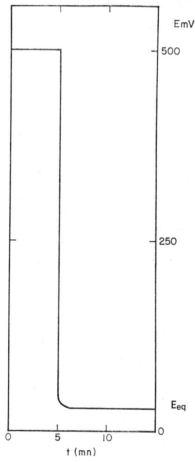

FIG. 3. Polarization curve of a platinum/platinum chloride reference electrode (after Hladik, 1966).

The imposed potential is 500 mV from the equilibrium potential and is applied during 5 mn. After removing the imposed potential, the electrode potential falls about 20 mV from its equilibrium potential E_{eq}, then it slowly decreases and returns to E_{eq} after 3 mn.

C. Gas and Metal Electrodes

Reference electrodes may be classified in two main groups: gas electrodes and metal electrodes. The former are constituted of a metallic conductor in contact with a solid electrolyte, a gas flowing through the electrode and bathing the contact metal/electrolyte; such electrodes exhibit therefore a three-phases boundary electrochemical process. In the case of liquid metals the gas can be dissolved in the liquid which is in contact with the electrolyte.

The metal electrodes are made of a metal embedded in, or simply in contact with, a solid electrolyte.

Many experiments have been performed with cells where one or two of the electrodes could be used as reference electrodes. These cells being reviewed in the other chapters of this book, we shall only give some examples of cells where one electrode is used as reference electrode.

Wagner (1953) has given a general discussion of the various situation that may arise if a mixed conductor is placed between various combinations of electrodes. The various cells that are possible are summarized in Table I (only reversible electrodes).

TABLE I.   Electrolytic cells with various types of electrodes

Type of cell	Examples
Two identical reversible electrodes	$Ag/AgI/Ag$ $Pt_{(O_2)}/Cu_2\,O_{(O_2)}/Pt$
Different reversible electrodes	
(a) electronic probes	$C/Ag\,Br/C$
(b) ionic (Ag^+) probes	$Ag/AgCl/Ag_2S/AgCl/Ag$
Reversible electrode (electronic) versus electronic conductor	$Ag/Ag_2S/Pt$
Reversible electrode (electronic) versus ionic conductor	$Ag/Ag_2S/AgI/Ag$

The problem of the comparison between electrode potentials in liquid and solid electrolytes has been treated by Hladik *et al.* in a separate chapter of this book.

II. Gas Electrodes

A. Three Phase Boundary Reactions

1. Fuel Cells Reference Electrodes

Reference electrodes have been used to establish the current-potential curves of many fuel cell electrodes.

A gas reference electrode has been used by Takahashi *et al.* (1965) in a fuel cell with the CeO_2–La_2O_3 electrolyte. Figure 4 shows a scheme of the cell structure. Onto the tested electrolyte was sintered 150 mesh platinum powder on both sides covering a diameter of 8 mm to make them electrodes. Each of the electrodes was also covered with a platinum net for collection of electric current. The tested electrolyte was held by ceramic tubes of internal diameter of 9 mm on both sides via platinum packing, thus separating both electrodes gas tight. The fine tube which led the gas to the electrode was inserted into these tubes and through it the current lead wire of platinum was taken out

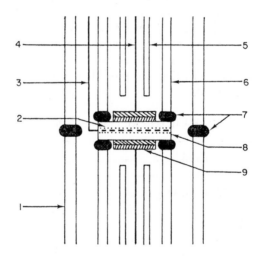

1   Outer Alumina tube internal dia. 24mm
2   Reference electrode
3   Pt lead wire from reference electrode
4   Pt lead wire from cathode
5   Alumina tube internal dia. 4mm
6   Alumina tube internal dia. 9mm
7   Pt packing
8   Electrolyte
9   Electrode

Fig. 4. Gas reference electrode. Scheme of a fuel cell (after Takahashi *et al.*, 1965).

to the measurement equipment. In this case, care was taken that the lead wire and the electrode were not in contact with the platinum packing to prevent the formation of a local cell. The cell was situated in the hinged electric furnace as shown in Fig. 4.

A reference electrode was provided by fixing platinum powder backing in a fine circle at the circumference of the electrolyte and was welded with a platinum lead wire. This electrode was exposed to air or oxygen gas stream. The reversibility of the reference electrode was so good that at 1000°C no polarization appeared even when a current of several milliamperes was taken out. The internal impedance of the measurement circuit was higher than 1 MΩ, so that the current taken out of the reference electrode was 1μA at the most. Its potential change, therefore, was negligibly small.

2. Reference Atmosphere

An arrangement with two electrodes, with reference atmosphere, has been used by Kleitz (1968) during studies concerning electrode reactions in solid electrolyte oxides. The principle of the experimental arrangement consists of a solid electrolyte (disc, tube bottom) which has one of its surfaces covered with a porous platinum layer which composes the counter-electrode. The working electrode is made up of a metal rod pressed on the electrolyte oppo-

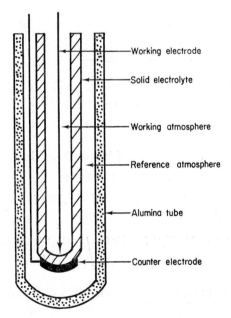

FIG. 5. Scheme of a gas electrode with a reference atmosphere (after Kleitz, 1968).

site surface (Fig. 5). The ratio between the two electrodes surfaces is greater than 10^3. In this case the absence of polarizability of the counter-electrode is fully satisfied by the gas electrode.

The whole system (electrodes electrolyte) is put inside an alumina protection tube. The working atmosphere is situated inside the electrolyte tube while the constant reference atmosphere is between the electrolyte and the alumina tubes. In order to choose a reference atmosphere pressure, one should be particularly aware of the gas permeability of the electrolyte tube. In this case, the oxygen flow increases with the difference between the working and reference oxygen partial pressures. The ideal condition corresponds to the pressures equality, but the perturbations are only important when the oxygen quantity in the working atmosphere is small.

The influence of the oxygen partial pressure of the reference atmosphere on the counter-electrode potential should be taken under consideration. When the oxygen partial pressure is much lower than 1mm Hg the Nernst law is not necessarily valid. Moreover, when the partial pressure of one of the electroactive species (O, CO, CO_2) is small, the electrode does not have a well defined potential and could be polarizable.

The three following reference atmospheres have been used by Kleitz (1968):

oxygen or air at atmospheric pressure.

pure oxygen at 1mm Hg pressure (in the case of very permeable electrolyte tubes).

the mixture CO/CO_2 at the atmospheric pressure.

A similar system has been employed by Guillou *et al.* (1968) using an argon–oxygen gaseous mixture corresponding to a partial pressure of 2×10^{-1} atm.

B. Gas Dissolved in the Electrode

1. Oxygen Electrodes

Masson and Whiteway (1967) have studied the diffusion of oxygen in liquid silver by means of the cell.

$$\text{Stainless Steel} \left/ \begin{array}{c} Ag(liq) \\ [O]\ ref \end{array} \right/ \begin{array}{c} 0.3\ Zr\ O_2 \\ 0.1\ Ca\ O \end{array} \left/ \begin{array}{c} Ag(liq) \\ [O] \end{array} \right/ \text{Stainless Steel}$$

illustrated in Fig. 6. Liquid silver in equilibrium with pure oxygen at barometric pressure was held in the alumina crucible K and served as a reference electrode. Diffusion was allowed to occur in a column of silver H, approximately 5 mm in diameter, held in an impervious lime–zirconia tube D (wall thickness $= 1$ mm). This tube also served as the electrolyte. The lower end of tube D dipped approximately 2 mm below the surface of the silver in the crucible. Electrical contact with the silver in each half of the cell was made by stainless steel wires A.

The electrochemical measurements of oxygen in liquid silver were reported by Diaz *et al.* (1966). In these experiments, the reference electrode is

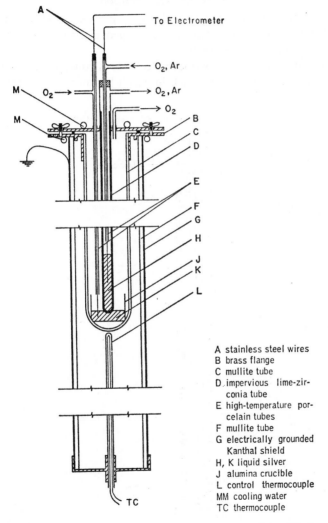

A stainless steel wires
B brass flange
C mullite tube
D impervious lime-zir-
conia tube
E high-temperature por-
celain tubes
F mullite tube
G electrically grounded
Kanthal shield
H, K liquid silver
J alumina crucible
L control thermocouple
MM cooling water
TC thermocouple

FIG. 6. Oxygen reference electrode. Experimental arrangement of the diffusion cell. (after Masson and Whiteway, 1967).

made from a tube of zirconia +15 mole per cent lime. This tube contained a small amount of liquid silver into which a stainless steel conducting lead was dipped. Oxygen gas at atmospheric pressure was passed down the inner alumina tube and over the silver. The experimental arrangement is shown in Fig. 7. This oxygen reference electrode was dipped into the liquid silver of a cell.

2. Hydrogen Electrode

The chronoamperometric studies of the hydrogen oxidation the LiCl–KCl–BaO solid electrolyte show that the hydrogen electrode can be used as reference electrode (Hladik, 1969).

In these experiments, the hydrogen electrode is constituted of a thin palladium tube, closed at its lower end, in which hydrogen flows (Fig. 8). The

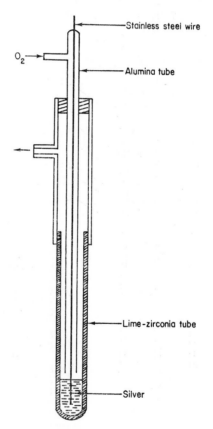

FIG. 7. Oxygen reference electrode. Electrochemical measurements of oxygen in liquid silver (after Diaz *et al.*, 1966).

tube is partly immersed in the molten electrolyte, the system is then cooled below the melting point of the mixture, and the palladium tube is thus embedded in the solid electrolyte. A known hydrogen pressure is then established in the palladium tube fixing the hydrogen electrode potential.

The chronoamperometric curves of the hydrogen oxidation reveal the existence of two oxidation peaks (Fig. 9). The currents values indicate important variations with decreasing temperature.

The first oxidation peak may be attributed to the hydrogen oxidation reaction by $O^=$ ions, like in molten electrolyte (Hladik, 1966). The theoretical values, calculated from the thermodynamic data, indicate that the activation polarization is weak. Such a reversible system can be used therefore like a gas reference electrode. The polarizability curves show a fast return to the equilibrium potential.

The results obtained for the hydrogen oxidation on gold-emerging electrode (Hladik, 1970) indicate a slight polarization.

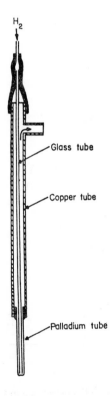

FIG. 8. Hydrogen/palladium electrode (after Hladik, 1969).

C. Oxide Reference Electrodes

Oxide reference electrodes can be used in oxygen concentration cells involving solid oxide electrolytes.

For example, the galvanic cell technique involving the $Zr_{0.85}$ $Ca_{0.15}$ $O_{1.85}$ solid electrolyte was introduced to determine the standard free energies of oxide formation by Kiukkola and Wagner (1957). Under the conditions of exclusive ionic conduction, the voltage E of a galvanic cell involving solid oxide electrolyte, of the type:

$$M, MO/\text{solid oxide electrolyte}/M'O, M'$$

may be related to the free energy change of the virtual cell reaction by the equation (Kiukkola and Wagner, 1957).

$$\mu_{O_2}'' - \mu_{O_2}' = 4\,EF$$

where μ_{O_2}' and μ_{O_2}'' are the chemical potentials of oxygen of the cathode

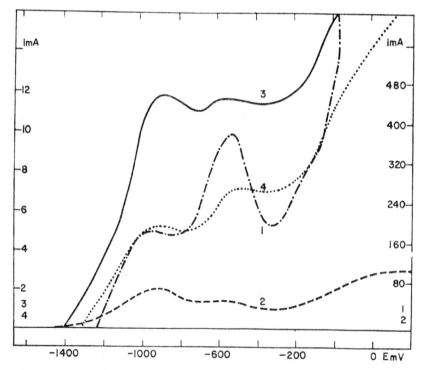

Fig. 9. Hydrogen oxidation on a palladium electrode in LiC–KCl–BaO 1·324°C; 2·302°C; 3·290°C; 4·284°C (after Hladik, 1969).

and anode, respectively, and F is Faraday's constant. The e.m.f. of the cell is, then:

$$E = \frac{RT}{4F} \ln \frac{P_{O_2}{''}}{P_{O_2}{'}}$$

the partial pressures of oxygen being fixed by the equilibrium between M and MO, $P_{O_2}{'}$ and between the M' and $M'O$, $P_{O_2}{''}$.

The chemical potentials of oxygen in the two phase mixtures $Ni + NiO$ and $Fe + Fe_xO$ are both well known. Each of these mixtures may, therefore, be used as a reference electrode in a galvanic cell involving solid oxide electrolytes with exclusive ionic conduction.

These reference electrodes have been used by Rapp (1963) for determining the free energy of formation of molybdenum dioxide. In his investigation, Rapp used the two cells:

$$\text{Mo, } MoO_2 / Zr_{0.85} \, Ca_{0.15} / Fe, \, Fe_xO$$

$$\text{Mo, } MoO_2 / Zr_{0.85} \, Ca_{0.15} \, O_{1.85} / Ni, \, NiO$$

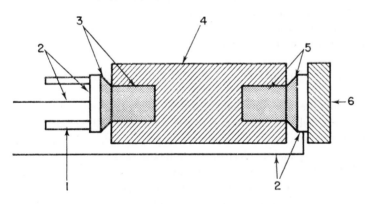

1	Quartz tube applying spring pressure
2	Pt
3	Fe + FeO or Ni + NiO
4	$Zr_{0.85} \, Ca_{0.15} \, O_{1.85}$ Electrolyte
5	Mo + Mo O_2
6	Fixed ceramic block

FIG. 10. Ni + NiO (or Fe + FeO) reference electrode. Determination of the free energy of molybdenum dioxide formation (after Rapp, 1963).

The experimental arrangement is shown in Fig. 10. The Ni + NiO electrode was made of pure NiO powder, heated in a vacuum for 12 hours at 1050°C to remove excess absorbed oxygen. This powder was then mixed with nickel powder. The Fe + Fe_xO electrode was prepared with iron powder mixed with Fe_2O_3 powder, pressed into large pills, and heated in a vacuum for 2 days at 1000°C. The resulting sintered pellet was reduced to a large grain powder by cutting it on a lathe. This powder was then finely ground with a mortar and pestle.

Similar cells involving solid oxide electrolytes can be used for the determination of activities in solid metallic systems (Kubik and Alcock, 1968). Cells of the type as:

$$M, MO/\text{solid oxide electrolyte}/M-M' \text{ alloy}, MO$$

are set up in which the cell reaction involves the transfer of oxygen ions from the side of high oxygen partial pressure, P_{O_2}'', to that of lower pressure, P_{O_2}'. In the reference electrode the activities of M and MO may be made equal to unity, so long as the oxygen solubility in the metal M is negligible in the temperature range under investigation.

For example, Rapp and Maak (1962) determine the activity of Ni in Cu–Ni alloys by using the following-cell.

$$\text{Ni, NiO}/Zr_{0.85}\,Ca_{0.15}\,O_{1.85}/\text{Cu–Ni alloy, NiO}$$

They discuss the processes that control the final attainment of the equilibrium e.m.f. of the cell. A steady e.m.f. of the whole cell will be reached after electrochemical equilibrium at all interfaces between zirconia and the adjacent phases involving mainly electronic conduction has been established. For the Ni + NiO tablet, this condition is satisfied only if the chemical potentials of oxygen in Ni and NiO (adjacent to the zirconia tablet) are equal. Primarily, the chemical potential of oxygen is determined by an equilibrium at the Ni–NiO interface. From this interface, oxygen diffuses into the interior of the nickel phase in order to establish a uniform O/Ni ratio in each individual phase which determines the local chemical potential of oxygen. Hence it follows that diffusion of oxygen in nickel and diffusion of nickel ions and electrons in nickel oxide are essential for reaching the complete equilibrium. Fortunately the attainment of equilibrium is supported by local cell action, i.e. migration of electrons in Ni or NiO and migration of oxygen ions in zirconia in direction essentially parallel to the interfaces. Thus a uniform chemical potential of oxygen along the boundary between a Ni–NiO tablet and a zirconia tablet is approached even faster than in the interior of a Ni–NiO tablet not in contact with ZrO_2.

Diaz and Richardson (1967) used cells with Ni–NiO as reference electrode for electrochemical measurements of oxygen in molten copper.

The nickel–nickel oxide pellet used as a reference electrode was maintained in close contact with the solid electrolyte tube and with a platinum lead ending in a platinum disc by pressure from a spring-loaded alumina tube. This same alumina tube carried the argon used to flush the reference electrode and the gas escaped through small slits cut into the end of the tube (Fig. 11). The Ni–NiO pellets were prepared by mixing "Analar" nickel and nickel oxide powders in equimolecular proportions and pressing the mixture at 10 ton in^{-2}. The pellets were sintered in purified argon at 900°C for 6 hours. The solid electrolytes were commercial grades of ZrO_2 0·7 mol per cent CaO and $ZrO_2 . 4·5$ mol per cent Y_2O_3. They consisted of tubes 6 mm in outside diameter, with a wall thickness of about 1 mm. Diaz and Richardson discuss the processes that control the polarization of the reference electrode, and

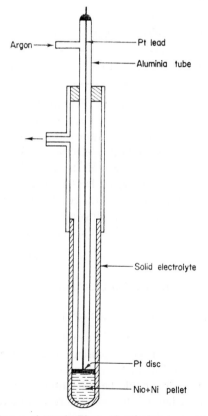

FIG. 11. Ni + NiO reference electrode. Electrochemical measurements of oxygen in molten copper (after Diaz and Richardson, 1967).

conclude that Pt/O_2 reference electrode is a better reference than the $Ni + NiO$ in the temperature range under investigation.

Pastorek and Rapp (1969) used a $FeO + Fe_3O_4$ reference electrode in their experiments on solubility and diffusivity of oxygen in solid copper. They used two cells of the type:

$$FeO + Fe_3O_4/Zr_{0.85} Ca_{0.15} O_{1.85}/Cu + (O\ dissolved)$$

for chronoamperometric and chronopotentiometric studies. The reversible electrode were prepared from reagent quality materials to yield two-phase equilibrium mixtures of $FeO-Fe_3O_4$ in a $1:1$ weight ratio. The equilibrium oxygen activity values for the $FeO-Fe_3O_4$ electrode were measured in their work using a galvanic cell with the zirconia elctrolyte and a $Co-Co\ O$ reference electrode. The equilibrium oxygen activities for the $FeO-Fe_3O_4$ reference electrode are:

$T°, C$	1030	1000	950	900	850	800
$-\log P_{O_2}$	12·21	12·81	13·89	15·05	16·30	17·70

In some preliminary experiments a $Fe-FeO$ mixture was used as a reversible electrode. However, the $Fe-FeO$ electrode contaminated the $Zr_{0.85}$ $Ca_{0.15} O_{1.85}$ electrolyte at high temperatures, $1000°C$ or above, and thus was not a suitable choice. A $FeO-Fe_3O_4$ electrode was used in attempt to eliminate polarization of the reference electrode at the electrode/electrolyte interface. The high defect structure of FeO should allow adequate volume diffusion of iron to maintain $FeO-Fe_3O_4$ equilibrium at the electrode/electrolyte interface while oxygen is withdrawn from the electrode. This electrode should then be able to function simultaneously as an oxygen source and a reversible electrode.

III. Metal Electrodes

A. Pure Electrolytes

The metal electrodes are constituted of a metal in contact with a solid electrolyte, the cations of this metal being present. The electrolyte can be a pure compound or a mixture of the metal salt in other electrolytic species. Many examples of reversible system could be cited among the electrochemical cells but we will only give a few of the possible reversible redox systems. It is particularly interesting to cite some electrochemical experiments using three electrodes: reference, working and counter-electrodes.

The measurement of the chemical potential of sulphur in Ag_2S has been obtained by Rickert (1964) with the aid of the solid-state galvanic cell:

$$Pt/Ag/AgI/Ag_2S, Pt \qquad\qquad (A)$$

where silver iodide is a practically pure ionic conductor. This cell has been used for investigations on silver sulphide by Reinhold (1934), Wagner (1953) and Kiukkola and Wagner (1957).

The e.m.f. "E" of this cell is related to the chemical potential of silver μ_{Ag} in the silver sulphide probe by:

$$\mu_{Ag} - \mu_{Ag}^{\circ} = -EF$$

where μ_{Ag}° is the chemical potential of silver in the standard state, which is pure metallic silver.

The electrochemical study of vaporization of sulphur from silver sulphide has been performed by Rickert with two separate type (A) cells (Fig. 12). Current is passed through the lower half of one. The magnitude of this current controls the rate of sulphur evaporation. Voltage measurements on the second cell (top of the cell) give values for the chemical potential of silver and, hence, of sulphur in Ag_2S. In this way, the current and potential measurements are carried out separately and errors due to polarization effects in the potential values are reduced to a minimum.

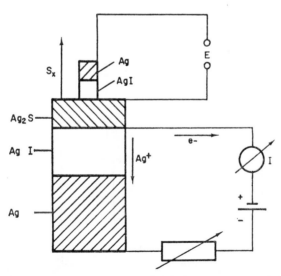

FIG. 12. Ag/AgI reference electrode. Electrochemical measurements of the vaporization rate of sulphur from solid silver sulphide (after Rickert, 1964).

The reaction between copper and liquid sulphur has been tested by Rickert (1968). The experimental arrangement is depicted by Fig. 13. Liquid sulphur is in a glass tube, closed at its lower end by the solid Cu_xS. It is then set up on a copper cylinder. The copper chemical potential in Cu_xS, near the phase boundary Cu/Cu_xS, is measured by means of a probe $Cu/CuBr$. In the galvanic cell:

$$Pt, Cu/CuBr/Cu_xS, Pt$$

the copper chemical potential μ_{Cu} in Cu_xS is related to the e.m.f. of the chain by the relation:

$$\mu_{Cu} - \mu_{Cu}^{\circ} = -EF.$$

B. Electrolyte Mixtures

Silver has been frequently used as a reference electrode in molten electrolytes (Morand and Hladik, 1969) and it seems to have the proper properties to be used in numerous solid electrolytes.

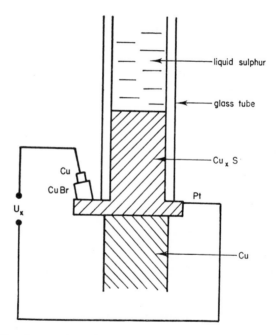

FIG. 13. Cu/CuBr reference electrode. Measurement of copper chemical potential in Cu_xS near the phase boundary Cu/Cu_xS (after Rickert, 1968).

The Ag/Ag(I) reference electrode has been employed in solid NaCl–KCl equimolecular mixture (Hladik *et al.*, 1970). The silver chloride is formed in the molten mixture, at 750°C, by anodic oxidation of a silver lead. The cathode is a graphite rod inside of a porous alumina tube in order to avoid contamination of the molten electrolyte by reduction compounds. The silver chloride concentration in the NaCl–KCl mixture is 10^{-1}M.

The reference electrode consists of a silver lead plunging in the NaCl–KCl–AgCl electrolyte. A quartz envelope, closed at its lower end, contains the solid electrolyte and allows the junction between the reference electrolyte and the outside solid electrolyte. The experimental arrangement is depicted by Fig. 14. We used this electrode in solid NaCl–KCl; for this purpose, the reference electrode was partly immersed in the molten salt and then the whole system was cooled. The lower part of the reference electrode is then embedded in the solid NaCl–KCl electrolyte.

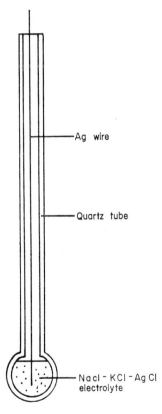

Ag wire

Quartz tube

Nacl – KCl – Ag Cl electrolyte

Fig. 14. Silver reference electrode (after Hladik *et al.*, 1970).

858   J. Hladik

This solid state Ag/Ag(I) electrode exhibits proper characteristics for a reference electrode in a NaCl–KCl medium. The current potential curves show that under such conditions the Ag/Ag(I) system is reversible. Figure 15 indicates the presence of silver chloride reduction waves whose height changes according to the electrolyse duration.

The resistance of the reference circuit is composed of three solides: NaCl–KCl–AgCl mixture, quartz membrane and outside electrolyte, the resistivity of each part being different. At 550°C, this resistance is about 3 to 5 MΩ, varying according to the experimental arrangements.

The use of a separating membrane introduces a junction potential which is difficult, actually, to calculate theoretically. However, this junction potential is weak when the supporting electrolyte is identical on the two sides of the quartz membrane and if the foreign ions inside and outside of the reference electrode are present in small quantities.

A similar study of silver electrodes was performed in solid LiCl–KCl and KNO_3 electrolytes. The current-potential curves of silver oxido-reduction indicate reversible systems in these electrolytes.

A platinum/platinum chloride system has been used, mainly in molten LiCl–KCl as a reference electrode (Morand and Hladik, 1969). Such an electrode can be also used in the solid state (Hladik, 1966). Platinum chloride is prepared by anodic oxidation of platinum lead in molten LiCl–KCl at 450°C. The reference electrode is quite similar to the one previously des-

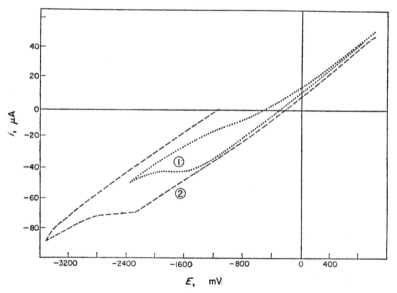

FIG 15. Silver oxido-reduction in NaCl–KCl at 540°C (after Hladik *et al.*, 1970).

cribed for the Ag/Ag(I) system (Fig. 14) except that the separating membrane is a Pyrex glass tube. The platinum/platinum chloride system is quite reversible.

Similar studies have been performed in solid equimolecular NaCl–KCl for the platinum and gold electrodes. Figure 16 shows the oxido-reduction current-potential curves of the gold electrode in NaCl–KCl at 540°C. and the influence of the oxidation maximum potential. Figure 17 shows the platinum oxido-reduction and the influence of sweeping rate.

In conclusion, we see that electrodes similar to the ones used in fused salts can be employed in solid electrolytes. Although the temperature range in solids is 100 to 300°C below the melting point, (according to the preceding considered electrolyte), the behaviour of many redox systems is quite analogous in the solid and liquid states.

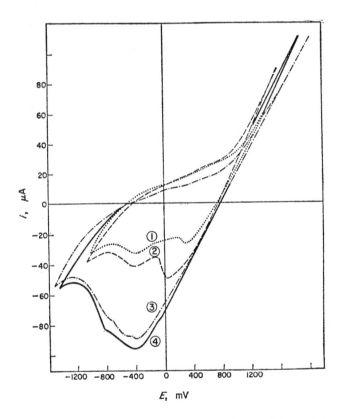

FIG. 16. Gold oxido-reduction in NaCl–KCl at 540°C. Influence of the oxidation potential. 1·1400 mV; 2·1600 mV; 3·2000 mV; 4·2050 mV (after Hladik *et al.*, 1970).

IV. Comparison Electrodes

A. Unpolarizable Counter-Electrodes

The principle of the unpolarizable electrode consists of using a comparatively large-surfaced counter-electrode and a smaller working electrode. Since current density is very small on the counter-electrode, its potential can be considered as a constant during the electrolysis between working and counter-electrodes.

This principle is applied in standard polarography where a mercury pool is used as an unpolarizable electrode with the dropping mercury electrode. Electroanalytical studies of molten salts have also been performed with working microelectrodes and unpolarizable counter-electrodes, the latter being made up on a long, spiral lead or a large surface metallic plate (Morand and Hladik, 1969). A similar technique can be used in solid media.

Filayev *et al.* (1961, 1963) studied oxygen electrode polarization in $Zr_{0.85} C_{0.15} O_{1.85}$ electrolyte, using a large, unpolarizable platinum electrode. They used cylindrical electrolyte samples with coaxially distributed electrodes (Fig. 18). The working electrode was a platinum wire of 0·1 mm diameter moulded symmetrically along the cylinder axis into the electrolyte. The apparent area of the contact surface between the electrode and the solid electrolyte was 1·2 mm². The second electrode was made of pulverized platinum deposited on the external surface of the cylindrical sample. The ratio of areas between the test and external electrode was 1 : 50. Therefore, polarization of the external electrode was neglected.

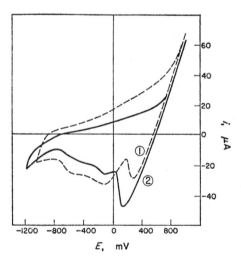

FIG. 17. Platinum oxido-reduction in NaCl–KCl at 540°C (after Hladik *et al.*, 1970).

The electrochemical reactions of oxygen exchange between a semi-conductor oxide and non-ionized or slightly ionized gases, have been studied by Guillou *et al.* (1968) using a large, unpolarizable counter-electrode made of Rhodium/platinum alloy. The galvanic cell is formed by ZrO_2 (0·85) CaO(0·15) sample pressed between two alumina tubes of similar interior diameter (Fig. 19). In the upper compartment, a platinum-rhodium rod (Pt–Rh 10%) with a pointed tip constitutes the working electrode. In the lower compartment, a large-surfaced Pt–Rh grate (Pt–Rh 10%) forms the counter-electrode. Swept by a gaseous oxygen–argon mixture, corresponding to a $2·10^{-1}$atm partial pressure, this electrode is used as an unpolarizable comparison electrode.

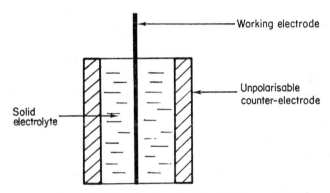

FIG. 18. Cylindrical counter-electrode (after Hladik *et al.*, 1970).

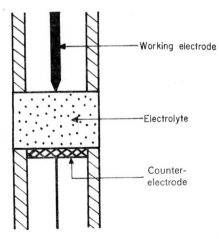

FIG 19. Scheme of a plane unpolarizable counter-electrode (after Guilliou *et al.*, 1968)

B. Pseudo-Reference Electrodes

In order to compare separately the anodic and cathodic potentials during electrolysis, it is possible to use a pseudo-reference electrode in contact with the solid electrolyte. This third electrode is generally a metallic wire but should be eventually replaced by graphite.

In this kind of study, anodes and cathodes generally have approximately equal surfaces. The third electrode does not participate in the electrolysis circuit and its potential, under particular conditions may be considered as practically stable.

There are three main arrangements for these pseudo-reference electrodes (Fig. 20):

a metallic wire partly embedded in solid electrolyte (a).

a rod in contact with the surface electrolyte (b).

a metallic film deposited on a part of the electrolyte (c).

The first type of electrode has been used, for example, by Takahashi and Yamamoto (1966) during their studies of the $Ag/Ag_3 SI/I_2$ solid electrolyte cell. A cell was prepared as in Fig. 21. The electrodes of cell are a silver plate (0·005 cm thick) which serves as the anode and an iodine and graphite mixture as the cathode. The thickness of the electrolyte is about 0·15 cm, and the diameter is 1.2 cm. To measure the anodic and cathodic polarizations independently, a silver wire reference electrode, diameter 0·05 cm, was inserted into the electrolyte. The reversibility of the reference electrode is good when the current passed through it is maintained below 0·1μA. An automatic voltage recorder with a high internal impedance (2MΩ) was used to measure the voltage drop between the reference electrode and the anode.

An analogous cell has allowed Takahashi and Yamamoto (1970) to study the characteristics of the solid electrolyte battery $Ag/Rb Ag_4I_5/Ag$.

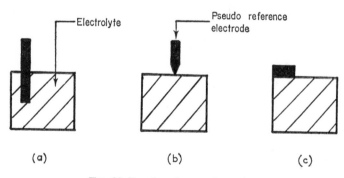

FIG. 20. Pseudo-reference electrodes.

In this experiment, the solid electrolyte $Rb\,Ag_4I_5$ has an ionic conductivity high enough to produce only a very small ohmic potential drop at current densities of the order of a milliampere per square centimeter. However, large polarization at the boundary between the electrode and the electrolyte would make a solid electrolyte cell unemployable. The anodic polarization of the cell, $Ag/Rb\,Ag_4I_5$, was measured, using a silver plate reference electrode placed in the electrolyte on the cathode side of the cell.

Karpachev and Obrosov (1969) have also used a silver wire embedded in the solid electrolyte NaCl–AgCl during a chronoamperometric, study at constant potential. In their experiment, the cathode was a thin silver wire with a diameter of 0·2 mm embedded along the axis of a cylindrical sample of the solid electrolyte. The comparison electrode was a silver wire embedded in the electrolyte, and the auxiliary electrode was deposited on the side surface of the cylinder in the form of silver powder.

Perfilev and Palguev (1966) have studied the platinum electrodes polarization in a CaO(0·15) CeO(0·85) solid electrolyte. The electrodes (two working electrodes and one reference electrode) were prepared by depositing fine platinum powder on compressed, but not yet sintered, tablets of the solid electrolyte. The electrodes were sintered for 2h in air at 1500–1530°C at the same time as the electrolyte was annealed. The results were dense, well sintered electrodes. They wanted to discover whether the potential of the reference electrodes is affected by the current passing from one working electrode to the other. For this purpose, two auxiliary reference electrodes were deposited on the electrolyte tablet at different distances from the working electrodes (Fig. 22). The difference of potentials between them was determined

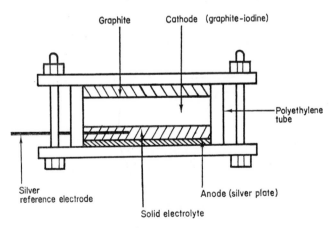

FIG. 21. Scheme of the $Ag/Ag_3Sl/I_2$. Solid electrolyte cell (after Takahashi and Yamamoto, 1966).

as a function of the density of current passing through the electrodes. Within the range of current densities investigated (up to 5mA/cm^2) the difference of potentials between the reference electrodes did not exceed 1 mV, or 1% of the value of the polarization of the electrodes. The equilibrium potentials of the working electrodes and the reference electrodes usually vary by no more than 0·1–0·3 mV.

C. Advantages and Disadvantages of Comparison Electrodes

The technical simplicity of the unpolarizable counter-electrode is its main advantage, since a two-electrode cell is easier to make than a three electrode model. In many studies, such simplicity will be satisfactory.

However, the comparison electrode potential can vary under particular conditions. Thus, during the chronoamperometric studies of a solute at various concentrations in a supporting electrolyte, the counter-electrode potential varies from one experiment to another.

The pseudo-reference electrodes are also easier to make than a true reference electrode, although reference electrode technology is relatively simple.

The pseudo-reference electrode exhibits the same inconveniences as the unpolarizable electrodes, that is to say a variable and not well defined potential. In the case of relatively large currents during electrolysis, concentration gradients can occur in the neighbourhood of the pseudo-reference electrode, involving a potential variation of this electrode. When solutes of different concentrations are used in a series of experiments, the pseudo-reference electrode potential also changes with concentration.

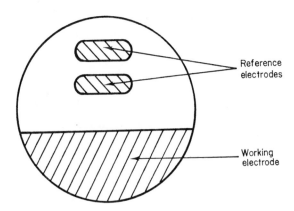

Fig. 22. Position of the electrodes on the tablet of CeO–CaO solid electrolyte (after Perfilev and Palguev, 1966).

References

Charlot, G., Badoz-Lambling, Mrs. J. and Tremillon, B. (1959). "Les réactions electrochimiques". Masson et Cie, Paris.

Diaz, C. and Richardson, F. D. (1967). *Trans. Instn. Min. Metall.* **76-C**, 196.

Diaz, C. Masson, C. R. and Richardson, F. D. (1966). *Trans. Instn. Min. Metall.* **75-C**, 183.

Filayev, A. T., Palguev, S. F. and Karpachev, S. V. (1961). *Tr. Inst. Electrokhim, Akad. Nauk. SSSR.* **2**, 199.

Filayev, A. T., Karpachev, S. V. and Palguev, S. F. (1963). *Akad. Nauk. SSSR.* **149**, 909.

Filayev, A. T., Karpachev, S. V. and Palguev, S. F. (1967). *In* "Electrochemistry of molten and solid electrolyte", Tome 4, pp. 161–165. Consultants Bureau, New York.

Guillou, M. ,Millet, J. and Palous, S. (1968). *Electrochimica Acta.* **13**, 1411.

Hladik, J. (1966). Thesis, University of Paris.

Hladik, J. (1969). *Comptes Rendus Acad. Sciences* **268**, 1019.

Hladik, J. (1970). *Comptes Rendus Acad. Sciences* **270**, 1504.

Hladik, J., Pointud, Y., Bellissent, M. C. and Morand, G. (1970). *Electrochimica Acta.* **15**, 405.

Ives, D. J. G. and Janz, G. J. (1961). "Reference Electrodes", Academic Press, London.

Karpachev, S. V. and Obrosov, V. P. (1969). *Soviet Electrochem.* **5**, 445.

Kiukkola, K. and Wagner, C. (1957). *J. Electrochem. Soc.* **104**, 379.

Kleitz, M. (1968). Thesis, University of Grenoble.

Kubik, A. M. and Alcock, C. B. (1968). *In* "Electromotive Force Measurements in High-Temperature Systems" (Alcock, G.B., ed). The Institution of Mining and Metallurgy, London, pp. 43–49.

Masson, C. R. and Whiteway, S. G. (1967). *Can. Metall. Quarterly* **6**, 199.

Morand, G. and Hladik, J. (1969). "Electrochimie des sels fondus", Tome II, Masson, Paris.

Pastorek, R. L. and Rapp, R. A. (1969). *Trans. Metal. Soc. AIME.* **245**, 1711.

Perfilev, M. V. and Palguev, S. F. (1966). *In* "Electrochemistry of Molton and Solid Electrolytes", Tome 3, pp. 97–104. Consultants Bureau, New York.

Rapp, R. A. (1963). *Trans. Metall. Soc.* **227**, 371.

Rapp, R. A. and Maak, F. (1962). *Acta Metallurgica.* **10**, 63.

Reinhold, H. (1934). *Z. Electrochem.* **40**, 361.

Rickert, H. (1964). *In* "Condensation and Evaporation of Solids", pp. 202 to 242. Gordon and Breach Science Publ., New York.

Rickert, H. (1968). *Werstoffe und Korrosion* **19**, 869.

Steele, B. C. H. (1970). *In* "Heterogeneous Kinetics at Elevated Temperatures", Plenum Press.

Takahashi, T. and Yamamoto, O. (1966). *Electrochimica Acta.* **11**, 779.

Takahashi, T. and Yamamoto, O. (1970). *J. Electroch. Soc.* **117**, 1.

Takahashi, T., Ito, K. and Iwahara, H. (1965). *Proc. Internat. J. Fuel Cells.*

Wagner, C. (1953). *J. Chem. Physics* **21**, 1819.

21. CHRONOAMPEROMETRY, POTENTIAL-SWEEP CHRONOAMPEROMETRY, AND CHRONOPOTENTIOMETRY

J. Hladik

Université de Dakar, Sénégal, West Africa

I. Electroanalytical Techniques

A. Classification

1. Definitions

The oxidation and reduction reactions, corresponding to the exchange of electrons between an electrode and a solid electrolyte, may be studied by means of many electroanalytical techniques. Delahay *et al.* (1960) have suggested a classification of electroanalytical methods. They divided these techniques into three groups depending on the electrode processes. The techniques that concern us are classified in the third group. We suppose, in this chapter, that diffusion is the only means of mass transport, that the electrode area is constant and that the concentrations of the electroactive species in the supporting electrolyte are small.

According to the nomenclature suggested by these authors, we shall study in the present chapter, chronoamperometry, potential-sweep chronoamperometry and chronopotentiometry.

Chronoamperometry involves the sudden application of a potential of sufficient magnitude to produce an electrode reaction and the measurement of the resulting current as a function of time, $i = f(t)$.

During potential–sweep chronoamperometry, the working electrode changes linearly with time, $E = E_i + vt$, and one measures the resulting current versus potential, $i = f(E)$.

In chronopotentiometry, dependence of the potential of a working electrode is measured during electrolysis at constant current, $E = f(t)$.

2. Distribution of the Electroactive Species

There are four possible types of distribution for the electroactive species. The solute can be dissolved in the electrolyte or in the electrode, it can be located at the electrode/electrolyte boundary or situated outside the electrode–electrolyte system.

(*a*) *Electroactive species dissolved in the supporting electrolyte.* In this case the electroactive species is scattered throughout the electrolyte. The mass transport occurs from the electrolyte towards the electrode/electrolyte boundary. We then obtain the following scheme (Fig. 1a) in which the arrows indicate the direction in which the electroactive particles are moving. This case corresponds to solid mixtures; for example, $CdCl_2$ dissolved in LiCl–KCl (Hladik, 1966a). Other types of solid compounds can be prepared by diffusion of gas through the solid electrolyte.

Electroactive compounds can also be introduced into a solid electrolyte by electrolysis *in situ*. For example, the anodic oxidation of a platinum electrode in solid NaCl–KCl gives platinum chloride which diffuses into the supporting electrolyte (Hladik *et al.*, 1967).

(*b*) *Electroactive species dissolved in the electrode.* Here the electroactive substances are dissolved in the electronic conductor used as an electrode. The mass transport from the electrode/electrolyte interface proceeds from the electrode to the phase boundary (Fig. 1b). The electronic conductor can be liquid or solid.

For example, the diffusion of oxygen out of liquid silver electrodes has been studied in fuel cells (Hladik, 1965). Electroanalytical studies have been performed with such electrodes by Rickert and El Miligy (1968). The diffusion of hydrogen through thin paladium membranes in contact with solid electrolytes may be studied by means of potential-sweep chronoamperometry (Hladik, 1969a).

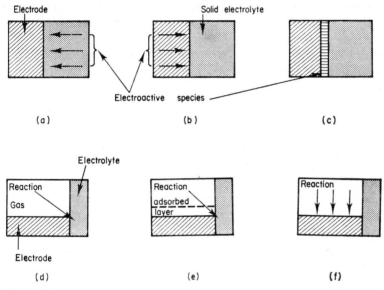

FIG. 1. Repartition of electroactive species. (a) electroactive species dissolved in the supporting electrolyte. (b) electroactive species dissolved in the electrode. (c) thin films of electroactive species. (d, e and f) reactions at the three-phase boundary.

(c) *Thin films of electroactive species.* The electroactive compounds can be deposited in thin layers at the electrode/electrolyte interface (Fig. 1c). The electrode can be covered with the electroactive species before placing it in contact with the electrolyte, or the thin film can be prepared by electrolysis in the solid state. For example, nickel has been deposited on a gold electrode by the electrolysis of nickel chloride in solid LiCl–KCl (Hladik, et al., 1967).

(d) *Reactions at the three-phase boundary.* The electroactive species may be a fluid situated outside the electrode–electrolyte system (Fig. 1d, e, f). The electrochemical reaction occurs either on the contact line between the three phases or at the metal/gas interphase. A third case may be distinguished by the possible adsorption of the electroactive species on the electrode.

The first case corresponds to the insolubility of the gas both in the electrode and the electrolyte; the second case to the formation of an adsorbed layer on the electrolyte and the third to the gas dissolution in the electrode metal, the dissolved gas being in an ionised state.

3. Theoretical Aspects

Electric currents, measured during chronoamperometric experiments, correspond to the flow of the electroactive particles arriving at the electrode surface. The theoretical expressions for the electrolysis currents are obtained, therefore, by solving transport equations for the electroactive species.

Electrode potentials measured during chronopotentiometric studies may be calculated theoretically from the concentration of electroactive species at the electrode surface. The numerical values of these concentrations may also be obtained by solving the diffusion equations for the electroactive compounds.

In both cases, the solution of mass transport equations can be obtained if the boundary conditions of a given system are known. The Nernst equation for reversible systems, or electrochemical process equations for irreversible systems, will generally determine the boundary conditions at the electrode surface. Other conditions, depending on the experimental arrangement, will complete the hypotheses, relative to the given mass transport problem.

Concerning substances dissolved in the supporting electrolyte, several particular problems have been solved for both reversible and irreversible systems, with various boundary conditions (Delahay, 1954). The case of electroactive species dissolved in thick electrodes is analogous to the case of its dissolution in the electrolyte.

Carslaw and Jaeger (1965) have treated many theoretical problems of heat conduction whose results may be used for gas diffusion through thin membrane electrodes.

We have solved the problem of thin-layered electrolytes studied by potential-sweep chronoamperometry (Hladik, 1966) and the results may be used for thin membrane electrodes.

The problem of thin layers deposited on an electrode has only been treated by supposing that the activity coefficient of the electroactive substance is equal to unity (Delahay, 1954). The problem of three-phase boundary reactions has been treated in a limited number of cases by Kleitz (1968) and Perfilev and Palguev (1967).

B. Electrolytes and Electrodes

1. Decomposition Potentials

During chronoamperometric and chronopotentiometric studies, the existence of the electrochemical reactions can only be made evident in the electrolyte electroactivity range. This range is definited by the oxidation and reduction potentials of the electroactive species of the supporting electrolyte.

The theoretical decomposition potentials of solid electrolytes have been calculated for numerous compounds; chlorides (Hamer *et al.*, 1965), bromides, iodides and fluorides (Hamer *et al.*, 1965), and oxides (Hamer, 1965). These values give the theoretical potentials between the anodic and cathodic limits of solid electrolytes.

The current-potential curves of the solid electrolyte decomposition allow one to determine the experimental range of electroactivity. Figure 2 shows the anodic and cathodic limits of LiCl–KCl solid electrolyte at 300° C (Hladik, 1966a). The oxidation of the Cl⁻ anions is obtained with graphite electrode and the reduction of the Li⁺ cations with gold electrode, while platinum/platinum chloride is used as the reference electrode.

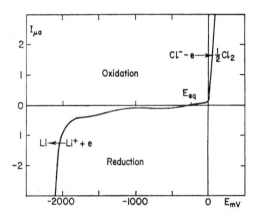

FIG. 2. Potential–sweep chronoamperometric curve. Oxidation and reduction limits of the LiCl–KCl eutectic at 300°C (After Hladik, 1966a).

The ZrO_2 (0·85) CaO (0·15) electrolyte has been investigated by Kleitz (1968) and Guillou *et al.*, (1968). They obtained the decomposition potentials by tracing the current-potential curves with platinum–rhodium electrodes. The electrolyte was maintained under argon pressure at $P_{O_2} < 21 \times 10^{-5}$ atm. The oxidation and reduction limits respectively correspond to the $O^=$ anion reactions and to the black zirconia formation (Fig. 3). After measuring the ohmic resistance of the sample, the authors calculated the current–potential curves corrected for the ohmic contribution.

2. Oxidation of the Electrodes

The existence of oxidation current–potential curves of an electrode can only be made evident when the electrode oxidation potential is situated inside the electrolyte electroactivity range. Very often the electrolyte oxidation limit cannot be obtained by using metallic electrodes.

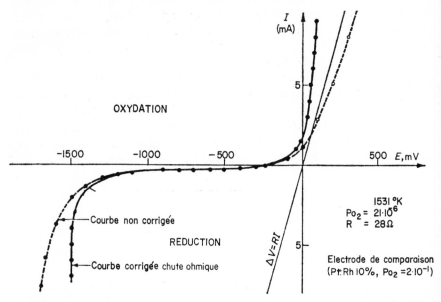

Fig. 3. Oxidation and reduction limits of the $(ZrO_2)_{0.85}$ $(CaO)_{0.15}$ electrolyte. (After Guillou et al., 1968).

Even for noble metals, the electrode oxidation occurs in solid alkali chloride electrolytes. Figure 4 shows the oxidation of the platinum electrode in NaCl–KCl, at 540°C, and the reduction of the compounds formed during the oxidation (Hladik et al., 1967). The oxidation is performed during various lengths of time and the height of the reduction peak increases with time.

Similar studies have been performed in solid LiCl–KCl with various metals (Hladik, 1966a). Figure 5 shows the oxide-reduction of silver, platinum, gold and tungsten electrodes in LiCl–KCl, at 240°C. At this temperature, the reduction waves are gradually decreasing.

II. Electroactive Species Dissolved in the Supporting Electrolyte

A. Supporting Electrolytes

1. Introductory Remarks

Numerous foreign cations and anions may be introduced into liquid electrolytes and the disordered structure of liquids allows these ions to be accomo-

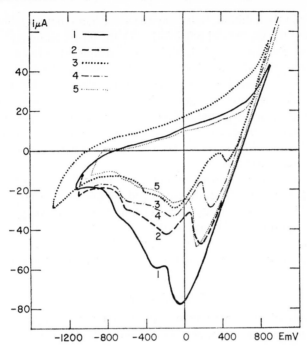

FIG. 4. Potential–sweep chronoamperometric curves. Oxidation of a platinum electrode and reduction of the platinum chlorides in solid NaCl–KCl at 540°C (After Hladik *et al.*, 1967).

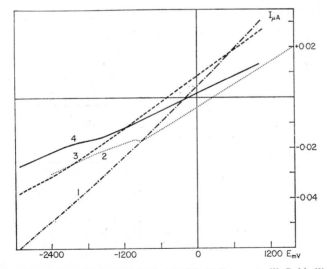

FIG. 5. Oxidation of electrodes in LiCl–KCl at 240°C. (1) Tungsten; (2) Gold; (3) Platinum; (4) Silver. (After Hladik and Rovetto, 1970.)

dated easily in the solution. Solids, on the other hand, have a more compact structure and the theoretical possibilities for the introduction and migration of foreign ions within a solid electrolyte are much less than in a liquid.

For the crystallized solids, having a well-defined structure, it is necessary to consider solvent and solute parameters characterizing crystalline solid electrolytes. The electrolyte lattice will be progressively distorted to give amorphous electrolytes as the solute concentration increases. On the other hand, diffusion is the essential mechanism of mass transport and it is important to know the values of heterodiffusion coefficients in solid state supporting electrolytes.

The rate of diffusion and the solubility of impurities in solids are closely interrelated. Both processes are controlled by the concentration of vacant sites available for the impurity atoms or ions and by the energy with which the impurity particles are bound to the atoms of the solvent. It follows that diffusion and solubility studies are jointly needed to give a full understanding of the interactions between impurity atoms and the atoms of the host material.

2. Solid Solutions

(a) *Classification*: The study of equilibrium phase diagrams of binary systems indicates that an element can be introduced into the crystal lattice of another, up to a certain concentration, without any loss of homogeneity. These homogeneous regions of the phase diagram are called solid solutions, of which there are three distinct kinds (Fig. 6): substitutional, interstitial and defect solid solutions.

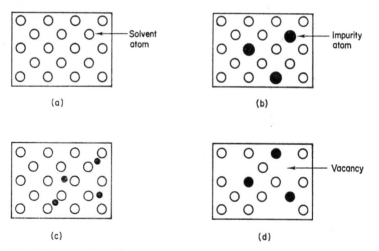

FIG. 6. Solid Solutions. (a) Solvent (b) Substitutional solid solution (c) Interstitial solid solution (d) Defect solid solution.

In substitutional solid solutions, some of the normal lattice sites in the solvent crystal are occupied by solute atoms, and the structure of the solvent remains unchanged (Fig. 6b). Thus KCl and KBr give solid solutions of any composition between the two extremes.

Interstitial solid solutions are formed when the solute atoms occupy positions in the interstices of the crystal lattice of the solvent. Generally, it is the atoms of small atomic radius (Li, B, H, N, C, O) which are involved (Fig. 6c).

Solid solutions CaF_2–YF_3 and SrF_2–LaF_3 provide examples of crystals containing interstitial ions. Density measurements show that of the two possibilities, $2M^{3+}$ replacing $3M^{2+}$ leaving vacant cation positions, or interstitial F^- ions, the latter is realized in these crystals. The F^- ions presumably occupy statistically the largest holes in the CaF_2 structure.

In defect solid solutions, some sites in the lattice of one of the components remain vacant (Fig. 6d). Thus in crystals of AgCl and AgBr containing small amounts of $CdCl_2$, $CdBr_2$, $PbCl_2$ and $PbBr_2$ there is a vacant Ag position for every Cd^{++} or Pb^{++} ion. Solid solutions of this type are nearly always semi-conductors. Defect solid solutions are formed typically in chemical compounds of transition elements, as well as in sulphides, selenides, tellurides and some oxides.

There is as yet no complete physical theory of solid solutions. However, from the work of Hume–Rothery and Raynor (1938) and Jones (1934), one can formulate a number of quantum mechanical conditions governing the formation and the stable existence of various phases when two substances are mutually dissolved.

(b) *Conditions of formation*—(i) Compounds of the same structure. The ability of two substances to form solid solutions is determined primarily by geometrical considerations. If the two structures are the same, and if the radii of the substituent atoms differ by not more than about 15% from that of the smaller one, a wide range of solid solutions can be expected at room temperature.

Thus, from solutions containing KCl and Br, it is possible to grow homogeneous crystals of any composition between the two extremes. In this case, the radii of the Cl^- and Br^- are respectively 1·81 and 1·95 Å and differ only by about 8·5%. These crystals all have the sodium chloride structure with the anion sites occupied by Cl^- and Br^- distributed at random. The cation sites are occupied by K^+.

At higher temperatures a somewhat greater degree of tolerance between the radii is permitted, but if the radius difference exceeds the limit of 15%, only restricted solid solution occurs. Thus NaCl and KCl form solid solutions only at high temperatures; KCl and Ki are mutually soluble only over a limited range of composition.

(ii) Types of binding. The formation of a solid solution requires that the bonds in the two crystals be of similar type; KCl forms solid solutions with KBr but not with PbS which has the same crystal structure. Even if the dimension of the unit cells of two crystals are sufficiently close to permit the formation of solid solutions, it still might not be possible to grow them. This will be the case if one of the salts is insoluble or is decomposed before melting, since mixed crystals are made from a solution or molten mixture of the two salts.

Thus the cell constants of calcite and sodium nitrate are very nearly the same,

$$CaCO_3 \quad a = 6 \cdot 361 \text{ Å} \quad \alpha = 46°07'$$

$$NaNO_3 \quad a = 6 \cdot 32 \text{ Å} \quad \alpha = 47°15'$$

but mixed crystals cannot be formed.

(iii) Compounds of different structures. Many examples are known of partial solid solution between compounds with different crystal structures and, also, between compounds of quite different chemical compositions.

The system AgI–AgBr gives a solid solution between compounds of similar composition but different structure. Silver iodide has the zincblend structure at room temperature and silver bromide has the sodium chloride structure. The silver bromide shows little tendency to assume the zincblend arrangement and only very limited solution of AgBr in AgI is possible. On the other hand, AgBr can take up in solid solution as much as 70% of AgI.

The analogous system AgBr–CuBr behaves differently. Copper halogen bonds are more covalent than the corresponding silver-halogen bonds, and consequently the copper atoms in CuBr retain their characteristic tetrahedral coordination even in solid solution. As CuBr is taken up in AgBr, the copper atoms do not occupy the sites of the silver atoms, but, instead enter the previously vacant tetrahedral interstices between the bromine atoms, so that for every tetrahedral site thus occupied, one empty octahedral site is created. The extent of the solid solution is limited.

The compounds $MgCl_2$ and LiCl, which differ both in crystal structure and empirical formula, give solid solutions. Lithium chloride has the sodium chloride structure whereas $MgCl_2$ has the cadmium chloride layer structure. Only a limited degree of solid $MgCl_2$ solution in LiCl and of LiCl in $MgCl_2$ is possible. Although the cadmium chloride and sodium chloride structure are quite distinct there is, nevertheless, a close geometrical relationship between them.

(c) *Local strains in the atomic structure of substitutional solid solutions.* Differences in the atomic sizes of the components lead to local strains in the lattice. These have important effects on the properties of the solid solutions; in particular they lead to changes of the mean lattice parameter.

Local strains are shown schematically in Fig. 7 for the two cases where the atoms of the solute are respectively larger and smaller than the atoms of the solvents. The lattice positions occupied by the solute atoms are centres of maximum strains. This is the location for the major part of the strain energy concentration.

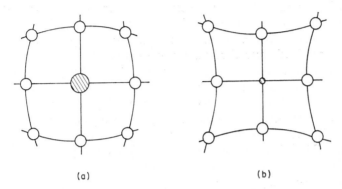

(a)   (b)

FIG. 7. Local strains in the lattice of a solid solution. Solvent atoms: (a) smaller, (b) larger than those of the solute.

The amount of lattice strain in a solid solution is determined by the magnitude of the mean square static displacement of the atoms from the equilibrium position $\overline{u_{st}^2}$. From the elastic model of a solid solution it follows that:

$$\overline{u_{st}^2} = \gamma c \, \Delta R^2$$

where γ is a numerical factor (equal to 7·8 for a f.c.c. lattice and 7·3 for a b.c.c.); c is the concentration of solute (assumed small) and ΔR is the difference in the atomic radii.

To correct the magnitude of the static stresses, one must take into account the dynamic displacement u_d^2 of the atoms from their equilibrium sites as a result of the thermal motion in the crystal.

In a solid solution, the magnitude of the dynamic displacements can change either because of changes in the masses of the atoms or of changes in the forces of interaction. The total value of the mean squared displacement of atoms in a solid solution is determined by the sum:

$$\overline{u^2} = \overline{u_{st}^2} + \overline{u_d^2}$$

and can be measured by intensity of X-ray scattering. Investigation of scattering at different temperatures enables us to separate static and dynamic strains. Experimental data on the static and dynamic displacements in solid solutions show that these displacements are of comparable size.

3. Diffusion

Diffusion in ionic crystals, glasses and ion-exchange resins is reviewed elsewhere in this book, therefore, we only give some indications concerning the heterodiffusion in supporting electrolytes. Experiments show that small amounts of impurities can have profound effects on the conductivity of ionic crystals, either decreasing or increasing it. For example, the addition of $CdCl_2$ to the NaCl crystal requires that each Cd^{++} ion replace two Na^+ ions, simply to maintain neutrality. The Cd^{++} ion, however, can fill only one of the remaining sites left by the two Na^+ ions; the other site must remain vacant. Therefore adding positive ions of greater valence than Na^+ increases the vacancy concentration. When $CdCl_2$ was added to sodium chloride, it was found that the conductivity was considerably increased. Since the Cd^{++} ions increase the positive ion vacancy concentration, we conclude that the primary mechanism for migration is the positive ion vacancy mechanism.

The transport number for Ag^+ in AgBr is unity. When $CdBr_2$ is added in trace amounts to AgBr, the conductivity decreases below its normal value. In this case, the Cd^{++} ions increase the vacancy concentration, but if the crystal contains Frenkel defects, the positive ion interstitial concentration is decreased. The decrease in conductivity is, therefore, accounted for if the diffusion mechanismes involve primarily cation interstitials.

The substances whose diffusion coefficients in glass have been measured can be placed into three groups, according to Doremus (1962):

Molecules: oxygen, hydrogen, helium, neon, gold, nitrogen.

Ions: silver, lithium, sodium, potassium, calcium.

Bound covalently in the glass network: silicon, aluminium, phosphorus, water, oxygen . . .

Doremus has shown that gold can diffuse in glass as either an atom or an ion. The classification of oxygen requires a more extensive discussion because there are two possible types of oxygen diffusion in silicate glasses: diffusion of oxygen that is bounded in the silicate network and diffusion of dissolved oxygen, which is distinct from network oxygen.

The univalent ions can, more or less, substitute one another in the glass network. Thus, silver ions exchange readily with sodium ions in a silicate glass, below the transformation temperature, without any substantial effect on the glass structure. Lithium and potassium ions also exchange with the sodium in a silicate glass, but if the concentration of these ions in the glass becomes too high, it will crack.

The ionic diffusion coefficients in glasses which are cooled rapidly to below the transformation temperature, are generally higher than those in annealed glass. Ionic diffusion coefficients in glass increase much more slowly with temperature than does the fluidity of the glass.

B. Experimental Techniques

Solid electrolytes with small solute concentrations may be prepared by the fusion of the solute and the supporting electrolyte, and then cooling the mixture until solidification. Electroactive species may be prepared *in situ* by anodic oxidation of electrodes. Some electroactive compounds may also be introduced by diffusion through certain electrolytes.

The working electrode can be embedded wholly or partially in the electrolyte, or simply placed in contact with the electrolyte surface (Fig. 8). The working electrodes may be cylindrical wires, with the cross section or the lateral surface in contact with the electrolyte (Hladik, 1965, 1966); hemispherical electrodes may also be used (Karpachev and Obrosov, 1969). Rods in contact with the surface electrolyte, metallic films deposited on a part of the electrolyte and liquid metal electrodes have also been used.

The classical potentiostatic apparatus may be used it if is sufficiently sensitive, the electroanalytical currents being generally small and the electro-

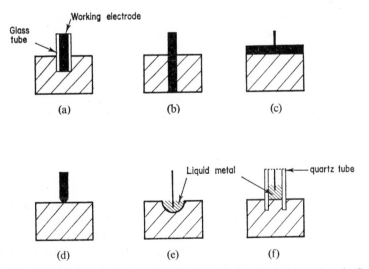

FIG. 8. Working electrode arrangements: (a, b) wire embedded in the electrolyte. (c, d) solid. electrode in contact with the electrolyte. (e, f) liquid electrode in contact with the electrolyte;

lyte having a high resistance. The measurements of the working electrode potential also require an apparatus adapted to the high resistance of the circuit involving the reference and working electrodes.

Figure 9 depicts the electrical arrangement using three electrodes, for potential–sweep chronoamperometry. A potentiostat, P_O, controls the well-defined potential of the working-electrode, the variation of the tension being realized by means of a pilot, P_i. The potential–current curve is recorded by means of an XY recorder.

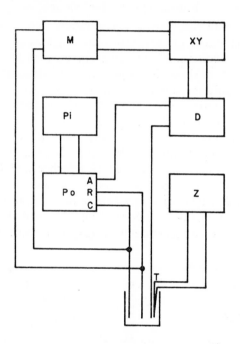

FIG. 9. Electrical arrangement fro potential–sweep chronoamperometry: P_o, Potentiostat; P_i, Pilot; XY, recorder; M, electronic voltmeter; C, working electrode; A, auxiliary electrode; T, thermocouple; Z, "zerostable"; R, reference electrode.

C. Chronoamperometry

1. Theory

Chronoamperometry involves the sudden application of a potential of sufficient magnitude to produce an electrode reaction and the measurement of the resulting current as a function of time. If an electroactive species is transported solely by a diffusion process. without any contribution from convection or migration, the diffusion coefficient can be evaluated from the observed current–time relationships.

In many constant potential chronoamperometric experiments, the electro-active species is reduced (or oxidized) at a rate which is sufficiently high so that the concentration of this cation (or anion) is equal to zero during all of the electrolysis.

In this case, the boundary conditions for the partial differential equations of the mass transport in the electrolyte are the simplest. Therefore, we will only consider the theoretical chronoamperometric curves with such an electrode potential for the two experimental arrangements corresponding to linear and cylindrical diffusion.

(a) *Semi-finite linear diffusion.* A plane electrode is immersed in a solid solution of substance O, which is reduced to another substance R at the electrode surface. The solution is supposed to extend infinitely in the direction perpendicular to the electrode. According to Fick's second law we have for the substance O:

$$\frac{\partial C(x,t)}{\partial t} = D \frac{\partial^2 C(x,t)}{\partial x^2}.$$

If the substance O is reduced at a rate which is so high that the concentration of this substance at the electrode surface is equal to zero as soon as electrolysis is started, the boundary condition is:

$$C(0,t) = 0 \qquad \text{for } t > 0.$$

When the concentration $C(x,0)$ is constant in the electrolyte, before electrolysis, we have:

$$C(x,0) = C°; \qquad t = 0.$$

We assume that $C(x,t)$ approaches $C°$ when x tends towards infinity for all values of t.

The solution of this equation is then well-known:

$$C(x,t) = C° \operatorname{erf}\left(\frac{x}{2D^{1/2}t^{1/2}}\right).$$

We used the two-variable Laplace transform as a fast method of resolution (Hladik, 1966a, 1969c). The notation erf (α) represents the error integral defined by:

$$\operatorname{erf}(\alpha) = \frac{2}{\pi^{1/2}} \int_0^\alpha e^{-y^2}\, dy.$$

The definition of the flux,

$$\phi(x, t) = D \frac{\partial C(x, t)}{\partial x}$$

shows that the electrolysis current is equal to the product of the charge involved in the reduction of one mole of electroactive species O by the flux of this substance at the electrode surface. Thus, for a plane electrode of area S, the current is:

$$i = nFS \left[\frac{\partial C(x, t)}{\partial x} \right]_{x=0}$$

n being the number of electrons involved in the reduction of substance O. and F being the Faraday. By differentiating the expression $C(x, t)$ with respect to $x = 0$ in the resulting formula, the value of the current is obtained:

$$i = \frac{nFS \, D^{1/2} \, C^\circ}{\pi^{1/2} \, t^{1/2}}.$$

If i is in amperes, the units in this equation are: S in square centimeters, F in coulombs, D in cm^2 sec^{-1}, C° in moles per cubic centimeter, t in seconds.

(b) *Semi-infinite cylindrical diffusion.* In this case, the electrode is a circular cylinder and the field of diffusion is limited by two planes perpendicular to the axis of the cylinder. The radius of the electrode is small in comparison with the distance between these planes and there is no interference with the diffusion process. Under such conditions, the diffusion field is symmetrical and the diffusion problem can be discussed in terms of the variables r, distance from the axis of the cylinder and t, time elapsed since the beginning of electrolysis. The partial differential equation for symmetrical cylindrical diffusion is: (Delahay, 1954)

$$\frac{\partial C(r, t)}{\partial t} = D \frac{\partial^2 C(r, t)}{\partial r^2} + \frac{D}{r} \frac{\partial C(r, t)}{\partial r}.$$

The boundary conditions are as follows:

$$C(r_0, t) = 0$$

r_0 being the radius of the cylinder; $C(r, t) \to C^\circ$ when $r \to \infty$. The initial condition is:

$$C(r, 0) = C^\circ; \qquad t = 0.$$

We have used a new functional transformation for rapidly solving this and other symmetrical cylindrical problems (Hladik, 1966b).

The proposed functional transformation is a two-variable Laplace–Bessel transformation:

$$u(q, s) = \int_0^\infty \int_{r_0}^\infty e^{-st} r K_0(qr) U(r, t) \, dr \, dt$$

where $K_n(z)$ is the Bessel function (Petiau, 1965)

$$K_n(z) = \lim_{x \to n} \frac{\pi}{2} \frac{\dot{I}_{-x}(z) - \dot{I}_x(z)}{\sin \pi x}$$

$\dot{I}_x(z)$ being the modified Bessel function of the first kind.

The Laplace–Bessel transformation is applied to the differential equation. We directly obtain the solution by using the inverse Laplace transformation:

$$\left[\frac{\partial C(r, t)}{\partial r}\right]_{r=r_0} = \frac{4C^\circ}{r_0 \pi^2} \int_0^\infty \frac{e^{-Dx^2 t} \, dx}{x[J_0^2(r_0 x) + Y_0^2(r_0 x)]}$$

J_0 and Y_0 being Bessel functions of zero order and of the first and second kinds respectively. This last integral was evaluated by Jaeger and Clarke (1942). The chronomaperometric current is always obtained by the same relation:

$$i = nFSD \left[\frac{\partial C(r, t)}{\partial r}\right]_{r=r_0}$$

$S = 2\pi r_0 h$ being the electrode surface and h the length of the cylinder. Itc an be demonstrated that the current approaches zero very slowly when t increases and no steady state is observable.

2. Experimental Results

Hladik and Morand (1965) investigated the transport properties of potassium nitrate in the solid electrolyte LiCl–KCl.

The solid solution is prepared by dissolving KNO_3 im nolten LiCl–KCl eutectic; the $pO^=$ is fixed by adding BaO. Leroy (1963) established the stability of the system NO_3^-/NO_2^- if the $O^=$ concentrations are sufficiently high. Preliminary studies of KNO_3 (Hladik and Morand, 1965) in solid electrolyte indicated the possibility of electrochemical reduction for NO_3^- ions according to the reaction:

$$NO_3^- + 2e \rightleftharpoons NO_2^- + O^=.$$

The construction of the cell is shown in Fig. 10. The working electrode is a gold wire of 2 mm diameter sealed in a "Pyrex" tube, the cross section of the wire being in contact only with the solid electrolyte. The auxiliary electrode is a gold wire, 2 cm long, and the reference electrode is a platinum/platinum chloride electrode. The three electrodes are immersed in the molten solution and the cell is frozen at a temperature below the melting point of the solution. Thus the lower extremities of the electrodes are embedded in the solid electrolyte and the contact between the electrodes and the solid electrolyte is very good.

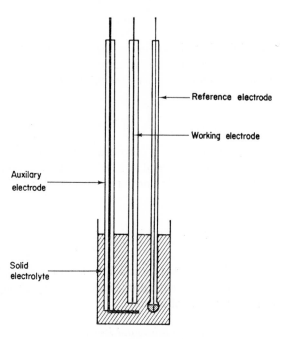

Fɪɢ. 10. Cell for chronoamperometric and potential–sweep chronoamperometric experiments.

A potential of $-1400\,mV$ is applied to the working electrode by means of a potentiostat. The current–potential curves show that this potential corresponds to the maximum current for the reduction of KNO_3. Thus the concentration of $KNO_3{}^-$ anions, near the working electrode surface, is zero.

The current–time curve obtained from the experiments with a concentration $4 \times 10^{-2}\,mol/kg$ of NKO_3 is given in Fig. 11.

Karpachev and Obrosov (1969) investigated the transport properties of silver in solid sodium chloride.

Thermodynamic investigations of the AgCl–NaCl solid solution which they carried out show that the activity coefficient of silver chloride barely changes over the concentration range from 0·1 to 1 mol % AgCl. This makes it possible to follow the change in the diffusion current in relation to the silver chloride concentration. The investigations were carried out with solid and liquid electrodes. The experimental arrangement is shown in Fig. 12. In the case of the liquid electrode, the cathode was a liquid drop of 40% Ag + 60% Bi alloy, 0·8 mm in diameter. The comparison electrode was a silver wire and the auxiliary electrode was a liquid alloy into which the sample was placed.

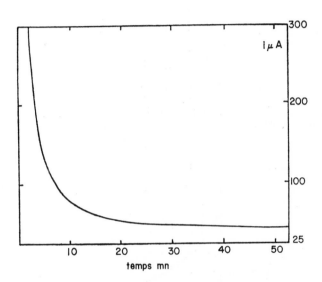

FIG. 11. Current–time relation obtained with a solid electrode. Reduction of potassium nitrate in solid LiCl–KCl at 320°C. (After Hladik and Morand, 1965).

In the case of the solid electrode, the cathode was a thin silver wire, 0·2 mm in diameter, embedded along the axis of a cylindrical sample of solid electrolyte. The comparison electrode was a silver wire embedded in the electrolyte and the auxiliary electrode was deposited on the surface of the cylinder in silver powder form. The ratio of the areas of the surfaces of the cathode and the auxiliary electrode was 1 : 50, making it possible to neglect anode polarization.

A cathode potential of 700 mV relative to the comparison electrode was applied to the liquid metal drop by means of a potentiostat. The pure cathod polarization of the drop then equalled 500–530 mV, guaranteeing that the concentration of discharging silver cations close to the drop surface was near to zero.

886   J. Hladik

The characteristic current–time curves obtained working with solid and liquid electrodes are given in Fig. 13.

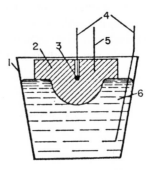

FIG. 12. Cell with liquid electrodes: (1) alundum crucible; (2) sample of AgCl–NaCl solid solutions (3) cathode, liquid alloy of 40% Ag + 60% Bi; (4) nickrome conductors; (5) comparison electrode, silver wire; (6) anode, liquid alloy of 40% Ag + 60% Bi. (After Karpachev and Obrosov, 1969).

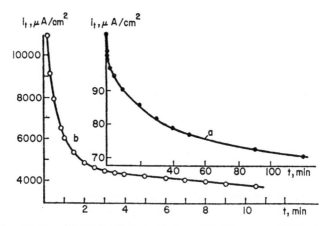

FIG. 13. Current–time relation in NaCl–AgCl electrolyte (a) on a solid electrode (cathode: silver wire $\phi0.2$ mm; mole fraction of AgCl 0.13×10^{-2}; 603°C). (b) on a liquid electrode (cathode: liquid drop of 40% Ag + 60% Bi; mole fraction of AgCl, 10^{-2}; 731°C) (After Karpachev and Obrosov, 1969).

Figure 14 gives the relations of i to $i^{-1/2}$ obtained by using liquid electrodes for different concentrations of silver chloride in the solid solution. As the figure shows, over the range of times from 5 to 210 sec, the time dependence of the diffusion current obeys the diffusion law for a spherical electrode. It should be noted that the diffusion current at a particular moment of time is proportional to the concentration of the silver ion discharged. Figure 15 shows the relation of the diffusion current to the concentration. It is concluded that the area of contact between the liquid electrode and the electrolyte was constant.

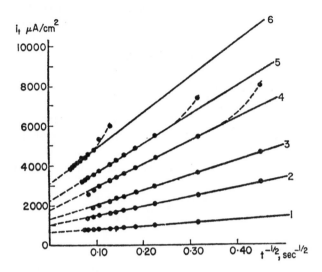

FIG. 14. Relations $i_t - 1/\sqrt{t}$ obtained with liquid electrodes for different concentrations of AgCl in AgCl–NaCl solid solution at 731°C. (1) 0.45×10^{-4}; (2) 1.04×10^{-4}; (3) 1.72×10^{-4}; (4) 2.24×10^{-4}; (5) 2.76×10^{-4}; (6) 3.45×10^{-4} mol/cm³ After Karpachev and Obrosov, 1969).

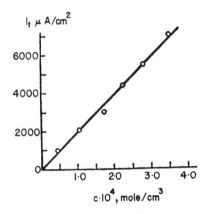

FIG. 15. Relation of the diffusion current to the concentration of the Ag^+ ion discharged at $t = 20$ sec (After Karpachev and Obrosov, 1969).

D. Potential-Sweep Chronoamperometry

1. Theory

(a) *Semi-infinite linear diffusion.* (i) Reversible process involving soluble substances. Current–potential curves are determined by using a stationary electrode whose potential varies continuously during electrolysis. We will consider the case where the potential varies linearly with time:

$$E = E_i - vt.$$

Where E_i is the initial potential of the working electrode and v is the rate of potential change in volt sec^{-1}.

Concerning the electrochemical reduction of a substance O to another substance R, we assume that the electrochemical process is reversible. Potential E_i is adjusted to a value at which substance O is virtually not reduced. According to Fick's second law, we have for the substance O:

$$\frac{\partial C_O(x, t)}{\partial t} = D_O \frac{\partial^2 C_O(x, t)}{\partial x^2}$$

and for the substance R:

$$\frac{\partial C_R(x, t)}{\partial t} = D_R \frac{\partial^2 C_R(x, t)}{\partial x^2}$$

The initial conditions of electrolysis are:

$$C_O(x, 0) = C^\circ$$

$$C_R(x, 0) = 0$$

and for $t \to \infty$

$$C_O(x, \infty) = C^\circ; \qquad C_R(x, \infty) = 0$$

It is supposed that the Nernst equation for the electrode potential will hold up to a certain rate of change of E. Above this rate, it will be affected by the velocity of the electrode reaction.

A first boundary condition is thus obtained by applying the Nernst equation:

$$\frac{C_O(0, t)}{C_R(0, t)} = \frac{f_R}{f_O} \exp\left[\frac{nF}{RT}(E_i - E^\circ)\right] \exp\left(\frac{-nF}{RT} vt\right).$$

The second boundary condition is obtained by saying that the sum of the fluxes of the species O and R, is equal to zero at the electrode surface. Thus:

$$D_O \left[\frac{\partial C_O(x, t)}{\partial x} \right]_{x=0} + D_R \left[\frac{\partial C_R(x, t)}{\partial x} \right]_{x=0} = 0.$$

This boundary value problem was solved by Sevcik (1948). The present author used a faster method for solving this type of equation, the double-Laplace method (Hladik, 1966a). The current–potential equation was calculated by Sevcik (1948) and by De Vries and Van Dalen (1963). The latter found a more accurate equation.

The results are summarized in the following current–potential equation:

$$i = \frac{n^{3/2} F^{3/2}}{R^{1/2} T^{1/2}} SC^\circ v^{1/2} \left[\sqrt{D_O} + \frac{\sqrt{D_R}}{\theta} \right] P(\sigma t)$$

where S is the electrode surface;

$$\theta = \frac{f_R}{f_O} \exp \left[\frac{nF}{RT} (E_i - E^\circ) \right]$$

$$\sigma = \frac{nFv}{RT}$$

$$P(\sigma t) = \frac{1}{4\sqrt{\pi}} \int_0^{\sigma t} \frac{1}{\sqrt{\sigma t - \sigma \alpha}} \frac{1}{\cosh^2 \frac{1}{2}(\sigma t_{1/2} - \sigma \alpha)} \, d\sigma \alpha$$

$$t_{1/2} = \frac{E_i - E^\circ}{v} + \frac{1}{\sigma} \ln \left(\frac{f_R}{f_O} \right) \left[\frac{D_O}{D_R} \right]^{1/2}.$$

The computed P-values are given in Fig. 15. It appears that P is practically independent of the value of the parameter $\sigma t_{1/2}$. Only the very first part of the curve is appreciably dependent on $\sigma t_{1/2}$. The given figure is constructed for $\sigma t_{1/2} = 10$. The origin of the potential axis is at $\sigma t_{1/2}$, which corresponds with the chronoamperometric half-wave potential of the redox system.

For the maximum of P, the authors found (De Vries and Van Dalen, 1963)

$$P_{\max} = 0 \cdot 446$$

at

$$E = E_{1/2} - 1 \cdot 1 \frac{RT}{nF}.$$

They assumed that their P-values were accurate to well within $0 \cdot 5\%$.

(ii) Reversible deposition of an insoluble substance. Now we will consider the reversible reduction of a substance O to another substance R. We assume that the electrochemical process is reversible and that the electrolysis product R is insoluble, as, for example in the deposition of a metal on a solid electrode. We assume that the activity of the deposit is equal to unity. Thus the Nernst equation is:

$$E = E^\circ + \frac{RT}{nF} \ln f_O\, C_O(0, t)$$

the assumption being same as for the preceding case (i).

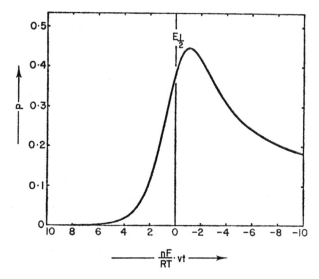

FIG. 16. The function $P(\sigma t)$ computed with Sevcik's modified convolution integral (After De Vries and Van Dalen, 1963).

This boundary value problem was solved by Berzins and Delahay (1953). The current–potential curve is:

$$i = \frac{2n^{3/2}\, F^{3/2}}{\pi^{1/2}\, R^{1/2}\, T^{1/2}}\, SC^\circ D_O^{1/2}\, v^{1/2}\, \phi\big[(\sigma t)^{1/2}\big]$$

where the function $\phi(\alpha)$ is:

$$\phi(\alpha) = \exp(-\alpha^2) \int_0^\alpha \exp(z^2)\, dz.$$

Variations of $\phi(\alpha)$ are represented in Fig. 17. Extensive tables of this function were preparated by Miller and Gordon (1931).

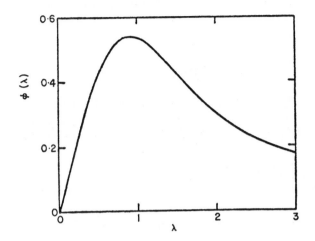

FIG. 17. Reversible deposition of an insoluble substance. Variation of the function $\phi(\lambda)$. (After Delahay, 1954).

The function $\phi(\alpha)$ exhibits a maximum equal to 0·541 when the argument is 0·924.
The corresponding peak current is readily calculated from the current–potential curve.

2. Finite Linear Diffusion

Hladik (1967a) obtained an equation for a reversible process using thin electrolyte layers. This is a very useful technique when working with solid electrolytes with high resistivity. A short distance between the working and auxiliary electrodes allows the attainment of sharper peaks than can be had in semi-finite linear diffusion.

The assumption of the reversible electrolysis of the substance O to R are the same as in the preceding case. The thickness of the electrolyte is L, and we suppose the $D_O = D_R = D$.

We used the two-variables Laplace transformation:

$$f(p, q) = \int_0^\infty \int_0^\infty e^{-px - qt} F(x, t)\, dx\, dt$$

for solving the mass transfer equation. This method allows one to obtain the relationship between the transform of the boundary and initial conditions.

The solution for the flux at the electrode surface is:

$$D\left[\frac{\partial C_o(x,t)}{\partial x}\right]_{x=0} = C^\circ D \frac{d}{dt}\left[\frac{1}{Ke^{\sigma t}+1}\right] \overset{t}{\star} \theta_2\left(0, \frac{Dt}{L^2}\right)$$

where:

$$\sigma = nFv/RT$$

$$K = \frac{f_a}{f_b}\exp\left[nF(E^\circ - E_i)/RT\right].$$

The function $\theta_2(x,y)$ is the Jacobi theta-function of order two:

$$\theta_2(x,y) = (\pi y)^{-1/2} \sum_{k=-\infty}^{\infty} (-1)^k \exp\left[-\frac{(x+k)^2}{y}\right].$$

The symbol $\overset{t}{\star}$ indicates a convolution integral with respect to the variable t. In the case of $K=1$, the solution for the current–potential curve is:

$$i(t) = \frac{D^{1/2}C^\circ nF}{4\pi^{1/2}} SI(t)$$

with:

$$I(t) = \sigma \cosh^{-2}\left(\frac{\sigma t}{2}\right)\overset{t}{\star} t^{-1/2} \sum_{-\infty}^{\infty}(-1)^k \exp(-k^2L^2/Dt).$$

This series converges rapidly when the values for L^2/D are sufficiently large. Figure 18 shows the current potential curves for various values of the parameter L^2/D.

3. Cylindrical Diffusion

Diffusion currents at cylindrical electrodes were studied by a numerical method of approximation by Nicholson (1954). The method of resolution consists of transforming the mass transfer equation in the corresponding finite difference equation. Hladik (1966b) used the two-variables Bessel–Laplace transformation for solving the mass transfer equation with cylindrical symmetry.

Concerning the deposition of an insoluble substance on the electrode, the solution derived by Hladik (1966b) with the aid of a Bessel–Laplace transformation is:

$$\left[\frac{\partial C(r,t)}{\partial r}\right]_{r=r_0}$$

$$= \frac{4nFv}{r_0\pi^2 RT}\exp\left[\frac{nF}{RT}(E_i - E^\circ - vt)\right]\overset{t}{\star}\int_0^\infty \frac{e^{-Dx^2t}\,dx}{x[J_0^2(xr_0)+Y_0^2(r_0x)]}$$

where r_0 is the radius of the electrode, J_0 and Y_0 are Bessel functions of zero order of the first and second kind, respectively; the convolution integral is calculated with respect to the variable t.

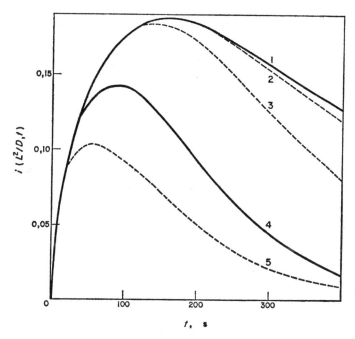

FIG. 18. Potential–sweep chronoamperometric curves. Finite linear diffusion. Variations of L^2/D: (1) 10^4; (2) 900; (3) 400; (4) 100; (5) 40. (After Hladik, 1967a).

When the quantity $(1/r_0)(D/nv)^{1/2}$ is smaller than 2, the current potential curves remain virtually identical and we find the case of linear diffusion. Thus, in slow recordings, the effect of the electrode curvature should be taken into account.

4. Experimental Results

(a) Electroactive species in solution. The first potential–sweep chronoamperometric studies were carried out in the solid LCil–KCl eutectic, with the temperature varying from 250–300° C (Hladik and Morand, 1965; Hladik, 1966a, 1970; Hladik et al., 1967, 1969). These studies have been continued in equimolecular NaCl–KCl mixture (Hladik et al., 1967, 1970).

The current–potential curves generally show one or several characteristic waves according to the nature of the ionic species, with a specific maximum for each electrochemical reaction. The heights of these maxima are proportional to the solute concentration when the latter are sufficiently small. This proportionality corresponds to the theoretical chronoamperomtric curves.

The reversibility of an electrochemical process is generally preserved in the neighbourhood of the melting point, when the system changes from the molten state to the solid state. A displacement of the reduction (or oxidation) potentials may be obtained when the reaction products are transformed from the gaseous state to the liquid or solid state, with lowering of the temperature.

Figure 19 shows the current–potential curves obtained from the reduction of potassium nitrate (Hladik and Morand, 1965; Hladik, 1966a). The influence of the temperature on the height of the reduction maxima is very large.

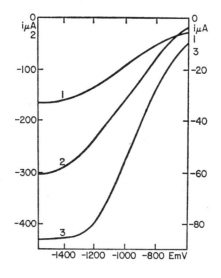

Fig. 19. Potential-sweep chronoamperometric curves of the potassium nitrate reduction. ($KNO_3 = 4 \times 10^{-2}$ mol/kg). Temperatures (1) 220°C; (2) 320°C; (3) 270°C. (After Hladik and Morand, 1965.)

(*b*) *Electroactive species made up in the electrolyte.* Numerous electroactive compounds may be made up *in situ* by solid state electrolysis in bulk (Hladik *et al.*, 1967, 1969, 1970). We used the anodic oxidation of a metal to establish a concentration gradient of the reaction products in the neighbourhood of the working electrode. This gradient is relatively stable, because of the small values of the diffusion coefficient.

5. Experimental Parameters

(a) *Electrochemical reaction.* The first purpose of the potential–sweep chronoamperometry is to provide evidence that one or several electrochemical reactions exist for a given solution. Then the current–potential curves show one or several waves corresponding to the possible electrochemical reactions.

When the chronoamperograms show several waves for only one ionic species in solution, this is an indication that the electrochemical process occurs in several steps. For example, when reducing Cu(II) we first obtained the reduction to Cu(I), then Cu(O) (Hladik *et al.*, 1966).

When the solid electrolyte contains several ionic species and the reduction (or oxidation) potentials are sufficiently far from one another, one finds characteristic maxima for each species.

Figure 20 shows a chronoamperogram obtained at 292°C, in the LiCl–KCl–BaO mixture ($pO^= = 1$). The reduction curve shows the reduction of gold chloride made up by the electrode oxidation. This reaction is followed by the oxygen reduction, the Ba^{++} reduction and finally the solvent limit corresponding to the alkali cation reduction. The oxidation curve,

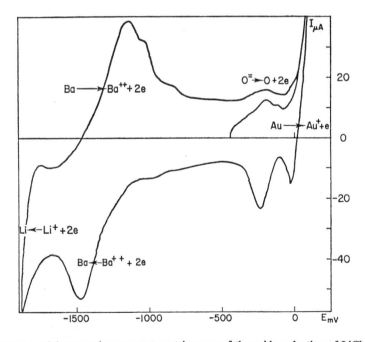

Fɪɢ. 20. Potential–sweep chronoamperometric curve of the oxido-reduction of LiCl–KCl–BaO($pO^= = 1$) at 292°C. (After Hladik, 1966).

from the negative potentials to the positive, indicates the oxidation of metallic lithium and barium, followed by O^{2-} and by the gold working electrode.

(b) *Successive linear sweeps.* When a great number of successive potential-sweeps are carried out, the chronoamperometric peaks decrease rapidly. This is explained by the impoverishment in electroactive species near the working electrode due to the slowness of the diffusion process. The original height of the peaks will only be recovered after melting the solid electrolyte and repolishing the working electrode.

A series of reduction curves and successive depolarizations have been carried out in LiCl–KCl–NiCl$_2$ electrolyte, at 256°C (Stein, 1969). The reduction potential is always stopped at the same value -1350 mV. After the reduction, the electrolysis is stopped and the electrode takes an equilibrium potential at -725 mV, after a certain delay which grows with each sweep. Figure 21 shows the diminution of the chronoamperometric maxima thus obtained. We have the following results (Table I).

TABLE I.   Successive linear sweeps—Reduction of Ni(II)

Curve	Maximum reduction current (μA)	Depolarisation duration (seconds)
1	0·85	106
2	0·57	191
3	0·49	270
4	0·45	315
5	0·42	415

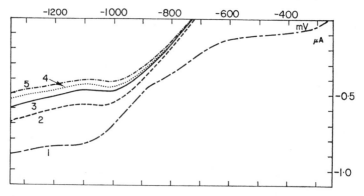

FIG. 21. Reduction of Ni(II) in LiCl–KCl at 256°C. Successive linear sweeping after depolarisation (After Stein, 1969).

The first equilibrium potential (-270 mV) corresponds to the potential of the gold electrode in the solid electrolyte; the following equilibrium potentials correspond to an electrode covered with nickel, deposited during the reduction. Although the equilibrium potentials are the same (-725 mV), the concentration gradients in the neighbourhood of the working electrode are not the same, since the heights of the peaks are decreasing.

(c) *Rate of linear sweep.* The rate of sweep has been varied for a series of chronoamperograms. We obtained the following values (Table II) for the reduction of Cd(II) in LiCl–KCl, the solute concentration being 5×10^{-2} mol/kg, at 310°C (Fig. 22) (Hladik *et al.*, 1969).

TABLE II.   Influence of the linear sweep rate.

Curve	Rate of sweep (mV/min)	Heights of the reduction peaks (µA)	Heights of the oxidation peaks (µA)
1	1,000	13·5	14
2	500	10	9·5
3	250	6·8	7·4

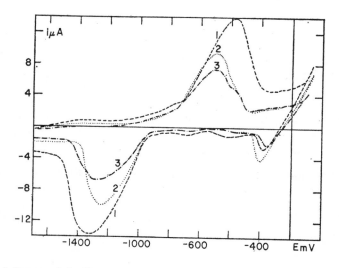

FIG. 22 Influence of the linear-sweeping rate on the heights of peaks. Oxido-reduction of CdCl$_2$(5×10^{-2} mol/kg) in LiCl–KCl at 310°C (After Hladik *et al.*, 1969).

According to the theory of reversible systems, the chronoamperometric current increases in proportion to the square root of the rate sweep. In the example given, this proportionality is well verified as far as the reduction is concerned (Fig. 23).

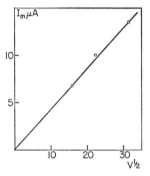

FIG. 23. Relation of the maximum current, i_M, to the sweeping rate (v, mV. sec^{-1}) (After Hladik et al., 1969).

(d) *Proportionality between current and concentration*. The theoretical equations which were calculated by using Fick's law indicate that the limiting diffusion current, i_M, is proportional to the initial concentration $C°$ of the electroactive substance, when the latter is uniformly dispersed within the electrolyte. We therefore have the equation:

$$i_M = KC°.$$

This proportionality is verified when the solute concentration is sufficiently small, i.e. when migration can be neglected. This relationship was verified for the reduction of $HgCl_2$ in LiCl–KCl at 300°C (Fig. 24) (Hladik, 1966a). The maximum values are well-defined, the chronoamperograms showing very sharp peaks. Moreover, the currents are perfectly stable contrary to the results obtained in the molten electrolyte, where mercury is in the gaseous state (Hladik, 1966a; Hladik et al., 1966). The reduction of potassium nitrate also permits the verification of this proportionality (Hladik, 1966a).

(e) *Temperature dependence*. The influence of temperature on the values of chronoamperometric currents is very important. This is a consequence of the rapid variation of diffusion coefficient with temperature.

In the case of linear diffusion, and for a reversible system, the relationship between the maximum-peak current i_M and the temperature is:

$$i = A(D/T)^{1/2}.$$

A being a parameter independent of the temperature.

The variation of the coefficient diffusion, with temperature is given by:

$$D = D_o\, e^{-H/kT}$$

H being the activation energy of the diffusion process. We can, therefore, write the equation:

$$i = AD_o\, e^{-H/2kT}\, T^{-1/2}$$

and in logarithmic form this becomes:

$$\ln(i_M\, T^{1/2}) - \ln(AD_o) = -\,H/2kT.$$

Thus a plot of $\ln(i_M\, T^{1/2})$ versus $(1/T)$ gives a linear relationship.

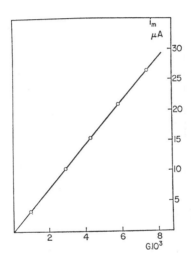

FIG. 24. Proportionality between current, i_M, and concentration. Reduction of $HgCl_2$ (mol/kg) in LiCl–KCl at 300°C (After Hladik, 1966a).

(f) *Number of electrons.* In the case of a reversible system, when the oxidant is the only electroactive species in the bulk, an approximate equation for the current–potential curve is:

$$E = E_{1/2} + \frac{2\cdot3\,RT}{nF} \log\left(\frac{i_a - i}{i}\right).$$

Such an equation can be used for understanding the chronoamperometric curves obtained in solid electrolytes, but only if it is applied to that part of the curve situated before the peak. The differentiation of the preceding equation gives:

$$\frac{2\cdot3\,RT}{nF} = \frac{\Delta E}{\Delta \log{(i_d - i)}/i} \; .$$

Identification of i_d (diffusion current) with i_M (height of the peak) enables a rough approximation to be made to the number of electrons involved in the electrochemical reaction.

For the reduction of cobalt chloride at 298°C in LiCl–KCl, Fig. 25 shows that the relationship between the potential E and $\log{(i_d - i)}/i$, is linear. The slope gives a value of $n = 1\cdot94$. Therefore, during the reduction of Co(II), two electrons are captured by each atom leading to the deposition of cobalt metal.

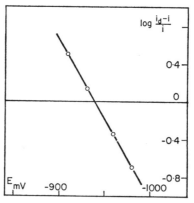

Fig. 25. E plotted versus $\log{(i_d - i)}/i$. Reduction of $CoCl_2$ in LiCl–KCl at 298°C. (After Hladik and Rovetto, 1970).

III. Electroactive Species Dissolved in the Electrode

A. Types of Electrodes

Two kinds of studies may be considered in the case of electroactive species dissolved in an electrode. On the one hand, the electrode may be made of a thin membrane through which an electroactive substance diffuses; this electroactive compound is on one side of the membrane and the electrolyte is on the opposite side. On the other hand, the electrode may be made of an alloy or it may contain gaseous molecules previously dissolved in the metal.

When the electrode is thick, with a given solute concentration, the theoretical chronoamperometric and potential–sweep chronoamperometric curves are identical to those calculated previously in the case of substances dissolved within the electrolyte, but, of course, the diffusion coefficient applies to the electroactive species in the metal of the electrode.

In the case of thin membranes, the flow of the electroactive compound through the electrode may be calculated from the theoretical results given by Carslaw and Jaeger (1965). For thin electrodes, without external contribution after the beginning of the electrolysis, it is possible to use the theoretical results tabulated in the case of thin electrolyte layers (Hladik, 1967b).

B. Chronoamperometry

Rickert and Steiner (1966) investigated the electrochemical measurement of oxygen diffusion in metals at high temperatures using a zirconia-based solid electrolyte. The principle of this electrochemical measurement consists of first bringing the metal into equilibrium with a fixed oxygen partial pressure in such a way that a known oxygen concentration is established in the metal. Then the metal sample is placed on one side of a solid electrolyte and functions as the electrode of a galvanic cell. On the other side of the electrolyte is an electrode which is practically unpolarizable, such as porous platinum in contact with air or an Fe–FeO electrode which has an oxygen partial pressure of about 10^{-10} atm at 800°C. Figure 26 shows an example where the cell has an Fe–FeO electrode on one side and silver metal containing oxygen in solution on the other.

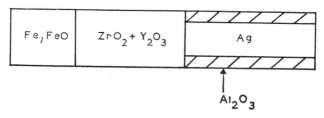

FIG. 26. Cell for electrochemical measurements of oxygen diffusion in silver (After Rickert and Steiner, 1966).

Figure 27 indicates the current density i as a function of $t^{-1/2}$ for the diffusion of oxygen in silver. The values obtained for the diffusion coefficients of oxygen in silver at 800°C were 1.9×10^{-5} cm^2 sec^{-1} and 1.8×10^{-5} cm^2 sec^{-1}, corresponding to values given in the literature, which were obtained using a different technique.

Rickert (1968) has shown that the diffusion coefficient of a gas in metal can be calculated without knowing the solubility, by means of a cylindrical geometry of the working cell.

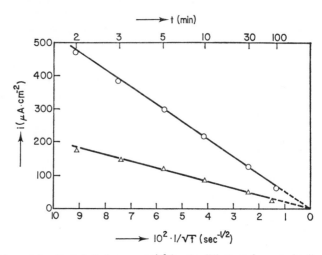

FIG. 27. Current density i plotted versus $t^{-1/2}$ for the diffusion of oxygen in silver at 800°C. ◯ values of i for an initial concentration. $C = 11.5 \times 10^{-6}$ moles oxygen cm^{-3}Ag. △ $C = 3.2 \times 10^{-6}$ (After Rickert and Steiner, 1966).

Raleigh and Crowe (1969) used a potentiostatic technique to measure diffusion coefficients of silver in solid silver$\frac{1}{2}$gold alloys. The working cell consisted of a single crystal AgBr pellet spring-loaded between a square of Ag foil and the appropriate gold electrode. For this, gold single crystal and Ag–Au alloy single crystals were employed in different runs. The electrodes were shaped from larger crystals and provided with a smooth, planar surface by spark cutting, followed by mechanical polishing and electropolishing. Due to the short diffusion length involved, $((Dt)^{-1/2} \simeq 10^{-6}$ cm), it was critically important to have the smoothest possible electrode surface with a minimum of surface damage. In general, the aqueous polish gave more consistent results and less surface damage than electropolishing in fused NaCl–KCl. The chronoamperometric method was used to measure diffusion values corresponding to both inward and outward diffusion into a phase. Plots of i versus $t^{-1/2}$ were linear over a wide time range. This was the case for both forward and reverse steps. Figure 28 shows a typical plot for a forward step, showing the time range 6–4500 sec. During time periods of less than 10 sec, currents showed a general upturn from linearity, attributable to residual double-layer charging. During periods of time in excess of several hundred seconds, there were often small drifts from linear deviation,

0·1–0·3 μA at most. Since these were larger than expected from the calculated nonlinearity and did not show a consistent direction, they probably represent small furnace temperature drifts. In general, the time range, 10–100 sec was used for the diffusion coefficient determination.

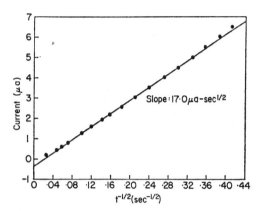

Fig. 28. Plots of i versus $t^{-1/2}$. Chronoamperometric measurements of the diffusion of silver in solid silver–gold alloys, at 394°C (After Raleigh and Crowe, 1969).

The solubility and diffusivity of oxygen in solid copper have been investigated by Pastorek and Rapp (1969). Solid-state electrochemical measurements by three alternative experimental procedures were made with the cell:

$$\text{FeO, Fe}_3\text{O}_4 \,/\, \text{Zr}_{0.85}\text{Ca}_{0.15}\text{O}_{1.85} \,/\, \text{Cu} \,/\, \text{Zr}_{0.85}\text{Ca}_{0.15}\text{O}_{1.85} \,/\, \text{FeO, Fe}_3\text{O}_4$$

in the temperature range 800–1030°C.

The authors showed that a plot of $\log i$ versus t should approach linear behaviour for large values of t. Thus the diffusion coefficient is obtained from the slope of the linear curve.

A typical plot is shown in Fig. 29. The values of the diffusivity were calculated from the slopes and the saturation solubility from the intercepts of $\log i$ versus t plots. In preliminary experiments, low voltages, i.e. very low oxygen activities close to FeO–Fe$_3$O$_4$ equilibrium were used; under these conditions inconsistent values for the saturation solubility were obtained. results were more consistent in the higher voltage ranges, i.e. working with higher oxygen contents at higher oxygen pressure. Despite the precautions which were taken (coating the copper electrode with an Al$_2$O$_3$ layer), a simple chemical exchange of oxygen between the copper electrode and the surrounding gas probably could not be completely avoided. This effect was

probably more influential under conditions corresponding to low oxygen
contents at the non-reversible electrode/electrolyte interface and, therefore
this condition was avoided for the potentiostatic experiments.

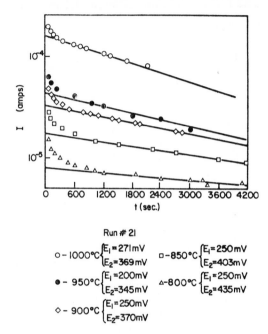

Run # 21

o – 1000°C $\begin{cases} E_1 = 271\,mV \\ E_2 = 369\,mV \end{cases}$ □ – 850°C $\begin{cases} E_1 = 250\,mV \\ E_2 = 403\,mV \end{cases}$

● – 950°C $\begin{cases} E_1 = 200\,mV \\ E_2 = 345\,mV \end{cases}$ △ – 800°C $\begin{cases} E_1 = 250\,mV \\ E_2 = 435\,mV \end{cases}$

◇ – 900°C $\begin{cases} E_1 = 250\,mV \\ E_2 = 370\,mV \end{cases}$

FIG. 29. Current–time plot for chronoamperometric determination of the solubility and
diffusivity of oxygen in solid copper at 800° to 1000°C (After Pastorek and Rapp, 1969)

C. Potential-sweep Chronoamperometry

The oxidation kinetics of hydrogen on a palladium electrode has been studied
in solid LiCl–KCl–BaO (Hladik, 1969). The working electrode is made of
a palladium cylinder, closed at its lower end, into which flows the hydrogen
(Fig. 30). The electrodes are plunged into the molten electrolyte and, after
freezing, the electrodes are embedded in the solid mixture.

The chronoamperograms show the existence of two oxidation peaks of
which the maxima are, respectively, about −110 mV and −750 mV, at
$pO^= = 2$ (Fig. 31). The first peak can be attributed to the oxidation of
hydrogen by $O^=$ ions, and the second to the oxidation by Cl^-. The theoretical
potential calculated from the thermodynamic data for the reaction leading
to the formation of H_2O, shows that the activation polarization is small,
which is suggested by the current potential curves. When successive sweeps
are performed for a constant flow of hydrogen, the oxidation currents
decrease rapidly.

IV. Electroactive Species Reacting at the Three-Phase Boundary

A. Gas Electrodes

Research on fuel cells has made a great contribution to the electrochemistry of gases in contact with solid electrolytes (Institut Francais du Pétrole, 1965; Mitchell, 1963; Young, 1960). Solid electrolytes with ionic conduction by O^{2-} have been principally used at high temperatures. For low-temperature fuel cells, ion-exchange membranes have mostly been used.

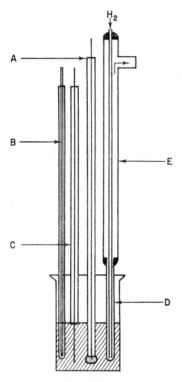

Fɪɢ. 30. Hydrogen/palladium cell with electrodes embedded in the solid electrolyte (After Hladik, 1969a). (A) reference electrode; (B) thermocouple; (C) counter-electrode; (D) palladium tube; (E) copper tube.

An experimental arrangement for fuel cells using ion-exchange membranes is depicted in Fig. 32. The two membranes are separated by an aqueous electrolyte and the reference electrode is situated in the middle of the cell. The electrodes are covered with a catalyst and are applied as a coating on the ion-exchange resins. The gaseous fuel and combustion air react where the electrolyte and the electrode are in contact. This experimental apparatus allows the polarization curve of each electrode to be measured separately.

We do not intend to give here an extensive survey of fuel cells, which are reviewed in another chapter of this book, since most of the polarization curves for the fuel cells were not obtained by potential–sweep chrono-amperometry.

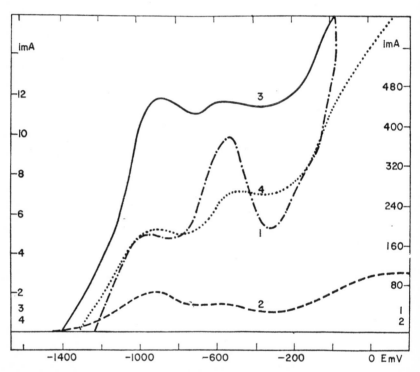

FIG. 31. Hydrogen oxidation in LiCl–KCl ($pO^= = 2$). Palladium electrode, thickness, 0·25 mm. Temperatures: (1) 324°C; (2) 302°C; (3) 290°C; (4) 284°C. (After Hladik, 1969a).

Some further studies have been concerned with gas–electrode polarization in contact with solid electrolytes but current–potential curves were generally obtained in stationary conditions or with a technique different from potential–sweep chronoamperometry. The reader is referred to the original studies: (Filayev *et al.*, 1967; Karpachev and Ovchinnikov, 1969; Perfilev and Palguev 1966, 1967; Karpachev and Filayev, 1968).

B. Partially Immersed Electrodes

Hydrogen oxidation reaction has been studied by Hladik (1970) using a gold electrode partly immersed in the electrolyte LiCl–KCl–BaO. As a consequence of the insolubility of hydrogen in the metallic phase as well as in the electrolyte, an electrochemical reaction occurs at the three phase boundary.

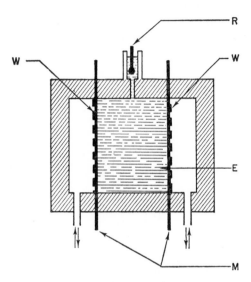

FIG. 32. Experimental arrangement for fuel-cells using ion-exchange membranes. R, reference electrode; W, working electrode; E, electrolyte; M, membranes.

The experimental cell is depicted in Fig. 33. The working electrode is made up of a thin gold plate partly embedded in the solid electrolyte near the reference- and counter- electrodes. Hydrogen flows in the glass tube surrounding the working electrode, altering the equilibrium potential of the electrode as it begins flowing.

The equilibrium potential of the working electrode varies from -650 mV to -450 mV, for a temperature range of 350°C to 240°C, the concentration in BaO giving $pO^{--} = 1$. When hydrogen, at a pressure of 76 cm of mercury is introduced, the equilibrium potentials are changed as follows:

	Equilibrium without H_2	Equilibrium with H_2
348°C	-650 mV	-1350 mV
238°C	-450 mV	-1120 mV

The equilibrium potential in a hydrogen atmosphere is even slower to establish itself as the temperature is lowered. For temperatures of the order of 240°C, however, the time necessary to reach this equilibrium does not exceed 30 min. Chronoamperometric curves show two successive waves of hydrogen oxidation (Fig. 34). The half-wave potentials are difficult to determine because the shape of the curves no longer corresponds to the classical theories of the diffusion of a solute in an electrolyte. The first wave presents a very steep slope at the origin, the half-wave potential being around 20–30 mV from the equilibrium potential. The second wave has a half-wave potential at 350 mV from the equilibrium potential.

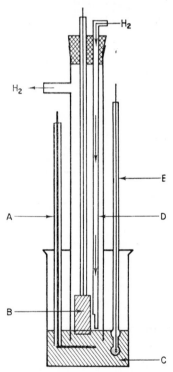

FIG. 33. Emergent electrode cell for the hydrogen oxidation. (After Hladik, 1970). (A) auxiliary electrode; (B) working electrode; (C) electrolyte; (D) glass tube for hydrogen; (E) reference electrode.

The first oxidation peak can be attributed to the oxidation of hydrogen by O^{--} ions, as it is in fused electrolyte (Hladik, 1966a). The second step, as in the case of the fused electrolyte, would be due to the oxidation reaction of the chloride ions.

We point out in the figure the influence of successive sweeps on the second step of oxidation, the first one staying invariable. The sweeps are carried out at 1 V/min between fixed limits: equilibrium potential and zero potential.

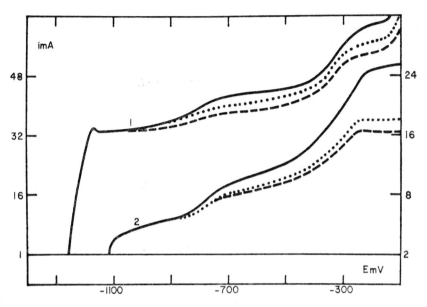

FIG. 34. Potential–sweep chronoamperometric curves. Hydrogen oxidation on emergent electrode (LiCl–KCl–BaO; $pO^= = 1$). (1) 336°C; (2) 238°C (After Hladik, 1970).

C. Contact Electrodes

1. Apparatus

The working electrode consists of a metal rod, pressed on to the surface of the solid electrolyte. The auxiliary electrode is formed by applying a gauze of large working area to the opposite side of the electrolyte. Generally, this second electrode is unpolarizable and is used as a comparison electrode. Kleitz (1968) and Guillou *et al.* (1968) used this technique. Thus we obtain the scheme of Fig. 35 in which the compartments containing the working electrode and the auxiliary electrode are separated so that the latter can be exposed to a reference atmosphere and thus act as reference electrode.

In the apparatus described by Kleitz (1968) the working electrode is contained in an electrolyte tube of zirconia with the end of the rod in contact with the bottom of the tube. The lower surface of the tube is platinized and is used as a counter electrode. The whole cell is placed in an alumina tube through which the reference atmosphere flows.

2. Experimental Results

The oxygen reduction on an electrode in contact with the binary electrolyte ZrO$_2$ (0·85) CaO (0·15) has been studied by Guillou *et al.* (1968). Figure 36 shows the non-corrected current–potential curves corresponding to the oxygen reduction on the solid electrolyte. The partial pressures of oxygen in the flowing gas take the successive values of 2·1 × 10^{-5}, 10^{-3}, 8·2 × 10^{-3}, 10^{-1} and 2 × 10^{-1} atm.

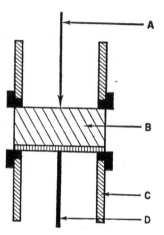

FIG. 35. Working electrode in contact with the solid electrolyte. (A) Working electrode; (B) electrolyte; (C) alumina tube; (D) counter electrode.

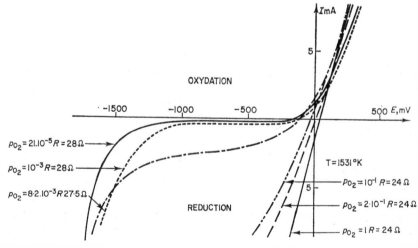

FIG. 36. Oxygen reduction on a platinum–rhodium electrode in contact with the binary electrolyte ZrO$_2$(0·85) CaO(0·15). (After Guillou *et al.*, 1968).

They all demonstrate the existence of a cathodic current at an electrode potential high above those corresponding to the reduction of the solvent. This is due to the oxygen reaction:

$$\tfrac{1}{2}O_2 + 2e + \square_{O^2} \rightarrow O_R^{2-}.$$

The redox couple O^{2-}/O_2 is a reversible system at high temperatures for such an electrolyte. In the anodic part, we find the different current potential curves practically superimposed. The oxidation step of the solvent is independent of the partial oxygen pressure.

Stationary polarization curves of reduction and oxidation of various gases at the surface of stabilized zirconia have been determined by Kleitz (1968) and Kleitz *et al.* (1967). The characteristics of the oxygen electrode have the shape indicated in the Fig. 37.

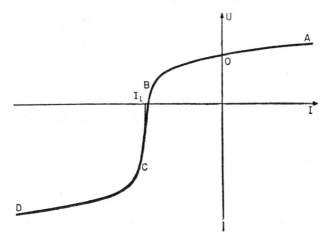

Fig. 37. Potential plotted versus *i*, for the oxygen reduction in stabilized zirconia (After Kleitz, 1968).

In the field *AB* of this characteristic, the oxygen electrode reaction occurs alone. It is preceded by an oxygen adsorption on the zirconia and the metal following the reactions:

$$O_2 \rightleftharpoons 2O_{ads} \qquad\qquad (I)$$

$$O_{ads} + |O|^{2+} + 2e^-_{metal} \rightleftharpoons zero$$

$|O|^{2+}$ being the oxygen ion vacancy. In the field CD, a second electrode reaction is superimposed on the preceding one. It is a reduction of zirconia that we can write in the form:

$$e^-_{metal} \rightarrow e^-_{(electrolyte)} \qquad\qquad (II)$$

in order not to prejudge the state of the electrons in the electrolyte (free or trapped). This reaction is followed by a chemical reduction of the electrolyte which is partially compensated by a third process which no longer directly interferes with the electron of the metal:

$$O_{ads} + |O|^{2+} + 2e^{-}_{electrolyte} \rightarrow zero. \tag{III}$$

The model schematized in the Fig. 38 summarizes the above phenomena.

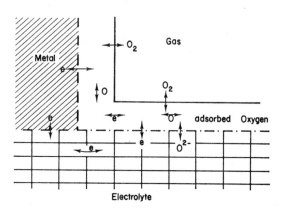

FIG. 38. Scheme of an oxygen electrode in solid oxide electrolytes, with electrochemical reduction of zirconia (I, I, III processes) (After Kleitz, 1968).

Stationary cathodic polarization curves of the oxygen electrode have been determined for various oxygen pressures pO_2. The author obtained "diffusion steps" of oxygen which show a limiting intensity dependent on the oxygen pressure pO_2.

The characteristics of the oxygen electrode may be described by an empirical equation of the form:

$$U = AI + \frac{RT}{ZF} \log \frac{I - I_0}{I_0}$$

in which A and I_0 are constants independent of the temperature and the oxygen pressure. The value I_0 is approximately equal, in absolute value, to the linear-intensity of the steps. The term A differs notably from the ohmic resistance of the cell.

3. Theoretical Current–Potential Curves

Experimental results allowed Kleitz (1968) to assume the following oxygen electrode model:

The electrode reaction is reversible and the greater part of the polarization is the result of the slowness of mass transport to the electrode.

The electrochemical reaction is preceded by an oxygen dissociation.

for small oxygen pressures the concentration variations in the gaseous phase becomes preponderant;

when the oxygen pressures are sufficiently high, the concentration variations within the gaseous phase can be neglected with only the concentration variation of the O species remaining.

the oxygen solubility within the zirconia is very small in the experimental conditions of oxygen pressure (20–600 mmHg).

For specifying the physical meaning of the coefficients of the experimental equation. Kleitz has proposed a set of equations describing the preceding electrode model. He assumed that the curvature of the three-phase boundary is great compared to the thickness of electrode diffusion layers.

The adsorbed oxygen on the metal is denoted O_M and that on the zirconia, O_Z. The variables x and y refer to the phase boundaries electrolyte/adsorbed oxygen and metal/adsorbed oxygen, respectively. At all points of the metal surface, the balance of the mass exchanges is expressed by the following equation (under stationary conditions).

$$-\frac{dJ_{O_M}}{dy} + F_1(P_{O_2}) - F_2(O_M) = 0 \tag{1}$$

J_{O_M} represents the superficial flow of adsorbed oxygen (ionized or not) at point y, $F_1(P_{O_2})\,dy$ is the quantity of oxygen which is adsorbed on the metal in unit time between points y and $y + dy$; $F_2(O_M)\,dy$ is the quantity which is desorbed.

On the electrolyte surface, we obtain the equation:

$$-\frac{dJ_{O_z}}{dx} + G_1(P_{O_2}) - G_2(O_z) - J_{O^{2-}} = 0. \tag{2}$$

The first three terms have the same significance as the preceding ones, and the adsorbed oxygen is represented by O_z, which is now in the atomic state. $J_{O^{2-}}\,dx$ indicates the quantity of O^{2-} ions which penetrate into the electrolyte in unit time between points x and $x + dx$. Always on the electrolyte surface, the electrical balance will be expressed by:

$$-\frac{dJ_e}{dx} + 2J_{O^{2-}} = 0. \tag{3}$$

J_e being the electron flow on the electrolyte surface at point x.

In spite of the simplicity of the Kleitz model, the set of equations obtained is too complicated. Kleitz has considered other simplifying hypotheses.

After an experimental study of the relative importance of electronic conduction and of adsorbed-oxygen diffusion on surfaces, Kleitz concluded that the superficial electronic conduction of zirconia may be considered as negligible. The initial system is then reduced to eqn (1) together with the simplifying equation:

$$-\frac{dJ_{O_z}}{dx} + G_1(P_{O_2}) - G_2(O_z) = 0.$$

Kleitz solved such a system after introducing other simplifying assumptions and has calculated an expression which can be verified experimentally.

V. Thin Films of Electroactive Species

A. Formation of Thin Layers

1. Introduction

There are many techniques which may be used to deposit films of electroactive species on electrodes, particularly by electrolysis in solid electrolytes. We may proceed by reduction either of the supporting electrolyte or of one electroactive substance in solution, or by the oxidation of the electrode.

One can study the formation rate of thin layers by chronoamperometry (Rickert, 1968). The potential characteristics of the deposition of such layers in solid electrolytes may be studied by potential–sweep chronoamperometry (Hladik, 1966a).

The oxidation (or reduction) of such films have been studied in solid electrolysis by means of potential–sweep chronoamperometry (Hladik *et al.*, 1967, 1969, 1970).

The deposition of a substance resulting from electrolysis is possible when the compound is stable and insoluble under experimental conditions.

Electrochemical studies of the formation of nickel sulphide in the solid state at high temperatures have been reported by Rickert (1968). The galvanic cell is shown in Fig. 39. The negative pole of a current source is connected on the left-hand side of Pt_1; the positive pole is connected to the lead Pt_2. The electric current which passes through these cells gives a measurement of the rate at which silver is removed from Ag_2S, as silver ions pass through AgI and electrons pass through the lead Pt_2. When the chemical potential of silver μ_{Ag} is less than $\mu_{Ag}{}^\circ$ i.e. $E > 0$, then the silver loss from Ag_2S, in the steady state, is practically equivalent to the sulphur loss from Ag_2S, since the latter has only a very narrow range of composition. When the vaporization of sulphur can be excluded, which is possible when its vapour pressure, i.e. potential, is maintained at low values, the rate of silver loss from Ag_2S is equivalent to the rate of formation of NiS. It is clearly necessary to establish that NiS has been formed as a result of the reaction between Ni and Ag_2S. Nickel ions and electrons pass through NiS, while an equivalent amount of silver ions and electrons pass from Ag_2S through the AgI and the lead Pt_2. This can be measured electrically.

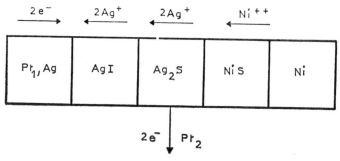

FIG. 39. Electrochemical cell for the formation of nickel sulphide in the solid state. (After Rickert, 1968).

A constant potential was applied and the current was measured as a function of time. If the NiS formed grows according to the parabolic law, and the phase boundaries are in thermodynamic equilibrium, we have, according to Tammann (1920) and Pilling and Bedworth (1923):

$$\frac{dX}{dt} = \frac{k}{X}$$

where X is the thickness of the NiS layer and t is the time. Since the observed current density i, flowing through the nickel sulphide layer of the cross-section is equivalent to the flux of nickel ions, it is a measurement of the growth rate of the NiS layer of molar volume V. With j as the flux of nickel ions in moles per unit area, per unit time, it follows that:

$$\frac{dX}{dt} = jV = \frac{iV}{2F}.$$

Integration of the differential equations gives:

$$X^2 = 2kt$$

and combination with the preceding equation yields,

$$i = \frac{F\sqrt{2k}}{V\sqrt{t}}.$$

The rate constant k can be calculated from the slope of the straight line i versus $t^{-1/2}$.

Figure 40 shows a typical current density–time diagram for a chrono-amperometric measurement. The current rises first up to a maximum value. During this incubation period, nucleation and the phase bounadryreaction may play an important role; thereafter, the current decreases.

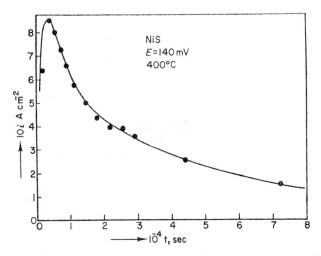

FIG. 40. Current density i plotted versus time. Chronoamperometric curve, $E = 140\,\text{mV}$, at 400°C (After Rickert, 1968).

A plot of i versus $t^{-1/2}$ gives, for large values of time, a straight line passing through the origin, which is in accordance with the assumption of a parabolic rate law used in the derivation of the theoretical equation.

We have studied the deposition of nickel, cobalt and cadmium on gold electrodes in LiCl–KCl supporting electrolyte (Hladik et al., 1967).

Figure 41 shows some chronoamperograms obtained with various reduction potentials followed by re-oxidation of the deposited compound.

The oxidation of a platimun electrode in solid chloride electrolyte shows a passivation peak from the beginning of the oxidation of chloride ions (Hladik, 1966a). The behaviour of a platinum electrode on anodic polarisation has been studied in LiCl–KCl–BaO and indicates oxide formation (Hladik et al., 1969).

3. Reduction of the Deposited Layers

The reduction of the anodic film formed on a platinum electrode shows a characteristic wave of reduction (Hladik et al., 1967). This passivation begins near $+1000\,\text{mV}$.

No similar passivation phenomenon has been observed in NaCl–KCl at 540°C, even under high polarisation.

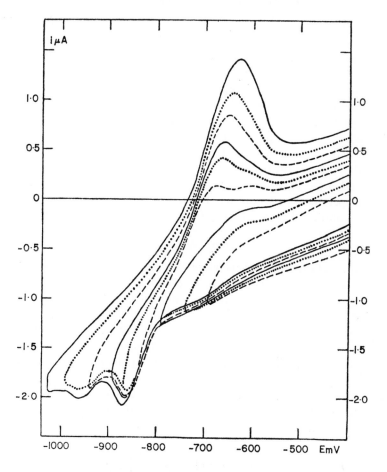

FIG. 41. Potential–sweep chronoamperometric curves. Oxido reduction of nickel on gold electrode in LiCl–KCl at 300°C. ($NiCl_2 = 2 \times 10^{-2}$ mol/Kg) (After Hladik *et al.*, 1967).

The deposition of thin metal layers on a noble metal electrode allows one to obtain an oxidation peak characteristic of the deposited metal (Hladik *et al.*, 1967, 1969). Figure 41 shows oxidation chronoamperograms of nickel deposited on a gold electrode. The height of the peaks is proportional to the amount of deposited metal.

VI. Chronopotentiometry

A. Theory

1. Chronopotentiometry Development

In chronoamperometry the time dependance of the potential of a working electrode is measured during electrolysis as a constant current intensity. Weber (1879) used this technique to calculate the ionic concentration in a solution. Chronopotentiometry was applied to molten salts by Laitinen and Ferguson (1957). We have shown that it is possible to extend this technique to solid electrolytes (Hladik, 1966a; Hladik et al., 1966).

Mathematical problems, concerning the variations of concentration of the electroactive species in the vicinity of the working electrode during chronopotentiometry have been solved by a number of authors. (Weber, 1879; Sand, 1901; Rosebrugh and Miller, 1910; Karaoglanoff, 1906). Dehalay (1954) and his collaborators have developed the mathematical analysis of chronoamperometric phenomena. The influence of the electric double-layer has been considered more recently (Gierst, 1952; De Vries, 1968).

2. General Assumptions

The theoretical expression of the working electrode potential $E(t)$ is deduced from the value of the concentration of electroactive substances on the electrode surface. These concentrations are calculated from the diffusion equation.

We consider chiefly the case of linear semi-infinite diffusion on a plane electrode. The mass transport under the action of the gradient of electric potential is neglected, which is a reasonable approximation when a large excess of supporting electrolyte is used.

Consider the electrolytic reduction of a substance O to another substance R. As the current is kept at the fixed density i_0, during electrolysis, we have the following relation for the flux at the electrode:

$$i_0 = nFD_O \left[\frac{\partial C_O(x, t)}{\partial x} \right]_{x=0}.$$

3. Simple Electrochemical Reaction: Reversible Process

If the substance R can diffuse in the supporting electrolyte, the sum of the flux of the products O and R at the electrode surface is nil. Thus we have the relation:

$$D_O \left[\frac{\partial C_O(x, t)}{\partial x} \right]_{x=0} + D_R \left[\frac{\partial C_R(x, t)}{\partial x} \right]_{x=0} = 0.$$

These two relations compose two boundary conditions of the diffusion equations. The initial conditions are, in the simplest case, the concentration R equal to zero at the beginning of the electrolytis and the concentration O, constant inside the electrolyte. Therefore:

$$C_O(x, 0) = C^\circ, \qquad C_R(x, 0) = 0.$$

On the other hand, we have, for $x \to \infty$, the conditions:

$$C_R(x \to \infty, t) \to 0; \qquad C_O(x \to \infty, t) \to 0.$$

The concentration gradient of O and R species has been calculated by Karaoglanoff (1906).

The time–potential curve is calculated from the Nernst equation, the concentrations at the electrode surface being obtained from the concentration expression sin which $x = 0$.

The equation of the chronopotentiometric curve may be established by introducing the transition time, τ, at which the concentration of the product O at the electrode surface is equal to zero.

$$E = E_{1/2} + \frac{RT}{nF} \ln \frac{\tau^{1/2} - t^{1/2}}{t^{1/2}}$$

with

$$E_{1/2} = E^\circ + \frac{RT}{nF} \ln \frac{f_O D_R^{1/2}}{f_R D_O^{1/2}}; \qquad \tau^{1/2} = \frac{C^\circ \pi^{1/2} nF D_O^{1/2}}{2i_0}.$$

The potential $E_{1/2}$ is reached when $t = \tau/4$. In the case of the deposit of an insoluble substance, and supposing that the activity of the deposit is equal to one, it is sufficient to introduce the value of $C_O(0, t)$ in the Nernst equation. The transition-time is then the same as in the case of a process with two soluble species.

4. Reduction Followed by an Oxidation Process

A compound is reduced to a substance R, assumed soluble, then the direction of the current through the cell is reversed at the transition-time τ corresponding to the reduction of O. The product R is then oxidized and the time–potential plot is obtained again for this oxidation process with a new transition time τ'. The initial conditions, for the second part of the curve corresponding to oxidation, are obtained by using the expressions $C_R(x, \tau)$ and $C_O(x, \tau)$ previously calculated during the reduction of the substance O.

When the current densities for the reduction and oxidation process are equal, the transition-time τ' is:

$$\tau' = \tau/3.$$

When there is a deposition of any soluble substance on the working-electrode, the reoxidation transition-time is equal to the reduction transition-time.

5. Deviations from the Transition-Time Equation

Various factors can cause deviation from the transition-time equation:

(*a*) *Surface roughness of solid electrodes.* The expressions for the concentrations of oxidized and reduced reactants and the transition-time were calculated for semi-infinite linear diffusion to a plane electrode. For a rough electrode, the diffusion is not strictly linear. For the steady-state conditions at the electrode, it has been shown that at distances large in comparison with the amplitude of a sinusoidal type of imperfection, the concentration profile is unaffected. In terms of the thickness of the diffusion layer, this limiting condition can be stated: When the thickness of the diffusion layer is large in comparison with the amplitude of the imperfections of the electrode the theoretical equations are applicable.

(*b*) *Non linearity of the diffusion field.* Deviations from the transition-time equation are very often noticed when different types of electrodes are used. Under certain conditions semi-infinite linear diffusion may be valid for spherical electrodes and cylindrical wire electrodes. When the dimensions of an electrode are large in comparison with the thickness of the diffusion layer the electrode surface approximates to an infinite plane.

Diffusion to an unshielded planar electrode occurs not only in directions normal to the surface but also in arbitrary directions at the edge. There are spherical contributions to diffusion to a planar, circular electrode. Non-linearity of the diffusion field around the electrode can be a cause of deviation from the theoretical equations. When the thickness of the diffusion-layer is large in comparison with the dimensions of the electrode, the diffusion field is, at first approximation, linear. As the thickness of the diffusion layer decreases, the departure from linearity increases. Therefore, since the thickness of diffusion layer is given by $\sqrt{D\tau}$, one would expect deviation, due to the non-linearity of diffusion for longer transition times, i.e. lower current densities.

(*c*) *Double-layer effect.* The effect of charging the double-layer on the transition-time can be qualitatively estimated by assuming that the charging of the double-layer occurs constantly during the time, τ, with the constant, capacitive current:

$$i_c = C_{DL} \, \Delta E / \tau.$$

The quantity of electricity Q_c, during the charging of the double-layer is:

$$Q_c = i_c \tau = C_{DL} \, \Delta E.$$

C_{DL} varies with the potential and an average value is introduced for the quantity Q_c.

The quantity of electricity Q_e involved in the electrochemical reaction at the current density i_0, and over the time, τ, is:

$$Q_e = i_0\tau.$$

This last equation is not quite exact, since the current density for the electrochemical reaction is slightly smaller than i_0 because of the capacity effect, but this error can be neglected here.

The order of magnitude of the error resulting from the existence of the double-layer, can now be evaluated by considering the ratio Q_c/Q_e as derived from the theoretical equations:

$$\frac{Q_c}{Q_e} = \frac{C_{DL}\,\Delta E}{i_0\tau}$$

or, after introducing the expression for $\tau^{1/2}$

$$\frac{Q_c}{Q_e} = \frac{4i_0\,C_{DL}\,\Delta E}{\pi n^2 F^2 (C^\circ)^2 D_0}.$$

One deduces from this equation that the error resulting from the capacity effect is minimized when the current density is as low as possible, and the concentration of the reacting species is as large as feasible.

Many authors have studied the problem of correcting the constant-current chronopotentiograms for the distortion by double-layer charging. A discussion of the performance of empirical correction methods for obtaining transition times from distorted chrnoopotentiograms is given by De Vries (1968).

(d) *Oxide formation and adsorption of electroactive species.* The formation or reduction of an oxide film may precede, follow or occur simultaneously with the electrode reaction. The effect of this complicated process is the same as in the adsorption of an electroactive species, that is, an increase in the chronopotentiometric constant with increasing current density.

(e) *Temperature gradient.* Deviations can be caused by a temperature gradient through the solid electrolyte. The ideal conditions for long transition times occur in solid electrolytes instead of liquid ones, where deviations can be caused by convection.

B. Experimental Results

1. Electroactive Species Dissolved in the Supporting Electrolyte

Until now, very few experimental chronopotentiometric studies have been performed in solid electrolytes. The experimental arrangements are generally quite similar to the ones used for chronoamperometric studies, and often the authors simultaneously used the two techniques.

The chronopotentiometric study of metallic ions in the solid LiCl–KCl electrolyte was undertaken by Hladik et al. (1966), Hladik (1967) and Pointud et al. (1969). The working electrode is a metallic lead sealed in a glass tube, with only the cross-section of the metal in contact with the electrolyte. Such an electrode is embedded in the solid electrolyte with the reference- and counter-electrodes. Inman (1970) obtained some results for silver diffusion in solid lithium sulphate using a similar technique.

Studies of cadmium chloride in solid LiCl–KCl electrolyte (Hladik, 1967) have been performed with large $CdCl_2$ concentrations in order to obtain easily measurable transition times.

Figure 42 shows the chronopotentiograms of the reduction of nickel chloride and the oxidation of nickel (Pointud et al., 1969). The transition times are identical for oxidation and reduction which conforms to the theoretical results for an insoluble substance deposited on the working electrode.

Figure 43 shows that the transition times for the reoxidation of cadmium are smaller than those for the reduction of cadmium chloride. The study of sub-chloride formation by potential–sweep chronoamperometry should lead

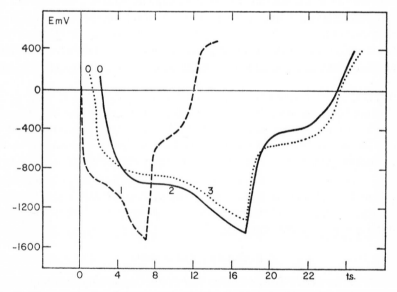

FIG. 42. Chronopotentiometric curves for nickel chloride reduction and nickel oxidation, in LiCl–KCl, at 255°C. $NiCl_2 = 5 \times 10^{-2}$ mol/kg; electrode ϕ 1 mm; (1) 2 μA; (2) 1·5 μA; (3) 1 μA. (After Pointud et al., 1969).

to the interpretation of this diminution of transition time. A smaller value of the transition time indicates the presence of a wholly or partly soluble compound.

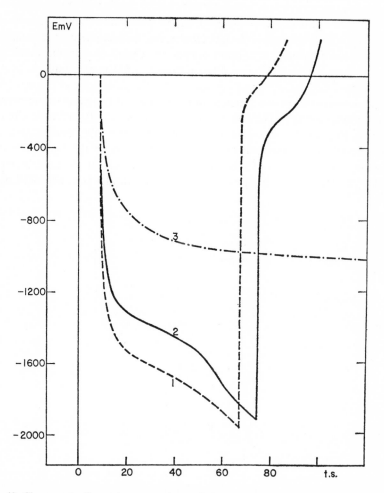

FIG. 43. Chronopotentiometric curves for cadmium chloride reduction and cadmium oxidation, in LiCl–KCl, at 288°C. $CdCl_2 = 10^{-2}$ mol/kg; electrode ϕ 1 mm; (1) 2 μA; (2) 1·5 μA; (3) 1 μA.

Chronopotentiometric studies have been performed with the hydrogen/palladium electrode already described for the chronoamperometric experiments (Hladik, 1969a). The LiCl–KCl–BaO electrolyte was used in the

temperature range 250–300°C. Characteristic two-wave chronopotentiograms were obtained corresponding, respectively, to the oxidation of hydrogen by the $O^=$ and Cl anions.

2. Electroactive Species Dissolved in the Electrode

The oxygen diffusion coefficient in liquid silver and liquid copper was
measured by Rickert and El Miligy (1968). The measurements were obtained
by using a CaO-doped zirconium dioxide as a pure oxygen ion conductor
in the solid state galvanic cell, Pt, air/ZrO_2 + CaO/metal (liquid) (+O in
solution).

The investigation was carried out by both chronoamperometry and
chronopotentiometry. The cell has a cylindrical geometry corresponding to
Fig. 44; the liquid metal is contained in a zirconia tube, and oxygen can
diffuse in a radial direction from the cylinder axis to the periphery. It is
necessary that the ratio of the length to the diameter of the tube be large,
so that the diffusion to the base of the zirconia cylinder is negligible. Linear

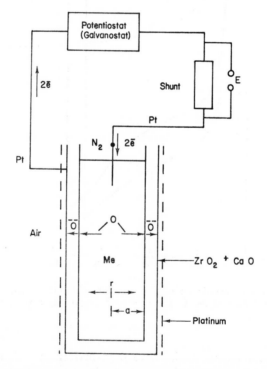

FIG. 44. Cylindrical cell for studying the oxygen diffusion coefficient in liquid silver and
liquid copper (After Rickert and El Miligy, 1968).

approximation can be used in the case of small transition times. Figure 45 shows three potential–time curves, recorded at various currents in liquid silver at 990° C. The equation for chronopotentiograms in the linear approximation is:

$$E = E° + \frac{RT}{nF} \ln \frac{\tau^{1/2} - t^{1/2}}{\tau^{1/2}}$$

τ being the transition times.

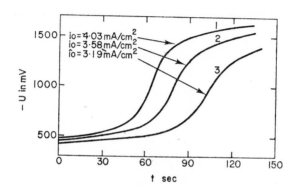

FIG. 45. Chronopotentiometric curves. Oxygen diffusion through liquid silver, at 990°C (After Rickert and El Miligy, 1968).

A plot of the log $(\tau^{1/2} - t^{1/2})/\tau^{1/2}$ versus E gives a linear relation which is verified as shown in Fig. 46. The transition time is related to the diffusion coefficient by:

$$\tau^{1/2} = \frac{nF \, \pi^{1/2}D^{1/2}C°}{2i_0} \, .$$

From this relationship, the authors have calculated the diffusion coefficient.

Pastorek and Rapp (1969) studied the solubility and the diffusion of oxygen in solid copper, in the temperature range 800° to 1030°C. A schema of the experimental arrangement is depicted in Fig. 47. The copper electrode and electrolyte discs were held together under light pressure at 1064°C for several hours to establish intimate $Cu/Zr_{0.85}Ca_{0.15}O_{1.85}$ contact.

To prevent the circumferential surfaces of the copper specimen from interacting with the surrounding helium gas phase, aluminium was first vacuum-deposited onto the circumference and then oxidized with a hot-air fan to form a tenacium Al_2O_3 coating. The reversible electrodes were prepared from reagent quality materials to yield two-phase equilibrium mixtures of $FeO–Fe_3O_4$ in a 1 : 1 weight ratio. The high defect structure of FeO allowed adequate volume diffusion of iron, in order to maintain $FeO-Fe_3O_4$ equilibrium at the electrode/electrolyte interface while oxygen was withdrawn from the electrode. This electrode was then able to function simultaneously as an oxygen source and a reversible electrode.

The galvanic cell and electrical circuit, illustrated in Fig. 47, were used in a series of chronopotentiometric experiments.

Initially, the copper crystal was equilibrated throughout to the oxygen activity corresponding to the coexistence of $FeO-Fe_3O_4$ at the diffusion temperature with the switch in position 1. After equilibration, the switch was moved to position 2 so that a predetermined constant current passed

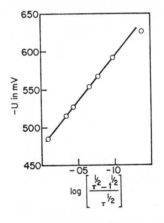

Fig. 46. Log $(\tau^{1/2} - t^{1/2})/\tau^{1/2}$ plotted versus E. Oxygen diffusion through liquid silver, at 990°C (After Rickert and El Miligy, 1968).

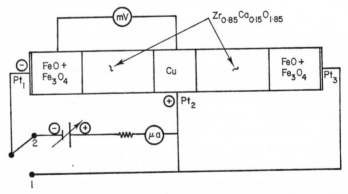

Fig. 47. Experimental arrangement for chronopotentiometric determination of diffusivity of oxygen in solid copper (After Pastorek and Rapp, 1969).

through the cell on the left. Thus, a constant flux of oxygen atoms entered the left-hand face of the cylindrical copper electrode. Simultaneously, the cell voltage E was measured continuously with a microvoltmeter. The duration of the experiment was limited to a time which was insufficient to form Cu_2O at the copper/electrolyte interface; in other words, the cell voltage was restricted to values lower than that corresponding to $Cu–Cu_2O$ equilibrium.

The authors have expressed the theoretical chronopotentiometric equation under the form:

$$E(t) = -i\Omega = \frac{RT}{2F}\left\{\ln f_0 + \ln\left[N_0(t_0) + \frac{V_{Cu}It^{1/2}}{FA\pi^{1/2}D_0^{1/2}}\right]\right\}$$

$$-\frac{RT}{4F}\ln[P_{O_2}''(FeO - Fe_3O_4)]$$

where Ω is the ionic electrical resistance of the electrolyte, I the cell current, N_0 the atom fraction of oxygen in copper, V_{Cu} is the molar volume of copper, A the cross-sectional area of the copper electrode, and f_0 the Henrian activity coefficient for oxygen atoms in copper.

Except for short periods of time (5 sec or less at 1000°C).

$$[VIt^{1/2}/FA\pi^{1/2}D_0^{1/2}] \gg N_0(t_0),$$

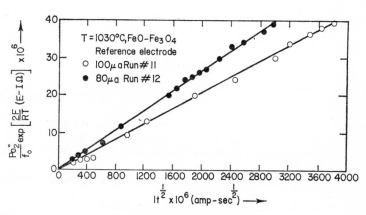

FIG. 48.

$$\frac{P_{O_2}''}{f_0}\exp\left[\frac{2F}{RT}(E - \dot{I}\Omega)\right]10^6$$

sveursplotted $\dot{I}t^{1/2}$. Diffusivity of oxygen in solid copper (After Pastorek and Rapp 1969).

so that a plot of $[E(t) - i\Omega]$ versus $\ln It^{1/2}$ should exhibit linear behaviour with a slope of $RT/2F$. A plot of

$$[P_{O_2}{}''^{1/2}/f_0]\left[\exp\left(\frac{2F(E - I\Omega)}{RT}\right)\right] \quad \text{versus} \quad It^{1/2}$$

should also be linear with a slope of $V_{Cu}/FA\,\pi^{1/2}D_O^{1/2}$ and an intercept of $N_O(t_0)$. The diffusivity D_O is then obtained from the measured slope. Figure 48 and 49 show that these relationships are well verified from experimental chronopotentiograms.

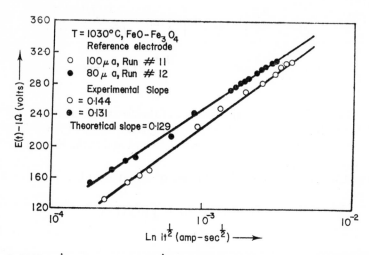

FIG. 49. $[E(t) - \dot{I}\Omega]$ plotted versus $\ln \dot{I}t^{1/2}$. Diffusivity of oxygen in solid copper (After Pastorek and Rapp, 1969).

References

Bard, A. J. (1961). *Anal. Chem.* **33**, 11.

Berzins, T. and Delahay, P. (1953a). *J. Am. Chem. Soc.* **75**, 555; (1953b) *Ibid.* **75**, 4205.

Butler, J. A. V. and Armstrong, G. (1933). *Proc. Roy. Soc.* **A139**, 406.

Butler, J. A. V. and Armstrong, G. (1934). *Trans. Faraday. Soc.* **30**, 1173.

Carslaw, H. S. and Jaeger, J. C. (1965). "Conduction of Heat in Solids", Clarendon Press.

Delahay, P. (1954). "New Instrumental Methods in Electrochemistry", Interscience Publishers.

Delahay, P. and Berzins, T. (1953). *J. Chem. Soc.* **75**, 2486.

Delahay, P. and Mamantov, G. (1955). *Anal. Chem.* **27**, 478.

Delahay, P. and Mattax, C. C. (1954a). *J. Am. Chem. Soc.* **76**, 874; (1954b) *Ibid.* **76**, 5313.

Delahay, P., Mattax, C. C. and Berzins, T. (1954). *J. Am. Chem. Soc.* **76**, 5319.

Delahay, P., Charlot, G. and Laitinen, M. A. (1960). *Analyt. Chem.* **32**. 103A.
De Vries, W. T. (1968). *J. Electroanal. Chem.* **18**, 472.
De Vries, W. T. and Van Dalen, E. (1963). *J. Electroanal. Chem. Nether.* **6**, 490.
Doremus, R. H. (1962). *In* "Modern Aspects of the Vitreous State" (J. D. Mackenzie, ed.) Vol. 2, Butterworths, London.
Evstropev, K. (1960). *In* "The Structure of Glass", (Consultants Bureau) Vol. II. p. 237.
Filayev, A. T., Karpachev, S. V. and Palguev, S. F. (1967). *In* "Electrochemistry of Molten and Solid Electrolyte", Vol. 4. pp. 161–165. Consultants Bureau.
Gierst, L. (1952). Thesis. University of Brussels.
Gierst, L. and Mechelynck, P. (1955). *Anal. Chim. Acta.* **12**, 79.
Guillou, M., Millet, J. and Palous, S. (1968). *Electrochimica Acta* **13**, 1411.
Hamer, W. J. (1965). *J. Electroan. Chem.* **10**, 140.
Hamer, W. J., Malmberg, M. S. and Rubin, B. (1956). *J. Electrochem. Soc.* **103**, 8.
Hamer, W. J. ,Malmberg, M. S. and Rubin, B. (1965). *J. Electrochem. Soc.* **112**, 750.
Hladik, J. (1965). "Les piles electriques" Presses Universitaires de France, Paris.
Hladik, J. (1966a). Thesis. University of Paris.
Hladik, J. (1966b). *Comptes Rendus Acad. Sc.* **263**, 1037.
Hladik, J. (1967a). *Electrochimica Acta.* **12**, 245.
Hladik, J. (1967b). *Comptes. Rendus Acad. Sc.* **265**, 1399.
Hladik, J. (1969a). *Ibid.* **268**, 1019.
Hladik, J. (1969b). "Electrochimie des Sels fondus" Tome II. Masson et Cie Paris
Hladik, J. (1969c). "La transformation de Laplace à plusieurs variables", Masson et Cie., Paris.
Hladik, J. (1970). *Comptes Rendus. Acad. Sc.* **270**, 1504.
Hladik, J. and Morand, G. (1965). *Bull. Soc. Chim. France* p. 828.
Hladik, J. and Rovetto, G. (1970). Unpublished results.
Hladik, J. Fromont, J. C. and Morand, G. (1966). *Comptes Rendus Acad. Sc.* **263**, 1106.
Hladik, J. Pointud, Y. and Morand, G. (1967). *Comptes Rendus Acad. Sc.* **265**, 691.
Hladik, J. Pointud, Y. and Morand, G. (1969). *J. Chime. Physique* **66**, 113.
Hladik, J., Pointud, Y., Bellissent, M. C. and Morand, G. (1970). *Electrochimica Acta.* **15**, 405.
Hume-Rothery, W. and Raynor, G. V. (1938). *Phil. Mag.* **25**, 335.
Inman, D. (1970). "Solid State Electrochemical techniques" Meeting at the atomic Energy Research Establishment, Harwell (unpublished results).
Institut Francais du Pétrole (1965). "Les Piles à combustibles" Editions Technip, Paris.
Jaeger, J. C. and Clarke, M. (1942). *Proc. Roy. Soc (Edinburgh).* **A61**, 229.
Jones, H (1934). *Proc. Roy. Soc.* **144A**, 225.
Karaoglanoff, Z. (1906). *Z. Electrochem.* **12**, 5.
Karpachev, S. and Filayev, A. (1968). *Z. für. Physikalische. Chemie.* **238**, 284.
Karpachev, S. V. and Obrosov, V. P. (1969). *Soviet Electrochem.* **5**, 445.
Karpachev, S. V. and Ovchinnikov, Yu, M. (1969). *Soviet Electrochem.* **5**, 181.
Kleitz, M. (1968). Thesis, University of Grenoble.
Kleitz, M., Besson, J. and Deportes, C. (1967). "Deuxièmes Journées internationales d'études des Piles à combustible", Bruxelles.
Latinen, H. A. and Ferguson, W. S. (1957). *Anal. Chem.* **29**, 4.
Leroy, M. (1963). Thesis, University of Paris.

930   J. Hladik

McMullen, J. J. and Hackerman, N. (1959). *J. Electrochem. Soc.* **106**, 341.
Mamantov, G. and Delahay, P. (1954). *J. Am. Chem. Soc.* **76**, 5323.
Miller, W. L. and Gordon, A. R. (1931). *J. Phys. Chem.* **35**, 2785.
Mitchell, W. (1963). "Fuel Cells". Academic Press, London.
Morand, G., Hladik, J. and Caby, F. (1964). *Comptes Rendus. Acad. Sci.* **258**, 329(
Nicholson, M. M. (1954). *J. Am. Chem. Soc.* **76**, 2539.
Nicholson, M. M. and Hkarchmer, J. (1955). *Anal. Chem.* **27**, 1095.
Pastorek, R. L. and Rapp, R. A. (1969). *Trans. Metall. Soc. AIME.* **245**, 1711.
Perfilev, M. V. and Palguev, S. F. (1966). *In* "Electrochemistry of Molten an
Solid Electrolyte" Vol 3 pp. 97–104. Consultants Bureau (1967). *Ibid* Vol. 4 p
153–159.
Petiau, G. (1965). "La theorie des fonctions de Bessel" Centre National de 1
Recherche scientifique Paris.
Pilling, N. B. and Bedworth, R. E. (1923). *J. Inst. Metals* **29**, 529.
Pointud, Y., Hladik, J. and Morand, G. (1969a). *Comptes Rendus Acad. Sci.* **26**
1423 (1969b) *Ibid.* **269**, 955.
Raleigh, D. O. and Crowe, M. R. (1969). *J. Electrochem. Soc.* **116**, 40.
Reilley, C. N., Everett, G. W. and Johns, R. H. (1955). *Anal. Chem.* **27**, 483.
Reinmuth, W. H. (1960). *Anal. Chem.* **32**, 1514; *ibid.* (1961). **33**, 485.
Rickert, H. (1968). *In* "Electromotive Force Measurements in High- temperatur
Systems" (C. B. Alcock, Ed.) The Institution of Mining and Metallurgy, London.
Rickert, H. and El Miligy, A. A. (1968). *Z. Metallkunde.* **59**, 635.
Rickert, H. and Steiner. R, (1966). *Z. Phys. Chem.* **49**, 127.
Rosebrugh, T. R. and Miller, W. L. (1910). *J. Phys. Chem.* **14**, 816.
Sand, H. J. S. (1901). *Phil Mag.* **1**, 45.
Sevcik, A. (1948). *Coll Czechoslov. Chem. Commun.* **13**, 349.
Steele, B. C. H. (1970). *In* "Heterogeneous Kinetics at Elevated Temperature"
Plenum Press.
Stein, B. (1969). Diplome d'Etudes Superieures-Paris.
Takahashi, T., Ito, K. and Iwahara, H. (1965). *Proc. Internat, J. Fuel Cells.*
Tammann, G. (1920). *Z. Anorg. Allg. Chem.* **111**, 78.
Wagner, C. (1954). *J. Electrochem. Soc.* **101**, 225.
Weber, H. J. (1879). *Wied. Ann. Physik.* **7**, 536.
Young, G. J. (1960). "Fuel Cells". Reinhold, New York.

22. COULOMETRY

J. H. Kennedy

University of California, Santa Barbara, California, U.S.A.

Introduction

The use of an electrolytic cell for a coulometer dates back to Faraday's experiments and, in fact, the silver coulometer is an accepted method for determining the value of the Faraday unit today (Richards and Anderegg, 1915; Craig *et al.*, 1960). Thus it is not too surprising that electrolytic cells have been used for applications requiring coulometers and, by using constant current, also for timers. The German V-2 Rocket during World War II employed a Ag/AgCl couple for timing and several such timers based on similar systems are available in the United States†. It is only natural that electrolytic cells containing solid electrolytes should also be studied as possible coulometers.

During the last several years extensive work has been carried out developing batteries using solid electrolytes (Foley, 1969; Hull, 1968; Takahashi 1968). Table I gives some examples of electrolytes used in solid state bat-

---

† Bissett-Berman, "E-Cell"; Bergen Laboratories, "t-Cell" and "Chronistor"; Gibbs Manufacturing and Research Corp., "Electrolytic Timing Coulometer"; General Electric, "Solion"; Curtis Instruments, "Indachron" and Fredericks Co., "Elapsed Time Indicator."

teries or fuel cells. Most solid electrolytes contain only one mobile ion (transference number for this ion equals one) and this serves as a method of classifying solid electrolytes. For room temperature operation the four most promising types are: (1) silver ion conductors, (2) fluoride ion conductors, (3) sodium or alkali ion conductors and (4) proton conductors. Of these, the most work, by far, has been carried out with silver ion conductors.

TABLE I.   Solid electrolytes for batteries

Electrolyte	Mobile Ion	Cell	Reference
AgI	Ag^+	$Ag/AgI/I_2$, Pt	Weininger, 1960
$RbAg_4I_5$	Ag^+	$Ag/RbAg_4I_5/RbI_3$	Argue et al., 1968
Ag_3SI	Ag^+	$Ag/Ag_3SI/I_2$, C	Takahashi and Yamamoto, 1965, 1966
β-Al_2O_3	Na^+	Na/β-$Al_2O_3/NaSx$	Weber and Kummer, 1967
ZrO_2/CaO	$O^=$	Pt, H_2/ZrO_2, CaO/O_2, Pt	Weissbart and Ruka, 1962
LiI	Li^+	$Li/LiI/AgI$	Liang et al., 1969

Basically, the coulometer consists of a material which is produced at an indicator (inert) electrode when the cell is "set" or accumulating "charge". Subsequently, the amount of this material is determined by weight, volume or electrical measurement. Historically, Tubandt used weight measurements to verify Faraday's Law for α-AgI and $PbCl_2$ solid electrolytes (Tubandt, 1932). However, most practical coulometers and timers today, especially those using solid electrolytes, make use of electrical "readout". This is accomplished by monitoring the voltage across the cell, and when the indicator electrode is completely "stripped" a large voltage occurs. A recording of voltage versus time will be referred to as the "discharge" or "stripping" curve and the final voltage rise when stripping is complete as the "cut-off" or "end-point". If one utilizes a silver ion conducting solid electrolyte the cell would operate in the following fashion:

$$
\begin{array}{llll}
\text{cell:} & \text{Ag} & \text{Ag}X & M \\
\text{"charge" or "set":} & \text{Ag} \to Ag^+ + e^- & Ag^+ + e^- \to \text{Ag} \\
\text{"discharge" or "strip":} & Ag^+ + e^- \to \text{Ag} & \text{Ag} \to Ag^+ + e^- \\
\text{"cut-off" or "end-point":} & Ag^+ + e^- \to \text{Ag} & M + nX^- \to MX_n + ne^- \\
& & \text{or } X^- \to X + e^-.
\end{array}
$$

The primary requirement is that the cell must operate at 100% current efficiency, i.e. the number of colombs necessary to strip the cell equals the number of colombs used in the initial setting. One could envisage a cell operating at <100% current efficiency, but at the same efficiency in both directions which would satisfy this primary requirement. However, such a current efficiency would depend on so many variables such as current density and temperature that the system would not be practical. A second requirement is that the end-point reaction must occur at a potential sufficiently different from the stripping reaction to be detected. In order for the system to qualify as a practical coulometer several operating characteristics must be investigated. These include electrolyte and electrode/electrolyte interface resistances, indicator electrode cut-off characteristics, coulometric accuracy including time delays between setting and stripping, and operating temperature range. This review will describe primarily the results for the few systems which have been studied in depth.

II. Silver Bromide

In solid electrolyte battery research there is a continuing quest for more highly conducting materials. A coulometer, on the other hand, may operate at low current densities which relaxes the requirements on conductivity. A coulometer can be designed using any material which exhibits an acceptable conductivity and meets the two requirements of 100% current efficiency and detectable end-point response.

The intrinsic conductivity of pure silver bromide is only 3×10^{-8} mho cm^{-1} calculated at room temperature, but polycrystalline silver bromide exhibits much higher conductivities from grain boundary conduction (Teltow, 1950; Kennedy et al., 1968). This high extrinsic conductivity was retained when cells were prepared by pressing silver bromide powder between pressed powder gold and silver electrodes at room temperature. The conductivity of the pressed powder was about $1-1.5 \times 10^{-5}$ mho cm^{-1}, (Kennedy et al., 1968) nearly three orders of magnitude higher than the intrinsic conductivity. An irreversible increase in resistance occurred when the pellet was heated to 100°C which was attributed to crystallite growth, decreasing the extent of grain boundaries.

The method of pressing powder together was developed to minimize electrode/electrolyte interface resistance. Since the contact resistance should be independent of pellet thickness the interface resistance could be checked by preparing pellets of varying thickness and plotting resistance versus thickness. Within experimental error the points lay on a straight line with the intercept at zero indicating little, if any, contact resistance.

After an amount of silver has been transferred electrolytically to the gold electrode (setting), this amount of silver can be determined by stripping, during which the silver is transferred back to the silver electrode. The voltage drop during the stripping cycle should remain equal to iR until the gold surface becomes depleted of silver, at which time the voltage will rise until it reaches the Ag/AgBr, AuBr/Au potential plus iR, or until Br_2 discharge occurs at the Au electrode. Figure 1 shows a stripping curve for a pellet cell ($1.27 \, cm^2$ area) charged for 1000 sec at $5 \, \mu A$. The voltage starts to rise somewhat before the "end-point" which in Fig. 1 was 0.5 V. The choice of an end-point potential is somewhat arbitrary, but by oxidizing the gold surface to this potential before charging the cell results will be consistent.

During the charge cycle, the system is a silver–silver couple, and the only voltage drop, in principle, should be from internal resistance. In practice, the voltage drop was somewhat higher than that predicted from a.c. conductivity measurements because of polarization effects. Polarization during both charge and strip cycles was measured using a 3-electrode technique.

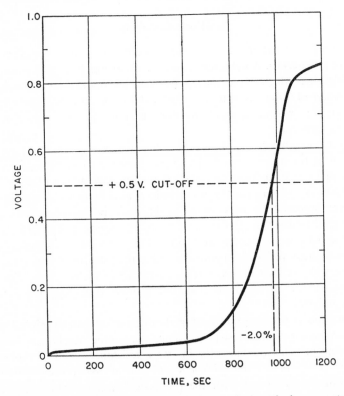

Fig. 1. Stripping curve for AgBr. Charge: 1000 sec at $5 \, \mu A$; stripping current: $5 \mu A$.

The silver layer was cut to form two silver electrodes. One was operated in the usual manner while the smaller second section was used as a reference electrode. Polarization at the silver electrode was pronounced at 75 µA during charge and some polarization was observed during stripping.

Coulometric accuracy for the silver bromide coulometers was checked at currents of 1–100 µA and for period of 50–200 sec. In the current range 5–50 µA the average error was less than $\pm 1\%$ in almost all cases (Table II). At 1–2 µA a positive error of about 25 µCoul was observed. Negative errors observed at high charge settings were shown to be predominantly silver remaining on the gold by stripping the gold at controlled potential, in which case the remaining µCoul were recovered at very low current. This difficult recovery of silver on gold has been observed with other coulometers and appears to be caused by silver atoms diffusing into the gold along grain boundaries.

TABLE II.   Timing accuracy for AgBr coulometers

Current, µA	Charge time (sec)					
	50		100		200	
	Average error, %	Average dev., %	Average error, %	Average dev., %	Average error, %	Average dev., %
1	+60·5	5·4	+26·2	2·4		
2	+17·8	2·4	+9·3	0·9		
5	+1·7	0·4	+0·9	0·6	−0·9	0·7
10	+1·7	0·8	0·0	1·2	+0·3	0·3
20	−0·2	0·3	−0·3	0·4	−0·1	0·6
50	−0·1	0·2	−0·1	0·3	−0·1	0·2
100	−0·2	1·0	−3·6	1·0	−10·3	0·3

Note: Results represent average of 8–10 determinations.

Although the AgBr coulometers showed high accuracy when stripping was carried out immediately, negative errors were observed when the cells were kept at open-circuit for a period of time between charging and stripping. The best results still showed a 13% loss with a 30 hour hold period.

Another practical consideration is the amount of charge which can be transferred to the gold and subsequently stripped without losing accuracy or shorting from silver growth. As can be seen from Table II, charges of 20 mCoul (5·56 µA h) could be readout with about a −10% error. Charges up to 50 μA h continued to yield low results although these errors could be kept below 20% by stripping at 1 μA current. No shorting was observed when the gold electrode was plated with a charge of 850 μA h, but shorting occurred during stripping. The tendency for silver growth to be dendritic when plated on silver was observed several years ago when Weininger studied thermocells (Weininger, 1964). He found the cell Ag/AgI/Ag to be unreliable because of silver dendrites and turned his attention to the cell I_2/AgI/I_2. Unfortunately, the high vapor pressure of iodine makes it difficult to contain at the high efficiency needed for coulometers.

In conclusion, the AgBr coulometer is suitable for charge capacities up to 10 mCoul/cm^2 stripped at currents of 4–40 μA/cm^2. Somewhat higher charges can be applied if stripping is carried out at a controlled potential or at a low current such as 1 μA/cm^2.

III. Silver Sulphide Iodide (Ag$_3$SI)

Several complex silver salts have been prepared which exhibit considerably higher conductivities than the silver halides (Takahashi et al., 1967; Reuter and Hardel, 1965; Bradley and Greene, 1967; Owens and Argue, 1967). A study carried out by Yamamoto and Takahashi (1968) investigated the coulometry of this system and describes an electrochemical coulometer similar to the one described above using AgBr.

The cell was Ag/Ag$_3$SI/Au and is shown in Fig. 2. Evaporated metal films were used in place of compacted powder electrodes, and the whole cell was encapsulated in epoxy resin. The electrode area was 0·1 cm^2. Silver sulphide iodide has a lower decomposition potential than AgBr and for this reason the end-point voltage used to indicate complete stripping should be kept low to prevent decomposition. Many of the curves presented by Yamamoto and Takahashi used 100–200 mV cut-off potentials, but a few results employing 300–500 mV, cut-off were also reported.

The high conductivity of Ag$_3$SI (0·01 mho cm^{-1}) allows the cell to be operated at higher currents than AgBr, and excellent results were obtained as high as 5 mA/cm^2. Voltage due to polarization was also reasonable over this current range reaching 100 mV at about 3 mA/cm^2 (Fig. 3). A reference electrode technique was not used so that silver and gold electrode polarization contributions could not be separated. However, in another study Takahashi and Yamamoto (1966) used a 3-electrode technique to study polarization in the cell Ag/Ag$_3$SI/I_2. Polarization at the silver plate anode was the major contributor to the total polarization for the cell. In fact, the silver plate could not be operated above 400–500 μA/cm^2 unless the silver electrode had been amalgamated with mercury. In the coulometer study, much less polarization was observed and this may be due to the evaporated silver film electrode in place of the silver plate giving better electrode/electrolyte contact.

When the current was between 20 μA (200 μA/cm^2) and 500 μA (5 mA/cm^2) and the total charge between 1 mCoul and 100 mCoul, coulometric errors were less than 5%. The cell was cycled over 100 times with no appreciable change in performance and was operable between $-20°$ and 60°C. In addition to constant current charging in which the cell operated as a timer, several runs were made with a varying current which tested the cell as a true coulo-meter. Results were somewhat less accurate, but still within 10%. Methods for improving coulometric accuracy when integrating a variable current will be discussed in a later section.

Of interest in light of the silver bromide results and other results to be presented later in this review was a comparison of discharge curves with immediate and a seven day hold period before stripping (discharge). There was less than a 2% difference in the results for a charge of 18 mCoul. This en-

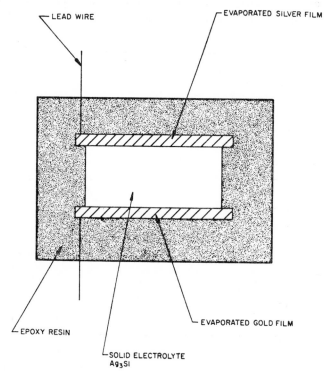

FIG. 2. Diagram of Ag$_3$SI coulometer .

couraging result may be due to the evaporated gold electrode offering less opportunity for silver diffusion into the gold.

IV. Silver Sulphide Bromide (Ag₃SBr)

During the same period of time that Yamamoto and Takahashi were investigating Ag_3SI coulometry, an independent study on the analogous compound, Ag_3SBr, was also initiated (Kennedy and Chen, 1969; Kennedy *et al.*, 1970). Both compounds can be prepared by mixing or precipitating together stoichiometric quantities of silver halide and silver sulphide followed by a heat treatment. The two compounds are quite similar in their electrochemical characteristics; the conductivity of Ag_3SBr reported in this study was 0.012 mho cm^{-1}, close to the value of 0.01 mho cm^{-1} reported for Ag_3SI. This value for Ag_3SBr was an order of magnitude larger than the value reported by Reuter and Hardel (1966), and the difference may be attributed to the extremely low contact resistance of compressed powder electrodes. The resistance of the pressed powder pellets was found to be proportional to the thickness, and the intercept at zero thickness showed no measurable contact resistance. Grain boundary conduction, so important to AgBr, was not an

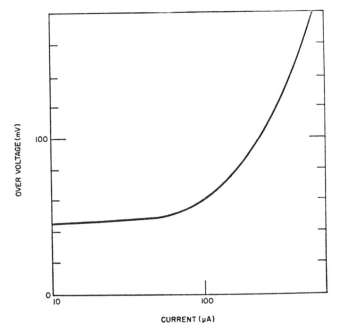

Fig. 3. Polarization curve of Ag/Ag₃SI/Ag–Au coulometer at 25°C.

important factor for Ag_3SBr. Annealing pellets at 150°C for seven days decreased the conductivity by only 10%.

Since Ag_2S is an electronic conductor which would be detrimental for coulometry, care was taken to avoid any excess Ag_2S in the Ag_3SBr. One way of accomplishing this was to add excess AgBr during the preparation. Addition of AgBr to Ag_3SBr increased the resistance of the pellets proportionally to the amount of AgBr added, up to 20 mole % excess AgBr. The resistivity of this mixture was 270 ohm cm, about three times higher than pure Ag_3SBr. Most of the Ag_3SBr used for coulometry contained only a small excess AgBr of 2–5%. If an electronic conducting impurity were present, a solid electrolyte would exhibit a residual current, i.e. a steady state current flows when a constant voltage below the decomposition potential is applied. By electrolyzing at a controlled potential (50–600 mV for 24–72 h) the cell reached a residual current of about 100–200 nA at 0·5 V. Assuming that no ionic contribution remains after a steady state current is reached, this residual current can be used to calculate the upper limit of electronic conductivity contribution. If the material contains conduction electrons a plot of current versus applied voltage will exhibit a plateau region where (Wagner, 1956):

$$i_0 = (RT/LF)\sigma_0$$

A reasonable plateau region was observed for pure Ag_3SBr and the electronic conductivity, σ_0, was calculated to be about $2·6 \times 10^{-7}$ mho cm^{-1} which is only 0·002% of the total conductivity. This probably represents the maxi-

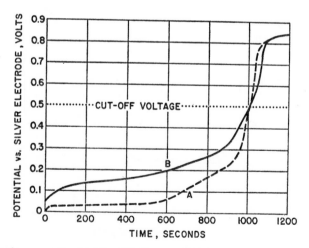

FIG. 4. Stripping curve for Ag_3SBr. (A) Charge: 1000 sec at 10 μA; stripping current: 10 μA; (B) Charge: 1000 sec at 36 μA; stripping current; 36 μA.

mum electronic contribution since the plateau was not strictly flat and may still have contained some ionic contributions. The residual current was an extremely sensitive test for electronic impurities as evidenced by a residual current 30 times higher for a pellet containing a small amount of metallic silver from decomposition and detected by X-ray diffraction.

One reason for using Ag_3SBr in place of Ag_3SI for coulometers is the higher decomposition potential of AgBr versus AgI. However, there remains the question of Ag_2S decomposition which should occur at 0·2 V versus silver. Cyclic voltammetry was used to examine the electrochemical reactions at the gold electrode, and a small peak at +0·2 V was observed. This peak was attributed to the oxidation of sulphide ion to sulphur but the reaction was highly irreversible since no cathodic peak was observed before reduction of silver ion. The fact that a small amount of sulphide could be oxidized at +0·2 V, while sulphur was not readily reduced formed the basis for a positive error mechanism whenever cut-off potentials greater than +0·2 V were employed. A stripping curve for Ag_3SBr, analogous to Fig. 1 is shown in Fig. 4. There was a definite voltage plateau at 0·2–0·4 V depending on current density. By using the 3-electrode technique described for AgBr coulometry, it was shown that the voltage arrests were occurring at the gold electrode during stripping (Fig. 5). During the charge cycle, polarization was predominantly at the silver electrode (Fig. 6), and after long periods this polarization increased to the decomposition voltage of the electrolyte. A current interruptor circuit was employed to show that almost no change

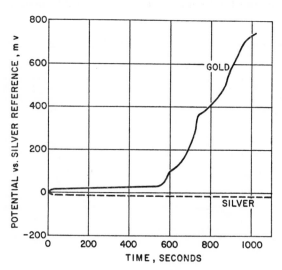

FIG. 5. Polarization during stripping cycle for Ag_3SBr. Charge: 800 sec at 25 μA; stripping current: 25 μA.

in resistance occurred during the charge cycle. The increase in silver electrode polarization was attributed to decreasing active area where silver oxidation could occur.

A similar voltage plateau for Ag_3SI was not reported by Yamamoto and Takahashi. However, many of their cut-off values were below $+0.2$ V and for cases in which higher voltages were used, the current density was over 1 mA/cm^2. At this high current density, no sulphide oxidation was observed for Ag_3SBr due to the slow electrokinetics of the sulphide reaction. This irreversible reaction would suggest that positive errors would be observed especially at low current densities. Results with Ag_3SBr at $10\,\mu A$ or less as well as with Ag_3SI showed positive coulometric errors. Higher currents with Ag_3SBr showed small negative errors which could arise from other causes, similar to the negative errors observed with AgBr coulometers. Timing accuracy for the silver sulphide bromide coulometers was checked for currents of 10–$360\,\mu A$ and periods of 50–250 sec and the results are given in Table III. The first few cycles on a fresh pellet gave relatively large negative errors and could arise from reduction of a species during charge that was not re-oxidized during stripping. For fresh pellets, the species could possibly be gold oxide or gold sulphide on the gold surface. In order to minimize this effect, pellets were conditioned by cycling 2–3 times before timing accuracy was reported. Negative errors observed for conditioned pellets may arise from the migration of silver into the gold via grain boundaries.

Positive errors were observed when the cell was held at open-circuit for a period of time before stripping. It appears that with aging, more sulphide

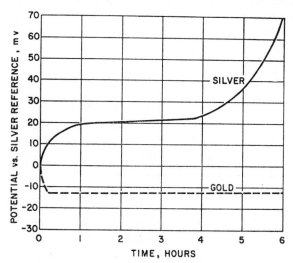

FIG. 6. Polarization during charge cycle for Ag_3SBr. Charge: 21,600 sec at 25 μA.

becomes available for oxidation at voltages below the 0·5 V cut-off used. A charge of 50 mCoul could be readout within 5% if the hold period was less than two days, but the error increased to 20% when stored for a month.

The capacity of Ag_3SBr coulometers for charge was considerably higher than with AgBr coulometers. However, unless the pellets were conditioned, 8–10% negative errors were observed for charges of 3–40 µA h. With conditioning, errors of less than 1% were reported. The largest charge reported was 417·6 µA h in which a −13% error was observed on stripping. Another possible mechanism for negative errors during stripping of such large charges is that the silver plate may be thick enough to allow under-cutting. That is, silver would be oxidized close to the gold during stripping leaving silver metal in the electrolyte region not connected electrically to the gold electrode. In support of this mechanism, charge lost for these pellets could not be recovered by controlled potential electrolysis, a technique which may recover silver that has migrated a small distance into the gold.

Most solid electrolyte coulometry studies have been carried out at ambient temperatures, and all of the results reported above for Ag_3SBr pertain to room temperature operation. Yamamoto and Takahashi did report that Ag_3SI coulometers could be used from −20 to +60°C and some data with Ag_3SBr at temperatures as low as −70°C have been published (Kennedy et al., 1970). At −70°C the cell resistance was still less than 100 ohms which probably indicates that the coulometer could be operated at even lower temperatures. Timing accuracy at constant current (Table IV) shows negative errors probably result from higher polarization voltages and iR drop leading to premature cut-off. Precision was within 1% and charge hold characteristics were better at low temperature. The cut-off curve was much sharper at low temperature which is related to the decreased sulphide oxi-

TABLE III.   Timing accuracy for Ag_3SBr coulometers

| Current, µA | Charge time, (sec) | | | | | |
| | 50 | | 100 | | 250 | |
	Average error, %	Standard dev., %	Average error, %	Standard dev., %	Average error, %	Standard dev., %
10	+4·6	2·3	+5·8	1·6	+3·2	0·26
36	−0·2	1·5	−0·1	0·8	−0·6	0·38
100	−1·8	1·3	−0·7	0·6	−0·6	0·30
360	−0·8	0·3	−0·7	0·2	−1·2	0·17

Note: Results represent average of eight determinations.

dation effects. In fact, this effect limited operation of Ag_3SBr coulometers to $<50°C$ for currents <100 $\mu A/cm^2$.

Results at constant current for all the coulometers described show that the charge recovered was a function of the current level. As the current increased, the coulometric error became more negative. Also, accuracy improved with succeeding cycles under the same operating conditions as the cell appeared to become "conditioned". For this reason, integration of a varying current could not be accomplished with the same accuracy as repeated cycles at constant current. The main reason for inaccurate readout is that the actual experimental function normally observed, applied voltage at the operating current, is not the function indicating the complete stripping of silver from the gold electrode, which is a gold electrode potential versus a reference. Although 3-electrode systems have been used for basic studies, current must be interrupted for each measurement to eliminate polarization. Applied voltage includes silver electrode potential, silver electrode/electrolyte polarization, electrolyte iR drop, and gold electrode/electrolyte polarization in addition to the desired gold electrode potential. Since polarization voltages and iR drop vary with current, the amount of charge required to reach a given applied voltage will vary with current level. A pulse discharge method was devised to measure, at least approximately, the gold electrode potential, and trigger a cut-off when a potential develops indicative of complete silver stripping (Kennedy et al., 1970). The method consists of pulsing the cell by a

TABLE IV. Low temperature timing accuracy (Constant current charge and discharge) Charge time: 200 sec

Current, μA	Temperature °C	Strip time* sec.	Average dev., %	Error %
36	0	200·4	0·1	+0·2
100	0	198·9	0·1	−0·6
36	−25	199·2	0·1	−0·4
100	−25	194·4	0·2	−2·8
36	−50	192.7	0·4	−3·7
100	−50	191·5	0·2	−4·2
10	−70	192·9	1·0	−3·6
36	−70	192·1	0·9	−4·0
100	−70	195·0	0·5	−2·5

* Average of eight cycles after two conditioning cycles.

capacitor discharge (0·1 mCoul) while the voltage across the cell is monitored. After the pulse, the voltage drops rapidly as the iR drop and polarization voltages decay, and when the cell voltage falls to a predetermined value (0·25 V) another pulse of charge is delivered. While silver remains on the gold electrode, the potential is essentially zero since it is being measured against the counter silver electrode. When all the silver has been stripped by the pulses of charge, a gold electrode potential develops which does not decay rapidly with time. Actually, since the measurement is still a 2-electrode case as opposed to a cell containing a reference third electrode, the auxiliary silver electrode will not be strictly a reference potential. However, polarization which develops during the pulse will have ten seconds to decay which is sufficient time to consider the auxiliary silver electrode as a reference potential. As soon as the cell potential remains above 0·25 V for 10 seconds, the circuit is inactivated, and the number of pulses which were delivered can be read on a counter. The instrument used in the Ag_3SBr study is available commercially for liquid electrolyte coulometers, and the values of 0·25 V, 10 sec time delay and 0·1 mCoul pulses may have to be adjusted for solid electrolyte coulometers.

TABLE V.   Pulse discharge results for constant current charge.

Charge conditions			Pulse Discharge* mCoul.	Average error, %	Average dev. mCoul.
Current µA	Time sec.	Charge mCoul.			
360	50	18·0	18·0	0·0	0·1
	100	36·0	35·9	−0·3	0·1
	200	72·0	71·5	−0·7	0·1
	500	180·0	180·4	+0·2	0·3
	1000	360·0	350†	−2·8	—
100	100	10·0	10·1	+1·0	0·1
	150	15·0	15·0	0·0	0·1
	200	20·0	19·5	−2·5	0·2
	500	50·0	49·8	−0·4	0·1
36	200	7·2	7·5	+4·2	0·1
	500	18·0	18·0	0·0	0·1
	1000	36·0	35·7	−0·8	0·1
10	1000	10·0	10·0	0·0	0·1

* Average of eight trials
† pellet broke after four trials, average of four

When the pulse discharge method was applied to Ag_3SBr coulometers, practically no "conditioning" effect was found and integration of a varying current could be performed with the same accuracy as constant current timing. Table V shows results obtained when the cells were charged at constant current and stripped with the pulse discharge while Table VI contains results for charges applied at various current levels. It would be of interest to use the pulse discharge method to readout charges applied by the triangular current program used by Yamamoto and Takahashi for Ag_3SI coulometers (Yamamoto and Takahashi, 1968). Again, Ag_3SBr cells could not be readout accurately at temperatures much above ambient even with the pulse discharge method, but results at low temperatures were excellent.

Although a coulometer based on a silver ion conducting solid electrolyte can be expressed as Ag/Ag electrolyte/M, the indicator electrode, M, has been gold in almost all studies. The requirement that the indicator electrode be more noble than the silver used for tabulating charge limits the choice to only a few materials such as gold, platinum group and carbon. A few experiments have been reported (Kennedy et al., 1970) for Ag_3SBr and

TABLE VI.   Integration results

Charging time (sec)				Total charge m. Coul.	Pulse-discharge readout m. Coul.	% error
10 μA	36 μA	100 μA	360 μA			
500	140	50	—	15·04	14·5	−3·3
500	140	50	—	15·04	14·6	−2·7
500	140	50	—	15·04	13·6	−10·5
500	140	50	—	15·04	14·8	−2·0
—	150	50	—	10·4	10·4	0·0
—	150	50	—	10·4	10·5	+1·0
—	150	50	—	10·4	10·3	−1·0
—	100	100	100	49·6	48·9	−1·4
500	—	—	50	23·0	23·2	+0·9
200	200	200	—	29·2	28·6	−2·1
500	200	100	50	40·2	40·3	+0·2
—	200	100	50	35·2	35·2	0·0
—	100	100	100	49·6	48·7	−1·8
—	150	50	—	10·4	10·1	−2·9*
—	150	50	—	10·4	10·1	−2·9*

* Operated at −70°C

Ag_3SI coulometers using palladium and graphite electrodes in place of gold (Fig. 7). Accuracy was quite poor, and the only desirable characteristic observed was that when using graphite the cut-off curve was extremely sharp. This, however, may have been due to a smaller true contact area between electrode and electrolyte which would exhibit less capacitance. On the basis of these few results, gold still appears to be the most suitable indicator electrode material.

V. Rubidium Silver Iodide ($RbAg_4I_5$)

The existence of a highly conducting solid electrolyte type, MAg_4I_5, was reported by Bradley and Greene (1967) and Owens and Argue (1967). The most stable member of the group is $RbAg_4I_5$ which exhibits the highest conductivity at room temperature of all solid electrolytes known to date. In fact, its conductivity value of 0.26 mho cm^{-1} at room temperature is quite close to the 0.7 mho cm^{-1} value for 30% KOH. In addition, the electrolyte still exhibits conductivity equal to 0.09 mho cm^{-1} at $-54°C$. Electronic conductivity was reported to be $<10^{-11}$ mho cm^{-1} which is less than 10^{-8}% of the total conductivity (Gould Ionics, 218P). The electrolyte is being used in a solid state battery:

$$Ag \ / \ RbAg_4I_5 \ / \ RbI_3$$

being developed by Gould Ionics (218P). A cell of the type:

$$Ag \ / \ RbAg_4I_5 \ / \ M \ (collector)$$

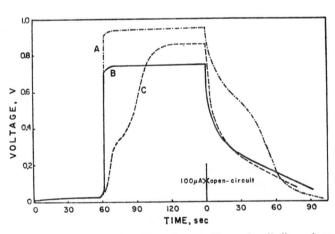

Fig. 7. Voltage rise and open-circuit decay for graphite and palladium electrodes: (A) C/Ag$_3$SBr/Ag, (B) C/Ag$_3$SI/Ag, (C) Pd/Ag$_3$SBr/Ag.

for coulometry has also been reported (Gould Ionics, 118–120). The coulometer, called "Set Cell" or "Coulistor" appears to have similar characteristics to those employing Ag_3SI or Ag_3SBr, but should have the advantage of possible higher current operation. Also, the absence of sulphide may eliminate one source of possible error. Extensive results have not been published

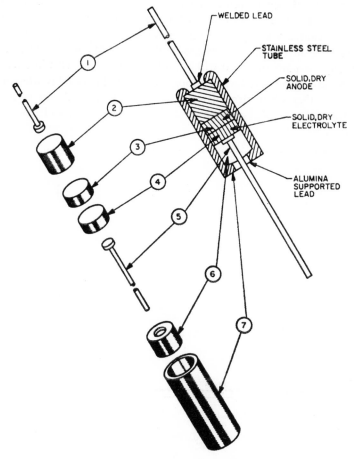

1. HEADED LEAD
2. METAL PLUG
3. ANODE
4. SOLID, DRY ELECTROLYTE
5. HEADED LEAD
6. ALUMINA BEAD
7. METAL TUBE

WELDED LEAD

STAINLESS STEEL TUBE

SOLID, DRY ANODE

SOLID, DRY ELECTROLYTE

ALUMINA SUPPORTED LEAD

FIG. 8. Cut-away diagram of $RbAg_4I_5$ coulometer (Gould Ionics, Inc.).

to date, but a diagram is shown in Fig. 8 and a discharge curve in Fig. 9. The lower decomposition potential for an iodide electrolyte compared to bromide is noted in the specifications which state that the cell should not be kept above 0·6 V, (Gould Ionics, 118–120). However, it is noted that the cell may not be permanently damaged and can re-establish good performance by cycling several times keeping below the 0·6 V, limit.

Timing accuracies of 5% are considered typical for charges up to 100 µA h and at temperatures as low as −54°C (Table VII) (Owens, private communication). Although most of the data reported are for charge settings < 100 µA h, a unit designated U–6000 is specified to be capable of storing 6000 µA h of charge. This is considerably larger than capacities reported for Ag_3SI and Ag_3SBr. The higher conductivity of $RbAg_4I_5$ may allow the silver plate plate to form evenly which would avoid shorting. Gould Ionics states that dendrite growth has not been observed for the "Coulistor".

Results for charge loss with storage or integration of varying currents are not available in the open literature at this time, but published results are specifically for units operated within an hour of setting. This infers that greater errors are encountered when a longer charge hold period is attempted similar to the other silver electrolyte systems. Another similarity to these systems is the higher accuracy after cycling the cell several times (Table VII).

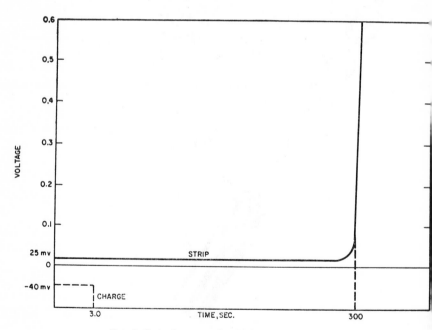

FIG. 9. Stripping curve for $RbAg_4I_5$ coulometer.

Hopefully, more results will be available soon to compare $RbAg_4I_5$ more critically with Ag_3SI and Ag_3SBr.

VI. Thin-Film Coulometers

Since one of the drawbacks for most solid electrolyte cells is the high resistance, some research groups have been investigating cells in the form of thin films. Several papers have been published attempting to develop thin-film batteries (Liang *et al.*, 1969; Vouros and Masters, 1969; Yamamoto and Takahashi, 1966) and a study has been devoted to thin-film cell coulometry (Kennedy and Chen, 1971). The system used was:

$$Ag \ / \ AgBr \ / \ Au$$

and a slide configuration containing twelve cells is shown in Fig. 10. The silver bromide thickness was about 100,000 Å since films less than 50,000 Å showed a high incidence of pinhole shorting. The primary requirement for the vacuum deposition was that the substrate temperature had to be maintained at ambient temperature. Under these conditions the silver bromide film showed small crystallite size (Fig. 11) which prevented pinhole shorting and also promoted high conductivity along grain boundaries. Resistivity of films produced under various conditions is given in Table VIII.

The thin-film cells were capable of the same accuracy as the pressed pellets as can be seen by the results in Table IX, yet these cells had resistances of about 200 ohm compared to the usual 5000 ohm for pressed powder AgBr cells. Cells of Ag_3SBr and Ag_3SI had resistances of about 5–20 ohm while

TABLE VII. Timing accuracy for $RbAg_4I_5$ coulometers 100 μA h. Capacity coulometer set at 1 ma and stripped at 10 μA within one hour of setting. Results are average of four cells run at each condition.

Set and operate temp. (°C)	Charge (μA h)	Average accuracy, % of set	
		Initial	After 5 cycles
−54	1	94·5	98·6
Room temp.	1	95·4	98·1
+71	1	95·9	97·4
−54	100	96	—
Room temp.	100	96	—
+71	100	96	—

those of $RbAg_4I_5$ are usually under 1 ohm. The AgBr thin-film cells exhibited long operating life as demonstrated by cycling the cells over 20,000 times without failing. The cells operated well from $-20°C$ to $+50°C$. This approach offers much promise for practical solid electrolyte devices both for improving electrochemical performance and offering production and packaging advantages over the pellet type cells.

TABLE VIII.   Conductivity of AgBr films

Deposition temp. (°C)	Conductivity (mho cm^{-1})
25°	960×10^{-7}
100°	14.7×10^{-7}
250°	7.5×10^{-7}

TABLE IX.   Coulometric accuracy of thin-film cells

Deposition conditions		Average error, %*	
Temperature (°C)	AgBr deposition rate (Å/sec)	3 μA	10 μA
25	870	$+2.6\%$	$+1.2\%$
25	870 baked gold	$+1.6\%$	-1.4%
25	60	$+1.6\%$	$+0.8\%$
25	60 baked gold	$+0.4\%$	-0.2%
100	770	$+1.4$	$+1.2\%$
100	770 baked gold	$+1.6\%$	$+0.8\%$
100	80	$+1.0\%$	$+0.2\%$
100	80 baked gold	$+1.8\%$	$+0.4\%$
250	1050	$+1.0\%$	-0.8%

* Results for three trials of 50 sec charge after two conditioning cycles.

VII. Coulometry With Other Solid Electrolytes

It was stated in the introduction that there are several types of solid electro lytes other than silver ion conducting materials. Some of these are bein developed primarily for battery application. One basic difference betwee a battery and the typical coulometer described here is that during discharge, a battery has a high voltage which drops at the end-point (complete discharge) while the coulometer has a very low voltage which increases at the end-point. There is no reason why coulometers could not be designed using the battery concept provided it demonstrated high current efficiency. Unfortunately, because of the high voltage when charged, there is usually a self-discharge which decreases the coulometric efficiency to $<100\%$. There have been some comments by researchers concerning coulometric efficiency of their devices and these are summarized here.

Mrgudich has been one of the pioneers in solid electrolyte batteries and other solid electrolyte devices, and a few years ago reported on rechargeable Ag/AgI/Pt batteries (Mrgudich *et al.*, 1965). The efficiency appeared to be high from the statement that 42 shallow charge-discharge cycles delivered 732 μA-minutes of charge with an input of only 727 μA-minutes. However, the initial condition was not defined enough to calculate a truly quantitative figure for efficiency. Other workers have not been as successful in achieving high current efficiency in a solid electrolyte battery. Weininger studied the Ag/AgI/I$_2$ cell and recovered 73% of the charge at 0°C (Weininger, 1958). He also pointed out two of the major problems for maintaining high current efficiency; high temperature and charge hold periods. With the same cell cycled at various temperatures and a 90 minute delay between charge and discharge only 73%, 63% and 15% of the charge was recovered at -210°C, 0°C and 25°C respectively. In another study in which a cell was operated in discharge only (primary cell), the efficiency was calculated from the con-

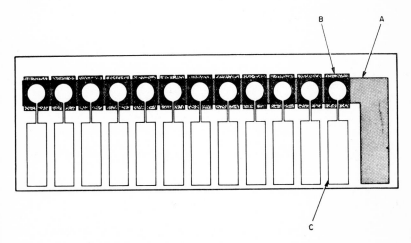

FIG. 10. Cell configuration for thin-film AgBr coulometer (12 cells on single substrate). (A) Silver strip electrode; (B) Silver bromide; (C) Gold electrode.

sumption of the silver wire electrode. One result of 85% efficiency was reported for a cell operated at 300°C (Weininger, 1959).

In an article describing electrochemical cells based on ion conductive ceramics (Hever, 1968), Hever concluded that power density as a battery was unacceptably low at room temperature, but the device might find utility as an electronic integrator or as a capacitor. Hever felt that the cell offered advantages in reliability and price, but that stability had not been fully investigated.

Durst and Ross reported that a fluoride-conducting membrane of LaF_3/EuF_2 could be used to generate fluoride ion at $>99\cdot2\%$ current efficiency in an aqueous medium (Durst and Ross, 1968). This demonstrates that the fluoride ion in the solid electrolyte has a transference number essentially equal to one and might be useful in a solid electrolyte coulometer. The same technique was used by Fletcher and Mannion (1970) to generate silver ion at 100% current efficiency with a silver sulphide membrane. An interesting feature here is that silver sulphide is partly an electronic conductor,

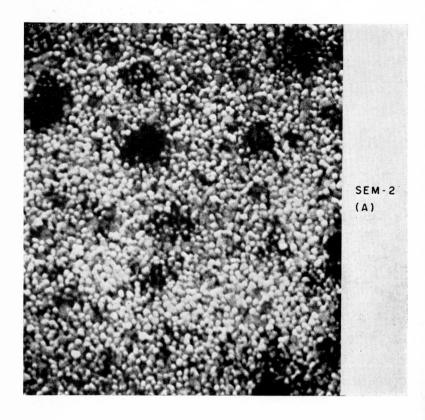

SEM-2
(A)

but the electronic contribution was blocked by the 100% ionically conducting aqueous solutions on both sides of the membrane. The use of silver iodide in solid electrolyte cells to perform the same blocking function has been reported (Kiukkola and Wagner, 1957).

VIII. Coulometric Titrations

One of the reasons the coulometers described above are finding application is their very sharp cut-off allowing for easy detection. Until silver is stripped completely from the indicator electrode, the silver activity is essentially the same on both electrodes and thus, the cell potential is zero. This behavior has the disadvantage that during this period there is no way of knowing how much silver remains on the indicator electrode. A cell in which the silver

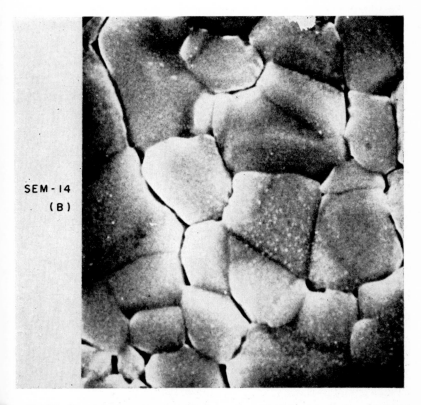

SEM-14
(B)

FIG. 11. Thin-film silver bromide crystallite formation. (A) AgBr deposited on 25°C. substrate, 10,000 X; (B) AgBr deposited on 250°C. substrate, 10,000 X.

or other metal formed a solid solution or compound with the indicator electrode would have a cell potential given by the equation:

$$E = - \frac{RT}{nf} \ln a_A$$

where a is the activity of A in the indicator electrode (C) for the cell:

$$A \mid AB \mid A(C)$$

The amount of A at the indicator electrode can then be determined by monitoring the cell potential. By passing a current through the cell, the activity of A in C can be varied, and the technique has been termed a coulometric titration.

There are severe limitations on the method since A and C must form a single phase (solid solution or compound) and reach equilibrium in a reasonably short time. The latter requires that A has a high diffusivity in C and C should usually be a very thin layer.

Wagner studied the Ag_2S stoichiometry by varying the Ag/S ratio over the range 2·000 to 2·002 (Wagner, 1953) for the cell:

$$Ag \mid AgI \mid Ag_2S \mid Pt$$

The limits on the system are the presence of sulphur on the sulphur-rich side to the presence of a silver metal phase at the silver-rich side. Since the cell $Ag/Ag_2S/S$ has a potential of 0·23 V while $Ag/Ag_2S/Ag$ would be zero the potential range is about 0·2 V. The cell reached zero at Ag/S = 2·0025.

A similar study was carried out by Rizzo and Smith (1968) on Wustite. The cell made use of calcia stabilized zirconia electrolyte which is an oxide ion-conducting electrolyte at high temperatures. The range in this system was Fe/Fe_xO at the iron rich side and Fe_yO/Fe_3O_4 at the oxygen-rich side. The reference electrode was Fe/Fe_xO, and the cell reached zero at 51·7 mol % oxygen. The highest potential was reached at 52·5–53·2% depending on temperature at which point the cell potential ranged from 60 mV at 765°C to 125 mV at 965°C. Equilibrium was reached in less than two hours. The cell potential was essentially linear with changes in oxygen content over this narrow range of the phase diagram.

The large change in potential with small changes in composition offers a sensitive method for determining the composition provided a reliable reference potential is available. Also, the ability to change the composition by adding or removing one component electrochemically (coulometrically) may make this technique important not only to the study of narrow range phase diagrams, but also to the development of practical solid state coulometric devices.

IX. Applications

There are a number of military applications which require timers having reasonable accuracy (5%) and are not prone to damage by shock, vibration, extreme temperatures and extended storage. It appears that solid electrolyte coulometers could meet the requirements for many of these applications.

As a coulometer, the cell can be used as an information storage device. One of the simplest modes is as a bistable device using the set condition (zero volts) as one state and the cleared condition (cut-off voltage) as the other. Andes and Manikowski studied AgI and AgCl between Ag and Pt electrodes and found that switching could be accomplished in 1msec (Andes and Manikowski, 1963). However, more information can be stored per cell when used in an analog fashion. A known amount of charge is put on the indicator electrode and when the system has been subject to usage corresponding to this charge value, a signal is given. In this mode, the device would be useful for maintenance monitoring or related applications. By using two cells, the initial known charge need not be lost. This is accomplished by connecting the cells in the following way:

$$\text{Ag} \mathbin{/} \text{electrolyte} \mathbin{/} M_1 \text{——} M_2 \mathbin{/} \text{electrolyte} \mathbin{/} \text{Ag}$$

One indicator electrode, M_1, is given an initial charge setting and as it is stripped, an equivalent amount of charge is deposited on M_2. At the endpoint, the cell is turned or current reversed so that the cell operates in the reverse direction, stripping M_2 and plating M_1. This "back-to-back" operation has been studied briefly (Kennedy et al., 1970) and results show that under limited conditions it may be possible to cycle a pair of cells hundreds or even thousands of times. A similar operation using two aqueous redox cells separated by an ion exchange membrane was described recently by Weininger (1969). Weininger's results illustrate that it must always be kept in mind that liquid electrolyte coulometers can also perform these functions, but the rugged nature and design possibilities especially as thin films may give the solid state systems a competitive advantage for certain applications.

Acknowledgement

The author expresses his gratitude to Dr. Owens of Gould Ionics, Inc. for information on the "Coulistor", Mr. Sueoka of Bissett–Berman Corporation for translation of Japanese articles, Mr. Goodwin of Bissett–Berman Corporation for figure drawings and to the many investigators whose work made this review possible.

References

Andes, R. V. and Manikowski, D. M. (1963). *J. Electrochem. Soc.* **110**, 66c.
Argue, G. R., Owens, B. B. and Groce, I. J. (1968). *Annu. Power Sources Conf.* **22**, 103.
Bradley, J. N. and Greene, P. D. (1967). *Trans. Faraday Soc.* **63**, 1023, 2516.
Craig, D. N., Hoffman, J. I., Law, C. A. and Hamer, W. J. (1960). *J. Res. Natl. Bur. Stand.* **64A**, 381. A more recent determination.
Durst, R. A. and Ross, J. W. (1968). *Anal. Chem.* **40**, 1343.
Fletcher, K. S. and Mannion, R. F. (1970). *Anal. Chem.* **42**, 285.
Foley, R. T. (1969). *J. Electrochem. Soc.* **116**, 13c. Hull, M. N. (1968). *Proc. Annu. Power Sources Conf.*, **22**, 106. Takahashi, T. (1968). *Denki Kagaku* **36**, 402, 481. Three of the many recent reviews.
Gould Ionics Inc., Bull. No. 218P.
Gould Ionics Inc., Bull. No. 118, 119, 120.
Hever, K. O. (1968). *J. Electrochem. Soc.* **115**, 830.
Kennedy, J. and Chen, F. (1969). *J. Electrochem. Soc.* **116**, 207.
Kennedy, J. and Chen, F. (1971). *J. Electrochem. Soc.* **118**, 1043.
Kennedy, J., Chen and F. Clifton, A. (1968). *J. Electrochem Soc.* **115**, 918.
Kennedy, J., Chen, F. and Willis, J. (1970). *J. Electochem. Soc.* **117**, 263.
Kiukkola, K. and Wagner, C. (1957). *J. Electrochem. Soc.* **104**, 308, 379.
Liang, C. C., Epstein, J. and Boyle, G. H. (1969). *J. Electrochem. Soc.* **116**, 1322, 1452.
Mrgudich, J. N., Schwartz, A., Bramhal, P. J. and Schwartz, G. M. (1965). *Proc. Annu. Power Sources Conf.* **19**, 86.
Owens, B. B. Private Communication.
Owens, B. B. and Argue, G. R. (1967). *Science* **157**, 308.
Reuter, V. and Hardel, K. (1965). *Z. Anorg. Allgem. Chem.* **340**, 158.
Reuter, V. and Hardel, K. (1966). *Ber. Bunsengesellschaft.* **70**, 82.
Richards, T. W. and Anderegg, F. O. (1915). *J. Am. Chem. Soc.* **37**, 7. A classical study.
Rizzo, F. E. and Smith, J. V. (1968). *J. Phys. Chem.* **72**, 485.
Takahashi, T. and Yamamoto, O. (1965). *Denki Kagaku* **33**, 518.
Takahashi, T. and Yamamoto, O. (1966). *Electrochim. Acta,* **11**, 779.
Takahashi, T. ,Yamamoto, O. and Mori, H. (1967). *Denki Kagaku* **35**, 181.
Teltow, J. (1950). *Ann. Physik* **5**, 63.
Tubandt, C. (1932) *Handbuch Der Experimentalphysik* **12**, (1), 383.
Vouros, P. and Masters, J. I. (1969). *J. Electrochem. Soc.* **116**, 880.
Wagner, C. (1953). *J. Chem. Physics* **21**, 1819.
Wagner, C. (1956). *Z. Electrochem.* **60**, 4.
Weber, N. and Kummer, J. T. (1967). *Proc. Annu. Power Sources Conf.* **21**, 37
Weininger, J. L. (1958). *J. Electrochem. Soc.* **105**, 439.
Weininger, J. L. (1959). *J. Electrochem. Soc.* **106**, 475.
Weininger, J. L. (1960). U.S. Patent, 2,933,546, April 19, 1960.
Weininger, J. L. (1964). *J. Electrochem. Soc.* **111**, 769.
Weininger, J. L. (1969). *J. Electrochem. Soc.* **116**, 1480.
Weissbart, J. and Ruka, R. (1962). *J. Electrochem. Soc.* **109**, 723.
Yamamoto, O. and Takahashi, T. (1966). *Denki Kagaku* **34**, 833.
Yamamoto, O. and Takahashi, T. (1968). *Denki Kagaku* **36**, 894.

E. ENERGETICS AND PHOTO-ELECTROCHEMICAL PHENOMENA

23. CELLS AND STORAGE BATTERIES

R. T. FOLEY

Chemistry Department, The American University,
Washington D.C., U.S.A.

The study of galvanic cells with solid electrolytes goes back to the turn of the century at which time Haber and Tolloczko (1904) and others made measurements of such precision that they were capable of being reproduced by other investigators decades later. The main use of the cells assembled by these early electrochemists appeared to be for the purpose of supporting thermodynamic arguments. However, it is known that these early electrochemists were very much aware of galvanic cells of all types as potential energy conversion devices. In the 1950's there was considerable effort by a number of industrial concerns to develop solid electrolyte batteries. There were a number of advantages to these cells which were recognized at that time and are still applicable as incentives for working in this field. Cells utilizing aqueous electrolytes have a definite temperature range limitation. At the lower temperature there is a point, above the freezing point, at which diffusion and migration processes become too slow to support realistic current flow. At higher temperatures, well below the boiling point of the electrolyte, the vapor pressure of the solvent or of certain cell materials becomes excessive. Solid electrolyte galvanic cells of the silver halide type are known to operate from $-20°C$ to $500°C$ and the oxide type of cell even higher. In the battery field the term "shelf-life" is used as a measure of the stability of the cell during storage without current drain. Usually this is limited in aqueous systems by the corrosion of the active anode material which, although inhibited, does occur to an appreciable extent over a period of several years. Solid electrolyte systems generally can be expected to be more stable because diffusion processes in solids are slower than those in liquids. In some systems the "shelf-life" has been extrapolated to 10 years and longer. A further problem of aqueous systems is cell leakage which has made necessary special sealing or fabrication techniques. This leakage would be minimized in solid electrolyte systems. Finally, it is noted that the present trend in electronic instrumentation is toward miniaturization. Conventional battery systems are severely limited in this regard because of the need for a minimum volume of electrolyte and for separators to prevent electrolytic action between the active components. It is possible to visualize miniature cells using only solid components as thin films in thicknesses in the micron range.

In the last decade there have been several developments that have led to the reconsideration of solid state cells as practical power sources. The first has been improved electrolyte compositions, mostly based on silver ion transport but bringing the desirable high temperature properties of AgI to room temperature. These improved electrolytes have specific conductances in the range of $0.2 \text{ ohm}^{-1} \text{ cm}^{-1}$, which, while not as high as conventional battery electrolytes, are several orders of magnitudes higher than the room temperature conductances of the silver halides. The second improvement has been in equipment and techniques for producing thin films by vacuum deposition. With a thin film it is possible to tolerate an electrolyte of higher specific resistivity because the electrode to electrode distance would be in the micron range.

This chapter deals with the employment of solid electrolyte galvanic cells as power sources. The very pertinent related problems such as ion transport, conductivity, and dielectric properties, which are covered in detail in other chapters, will only be touched on when they bear on the operation of the galvanic cell as a power producing system.

I. Historical Development

Haber and Tolloczko (1904), early in the development of electrochemistry, recognized that galvanic cells could be constructed with solid electrolytes.

They measured the electromotive forces of the cells

$$Pb/PbCl_2/AgCl/Ag \qquad \text{(I)}$$

and

$$Cu/CuCl/AgCl/Ag \qquad \text{(II)}$$

but their u-shaped cell configuration was such that no significant current could be drawn from the cell. Katayama (1908) confirmed Haber's results and extended the analogy between aqueous systems and solid electrolyte galvanic cells by showing that it was possible to construct *concentration* type cells, so-called "*chemical*" cells, and *Daniell* type cells. He reported e.m.f. measurements over a range of temperatures on lead amalgam *concentration* cells such as:

$$\text{amalgam I}/PbBr_2/\text{amalgam II}, \qquad \text{(III)}$$

chemical cells such as

$$Pb/PbBr_2/Br_2 \text{ and} \qquad \text{(IV)}$$

$$Ag/AgCl/Cl_2 \qquad \text{(V)}$$

and *Daniell type* cells such as

$$Pb/PbCl_2/AgBr/Ag \text{ and} \qquad \text{(VI)}$$

$$Pb/PbCl_2/AgCl/Ag \qquad \text{(I)}$$

Lorenz and Katayama (1908) discussed the thermodynamic treatment of these galvanic cells and showed that the electrical energy of these cells could be calculated from the Gibbs–Helmholtz relationship. The classical work of Tubandt and coworkers (Tubandt and Lorenz, 1912, 1914; Tubandt and Eggert, 1920) did much to clarify the mechanism of conduction in these cells and to furnish conductance data of value in the interpretation of galvanic cell measurements. They showed that Faraday's Law applied to solid electrolytes in the same manner that it applied to aqueous. Their measurements of the conductivity of AgCl, AgBr, AgI, TlCl, TlBr, and TlI over the temperature range of 100–700°C showed that solid electrolytes, given the proper crystal modification, possessed electrical conductivities of the same order of magnitude as aqueous electrolytes. This demonstrated that what appeared to be the main limitation in the utilization of these solid electrolytes in energy conversion devices, i.e. high electrical resistivity, was not a property inherent in all solid electrolytes. They also recognized that the current in solid electrolytes was carried not only by cations and anions but also by electrons. In some crystals, such as β–Ag_2S in the temperature range of 150–170°C, appreciable current was carried by electrons. In the operation of a galvanic cell electronic conduction cannot be tolerated in any significant degree because it will cause internal shorting of the cell.

Reinhold (1928) made significant measurements on power generation with thermogalvanic elements. Such cells as

$$Ag/AgI/Ag \qquad\qquad (VII)$$

$$Pb/PbCl_2/Pb \qquad\qquad (VIII)$$

$$Cl_2/AgCl/Cl_2 \qquad\qquad (IX)$$

wherein the electrical energy was derived from the anode being held at a different temperature than the cathode were investigated over the temperature range of 150–500°C. A cell such as IX, with gas electrodes, exhibited a higher voltage than the others, an observation confirmed in modern technology.

Reinhold (1928) also made measurements on the type of cell previously studied by Haber and Tolloczko (1904) and Katayama (1908). The e.m.f. measurements on the cell

$$Pb/PbCl_2/AgCl/Ag \qquad\qquad (I)$$

are compared as follows:

Temperature	Katayama	Haber	Reinhold
67°	0·480V	—	0·488V
151	0·471	0·480V	0·472
200	—	0·492	0·463
250	—	0·500	0·451

This comparison is presented to illustrate the good agreement among the early workers in this field. These results have been substantiated in more recent reports.

From these brief observations it is possible to conclude that the basic electrochemical theory for modern day work on solid electrolytes as energy conversion devices had been reported by the early 1930's. From the viewpoint of the theory of chemical cells and concentration cells, conduction mechanisms, equilibrium reactions and thermodynamics, solid electrolyte cells could be treated exactly like traditional aqueous cells. It was also evident that the total transference of current in a solid electrolyte was the sum of the cation, anion, and electron transference which meant that any new battery electrolyte would have to be evaluated not only in terms of the total specific conductivity but also with respect to the contribution from these three sources.

II. Solid Electrolytes Suitable for Galvanic Cells

The employment of a solid electrolyte in an energy conversion device implies certain specifications:

(a) the electrolyte conductance should be sufficiently high so that the internal resistance (ohmic drop) does not so lower the operating voltage that the energy density (on a mass or volume basis) be at an impractical low level. Thus, when electrolytes were observed with specific conductances in the range of $0.2\,ohm^{-1}\,cm^{-1}$, interest in this field was revived. Whereas *ionic* conductivity would be high, electronic conductivity should be low, probably less than $10^{-6}\,ohm^{-1}\,cm^{-1}$.

(b) the electrolyte should be stable over an appreciable temperature range in the sense of maintaining a low vapor pressure, exhibiting no phase transformations, or no reaction with other components in the galvanic cell. This temperature range would cover $-40°C$ to $70°C$ based on the possible range of operation or storage of many commercial and military batteries.

(c) the electrolyte should have a high ionic transference number with respect to species associated with the anode or cathode reaction. Thus AgI has been successfully used as an ionic conductor in the cell

$$Ag/AgI/I_2 \qquad\qquad (X)$$

because the transference number for Ag^+ in AgI is approximately unity.

van der Grinten (1968) has suggested that a logical basis for classifying and discussing solid electrolytes is in terms of their basis transport process. Thus, electrolytes and the galvanic cells resulting therefrom would be distinguished by those utilizing (a) Ag^+ transport, (b) Cl^- transport, (c) Na^+ transport, (d) Li^+ transport, (e) H^+ transport, and (f) other conducting species.

In a parenthetical manner a class of galvanic cells should be mentioned which are often referred to as "solid electrolytes". Grubb (1959) described cells of the type

$$Zn/sulphonated\ polystyrene\ resin/Ag \qquad\qquad (XI)$$

wherein the electrolyte was an ion exchange resin which allows the transport of Zn^{++} and Ag^+. The mechanism of conduction, however, is very likely identical to that in aqueous system, i.e. transport by means of hydrated species. A second "solid electrolyte" galvanic cell reported by Swindells (1969) was

$$\text{Zn/electrolyte complex/AgCl, Ag} \qquad \text{(XII)}$$

in which the electrolyte complex consisted of a paper separator prepared by immersing in a $ZnCl_2$, LiCl solution and then exposure to ammonia gas to produce a zinc chloride diammine complex. The useful life of this cell is related to the availability of a given partial pressure of water vapor and it is evident that the transport mechanism involves hydrated species.

A. Galvanic Cells Utilizing Solid Electrolytes Based on Ag^+ Transport

The earliest work on this type of galvanic cell was devoted mainly to utilizing AgI as an electrolyte. The reason becomes apparent when the properties of AgI are examined in the light of the three requirements mentioned above. The conductance of AgI (and other Ag halides) was thoroughly studied by Tubandt and Lorenz (1914). The low temperature, low conductance, beta form of AgI undergoes a transition to the highly conductive alpha form at about $145°$–$146°C$ (Tubandt, 1921; Jost et al., 1958; Lieser, 1954, 1955). This phase change is reversible with no decomposition occurring in the transition from one crystal modification to the other. The electronic conductivity of AgI is about 10^{-7} at an I_2 partial pressure of 3 mm Hg at $179°C$ (Jost et al., 1954). The current is carried almost exclusively by Ag^+, the cation transport number as reported by Tubandt (1921) is very close to unity. Thus a cell wherein Ag metal served as the anode appeared feasible.

The practical significance of the increased conductivity achieved by working above the transition point was demonstrated by Weininger (1960) in his measurements on batteries composed of a number of gas activated cells of the type

$$\text{Ag/AgI/I}_2 \text{ (Pt).} \qquad \text{(X)}$$

Temperature of Cell Operation	Open Circuit Voltage	Short Circuit Current
25°C	6V	1μA
200°C	6V	1000μA

Thus, it was established at an early date that AgI would be an excellent conductor for a galvanic cell intended to operate above $146°C$ but would be inadequate for low temperature operation. With this in mind a number of efforts were made to improve the low temperature conductivity of AgI by stabilizing the high temperature form of the crystal. These investigations might be separated into two categories dependent on the results. The earlier investigations effected modest improvements in conductivities or in the lowering of the transition temperature. On the other hand, several investigations reported since 1966 yielded results several orders of magnitude higher than the β-AgI measurement at room temperature. Among the early investigations, Weiss et al. (1958) measured the electrical conductivity and phase changes in the system β-AgI-CdI$_2$, β-AgI-PbI$_2$, β-AgI-Ag$_2$Se, β-AgI-KI, and α-AgI-PbI$_2$. The transition temperature (β to α form) was lowered from 144·6° to 100-101°C by the addition of 5 mol % PbI$_2$. The conductivity of the mixed crystal at 120° was about 18% that of the pure AgI. Lieb and DeRosa (1960) reported that additions of tellurium to silver halide reduced its resistance. The resistance of AgCl at 100°C was reduced from $7·7 \times 10^6$ ohms to $2·3 \times 10^3$ ohms by the addition of 5% Te. The addition of 5% Te to AgBr in the cell

$$Ag/AgBr/CuBr_2 \text{ (C)} \qquad\qquad \text{(XIII)}$$

decreased the resistance from 114 K to 18 K. Another patent, by Lehovec (1954b), claimed a galvanic cell with an improved electrolyte on the basis of increasing lattice defects and increasing conductance. An example of this cell was

$$Ag/AgCl (10\% PbCl_2)/Pb. \qquad\qquad \text{(XIV)}$$

Other electrolytes were formed with 10% CdCl$_2$ in AgCl, 15% CaCl$_2$ in Hg$_2$Cl$_2$, 10% SbCl$_3$ in PbCl$_2$, and 12% BiCl$_3$ in ZnCl$_2$. Lehovec and Smyth (1961, 1962) also cited improvements to be gained in the conductivity of silver halides by the addition of anions such as $Te^=$, $Se^=$, and $S^=$. Here the improvement in conductivity was about 10-fold but not of such a magnitude to make the cells applicable for high current applications.

Several cases are reported in the earlier literature (ante 1960) wherein attempts were made to overcome the deficiency of one electrolyte by using two electrolytes in a series arrangement. Lehovec and Broder (1954), for example, utilized AgI with its desirable high ionic conductivity in series with Ag$_2$S which exhibits appreciable electronic conductivity at 200°C. Their cell

$$Ag/AgI/Ag_2S/S \qquad\qquad \text{(XV)}$$

exhibited an open circuit voltage of 0·2V at 200°C which was close to the value calculated from the free energy for the reaction between Ag and S. This cell could be operated at 200°C with a short circuit current of 0·18 A/cm^2 and a shelf life of 100 minutes. The performance of this cell was compared to the cell

$$Ag/AgI/I_2 \text{ (C)} \qquad\qquad \text{(X)}$$

which operated at room temperature at current densities of several hundred microamperes and an e.m.f. of 0·7V. This again demonstrated the importance of the highly conducting α–AgI structure in galvanic cell work.

The Ag–I_2 cell (X) described in a patent disclosure by Broder (1954) was investigated in some depth by Smyth (1956) and coworkers. They recognized that the shelf-life of the battery was limited by the electronic current leakage in the cell which in turn was measured by the rate of tarnishing of Ag by I_2. The reaction between Ag and I_2 was studied in terms of the rate of consumption of the Ag anode and it was determined that the rate was controlled by the diffusion of cations and electrons through the reaction product film. The cell as written above (X) would not have sufficient storage life to be of practical importance in the battery field. To circumvent this difficulty Smyth (1956) used the configuration

$$Ag/AgCl/AgI/I_2 \qquad\qquad (XVI)$$

to overcome some of the electrochemical and mechanical problems. This system was only a partial solution to the problem because of its low e.m.f. (0·684V at 20°C) and the need to maintain a non-porous AgCl film over Ag to prevent reaction of the I_2 with the Ag. Weininger (1958b) also investigated the tarnishing reaction between bromine and iodine and silver in connection with a vapor activated cell utilizing AgBr and AgI electrolytes respectively. The rate determining step was established to be the diffusion of defect electrons which migrate in a direction opposite to that of silver ions. Water vapor was found to be important in the reaction mechanism, the reaction rate being reduced by a factor of two in the presence of water vapor. These seemingly unrelated studies were necessary to prescribe the conditions by which the proper electrolyte could be prepared and developing the knowledge that would make possible a prediction of storage life (without current drain).

During the 1960's, following what appeared to be the same scientific approach to the problem, a number of electrolytes were prepared that represent dramatic improvements in the solid electrolyte field. Reuter and Hardel (1961, 1965, 1966) reported on the preparation, properties, and phase behavior of Ag_3SBr and Ag_3SI. In their papers they described measurements of the electrical conductivity of α and β–Ag_3SI over a temperature range from about 25°C to 400°C (1961) and noted the high conductance of α–Ag_3SI. Their results agree approximately with those of Takahashi and Yamamoto (1964, 1965) who studied this system independently and were the first to utilize this electrolyte in a solid electrolyte galvanic cell. The Japanese workers (1966a) reported a conductivity of about $10^{-2}\,ohm^{-1}\,cm^{-1}$ for α–Ag_3SI at room temperature, several orders of magnitude higher than the room temperature form of AgI. Further, the conduction mechanism was similar to that of the high conductance form of AgI; the transference number for Ag^+ was measured as 0·977. There is no good agreement with respect to the value of the electronic conduction in this compound which would be critical in its employment as a battery electrolyte and in terms of its participation in solid state reactions. Takahashi and Yamamoto reported values of 10^{-8} (1966), 10^{-6} (1964), and 10^{-4} ohm^{-1} cm^{-1} (1965). From Reuter and Hardel's work (1966) it is possible to obtain an extrapolated value of between 10^{-4}–10^{-5} ohm^{-1} cm^{-1}. The value for electronic conduction is known to vary with the Ag/anion ratio so it is probable that all of these values were correct for the particular sample and was determined to a large extent by the method of preparation. It is concluded at this point that the value of 10^{-4} ohm^{-1} cm^{-1} is a reasonable value and it is too high considering requirements for long term stability and participation in tarnish reactions. Takahashi and Yamamoto (1966) investigated the galvanic cell

$$Ag/Ag_3SI/I_2 \text{ (C)}. \qquad \text{(XVII)}$$

At 20·0°C the experimental cell voltage was 0·675V in good agreement with 0·687V calculated from the free energy data and the iodine vapor pressure relationship. The internal resistance of a cell constructed with a thickness of 0·15 cm was 10 ohms and it was possible to maintain a current density of 100 µA/cm^2 at 25°C without appreciable lowering of this voltage. The authors concluded that the polarization observed at higher current densities was due to the anode reaction (the dissolution of Ag) and corrected this by amalgamation of the electrode. The amalgamated Ag electrode could sustain current densities up to 1000 µA/cm^2 at 25°C. The cell developed by these Japanese workers was capable of being discharged at current densities of 1–2 mA/cm^2 without appreciable polarization.

Owens et al. (1969) investigated the long term stability of the cell utilizing Ag_3SI as the electrolyte and I_2 as the cathode depolarizer. They found that the electrolyte degraded by reaction with the I_2 according to the reaction

$$I_2(g) + Ag_3SI(s) \rightarrow 3\,AgI\,(s) + S\,(s).$$

They experimentally found considerable reaction in times less than 100 hours at 65° and 70°C and concluded that the cell would transform to

$$Ag/Ag_3SI/AgI, S/I_2 \text{ (C)} \qquad \text{(XVIII)}$$

which is equivalent to interposing a high resistivity layer between the highly conducting electrolyte and the cathode. These results do not agree with those of Takahashi and Yamamoto (1966) who reported stability of the system for over one year. The difference might be due to the elevated temperature used by Owens *et al.*, but this is not an unrealistic temperature for the evaluation of the stability of battery systems. Thus, on the basis of this evaluation and on reported measurements on electronic conductance it must be concluded that this system has serious short comings.

The most significant recent improvement in the AgI electrolyte was reported independently by Bradley and Greene (1966, 1967a) and Owens and Argue (1967). An investigation of the AgI–RbI phase diagram revealed certain compounds such as $RbAg_4I_5$ with extremely high conductances as compared to other solid electrolytes. A plot of ionic conductivity versus mole percent of RbI in AgI shows a conductance maximum of $0\cdot20$ ohm^{-1} cm^{-1} at about 80–85 mol % of AgI, Argue and Owens (1969). Similar behavior is reported for the AgI–KI and AgI–NH_4I systems. These compounds transform from their high temperature, highly conductive, α-form to the β form at $-136°C$ (KAg_4I_5) and $-155°$ ($RbAg_4I_5$). The lower limits of stability, as reported by Bradley and Greene (1966, 1967), is 27°C for $RbAg_4I_5$, 38°C for KAg_4I_5, and 25°C for $NH_4Ag_4I_5$. This stability is established by the reaction

$$2\,MAg_4I_5 = M_2AgI_3 + 7\,AgI.$$

This decomposition reaction, which produces lower conducting species, is promoted by water vapor. The transport numbers for the ions were measured by Bradley and Greene (1966) and it was found that current was essentially carried by Ag^+ and further that there was no change in the composition of the anolyte and catholyte after electrolysis. Both Geller (1967) and Bradley and Greene (1967b) reported on the crystal structure of the compounds of MAg_4I where M may be K, Rb or NH_4. The high mobility of Ag^+ is due to the unusual crystal structure which was pictured to be analogous to Ag^+ in a liquid state inside of a rigid crystal lattice of I^- and M^+ ions. On the basis of the atomic size Cs^+ is too large and Na^+ is too small, so the series is limited to Rb^+, K^+, and NH^+_4. These considerations of atomic size versus the stability of the conducting crystal form get less straightforward in the light of new developments wherein substituted ammonium iodides are incorporated with AgI to form highly conducting systems.

The electrochemical performance of a cell of the type

$$Ag/RbAg_4I_5/RbI_3\ (C) \qquad\qquad (XIX)$$

was reported by Argue *et al.* (1968a). The open circuit voltage of 0·66V agreed with the theoretical value and the cell was operated to 90% of its theoretical capacity at an average current density of 0·41 mA/cm^2. The application of cells of this type will be discussed below but the data indicated at this point that solid electrolyte cells were capable of current densities of greater than the microampere range.

Rubidium salts are quite expensive and practical consideration directed investigators to the examination of other systems that would be inexpensive and still yield electrolytes with conductivities of the same order of magnitude as the silver rubidium iodide system, i.e. 10^{-1}–10^{-2} ohm^{-1} cm^{-1}. Recently several systems using substituted ammonium salts have been reported. Owens (1969b) showed a maximum in the conductivity—mole percent plot for the systems AgI–$(CH_3)_4$NI, AgI–$(C_2H_5)_4$NI and AgI–pyridinium iodide at about 80–90% mol fraction of AgI. Smyth *et al.* (1970) prepared cells with a 1:6 molar ratio of tetramethyl ammonium iodide and silver iodide which has an ionic conductivity of 0·04 ohm^{-1} cm^{-1} which is in agreement with data reported by Owens (1970). The electronic conductivity of this electrolyte is 10^{-11} ohm^{-1} cm^{-1} which would suggest long shelf-life (Smyth *et al.*, 1970). Cells constructed with this electrolyte with a Ag anode and an I_2 cathode yielded an open circuit cell voltage of 0·64V, had an internal resistance of 6–8 ohm, and discharged with current densities in the milliampere range. Owens *et al.* (1971) have extended the investigation of substituted ammonium salts and silver iodide and report a large number of compounds with conductivities as high as 10^{-2} ohm^{-1} cm^{-1}.

Mellors and Louzos (1971) have recently reported a new series of electrolytes based on the KCN–AgI and RbCN–AgI systems. Conductances at 25°C of 0·14 and 0·18 ohm^{-1} cm^{-1} were reported for the compounds KCN–4AgI and RbCN–4AgI respectively. X-ray studies of these compounds indicate that these are not mixtures but specific chemical compounds.

B. Solid Electrolytes with Cl⁻ Transport

There are a few systems, such as the salts of the bivalent metals $PbCl_2$ and $BaCl_2$, in which the anion transference number is high. In the temperature range of 200°–450°C the transference number for the Cl⁻ approaches unity (Sator, 1956). The transport data for these halides have been reported for elevated temperatures but there are very few room temperature data available. The total ionic conductivity of species like $PbCl_2$ is quite low at room temperature and extrapolation from the temperature dependence of the cation transport and anion transport leads to the inference that at room temperature current is carried by both the cation and anion to the same extent.

The cell

$$Pb/PbCl_2/AgCl/Ag \qquad (I)$$

involves Cl⁻ transport in its conduction mechanism. This cell has been studied extensively by Sator (1956) and is discussed below because until recently it was one of the few rechargeable cells.

C. Solid Electrolytes with Na$^+$ Transport

A number of alkali and alkaline earth compounds are good ionic conductors. Some of these compounds conduct by cation transport (Na$^+$, K$^+$, or Ca^{++}). This field has been recently reviewed by Kummer and Milberg (1969) who discuss the various mechanisms of ionic conduction and diffusion through solids. The classical case of a ionic conductor with Na$^+$ transport is "beta-alumina" characterized by Ridgway et al. (1936) as an alkaline aluminate rather than an allotropic modification of alumina. The approximate formula was given as Na$_2$O · 12Al$_2$O$_3$ or for the potassium version, K$_2$O · 12Al$_2$O$_3$. Kummer et al. write the empirical formula as Na$_2$O · 11Al$_2$O$_3$. The conduction is due to cation mobility and Yao and Kummer (1967) measured the self-diffusion coefficients of Na$^+$, Ag$^+$, K$^+$, Rb$^+$, and Li$^+$ in single crystals. Na$^+$ had the highest rate of the ions tested, about 1×10^{-5} cm^2/sec at 300°C and 40×10^{-7} cm^2/sec at 25°C. The proposed mechanism for cation diffusion which involves the movement of sodium ions along planes in a plate-like structure is discussed in detail by Kummer and Milbert (1969) in their review paper.

A secondary cell utilizing β-alumina was described by Weber and Kummer (1967). Whereas this cell used the solid sodium aluminate electrolyte, other components of the cell were liquids, the anode was liquid sodium and the cathode was liquid sulphur. The electrolyte used in the galvanic cell was a sintered product with an electrical resistance of 3·5 ohm–cm at 300°C and about 250 ohm–cm at 26°C. This required the operation of the cell at the elevated temperatures at which sodium and sulphur are in the molten stage. At 300°C the cell operated at about 2·0V and a current density of 170 mA/cm^2 of electrolyte surface and the voltage–current curves were reasonably linear indicating that the galvanic reaction was not limited by activation or concentration polarization. The ceramic electrolyte has been improved by Jones and Miles (1970) who report conductivities of 10^{-1} ohm^{-1} cm^{-1} at 300°C and the incorporation of these improved electrolytes into cells with power output exceeding 50W. A similar type of cell has been reported by Hever (1968). In one example the mobile species was Na$^+$ in an electrolyte consisting of an α-alumina–β-alumina eutectic and the electrodes were also ceramic, e.g. Na$_2$O · 5 (Fe$_{0.95}$ Ti$_{0.05}$ AlO$_3$). Cells of this type operating at 300°–500°C exhibited low power densities mainly due to the low operating voltage. Zeolite electrolytes were used by Freeman (1965) in a cell of the type

$$Zn/Na_2X/Cu(II)X \cdot Au. \qquad\qquad (XX)$$

This cell used a zinc anode, a sodium zeolite separator, a copper-exchanged zeolite catholyte, and a Au catholyte contact. Cells of this type operated at

0·10μA current and a cell voltage of 0·10V under load. Other compositions gave higher voltage—up to 0·80V on open circuit but the current under load was in the microampere range.

D. Solid Electrolytes with Li^+ Transport

The effort in Li^+ conducting solid electrolytes probably stemmed from recent research in the utilization of lithium as an anode in organic electrolyte cells. The high standard potential of Li, $-3·045V$, would make possible, at least on a theoretical basis, the development of a high energy cell utilizing the reaction with Li metal and a halogen. LiI is essentially a cation conductor and has a low specific conductance, $10^{-7} ohm^{-1} cm^{-1}$, at 25°C, Liang and Bro (1969) used thin films of LiI to overcome this deficiency in some measure. Their cell

$$Li(5 μ)/LiI(15 μ)/AgI(30 μ) \qquad \text{(XXI)}$$

produced an open circuit voltage of 2·1V in agreement with the free energy change calculated for the reaction

$$Li + AgI = LiI + Ag.$$

A typical polarization curve was linear from 2·09V on zero current drain to 0V at 120 μA/cm². The slope of the polarization curve gave a cell resistivity of $1·7 \times 10^7$ ohm–cm in agreement with the literature value for the LiI resistivity value. Schneider *et al.* (1970) described a cell of the type

$$Li/LiI/I_2 \qquad \text{(XXII)}$$

wherein the LiI is formed by exposing the metal to I_2 vapor. The cell reaction for this galvanic cell is the formation of LiI and the open circuit voltage is 2·80V. Discharge curves for this cell were linear over a temperature range of 60°C to $-40°C$. Practical currents for 1000h operation were 50 μA/cm² at 60°C and 10 μA/cm² at $-20°C$. In both of these applications the electrolyte was present as a thin film to overcome the high ohmic resistance of LiI. These cells will be discussed further under "Construction".

E. Solid Electrolytes with H^+ Transport

Electrolytes capable of proton conduction would be of great value and make possible low temperature galvanic cells based on the $H_2–O_2$ reaction. However, very little experimental work has been reported in strictly anhydrous systems. Proton conduction in ion exchange membranes has been described by Grubb and Niedrach (1960) in connection with hydrogen–oxygen fuel cell but the proton most likely moves in the hydrated form. The theoretical basis for proton conduction in solids is provided by the work of Eigen and de Maeyer (1958) in their study of conduction mechanisms in ice. These investigators found that proton mobility in ice was high, in fact, higher than that of other cations or anions in water. Specific conductances of water and ice (extrapolated values) at 0°C were reported by Eigen and de Maeyer as $1 \cdot 2 \times 10^{-8}$ and $0 \cdot 3 \times 10^{-8}$ ohm^{-1} cm^{-1} respectively. Proton concentration in ice is lower than that in water by a factor of two but the proton mobility is higher by one or two orders of magnitude. The ice crystal is described as a "protonic semi-conductor" because the observed ionic mo - bilities differ by only a factor of 50 from typical electronic mobilities in semi-conductors. It should be pointed out that the hydrogen bonded structure, as pictured by Eigen and de Maeyer, which is responsible for the exceptional proton mobility may be unique with water but, at least theoretically, it should be possible to design other proton conductors that would function at room temperature. This theoretical basis was used as the technical justification for solid electrolytes with proton conduction at room temperature (Compagnie Industrielle des Piles Electriques, 1965). The solid electrolyte cited in the French patent as an example was a mixture of urea or a urea derivative with an organic acid. Mixtures of urea and formic acid, thiourea and citric acid, or thiourea and toluene sulphonic acid were given as examples. A mixture of 100 g of urea and 8 g of acetic acid was claimed to have a conductivity of 5×10^{-3} ohm^{-1} cm^{-1} at ambient temperature. Thus, the conductivity of this mixture improved by a factor of 10^4 over the conductance of AgI and AgCl. Such electrolytes could be employed in galvanic cells or in fuel cells wherein a hydrogen electrode serves as the anode and the cathode reaction is one of oxygen reduction or depolarization with manganese dioxide. Attempts to reproduce the results of this patent have been unsuccessful (Thompson and Foley, 1970). The anhydrous condensation product obtained with urea and acetic acid had a specific conductance of 10^{-6} ohm^{-1} cm^{-1}. It would appear that until further work is done and published on this topic the concept of proton conduction at room temperature in anhydrous, solid electrolytes must remain in doubt.

F. Solid Electrolytes with Other Conducting Species

Several solid electrolyte galvanic cells have been described in which the conducting species is not apparent. Ruben (1955, 1957, 1958) reported cells capable of producing high potentials but were limited to low currents. These cells, such as

$$\text{Ni/tin sulphate/lead peroxide} \qquad \text{(XXIII)}$$

supplied 1·65V and "several microamperes". Other examples utilized bismuth sulphate and antimony sulphate which are noted as hydrolyzable salts. Tin sulphate was preferred because it was not hydroscopic. According to the patent claim cells with these electrolytes were capable of holding their potentials for a long time. Another version

$$Ni/tin\ sulphate/barium\ permanganate + carbon \qquad (XXIV)$$

included an improved cathode depolarizer and operated at 1·42V. These cells, as written, have no allowance for anode polarization, that is, there is no obvious anodic reaction for the high potential but "microampere" current seems reasonable. At any rate cells of this type would be of little value as energy conversion devices.

III. Anodic Reactions

Much of the solid electrolyte galvanic cell work has involved the utilization of a cell reaction such as

$$Ag + X \rightarrow AgX$$

wherein X was I^-, Cl^-, Br^-, or $S^=$. In such cases the anode must obviously be Ag metal, furnished as strip, foil, deposited layer, or pressed powder. Some effort has been expended toward improving the anodic characteristics of silver. Buchinski and coworkers (1960) observed that low alloying concentrations of Cu and Si improved the adherence of the AgI film which would serve as the electrolyte during the forming process. Takahashi and Yamamoto (1966b) found that anode polarization was a limiting process in the Ag_3SI cell and that this polarization could be reduced by amalgamation. The amalgamated Ag electrode sustained current densities as high as $1·0\ mA/cm^2$ at 25°C. Mrgudich (1960) also observed in his measurements of the conductivity of AgI that electrode polarization of the silver anode could be greatly reduced by amalgamation. In many cell designs the cell components are developed as thin films. The observation by Kostyshin et al. (1966), that certain compounds when vacuum deposited as thin films react with the substrate which functioned as an anode, is significant. Films of PbI_2 and CuCl when deposited as Ag, Pb, or Cu showed a greatly enhanced photographic sensitivity over that observed when the same material was deposited on a dielectric substrate.

Some of the more recent work in the solid electrolyte field has used the lithium anode to take advantage of its higher potential and the cell voltage corresponding to the free energy change of the reaction

$$Li + I_2 = LiI.$$

In the cell described by Schneider *et al.* (1970), Li foil was used and reacted with I_2 to form the electrolyte. Liang *et al.* (1969) vacuum deposited the Li anode on the LiI electrolyte, a typical thickness being 0.4×10^{-3} cm. One very serious problem in the employment of the Li electrode is its high reactivity. Assuming that proper experimental techniques can be developed to form the Li anode in the proper degree of purity, the highly reactive metal will, over a period of time, "scour" the assembled galvanic cell of such species as N_2, O_2, H_2O, and organic species with even very low partial pressures.

IV. Cathodic Reactions

The solid electrolyte cells investigated up until the last few years were composed in most part of a Ag anode and a silver salt electrolyte, so the cathode reaction involved the reduction of some species that would react with Ag yielding a free energy decrease proportional to a reasonable electromotive force. Cathode depolarizers have included sulphur (also Se and Te), the halogens, halogen-containing compounds, and certain oxidation-reduction couples. The earliest work in which experimenters attempted to develop practical power producing cells used a sulphur cathode. Lehovec (1954a) was issued a patent including a broad claim of a primary cell comprising......a solid crystalline electrolyte...... between two reactive electrodes which, in a sense, anticipated all subsequent patents in this field. His example was the cell

$$Ag/AgI/S \qquad\qquad (XXV)$$

with the "sulphur" electrode actually being a compressed mixture of 15% sulphur and 85% carbon and the AgI a thin film of 0.0025 mm produced by a tarnish reaction between Ag metal and I_2 vapor. This cell had an open circuit voltage of 0.2V and a short current of 0.22 amp/cm^2 per cell. A cell of this type becomes, upon the flow of current

$$Ag/AgI/Ag_2S/S \qquad\qquad (XV)$$

as Ag^+ migrate through the AgI electrolyte to react with sulphur. Solid electrolyte cells used Te and Se cathodes but the newer, better conducting electrolytes have been studied by Takahashi and Yamamoto (1970).

Cells of the type

$$Ag/RbAg_4I_5/Te \qquad\qquad (XXVI)$$

and

$$Ag/RbAg_4I_5/Se \qquad\qquad (XXVII)$$

yielded open circuit voltages of 0·217V and 0·265V respectively at 20°C. The advantage to these cells is that they may be discharged at current densities of several mA/cm^2 at room temperature and the polarization is small. However, as a generalization, the employment of S, Se, and Te as cathode materials in power producing cells has been discouraged by the low voltage of the galvanic cell and the production of a high resistance layer analogous to the Ag_2S in cell XVII.

In some of the earliest work in the solid electrolyte field cathode depolarizers similar to conventional depolarizers in commercial batteries were employed. Louzos (1959) described cells with V_2O_5, PbO_2, MnO_2, NiO_2, Ni_2O_3, AgO, and CeO_2 as cathode depolarizers. A typical configuration was

$$Ag/AgI/V_2O_5 \ (C) \qquad\qquad (XXVIII)$$

which yielded open circuit voltages in the range of 0·5–0·7V and sustained currents in the 1–10 microampere range at cell voltages of 0·45–0·55V over the temperature range of $-75°F$ and 200°F. Cells of this type have not been widely used because of their inability to sustain current densities in the milliampere range.

The halogens have been used extensively as cathode depolarizers with considerable ingenuity being employed to furnish a sufficient concentration for the electrochemical reaction yet separating the halogen from the reactive anode to prevent internal corrosion and short circuiting. Broder (1954) described a power producing cell somewhat similar to the Lehovec cell about but employing a cathode comprised of a pressed mixture of 80% carbon and 20% I_2. This cell exhibited an open circuit voltage of 0·67V and a short circuit current of 3 mA at room temperature.

Gas activated cells had been investigated by Reinhold (1928) but van der Grinten (1954, 1957) appears to have been the first to employ this technique in power producing units. In his cell he used a permeable solid structure with the object of developing a controlled partial pressure of the halogen. Such compounds as $CuCl_2$, $CuBr_2$, and $FeCl_3$ were used in addition to the elemental halogens. Typical cell performance reported by van der Grinten (1954) was as follows:

Cell Type	e.m.f. (volts)	Flask Current (mA)	Optimum Load Current (μA)
$Ag/AgBr/CuBr_2$ (C)	0·74	0·22	2
$Ag/AgBr/Br_2$ (C)	0·85	1·5	26

Obviously the limiting factor in these cells was the vapor pressure of the halogen developed by the inorganic halide. The vapor pressure over the

$CuBr_2(C)$ electrode was estimated as 10^{-4}–10^{-7} mm sufficient to yield a coulombic capacity of up to 3·6 amp sec. This capacity and current capability was improved by using Br_2 sorbed on carbon, the vapor pressure for which was in the 0·1–10 mm range.

The development of the gas activated cell was advanced by Weininger (1959b, 1960) who used the thermal decomposition of a polyiodide, e.g. CsI_4, or reaction with an oxidizing agent, e.g. ($KMnO_3 + CuI$) to produce the proper partial pressure of the halogen. Weininger's cell

$$Ag/AgI/I_2 \text{ (Ta)} \tag{X}$$

was designed to operate at elevated temperatures with current capability as indicated above. Smyth and coworkers (Smyth, 1956, 1959; Smyth and Shirn, 1959) extended the background in this area by a detailed investigation of cells of the type

$$Ag/AgCl/KICl_4. \tag{XXIX}$$

Potassium tetrachloroiodide is representative of polyhalide compounds that are capable of dissociating at slightly elevated temperatures to give iodine trichloride (ICl_3), I_2 and Cl_2. These compounds are relatively stable at room temperature. In the patent cited (Smyth and Shirn, 1959) there is an extensive discussion of the chemical and physical properties of these salts, in fact, the depth of technical discussion is quite unusual for the patent literature. The significant property of these polyhalides is that they should dissociate rapidly enough to furnish the active cathode compound (the halogen) to maintain sufficient current flow but slow enough so that the shelf life of the cell is not shortened by internal corrosion of the anode by the active cathode material. Smyth and Shirn (1959) describe a cell consisting of a silver anode shaped as a cup, a silver chloride electrolyte, 5–50 microns thick, prepared by tarnishing Ag with Cl_2 gas at 200–300°C for 30 minutes and a cathode composed of 30–50% $KICl_4$, 5–10% carbon black, and 40–60% grease, binder, e.g. Kel F. The e.m.f. of the cell corresponded to the theoretical value of 1·04V and was constant to ±5% over the temperature range of −40° to +75°C. This cell, as constructed, had an internal resistance of 10^5 ohm/cm² at 25°C and when a ten-cell battery was loaded with a resistance of 10^6 ohm a current drain of 1·0μA was sustained for 100 minutes with a flat discharge curve at 1·0V. In their rather extensive investigation lasting over several years Smyth (1956) and coworkers investigated four types of cells.

$$(1) \ Pb/PbCl_2/AgCl/Ag \tag{I}$$

$$(2) \ Ag/AgI/I_2 \tag{X}$$

(3) $Ag/AgCl/AgI/I_2$ (XVI)

(4) $Ag/AgCl/KICl_4$ (XXIX)

Of these only the fourth type appeared to be suitable for realistic power production.

The utilization of the high conductance electrolyte $RbAg_4I_5$ is discussed above. An early version of a cell with this electrolyte had the configuration (Argue *et al.*, 1968a)

$$Ag/RbAg_4I_5/RbI_3, C$$ (XIX)

In this cell the cathode mixture included RbI_3, carbon and $RbAg_4I_5$. The capacity of this mixture in the particular cell was nominally 30 mAh/gram. The use of RbI_3 has certain disadvantages including thermal stability, cost, and availability of reactive I_2 on a unit weight basis. Owens (1969a) proposed the use of organic ammonium polyiodides which includes a large number of possible materials. He cited tetramethylammonium heptaiodide $N(CH_3)_4I_7$ and tetraethyl ammonium triiodide, $N(C_2H_5)_4I_3$ as preferred compounds although in the patent 21 compounds with available I_2 ranging from 45–80 wt% were listed. Cells of the type

$$Ag/RbAg_4I_5/QI_n$$ (XXX)

wherein QI_n represented the organic ammonium iodide yielded open circuit potentials of about 0·66V. A cell utilizing a cathode mixture of tetrabutyl-ammonium iodide, and graphite was used by De Rossi, *et al.* (1969) in a reversible cell. Their cell was characterized by an open circuit voltage of 0·56V and a short circuit current 1·5 A/dm².

A significant recent contribution to the cathodic reaction field has been the employment of charge–transfer complexes as cathode depolarizers. Gutmann *et al.* (1967) described experiments with a cell of the general type

$$Mg/I_2 - complex/Pt, C, or Au.$$ (XXXI)

The I_2 complex consisted of approximately 50% by weight of I_2 complexed with perylene, phenothiazine, polyvinylpyridine, tetracyanoethylene (TCNE), tetracyanoquinodimethane (TCNQ), or other appropriate compounds. The important property of these charge–transfer complexes is their high electronic conductivity. The "electrolyte" in the conventional usage of the term is produced by the reaction of the active metal with the complex. A cell of the above type with perylene–iodine complex produced an open circuit voltage of 1·45–1·85V and a short circuit current of up to 16 mA. A typical polarization curve showed a flat plot at about 1·3V under a load of 10kΩ for 2·5 hours. Cells of this type were found to operate at room temperature but also were stable up to 130°C. The cell

$$Mg/Phenothiazine–I_2/Pt \qquad \text{(XXXII)}$$

was studied in some detail (Gutmann *et al.*, 1968) especially from the viewpoint of the effect of vapors of high permittivity liquids. It was observed generally that the presence of high permittivity liquids in their vapor state such as acetone, acetonitrile, or water, improved the performance of the cells, whereas complete removal of vapors or the introduction of benzene or CCl_4, had a detrimental effect on the short circuit current. The role of water (and other protonic compounds) has been a long standing unanswered question in the solid electrolyte field. van der Grinten (1954) observed that his vapor activated cells stored in a controlled humidity chamber (less than 1% relative humidity) dropped in voltage from 0·73V to 0·4–0·5V. The original voltage was regained when the cell was removed to the more humid laboratory atmosphere. Further, Smyth and Lehovec (1967) noted that the complete exclusion of all water vapor was detrimental and proposed adding a hydrated salt such as $MgCl_2 \cdot 6H_2O$ to their $KICl_4$- containing cathode. They explained their results in terms of the water being adsorbed on the cathode side and forming a layer which blocks the movement of electrons which would, in time, discharge the cell. Both Smyth and Lehovec (1967) and Weininger (1958) demonstrated that the rate of tarnishing of silver in bromine and iodine vapor is greatly reduced in the presence of water vapor. The tarnish reaction is understood to be the limiting factor in shelf life. Recently Friedel (1968) reported on experiments investigating the effect of water vapor on cells of the type

$$Mg/Perylene–I_2/Cu. \qquad \text{(XXXIII)}$$

His experiments definitely showed that water vapor was necessary for the operation of the cell. A well-desiccated battery was inoperable. Further, water vapor was a reactant, not a catalyst, and H_2 was evolved as the cell reacted. According to these findings the charge-transfer complex acts merely as an inert container for the molecular iodine restricting the vapor pressure of I_2 to the proper range. Whether the earlier observations made by van der Grinten, Smyth, and Weininger are related to the present work on charge-transfer complexes can not be said with certainty. It is quite clear, however, that in all these solid state cells, H_2O does play some role and this role has not been precisely elucidated as of this time.

In summary it appears that recent advances in cathode formulations have made for more stable galvanic cells. A number of reports would indicate that the combining or complexing of iodine with larger molecules restricts

he activity of the iodine and overcomes to a major extent the corrosion
eaction with the anode which has heretofore limited cell life.

V. Rechargeability and Reversibility

`he major effort in the solid electrolyte field has been in the development of
»rimary cells, very little work having been reported on secondary or recharge-
ıble cells. For a rechargeable cell, obviously, both electrode couples must be
eversible. The anodic reactions, e.g., the Ag–Ag$^+$ couple, are reversible,
»ut the cathode reactions, such as S–S$^=$ and I$_2$–I$^-$, do not exhibit this rever-
ibility.

Sator (1952, 1956) was the first to report experiments on a rechargeable
galvanic cell. Perrot and Sator (1952) described a method for measuring the
electrical resistance of ionic conductors at room temperature. They vacuum-
leposited thin films of PbCl$_2$ between silver layers and recorded voltage–
eurrent curves. Upon the passage of current they observed the formation of
AgCl at one electrode and Pb at the other, which suggested that it might be
»ossible to form a reversible galvanic cell. These investigators used vacuum
leposition techniques to prepare a number of cells of the type

$$Ag/Pb/PbCl_2/AgCl/Ag. \qquad (I)$$

These cells were charged at a constant current of 0·5 µA according to the
·eversible equilibrium

$$PbCl_2 + 2Ag \underset{\text{discharge}}{\overset{\text{charge}}{\rightleftarrows}} Pb + 2AgCl.$$

The e.m.f. of the cell was 0·523V after charging but stabilized at 0·460V
ıfter about one hour. Cells so constructed were carried through nine charge–
lischarge cycles and the reversible character of the cell was definitely demonstrated. Similar cells using bromides as well as chlorides were prepared by
Moulton et al. (1964) who also used vacuum deposition techniques. The
ϲhloride cell exhibited an open current voltage of 0·47–0·50V and the bromide
ϲell 0·33–0·35V which would be expected from the free energy change for
the galvanic cell reaction. These investigators were not successful in preparing
ϲhe iodide cell, but this has been accomplished in work reported during the
last few years and is described below. Further work was done on the PbCl$_2$–
AgCl cell by Smyth (1956) and coworkers who used different methods for the
preparation of the AgCl and PbCl$_2$ electrolytes. Their methods included
(1) evaporation, (2) tarnishing the metal electrode, and (3) anodic forming.
Of these, evaporation was the most successful, but it should be noted that

the thickness of the layers was much greater than those in the Sator cell
Typical cell dimensions were

$$Ag(0.150 \text{ mm})/AgCl(0.058 \text{ mm})/PbCl_2(0.050 \text{ mm})/Pb(0.001 \text{ mm}).$$

The evaporations were carried out from a graphite crucible which invest gators subsequently found to be a cause of contamination in this vacuum deposition process. In several cases the theoretical voltage for the cell, 0·46V was observed, but in a number of cases the voltage was lower. During the operation of the cell, Smyth and coworkers observed the deposition of free metallic silver at the silver chloride–lead chloride interface. This, of course would be a source of electronic conduction and would lead to the electrical shorting of the cell. For this reason the cell (I) was not investigated further by these investigators. As it now stands, the results reported by the three groups are in disagreement. The only obvious explanation is the thickness of the relative layers used by the investigators. Dendrite formation, however would be expected in the thinner layers rather than in the thicker, so the discrepancy remains.

Weininger (1958a) investigated the reversibility of the iodine activated cell

$$Ag/AgI/I_2(Pt). \qquad (X$$

While this cell cannot be considered completely reversible, it was possible to recover some cell capacity. The amount of capacity recovered depended on the time interval between discharge and recharge and the temperature. For example, a cell charged immediately after discharge yielded only 73% of the capacity, whereas cells stored for 90 minutes at $-210°$, $0°$, and $25°C$ yielded 73, 63, and 15% of capacity. The lack of rechargeability is definitely tied to the gas electrode, but attempts to correlate loss in capacity with loss of iodine by diffusion using literature values for the diffusion coefficients of I_2 and Br_2 in the appropriate halides were unsuccessful.

Mrgudich et al. (1965) described a thin film rechargeable solid electrolyte battery of the composition

$$Ag/AgI/Pt. \qquad (XXXIV$$

In some cells flash currents as high as $100 \, \mu A/cm^2$ and open circuit voltages of 0·370V were observed. These cells were subjected to a large number of charge–discharge cycles and it would appear from the reported experimental data that reversibility was demonstrated. It is difficult to accept the Author's explanation of the mechanism in terms of a silver concentration cell consid ering the low rate of diffusion of Ag in Pt. In this cell there is no evidence for cathode depolarization and yet higher currents were reported than for cells with iodine depolarization. Perhaps a more reasonable explanation for the performance of the Ag–Pt cell would be in terms of an air–depolarization mechanism as described by Hacskaylo and Foley (1966).

A further study of the reversibility of solid electrolyte cells using silver halides was made by Andes and Manikowski (1963). Their cells consisted of Pt wire and Ag wire electrodes in halide electrolytes, one of the preferred types being 80 mol % AgI and 20 mol % AgCl. The cells were charged with single pulses varying from a millisecond to more than a second and discharging was done with pulses of the same duration but opposite polarity. In a typical experiment, a charge voltage of 45V for a time of 10 milliseconds resulted in a current efficiency of 85% and an energy efficiency of 0·86%. The ability to charge with pulses of such short duration argues against any concentration cell mechanism based on intermetallic diffusion processes which are known to be quite slow at room temperature.

Recently, De Rossi et al. (1969) discussed the rechargeability of the cell

$$Ag(Hg)/RbAg_4I_5/I_2, TBAI, C \qquad (XXXV)$$

discussed above. A charge–discharge cycle at 25°C and with a low constant current of 0.2 mA yielded a high coulombic efficiency. Charging at higher current densities resulted in failure after the fifth cycle due to Ag dendrite formation. The introduction of some of the newer cathode materials such as the organic iodine compounds indicate that with the I_2 so stabilized, the employment of the reversible I^-–I_2 couple may be feasible. Previous efforts to use this couple were undoubtedly unsuccessful because of the high vapor pressure of the halogen, but if this were overcome, it would be possible to design highly reversible solid electrolyte cells.

VI. Construction of Solid Electrolyte Galvanic Cells

A number of investigators have described the fabrication of solid electrolyte cells in sufficient detail so that operating cells may be constructed in the laboratory. Many of the ingenious methods have stemmed from the need to use thin layers of high resistivity electrolytes and the need to prevent internal shorting when the conducting layers were in close proximity. van der Grinten and Mohler (1960) describe a miniaturized cell of the following dimensions:

$$Ag(0·019 \text{ mm})/AgCl(0·025 \text{ mm})/Ca(OCl)_2 + C(1·0 \text{ mm}). \qquad (XXXVI)$$

This cell produced an open current voltage of 0·9–1·0V and a short circuit current of 0·03 mA/in^2 at room temperature. To overcome the problem of internal electrical shorting an insulating annular washer was used to hold the cathode material in a fixed location. The anode which was usually Ag or Pb was also faced on the side exposed to the electrolyte with an impervious electrically conducting material.

The Louzos cell

$$Ag/AgI/V_2O_5 \; (C) \hspace{4cm} (XXVIII)$$

was described above in connection with the use of oxide cathode materials. Evans (1959) described the construction of this cell. A layer of AgI, 15 microns thick, was produced on a Ag strip by an elevated temperature tarnishing reaction (25 mm I_2 at 230°C). Annealing of this layer lowered the electronic conductivity. The cathode current collector was constructed from stainless steel. The cathode mix, applied to the AgI layer as a slurry or suspension, consisted of 5 parts by weight of AgI, 10 parts by weight of V_2O_5 and 2 parts by weight of micronized graphite. The cell thickness was about 0·625 mm. Cells so constructed gave stable voltages of 0·455V and flash currents of 50 $\mu A/cm^2$. Richter et al. (1961) developed the process for automatic manufacturing.

The gas activated cell with a solid AgI electrolyte was constructed by Weininger (1958a, 1960) to operate at elevated temperatures. In one configuration (1961) the electrolyte of 2–5 micron thickness was developed by chemically tarnishing Ag in I_2 vapor at 150°C. A bimetallic corrugated sheet was used to provide access space for gas diffusion to the active electrodes. When this cell was activated by I_2 vapor at 0·306 mm at 25°C a voltage of 0·67V and a short circuit current of 200 μA was produced.

Sator (1952) was the first to use vacuum deposition techniques and to examine other ways to produce thin films. For example, one method involved starting with a Ag electrode and forming a thin film of AgCl by chemical attack with Cl_2 gas. Upon this substrate a thin layer of $PbCl_2$ was vacuum deposited, then a layer of Pb, and finally a layer of Ag to complete the cell. All of these layers were in the range of 10^{-5}–10^{-4} cm.

Vouros and Masters (1969) discuss at considerable length methods of vacuum depositing the components of a solid electrolyte battery and some of the properties of the films. Films of both AgBr and AgI were vapor deposited at a pressure of 5×10^{-5} mmHg using a tungsten boat operating at 632°C and 560°C respectively for the two halides. These films were 5 to 12 μ thick. Ag electrodes were also deposited by evaporation from a tungsten boat and the platinum films were obtained by sputtering. The electrodes were prepared in thicknesses between 700 and 1000 Å. It was observed that the AgI that was deposited was in the form of grains with voids between individual grains. It was found that if a substrate film of AgI was formed on Ag by chemical reaction then a uniform layer could be vacuum deposited. The galvanic cell constructed in these experiments was of the type

$$Ag/AgI/Pt \qquad \text{(XXXIV)}$$

which unfortunately did not represent a good case for power production. However, these experiments did demonstrate the feasibility of constructing cells of controlled thickness by vacuum deposition techniques.

The fabrication by thin film techniques of the Li/LiI/AgI cell was discussed in detail by Liang *et al.* (1969). In their procedure the AgI electrolyte was formed by spraying on a Ag foil then melted in a three stage furnace. The LiI electrolyte was vacuum deposited by a process requiring close control to prevent a porous layer. The thickness of this film was required to be greater than 5 μ. The Li anode was also vacuum deposited. Typical thicknesses of the cell components were as follows: silver current collector, 25 μ; silver iodide cathode, 30 μ; lithium iodide electrolyte, 15 μ; and lithium anode, 4 μ.

Kennedy (1970) has made a detailed study of conditions for film deposition and properties of thin film electrolytes such as AgBr in the preparation of cells for coulometry.

Certainly the technology is now available to construct cells with thin films wherein the components are simple compounds, e.g. $PbCl_2$, AgI. Some of the better conducting compounds, such as $RbAg_4I_5$, decompose at elevated temperatures, and might require special techniques to prepare thin films. These techniques have not been reported on as yet.

In the preparation of practical cells one important problem has been the maintenance of low contact resistance. Thus Argue and coworkers (1968a) replaced the traditional silver foil anode with a mixture of powdered carbon, silver, and $RbAg_4I_5$. The powdered anode, electrolyte, and cathodes were pressed together, titanium foil served as an electronic contact on the cathode, and Cu, Ag, or Ti as a negative current collector. The completed cell was encapsulated in plastic.

VII. Application of Solid Electrolyte Cells

Solid state galvanic cells have been extensively and very productively used in developing physico-chemical concepts. As was mentioned above, it was established in the 1920's that solid electrolyte galvanic cells adhered to all the physico-chemical laws developed for aqueous systems, e.g. Nernst equation, Hittorf migration relationship, Gibbs–Helmholtz equation, etc. For this reason, galvanic cells utilizing solid electrolytes have proved valuable in gathering information that ordinarily might be difficult to obtain. A recent comprehensive review by Raleigh (1967) covers the utilization of solid electrolyte galvanic cells for the study of solid state reactions, electronic and ionic conductivity of compounds, and for the determinations of the free energies of compound formation and related thermodynamic properties.

The use of solid electrolyte cells of the type

$$Ag/AgBr/Au \qquad\qquad (XXXVII)$$

for coulometry has been described by Kennedy and coworkers (Kennedy and Chen, 1968, 1970; Kennedy et al., 1971) in several papers and is discussed in length in Chapter 22 of this work.

Solid electrolyte cells utilizing the high conductance, high temperature form of AgI have been employed in thermogalvanic cells (Weininger, 1964). The silver thermocell, i.e. based on the reversible Ag–Ag$^+$ couple,

$$T_1(Ag)/AgI/Ag(T_2) \qquad\qquad (VII)$$

wherein T_1 and T_2 represent two different temperatures preferably above 146°C was investigated and proved feasible for solar energy conversion (Foley, 1960). One of the shortcomings of this type of cell was the dendritic formation of Ag upon deposition on the cold electrode. This was remedied in the iodine thermocell

$$(T_1)(C)I_2/AgI/I_2(C)T_2 \qquad\qquad (XXXVIII)$$

which employed the same highly conducting AgI, but gas electrodes instead of Ag (Weininger, 1959a). In the configuration described in the patent, the electrodes were 10 cm^2 in area, the electrolyte thickness was 1·0 mm, the iodine vapor pressure was one atmosphere, and the electrodes were held at 350°C and 550°C. Under such conditions an open circuit voltage of 0·112V and a short circuit current of 27A was obtained. Weininger (1964) reports an e.m.f. of 0·347V generated in an iodine cell wherein the electrodes are maintained at a temperature difference of 280°C. The internal cell resistance of 91 ohms was measured at a current greater than one mA.

Mrgudich (1967) proposed the use of the symmetrical cell

$$Ag/AgI/Ag \qquad\qquad (VII)$$

as a pressure and temperature sensor. He demonstrated that putting one electrode in tension (and the other in compression) produced an electrical signal proportional to the applied torque, the ratio being about 0·1 mV/gram-centimeter. The electrode in tension became positive and the other electrode negative. He also proposed the application of the thermogalvanic effect in such a cell to temperature sensing. The heat generated by a person's fingertip approaching the silver electrode was measurable and varied positively and negatively for the individual. The temperature coefficient measured with his cells was about $0.8\,mV/°C$ (temperature difference between anode and cathode).

The application of the cell

$$Ag/RbAg_4I_5/RbI_3(C) \qquad\qquad (XIX)$$

to various applications requiring low, medium, and high current drain is reported by Argue *et al.* (1968b). In the low current drain battery five cells formed a $3.3V$ battery. The open circuit voltage was $3.30V$ and under a load of $100k\Omega$ supplied current for 144 hours to a $2.3V$ cutoff at $+25°C$. The cell operated over the temperature range of $-55°C$ to $+74°C$. It was demonstrated that the cells were capable of short circuit currents of $0.9A/cm^2$. At a current density of $50\,mA/cm^2$ it was possible to draw about 90% of the theoretical capacity. Improved cells of this type have been proposed for a number of electrochemical applications (Atomics International, 1970) such as in circuitry requiring capacitance, integrator, load controlling, etc.

VIII. Conclusions

Within the last decade there has been a tremendous growth in both the science and technology dealing with solid electrolyte galvanic cells. The major road block to progress was removed with the discovery of highly conducting electrolytes such as silver sulphur iodide, rubidium silver iodide, and potassium cyanide–silver iodide compounds. Some of these structures are stable over an extended temperature range. Vacuum deposition techniques have been developed as have other fabrication techniques to yield cells with low contact resistance and low internal resistance. The recent disclosure of new cathode materials, compounds incorporating iodine chemically or physically in organic molecules, have yielded cells of greater stability and higher current capability. Until recently the only reversible cell was the Sator type $PbCl_2$–$AgCl$ cell. However, it appears that the organic iodine compounds may furnish the equivalent of a stabilized iodine–iodide redox couple which will find application in a secondary cell.

References

Andes, R. V. and Manikowski, D. M. (1963). *J. Electrochem. Soc.* **110**, 66C.

Argue, G. R. and Owens, B. B. (1969). U.S. Patent 3, 443, 997, May 13, 1969.

Argue, G. R., Owens, B. B. and Groce, I. J. (1968a). "Solid State Batteries", 22nd Annual Power Sources Conference, 14–16 May 1968, Atlantic City, N. J.

Argue, G. R., Groce, I. J. and Owens, B. B. (1968b). Sixth International Power Sources Symposium, Brighton, England, September 24–26, 1968.

Atomics International (1967). "Solid State Electrochemical Batteries, Timers"; see also Gould Ionics (1970), Publication 70–SB–006A.

Bradley, J. N. and Greene, P. D. (1966). *Trans. Faraday Soc.* **62**, 2069.

Bradley, J. N. and Greene, P. D. (1967a). *Trans. Faraday Soc.* **63**, 424.

Bradley, J. N. and Greene, P. D. (1967b). *Trans. Faraday Soc.* **63**, 2516.

Broder, J. D. (1954). U.S. Patent 2,690,465, Sept. 28, 1954.

Buchinski, J. J., Foeching, R. M., Gluyas, R. E. and Powers, R. A. (1960). U.S. Patent 2,932,569, April 12, 1960.

Compagnie Industrielle des Piles Electriques (1965). French Patent 1,395,560, March, 8 1965.

De Rossi, M., Pistoia, G. and Scrosati, B. (1969). *J. Electrochem. Soc.* **116**, 1642–1646.

Eigen, M. and de Maeyer, L. (1958). *Proc. Roy. Soc. (London)*. **A247**, 505.

⟶ Evans, G. E. (1959). U.S. Patent 2,894,052, July 7, 1959.

Foley, R. T. (1960). "Solar Regenerated Fuel Cell", Contract DA–30–0690–ORD–2782, Final Report, November 22, 1960.

Freeman, D. C. (1965). U.S. Patent 3, 186,875, June 1, 1965.

Friedel, R. A. (1968). *J. Electrochem. Soc.* **115**, 614–615.

Geller, S. (1967). *Science* **157**, 310.

Grubb, W. T. (1959). *J. Electrochem. Soc.* **106**, 275.

Grubb, W. T. and Niedrach, L. W. (1960). *J. Electrochem. Soc.* **107**, 131.

Gutmann, F., Hermann, A. M. and Rembaum, A. (1967). *J. Electrochem. Soc.* **114**, 323–329.

Gutmann, F., Hermann, A. M. and Rembaum, A. (1968). *J. Electrochem. Soc.* **115**, 359–362.

Haber, F. and Tolloczko, St. (1904). *Z. anorg. Chem.* **41**, 407.

Hacskaylo, M. and Foley, R. T. (1966). *J. Electrochem. Soc.* **113**, 1231.

Hever, K. O. (1968). *J. Electrochem. Soc.* **115**, 830–836.

Jones, I. W. and Miles, L. J. (1970). Seventh International Power Sources Symposium, Brighton, England, Sept. 15–17, 1970.

Jost, W. and Weiss, K. (1954). *Z. physik. Chem. (N. F.)* (Frankfurt) **2**, 112.

Jost, W., Krug, J. and Sieg, L. (1952). Proc. Intern. Symposium on Reactivity of Solids. Gothenburg, pp. 81–2.

Jost, W., Oel, H. J. and Schniedermann, G. (1958). *Z. physik. Chem.* (Frankfurt) **17**, 175.

Katayama, M. (1908). *Z. physik. Chem.* **61**, 566.

Kennedy, J. H. and Chen, F. (1968). *J. Electrochem. Soc.* **115**, 918–924.

Kennedy, J. H. and Chen, F. (1970). *J. Electrochem. Soc.* **117**, 244C.

Kennedy, J. H., Chen, F. and Willis, J. (1970). *J. Electrochem. Soc.* **117**, 263–267.

Kostyshin, M. T. and Mikhailovskaya, E. V. and Romanenko, P. F. (1966). *Soviet Physics—Solid State* **8**, 451.

Kummer, J. and Milberg, M. E. (1969). C and EN, May 12, 90–99.

Lehovec, K. (1954a). U.S. Patent 2,689,876, Sept. 21, 1954.

Lehovec, K. (1954b). U.S. Patent 2.696,513, Dec. 7, 1954.

Lehovec, K. (1961). U.S. Patent 3,007,992, Nov. 7, 1961.

Lehovec, K. and Broder, J. (1954). *J. Electrochem. Soc.* **101**, 208.

Lehovec, K. and Smyth, D. M. (1962). U.S. Patent 3,036,144, May 22, 1962.

Liang, C. C. and Bro, P. (1969). *J. Electrochem. Soc.* **116**, 1322–1323.

Liang, C. C., Epstein, J. and Boyle, G. H. (1969). *J. Electrochem. Soc.* **116**, 1452–1454.

Lieb, H. C. and De Rosa, J. A. (1960). U.S. Patent 2,930,830, Mar. 29, 1960.

Lieser, K. H. (1954). *Z. physik. Chem. (N.F.)* **2**, 238.

Lieser, K. H. (1955). *Z. physik. Chem. (N.F.)* **5**, 125.

Lieser, K. H. (1956). *Z. physik. Chem. (N.F.)* **9**, 216.

Lorenz, R. and Katayama, M. (1908). *Z. physik. Chem.* **62**, 119.

Louzos, D. V. (1959). U.S. Patent 2,894,053, July 7, 1959.

Mellors, G. W. and Louzos, D. V. (1971). *J. Electrochem. Soc.* **118**, 846.

Moulton, C. W., Hacskaylo, M. and Feldman, C. (1964). *J. Electrochem. Soc.* **111**, 60C.

Mrgudich, J. N. (1960). *J. Electrochem. Soc.* **107**, 475.

Mrgudich, J. N. (1967). Proceedings of the 21st Annual Power Sources Conference, May 18, 1967, pp. 117–120.

Mrgudich, J. N., Schwartz, A., Bramhall, P. J. and Schwartz, G. M. (1965). Proceedings of the 19th Annual Power Sources Conference, pp. 86–88, May 18–20, 1965, PSC Publications Committee, Red Bank, N.J.

Owens, B. B. (1969a). U.S. Patent 3,476,605, Nov. 4, 1969.

Owens, B. B. (1969b). U.S. Patent 3,476,606, Nov. 4, 1969.

Owens, B. B. (1970). *J. Electrochem. Soc.* **117**, 1536.

Owens, B. B. and Argue, G. R. (1967). *Science.* **157**, 308.

Owens, B. B., Argue, G. R., Groce, I. J. and Hermo, L. D. (1969). *J. Electrochem. Soc.* **116**, 312–313.

Owens, B. B., Christie, J. H. and Tiedeman, G. T. (1971). *J. Electrochem. Soc.* **118**, 1144.

Perrot, M. and Sator, A. (1952). *Compt. rend.* **234**, 1883.

Raleigh, D. O. (1967). "Solid State Electrochemistry". *In* Progress in Solid State Chemistry. Vol. 3, pp. 83–134. H. Reiss, (ed.), Pergamon Press, New York.

Reinhold, H. (1928). *Z. anorg. Allgem. Chem.* **171**, 181.

Reuter, B. and Hardel, K. (1961). *Naturwissenschaften,* **48**, 161.

Reuter, B. and Hardel, K. (1965). *Z. anorg. Allgem. Chem.* **340**, 158–67.

Reuter, B. and Hardel, K. (1966). *Ber. Bunsenges. Physik. Chem.* **70**, 82.

Richter, E. W., Shellek, D., McMillan, H. E., and Evans, G. E. (1961). U.S. Patent 3,004,093, Oct. 10, 1961.

Ridgway, R. R., Klein, A. A. and O'Leary, W. J. (1936). *Trans. Electrochemical Soc.* **70**, 71.

Ruben, S. (1955). U.S. Patent 2,707,199, April 26, 1955.

Ruben, S. (1957). U.S. Patent 2,816,151, Dec. 10, 1957.

Ruben, S. (1958). U.S. Patent 2,852,591, Sept. 16, 1958.

Sator, A. (1952). *Compt. rend.* **234**, 2283.

Sator, A. (1956). *Publ. Sci. Univ. Alger, Ser. B. Sci. Phys.* **2**, 115.

Schneider, A. A., Moser, J. R., Webb, T. H. E. and Desmond, J. E. (1970). Proceedings of the 24th Annual Power Sources Conference, May, 1970, PSC Publications, Red Bank, N. J.

Smyth, D. M. (1956). "Final Report on Solid Electrolyte Battery Systems" Contract No. DA–36–039–SC–63151, Period: 1 June 1954 to 30 November, 1956.

Smyth, D. M. (1959). *J. Electrochem. Soc.* **106**, 635.

Smyth, D. M. and Lehovec, K. (1967). U.S. Patent 3,346,423, Oct. 10, 1967.

Smyth, D. M. and Shirn, G. A. (1959). U.S. Patent 2,905,740, Sept. 22, 1959.

Smyth, D. M., Tompkins, C. H. and Ross, S. D. (1970). Procedings of the 24th Annual Power Sources Conference, May, 1970, PSC Publications, Red Bank, N.J.

Swindells, F. W. (1969). Final Report on Contract No. NAS5–11548.

Takahashi, T. and Yamamoto, O. (1964). *Denki Kagaku.* **32**. 610.

Takahashi, T. and Yamamoto, O. (1965). *Denki Kagaku.* **33**. 346.

Takahashi, T. and Yamamoto, O. (1966a). *Electrochimica Acta.* **11**, 779.

Takahashi, T. and Yamamoto, O. (1966b). *Electrochimica Acta.* **11**, 911.

Takahashi, T. and Yamamoto, O. (1970). *J. Electrochem. Soc.* **117**, 1-5.

Thompson, C. T. and Foley, R. T. (1970). Unpublished data.

Tubandt, C. (1920). *Z. Elektrochem.* **26**, 358.

Tubandt, C. (1921). *Z. anorg. Allgem. Chem.* **115**, 105.

Tubandt, C. and Eggert, S. (1920). *Z. anorg. Allgem. Chem.* **110**, 196.

Tubandt, C. and Lorenz, E. (1912). Nernst's Festschrift (W. Knapp, Halle) pp. 446-58.

Tubandt, C. and Lorenz, E. (1914). *Z. physik. Chem.* **87**, 513.

van der Grinten, W. J. (1954). Final Report on Contract No. DA-36-039-sc-42716, January 1, 1954.

van der Grinten, W. J. (1957). U.S. Patent 2,793,244, May 21, 1957.

van der Grinten, W. J. (1968). Private communication.

van der Grinten, W. J. and Mohler, D. (1960). U.S. Patent 2,928,890, March 15, 1960.

Vouros, P. and Masters, J. I. (1969). *J. Electrochem. Soc.* **116**, 880-884.

Weber, N. and Kummer, J. T. (1967). "Sodium-Sulphur Secondary Battery". Proceedings of the 21st Annual Power Sources Conference, 16-18 May, 1967, PSC Publications, Red Bank, N.J.

Weininger, J. L. (1958a). *J. Electrochem. Soc.* **105**, 439.

Weininger, J. L. (1958b). *J. Electrochem. Soc.* **105**, 577.

Weininger, J. L. (1959a). U.S. Patent 2,890,259, June 9, 1959.

Weininger, J. L. (1959b). *J. Electrochem. Soc.* **106**, 475.

Weininger, J. L. (1960). U.S. Patent 2,933,546, April 19, 1960.

Weininger, J. L. (1964). *J. Electrochem. Soc.* **111**, 769.

Weininger, J. L. and Liebhafsky, H. A. (1961). U.S. Patent 2,987,568, June 6, 1961.

Weiss, K., Jost, W. and Oel, H. J. (1958). *Z. physik. Chem.* (Frankfurt) **15**, 429.

Yamamoto, O. and Takahashi, T. (1966). *Denki Kagaku.* **34**, 833.

Yao, Y. Y. and Kummer, J. T. (1967). *J. inorg. nucl. Chem.* **29**, 2453.

24. SOLID ELECTROLYTE FUEL CELLS (THEORETICS AND EXPERIMENTS)

Takehiko Takahashi

Department of Applied Chemistry, Faculty of Engineering, Nagoya University, Nagoya, Japan

I. Introduction

Fuel cells operative at above 300°C are called high temperature fuel cells. Chemical cells, including fuel cells, can convert chemical energies to electrical energies at constant temperature and they need not be operated at high temperature as a general rule. But, it is only an expensive fuel like pure hydrogen which can make the low temperature fuel cell operate with satisfactorily high current densities even when an active electrode catalyst like platinum is used.

Generally speaking, chemical reaction rate increases as temperature rises. It is also true in the case of electrode reaction, and a high temperature fuel cell can be operated at a high current density even when fuels which do not give a high current density in low temperature fuel cells are used.

Of the materials for high temperature fuel cells, the selection of the electrolyte is one of the most important problems. The electrolyte used for the high temperature fuel cell so far is the molten or solid electrolyte.

In this chapter, solid electrolytes for fuel cells and high temperature fuel cells are described.

II. Solid Electrolytes for Fuel Cells

A. General Considerations

As is well known, the electrolyte for fuel cells must always be a conductor, the charge carrier in which is the ion of the fuel or the oxidant. At high temperature, however, conductors have not been found in which the charge carrier is an ion of the fuel such as hydrogen ion. Accordingly, oxidant ion conductors—usually oxide ion conductors—have been utilized as the solid electrolytes for high temperature fuel cells.

The cathodic reaction of this sort of high temperature solid electrolyte fuel cell is,

$$cO_2 + 4e^- \rightarrow 2eO^{2-} \tag{1}$$

where the prefixes c and e represent the states at the cathode and in the electrolyte respectively. At the anode, though, several reaction mechanisms may be considered according to the kinds of fuel, the reverse reaction to eqn (1) can be regarded thermodynamically as the first electromotive reaction, that is,

$$2eO^{2-} \rightarrow aO_2 + 4e^- \tag{2}$$

where the prefix a represents the state at the anode. Fuels oxidized at the anode by aO_2 reduce the activity of aO_2 and thus promote the anodic reaction. Consequently, the voltage determining reaction of the cell is represented by eqns (1) and (2) as

$$cO_2 \rightarrow aO_2. \tag{3}$$

That is, the high temperature fuel cell is considered to be the same as the oxygen electrode concentration cell, and the electromotive force E is given by

$$E = (RT/4F)\ln(cP_{O_2}/aP_{O_2}) \tag{4}$$

where P_{O_2} is the partial pressure of oxygen at the electrodes. Thus, E is large when $cP_{O_2} \gg aP_{O_2}$ which can be realized by feeding the appropriate fuel to the anode in fuel cells.

For example, when carbon monoxide is fed to the anode as the fuel, the reaction

$$\tfrac{1}{2}aO_2 + CO \rightarrow CO_2 \tag{5}$$

will take place to reduce aP_{O_2}. This aP_{O_2} is given by eqn (6).

$$aP_{O_2} = (aP_{CO_2}/aP_{CO} \cdot K_5)^2 \qquad (6)$$

In eqn (6), K_5 is the equilibrium constant of the reaction (5). Substituting eqn (6) into eqn (4),

$$E = E^\circ + (RT/4F) \ln cP_{O_2} + (RT/2F) \ln (aP_{CO}/aP_{CO_2}) \qquad (7)$$

is obtained, where

$$E^\circ = (RT/2F) \ln K_5. \qquad (8)$$

That is, the open circuit voltage of the cell

$$CO_2, CO \,|\text{oxide ion conductor}|\, O_2 \qquad (9)$$

is given by eqn (7) and by the change of one mole carbon monoxide to one mole carbon dioxide, the electrical energy of $2EF$ is obtained. The cell reaction in this case is from eqn (3) plus eqn (5),

$$cO_2 + 2CO \rightarrow 2CO_2. \qquad (10)$$

At standard state, E is represented by

$$E = E^\circ = (RT/2F) \ln K_5 = -\Delta G_5^\circ/2F \qquad (11)$$

where ΔG_5° is the standard free enthalpy change of eqn (5).

Eqn (11) should be held with any kind of fuel and the following equation must be established for every fuel,

$$E^\circ = -\Delta G^\circ/nF$$

where $-\Delta G^\circ$ is the standard free enthalpy change of the combustion reaction of fuels and n the number of moles of oxygen required to oxidize one mole of the fuel multiplied by four. The maximum energy obtained in these cases is given by $-\Delta G^\circ$ and the ideal thermal efficiency ε° is represented by $(-\Delta G^\circ/ -\Delta H^\circ)$ where $-\Delta H^\circ$ is the standard enthalpy change of the reaction. For several kinds of fuels, ΔG°, ΔH°, E° and ε° are listed in Table I.

On discharging these high temperature fuel cells, deviations from the equilibrium always arise which depend on the current i, temperature T or electrode materials. These deviations are (a) the diffusion polarization which appears when the feeding velocity of the oxidant to the cathode or the fuel to the anode or the removing velocity of the cell reaction products from the electrodes is slower than that corresponding to the discharging current i; (b) the reaction polarization which is based on the rate of the cell reaction (eqn (5)) being slow, (c) the charge transfer polarization which is due to slow discharge reaction which may occur in the reaction such as eqn (11) or (12); and (d) the resistance polarization which is based on the internal resistance of the cell consisting mainly of the resistance of the electrolyte in the case of the solid electrolyte fuel cell.

TABLE I.   Maximum thermal efficiency of fuel cells.

Reaction	$T°K$	$\Delta G°$ kcal	$\Delta H°$ kcal	$E°$	$\varepsilon°$
$H_2 + \frac{1}{2}O_2 = H_2O\ (g)$	1000	$-46{\cdot}0$	$-59{\cdot}1$	$0{\cdot}997$	$0{\cdot}78$
	1250	$-42{\cdot}6$	$-59{\cdot}7$	$0{\cdot}924$	$0{\cdot}71$
$CO + \frac{1}{2}O_2 = CO_2$	1000	$-46{\cdot}7$	$-67{\cdot}7$	$1{\cdot}013$	$0{\cdot}69$
	1250	$-41{\cdot}4$	$-67{\cdot}4$	$0{\cdot}898$	$0{\cdot}61$
$CH_4 + 2O_2 = CO_2 + 2H_2O$	1000	$-191{\cdot}8$	$-191{\cdot}3$	$1{\cdot}039$	$1{\cdot}00$
	1250	$-191{\cdot}9$	$-191{\cdot}5$	$1{\cdot}039$	$1{\cdot}00$
$C + O_2 = CO_2$	1000	$-94{\cdot}8$	$-94{\cdot}7$	$1{\cdot}027$	$1{\cdot}00$
	1250	$-94{\cdot}8$	$-94{\cdot}8$	$1{\cdot}027$	$1{\cdot}00$

On these polarizations, a little study has been done by dealing with methods analogous to those used for the case of aqueous electrolytes, and their details will be described in the later sections. Generally speaking, since the problem of charge transfer polarization is not serious at high temperatures of about 1000°C which is usually used for operating solid electrolyte fuel cells, the main emphasis has been put on the improvement of solid electrolytes to reduce resistance polarization which is mainly due to the resistance of the solid electrolyte used and on the development of suitable cell designs by which diffusion polarization can be decreased.

As described above, the solid electrolyte used for fuel cells is the oxide ion conducting substance, i.e. solid oxide electrolytes.

The history of solid oxide electrolyte began when Nernst produced his "glower".[1] He proposed an ionic conductor such as ZrO_2 with small amounts of added Y_2O_3 or CeO_2 to use as a glower of the electric lamp. The ionic conductor which Nernst finally used contained 85 w/o ZrO_2 and 15 w/o Y_2O_3 and was called the "Nernst mass". The Nernst glower could be operated for long periods on direct current though electrolysis was found to occur. It was explained that any loss of oxygen librated at the anode was balanced by an equal amount of oxygen taken into the glower at the cathode.

This is a reverse phenomenon to the solid electrolyte fuel cell and the oxide electrolyte was applied to the high temperature fuel cell firstly by Baur and Preis in 1937.[2] The most important solid oxide electrolytes studied by them were ZrO_2, $(ZrO_2)_{0.9}$ $(MgO)_{0.1}$ and $(ZrO_2)_{0.85}$ $(Y_2O_3)_{0.15}$, but high performance of the cell was not obtained. At the present time, extensive investigations on solid oxide systems are being carried out in order to find high ionic conductivity solids and to establish techniques for obtaining thin films of them.

B. Properties of Oxide Solid Electrolytes

The solid oxide electrolytes extensively investigated so far have been the stabilized zirconias. ZrO_2 is stable in the monoclinic structure at room temperature[3] and is transformed into a tetragonal form at about 1150°C being accompanied by a volume contraction amounting to 9%. This undesirable thermal contraction hinders the wide use of ZrO_2 at high temperatures. The above stated two modifications both differ slightly from the cubic structure and it will be expected that ZrO_2 will be able to take a form of cubic structure after firing with other oxides which have cubic structures and can form solid solutions with ZrO_2 giving only small stresses to the ZrO_2 lattice. ZrO_2 thus treated is called "stabilized" and has usually a fluorite type structure[4] and shows no thermal contraction like that mentioned above.

The unit cell of the fluorite type oxide has so called an $8:4$ configurated M_4O_8 structure where M is a metal. This structure is shown in Fig. 1. In

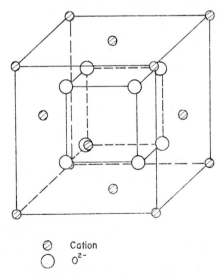

⊘   Cation
◯ O^{2-}

FIG. 1. Crystal structure of fluorite.

fluorite structure, each metal ion is surrounded by eight oxide ions forming a body centered cubic structure and each oxide ion is surrounded by four metal ions forming a tetrahedral arrangement. The limiting radius ratio of each ion to form this structure, that is $r_{\text{metal ion}}/r_{\text{oxide ion}}$ is 0·732. In the case of ZrO_2, this condition is not satisfied and the other oxide should be dissolved in ZrO_2 in order to obtain thermally stable ZrO_2. The added oxide for this object is called a stabilizer.

CaO and Y_2O_3 are the typical stabilizers and the phase diagrams of the systems ZrO_2–CaO[5] and ZrO_2–Y_2O_3[6] show that these systems have cubic structures in some composition and temperature ranges.[7] The lattice constants of the ZrO_2–CaO system change linearly against the concentration of CaO in some composition ranges of CaO as shown in Fig. 2 indicating that CaO can form a solid solution with ZrO_2 in this range.[7–14]

In this case, chemical defects are necessary in order to preserve the electrical neutrality in the solid solution. The probable structural defect models in such case are: (a) an oxide ion vacancy model, all metal ions being fixed at

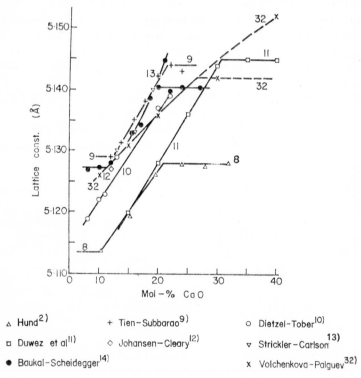

Fig. 2. Lattice constant of a solid solution of ZrO_2–CaO system as a function of CaO content.

he lattice points, (b) a cation interstitial model, all oxide ions being fixed nd (c) a mixed model of (a) and (b).

The density d of these solid solutions is given by

$$d = (xm + yn)/Na^2, \tag{12}$$

vhere x and y are the numbers of molecule in the unit cell of the host oxide nd those of the added oxide respectively, m and n their molecular weights, the lattice constant and N the Loschmidt's number. In the case of (a) $x + y = 4$ and (b) $2x + y = 8$ for MO_2–$M'O$ and $2x + 1·5y = 8$ for MO_2–$M'O_{1.5}$ systems, and x/y is obtained from the composition of the solid solu- ion and d can be calculated from eqn (12). By comparing the calculated lensity with that obtained by pycnometer measurement, the defect model can be determined.

Figure 3 shows the calculated and measured densities of the ZrO_2–CaO ystem sintered at 1460°C as a function of the CaO content and indicates hat the calculated values based on (a) coincide with the measured values vhich are shown by the circles in the figure.[8] This means that the solid solution with an imperfect fluorite lattice of the system ZrO_2–CaO contains vacant oxygen lattice sites. This has been confirmed by X-ray[9, 15] and neutron diffraction[16] and the measurement of the diffusion coefficient of he oxide ion in the solid solution[17]. Similar results have been obtained for he systems ZrO_2–Y_2O_3[8, 11, 13, 14, 18, 19] and CeO_2–La_2O_3.[20]

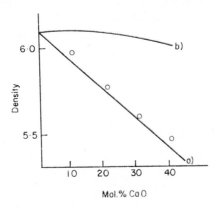

FIG. 3. Calculated and measured densities of the ZrO_2–CaO system as a function of CaO content.

The existence of the oxide ion vacancy suggests the appearance of oxide ion conduction by vacancy mechanism in these solid solutions. In this case, the oxide ion movement can be caused by the diffusion of the oxide ion.

From the Nernst–Einstein equation,

$$D_i = \mu_i \kappa T = \sigma_i \kappa T / n_i Z_i^2 e^2 \qquad (13)$$

where D_i is the diffusion coefficient of movable ions, μ_i the mobility, σ_i the conductivity, n_i the number of movable ions in unit volume, Z_i the ionic value and the other symbols have the usual meanings. The conductivity σ of $(ZrO_2)_{0.85} (CaO)_{0.15}$ sintered at 2000°C for 7 hr has been given, between 700°C and 1725°C, as

$$\sigma = 1 \cdot 50 \times 10^3 \exp(-1 \cdot 26/\kappa T). \qquad (14)$$

The diffusion coefficient of oxide ion $D_{O^{2-}}$ of this sample has been measured using ^{18}O, between 700°C and 1100°C, as

$$D_{O^{2-}} = 1 \cdot 0 \times 10^{-2} \exp(-1 \cdot 22/\kappa T). \qquad (15)$$

From the fact that the activation energies in eqns (14) and (15) have almost the same values, the electric carrier of this solid solution may be considered to be the oxide ion. By assuming $\sigma = \sigma_{O^{2-}}$, $D_{O^{2-}}$ was calculated by eqn (13) the value of which was confirmed as coinciding with the measured value as shown in Fig. 4 and as illustrating that the above assumption was correct. That is, the electrical carrier in $(ZrO_2)_{0.85} (CaO)_{0.15}$ is proved to be the oxide ion for the most part.

In this way, it is apparent that some stabilized zirconias are conducted electrically not by cations but by oxide ions. One more important problem is whether the electronic conduction can be intermixed in oxide ion conduction or not.

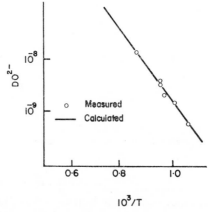

FIG. 4. Diffusion coefficient of oxide ion in $(ZrO_2)_{0.85} (CaO)_{0.15}$.

1. Oxide Ion Transference Number

It has been recognized that ionic conduction in oxides of fluorite type lattice is due mainly to oxide ions and the contribution of cationic conduction is negligibly small.[17, 21–23] The electronic conduction in the electrolyte, however, cannot be neglected in some cases.

When an external condition appears by which the oxide ion at the lattice point is converted to gaseous oxygen leaving an oxide ion vacancy $\square_{O^{2-}}$, an electron e^- may appear as

$$O^{2-} \rightarrow \tfrac{1}{2}O_2 + \square_{O^{2-}} + 2e^-. \tag{16}$$

In the reverse case, a positive hole h may be produced as

$$\tfrac{1}{2}O_2 + \square_{O^{2-}} \rightarrow O^{2-} + 2h. \tag{17}$$

Thus electronic conduction may appear.

These concentrations of the electron and positive hole must be affected by the external gas pressure. When these concentrations are low and the interactions between the species in eqns (16) and (17) are negligible, application of the mass action law to eqn (16) leads to

$$[e^-]^2 \cdot [\square_{O^{2-}}] \cdot P_{O_2}^{\tfrac{1}{2}}/[O^{2-}] = K \tag{18}$$

where P_{O_2} is the partial pressure of oxygen and K the constant. The brackets mean the concentrations. Since $[O^{2-}]$ and $[\square_{O^{2-}}]$ are fixed by the composition of the oxide solid solution, $[e^-]$ is given by

$$e^- = K_1 \cdot P_{O_2}^{-\tfrac{1}{4}} \tag{19}$$

where K_1 is a constant. Similarly,

$$h = K_2 \cdot P_{O_2}^{\tfrac{1}{4}}. \tag{20}$$

The total conductivity σ is expressed as

$$\sigma = \sigma_{O^{2-}} + F\mu_-[e^-] + F\mu_+[h] \tag{21}$$

where $\sigma_{O^{2-}}$ is the oxide ion conductivity, F the Faraday's constant and μ_- and μ_+ the mobilities of electrons and positive holes respectively. Substituting eqns (19) and (20) into eqn (21)

$$\sigma = \sigma_{O^{2-}} + K_1' \cdot P_{O_2}^{-\tfrac{1}{4}} + K_2' \cdot P_{O_2}^{\tfrac{1}{4}} \tag{22}$$

is obtained which indicates that the total conductivity should be dependent on the oxygen pressure when the electronic conductivity appears. In other words, when the total conductivity depends on the external oxygen pressure, it is a proof of the appearance of the electronic conduction in the solid solution.

By measuring such a dependence, electronic conductance was found somewhat in certain oxide systems, for example, ThO_2–CaO, ThO_2–Y_2O_3 and ThO_2–La_2O_3,[21, 24–26] whereas in the stabilized zirconias, the total conductivity was found to be independent of oxygen pressure over the range of 1 to 10^{-10} atmospheric pressures which is indicative of the fact that the conduction in the stabilized zirconias is fully ionic.[8, 17, 21]

This method, however, is rather qualitative and when only a small electronic conduction coexists, a quantitative measurement of the electronic conductivity should be carried out.

Let the conductivity of the electron in the solid solution be σ_e and the transference number of the oxide ion be $t_{O^{2-}}$, then

$$t_{O^{2-}} = \sigma_{O^{2-}}/\sigma \tag{23}$$

$$\sigma = \sigma_{O^{2-}} + \sigma_e. \tag{24}$$

Under this condition, the electromotive force of the fuel cell is not given by E in eqn (4) but by E'. The equivalent circuit of this case can be represented by Fig. 5. That is, the cell can be considered to be short circuited by the resistor, the resistance of which is C/σ_e and the following equations can be established where C is the thickness of the electrolyte,

$$E' = i_e \cdot C/\sigma_e \tag{25}$$

$$E' = E - i_{O^{2-}} \cdot C/\sigma_{O^{2-}} \tag{26}$$

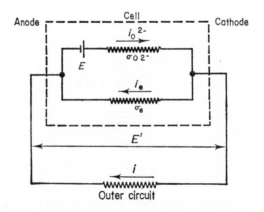

Fig. 5. Equivalent circuit of the fuel cell in operation.

where i_e and $i_{O^{2-}}$ represent the electronic and oxide ion current in the electrolyte respectively and at open circuit of the cell

$$i_e = i_{O^{2-}}. \tag{27}$$

By eliminating i_e and $i_{O^{2-}}$ in eqns (25), (26) and (27),

$$E' = E\sigma_{O^{2-}}/(\sigma_e + \sigma_{O^{2-}}) \tag{28}$$

is obtained and putting eqns (23) and (24) in eqn (28)

$$t_{O^{2-}} = E'/E \tag{29}$$

is established.

As has been described, the solid oxide electrolyte fuel cell is considered to be a sort of oxygen gas electrode concentration cell, E' was measured at $cP_{O_2} = 1\,\text{atm}$ and $aP_{O_2} = 2 \times 10^{-1}$ and $10^{-17}\,\text{atm}$ and the transference number of electron t_e in $(ZrO_2)_{0.85}\,(CaO)_{0.15}$ which is $1 - t_{O^{2-}}$ was calculated to be 0·002 and 0·006 at $1000°C$[22].

In order to form this concentration cell, oxygen produced by the dissociation of oxides at equilibrium has also been used. For example, in the cell

$$Fe \cdot FeO \mid (ZrO_2)_{0.85}\,(CaO)_{0.15} \mid Cu_2O \cdot Cu$$

the oxygen pressures at the anode and cathode are determined by the equilibrium pressures of oxygen of the reactions $FeO \rightleftarrows Fe + \frac{1}{2}O_2$ and $Cu_2O \rightleftarrows 2Cu + \frac{1}{2}O_2$ the values of which are $aP_{O_2} = 10^{-14.9}$ and $cP_{O_2} = 10^{-6.3}$ respectively. By this method, t_e in $(ZrO_2)_{0.85}\,(CaO)_{0.15}$ was measured as 0·003, 0·0004, $-0·0017$ and $-0·0041$ at 700, 800, 900 and $1000°C$.[23]

Strictly speaking, the values obtained above are merely mean values through the electrolyte, for the equilibrium states of the electrolyte may be considered to differ with the equilibrated oxygen pressure. In this case, $t_{O^{2-}}$ is not constant in the electrolyte and the oxygen electrode concentration cell is constructed of a series of minute unit concentration cells in which the difference of the pressure of the neighbouring oxygen electrode is so small that $t_{O^{2-}}$ in the unit cell may be considered to be uniform. Differentiating eqns (4) and (29), the electromotive force dE' of the unit cell is given by

$$dE' = t_{O^{2-}} (RT/4F)\,d \ln P_{O_2}. \tag{30}$$

Consequently, the measurable electromotive force E' of the concentration cell is

$$E' = t_{O^{2-}} (RT/4F) \int_{\ln a P_{O_2}}^{\ln c P_{O_2}} d \ln P_{O_2}. \tag{31}$$

Differentiating eqn (31) with respect to $\ln P_{O_2}$, the relation,

$$t_{O^{2-}} = (4F/RT)\, \partial E'/\partial \ln P_{O_2} \tag{32}$$

is obtained, thus, by plotting E' which is measured for the same value of cP_{O_2} and the different values of aP_{O_2} against $\ln aP_{O_2}$, $t_{O^{2-}}$ can be obtained as a function of aP_{O_2}.

By measurements of this kind, $t_{O^{2-}}$ has been found to be almost unity over wide pressure ranges in stabilized zirconia at elevated temperatures.[15]

The measurement of the rate of transfer of oxygen through the electrolyte has been carried out to determine $t_{O^{2-}}$. From eqns (23)–(26)

$$t_{O^{2-}} = 1 - i_{O^{2-}} \cdot C/\sigma E' = 1 - 2F v_{O^{2-}} \cdot C/\sigma E' \tag{33}$$

is obtained, where $v_{O^{2-}}$ is the velocity of oxide ion through the electrolyte at open circuit. $v_{O^{2-}} \cdot C$ is given by twice the number of moles of oxygen transferred from the cathode to the anode per second at open circuit, and $t_{O^{2-}}$ can be calculated by eqn (33). Using this method, t_e in $(ZrO_2)_{0.85}$ $(CaO)_{0.15}$ has been calculated as less than 0.02 at temperatures near $1000°C$.[22]

Moreover, t_e or $t_{O^{2-}}$ has been measured by the electrolysis method[27] or Wagner's polarization method.[28],[29] These experimental methods have not been used widely because of difficulties experienced at high temperatures.

Zirconium dioxide has a conductivity of $10^{-7}\,\Omega^{-1}\,cm^{-1}$ at $1000°C$, this low value making one suspect that the change of value of the transference number in the stabilized zirconias with the amount of the stabilizers is small. In the solid solutions, however, where the mother material has relatively high electronic conductivity like cerium dioxide, the change of the transference number due to the added oxides must be made clear.

Actually, the oxide ion transference number in the solid solution, the mother material of which is ceria or thoria, has been recognized to have a value below unity in some ranges of composition, oxygen pressure, and temperature. In Fig. 6, $t_{O^{2-}}$, measured by the oxygen gas concentration cell method, is plotted against the content of lanthana in ceria.[20]

2. Conductivity Changes by Compositions

For solid electrolyte fuel cells, the most widely studied system is the stabilized zirconia followed by ceria, thoria and perovskite solid solutions.

(a) ZrO_2–CaO System

The conductivities of this system measured by several investigators have not always been consistent.[9,12,16,17,30-35] Though the reason for this rediscpancy is not clear, it may be attributed to the differences of the temperature and duration of heat treatment of the specimens, since ageing effects have been found on measuring conductivities of this system.[9,16,34,36]

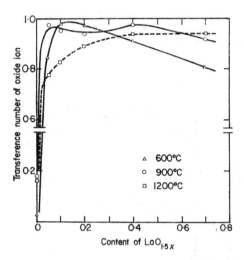

Fig. 6. Oxide ion transference number in $(CeO_2)_{1-x}(LaO_{1-5})_x$ system as a function of the content of $LaO_{1.5}$.

When the conductivities are plotted against the content of calcia, a maximum value of conductivity has been found between 13 and 15 m/o calcia. In Fig. 7, the result obtained by Carter and Roth[16] is shown. They found that the conductivities of the solid solution of this system sintered at 2050°C in a hydrogen stream and of the single crystal prepared by the Verneuil's method both decrease parabolically with time below the characteristic temperatures which were determined by the composition of the specimen, for example, 1150°C for 13 m/o CaO, 1325°C for 14·2 m/o CaO and 1280°C for 19 m/o CaO do not attain any constant value even after 3000 hr of measurement at 900 or 1000°C. These conductivity decreases have been attributed to the order–disorder transition.

Takahashi and Suzuki[36] have measured the conductivity of this system at 1000°C for over 1000 hr with the specimens sintered at 2000°C in air. After being annealed at 1600°C, the specimens were cooled to room temperature abruptly or slowly and heated again to 1000°C to measure the conductivity. The specimens used contained 11, 13, 15, 17 and 19 m/o CaO, the conductivities of which were found to decrease with time except the case of the specimen of 15 m/o CaO. The highest value of conductivity was observed by the specimen of 11 or 13 m/o CaO at the first stage of experiment, while, on and after 50 hr it was obtained by the specimen of 15 m/o CaO as shown in Fig. 8. In general, the resistivity $\rho(t)$ was expressed by an exponential function of time t in contrast with the Carter's result as

$$\rho(t) = A - B \exp(-Kt) \tag{34}$$

where A, B and K are constants.

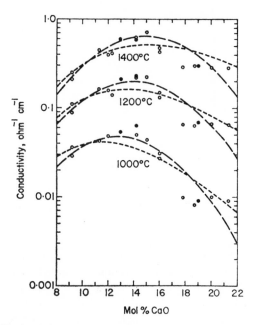

Fig. 7. Conductivity of ZrO_2–CaO system a a function of composition.

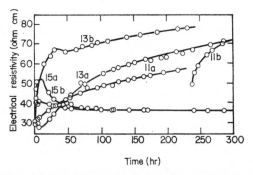

Fig. 8. Time dependence of the resistivity of the ZrO_2–CaO system at 1000°C. The numerals in the figure denote the contents of CaO in mol %. (a) slowly cooled; (b) rapidly cooled.

From the above, the concentration of calcia in stabilized zirconia which gives a highest conductivity will be said to vary with time elapsed before the measurement of conductivity.

Although little is known about the ageing effects of calcia stabilized zirconia, it is probable that these effects are incurred by the order–disorder transition, from the observation that a new phase has been discovered by neutron diffraction in the calcia stabilized zirconia containing CaO above 16 m/o which was annealed at 1010°C for 336 hr before the diffraction measurement.[16] Thus, the calcia stabilized zirconia must be used at above the characteristic temperatures which were determined by Carter in order to obtain a result which shows no ageing effect.

This system has been shown to have an impurity effect when SiO_2, TiO_2, Al_2O_3 or Fe_2O_3 coexist in the system to some extent. These impurities obstruct the stabilization of zirconia as a result of forming some compounds with CaO.[37]

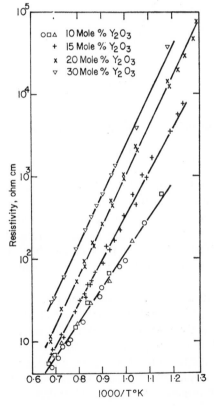

FIG. 9. Temperature dependence of the resistivity of the ZrO_2–Y_2O_3 system.

(b) $ZrO_2-Y_2O_3$ System

This system shows a higher conductivity[13,33,36,38,39] than that of ZrO_2-CaO system and is applicable to fuel cells.

In Fig. 9,[33] the temperature dependence of the resistivity of this system at above 10 m/o of Y_2O_3 is shown, and the resistivity at 1000°C is illustrated in Fig. 10 against the content of Y_2O_3.[36]

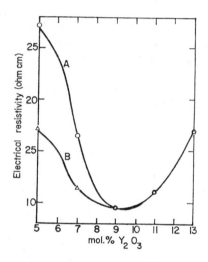

FIG. 10. Resistivity of the $ZrO_2-Y_2O_3$ system at 1000°C.
$\rho(\infty)$–curve A,
$\rho(1\ hr)$–curve B.

It will be found in Fig. 10 that the highest conductivity is at about 9 m/o Y_2O_3. In Fig. 10, $\rho(\infty)$ is calculated by eqn (34). When the content of Y_2O_3 is over 9 m/o the resistivity has been found to be almost unchanged with time at 1000°C showing that in this range the rate of the order-disorder transition will be faster than that for calcia stabilized zirconia, while the ageing effect has been shown (Fig. 10) when the content of Y_2O_3 was below 9 m/o.

The phase diagram shows that in the region where the concentration Y_2O_3 is below 9 m/o, the tetragonal structure appears at 1000°C. The time dependence of the resistivity in this region might be considered to be partly due to the separation of tetragonal phase from cubic structure.

Considering the fact that the ageing effect has been found in the specimens containing 9 m/o Y_2O_3 at 800°C and 600°C as shown in Fig. 11, it may also be necessary to take into account the order-disorder transition, as in the case of calcia stabilized zirconia.

Baukal measured the electrical conductivity of $(ZrO_2)_{0.91}$ $(Y_2O_3)_{0.09}$ at 800°C in air and hydrogen as a function of time,[38] and found that the systematic decrease in conductivity could be described by first-order kinetics. From the fact that the activation energy of conduction remained unchanged by the ageing process, an order–disorder transition in the fluorite lattice seemed likely as a possible mechanism of the ageing process.

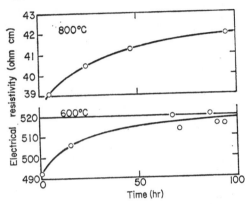

Fig. 11. Time dependence of the resistivity of 9 m/o Y_2O_3–ZrO_2 at 800 and 600°C.

(c) ZrO_2–Yb_2O_3 *System*

The conductivity of this system is shown in Fig. 12 against the content of Yb_2O_3 taking the measuring temperature as a parameter.[40] In this figure, the highest conductivity will be found at 8–9 m/o Yb_2O_3 at every temperature.

In this system, ageing effects similar to those in the ZrO_2–Y_2O_3 system have been found[36] as shown in Figs. 13 and 14. As the shape of the phase diagram of the ZrO_2–Yb_2O_3 system resembles that of the ZrO_2–Nd_2O_3 system[41] in the higher concentration region of ZrO_2, the ZrO_2–Yb_2O_3 system may possibly be considered also to be similar to them, the time dependence of this system may also be attributed to the order–disorder transition and/or to the cubic–tetragonal phase separation.

(d) ZrO_2–M_2O_3 *System*

As M, Sc,[33,38] Gd,[40] Sm[38] and Nd[33,40] have been taken into account besides Y and Yb. These M_2O_3 can form solid solutions with ZrO_2 taking the fluorite structure in certain composition range, thus stabilizing ZrO_2.

The conductivities of these systems show the highest values when the minimum amounts of M_2O_3 necessary to stabilize ZrO_2 in cubic structure are dissolved in ZrO_2, and the smaller the ionic radius of M^{3+}, the higher the conductivity. The values of the ionic radius of M^{3+} are 0·81Å for Sc^{3+}, 0·86Å for Yb^{3+}, 0·92Å for Y^{3+}, 0·97Å for Gd^{3+}, 1·00Å for Sm^{3+} and 1·04Å for Nd^{3+}. In Fig. 15 conductivities are shown against the concentration of M_2O_3. It is difficult to explain quantitatively why M_2O_3, in which M^{3+} has a smaller ionic radius, makes the conductivity of the stabilized zirconia high.

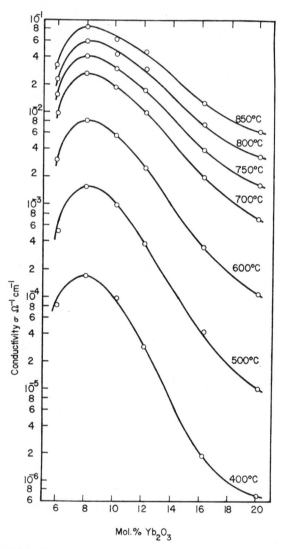

FIG. 12. Conductivity of the ZrO_2–Yb_2O_3 system as a function of the content of Yb_2O_3 at various temperatures.

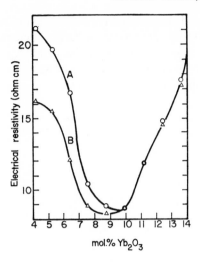

FIG. 13. Resistivity of the ZrO_2–Yb_2O_3 at 1000°C.
$\rho(\infty)$–curve A,
$\rho(1\ hr)$–curve B.

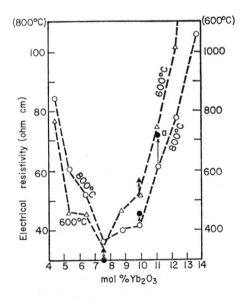

▲ : $\rho(\infty)$ at 600°C
● : $\rho(\infty)$ at 800°C (a : after 117 hours)

FIG. 14. Resistivity of the ZrO_2–Yb_2O_3 system at 800 and 600°C.

But it may be assumed that when Zr^{4+}, the ionic radius of which is 0·79Å, is to be substituted by M^{3+}, the more similar the ionic radius of M^{3+} to that of Zr^{4+}, that is, the smaller the ionic radius of M^{3+}, the smaller the strain of the lattice and the fewer the associations of M^{3+} with oxide ion vacancy, this makes the mobility of oxide ions greater.

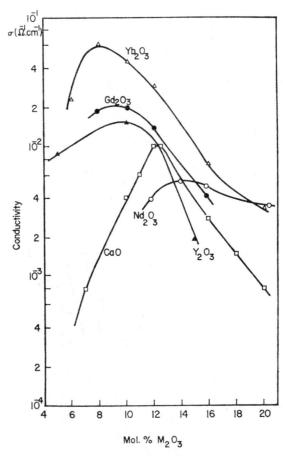

FIG. 15. Conductivities of the ZrO_2–M_2O_3 system at 800°C

Though the reason why the conductivity always shows maximum when plotted against the content of stabilizer is not clear, it is probably due to the concentration of oxide ion vacancy being proportional to the amount of the stabilizer added, when the concentration of the stabilizer is small and oxide ion vacancy in the crystal lattice is distributed statistically making the conductivity higher as the concentration of the stabilizer is made larger. When the amount of the stabilizer reaches a high level, the number of oxide ion vacancies increase and cause the interaction of oxide ion vacancies which makes the distribution be ordered and the mobility of oxide ion vacancy be inferred. As a result, maximum conductivity will appear at certain concentrations of the stabilizer.

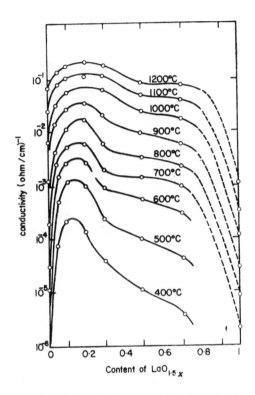

FIG. 16. Conductivities of $(CeO_2)_{1-x}(LaO_{1.5})_x$ as a function of x at various temperatures.

The three-component systems such as ZrO_2–CaO–Y_2O_3,[13] ZrO_2–CeO_2–Y_2O_3, ZrO_2–Ta_2O_5–Y_2O_3[42] and ZrO_2–Fe_2O_3–Y_2O_3,[43] have also been studied without finding any special merit.

(e) CeO_2–La_2O_3 System

As CeO_2 has the crystal structure of fluorite, an oxide ion conductor should result when M_2O_3 is incorporated. Instead of M_2O_3, La_2O_3 has been successfully chosen. But, as CeO_2 is an n-type semiconductor, the content of La_2O_3 in CeO_2 should be fixed so that the oxide ion transference number be high to say nothing of its high value of conductivity. In Fig. 16,[44] the conductivity of this system sintered at 1600°C is shown against the added amount of $LaO_{1.5}$ and the measuring temperature, and the oxide ion transference numbers measured by the electromotive force method indicate that the change of the oxide ion transference number with the amount of dissolved $LaO_{1.5}$ in this system is relatively large, as illustrated in Fig. 6. In this measurement, cP_{O_2} was fixed at 1 atm and the ratio of cP_{O_2}/aP_{O_2} was selected as 50.

In this system, oxide ion vacancy is represented by the following reaction,

$$(1 - x)\, CeO_2 + xLaO_{1.5} = Ce_{1-x}\, La_x\, O_{2-0.5x} + 0.5x \,\square_{O^{2-}} \qquad (35)$$

where x is the number of La^{3+} substituting for Ce^{4+} in the lattice. The larger the value of x, the higher the concentration of oxide ion vacancy as in the case of stabilized zirconias, making the oxide ion conductivity high. But when the concentration of oxide ion vacancy becomes so high as to have an interaction between oxide ion vacancies, the statistical distribution of oxide ion vacancy loses its disordered state and transfers to the ordered state making the conductivity of the system low.

As shown in Fig. 6, the oxide ion transference number is not always unity and it decreases under low partial pressure of oxygen as illustrated in Fig. 17. This is thought to be due to the partial reduction of ceria and, in order to avoid this, the addition of thoria to this system has been proposed to demonstrate the increase of oxide ion transference number as shown in Fig. 18.

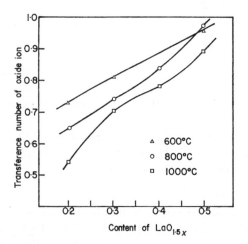

FIG. 17. Transference number of oxide ion of $(CeO_2)_{1-x}\,(LaO_{1.5})_x$ as a function of x at various temperatures calculated from the ratio of the measured and theoretical e.m.f. of the cell O_2(1 atm), $Pt|(CeO_2)_{1-x}(LaO_{1.5})_x|Pt$, $H_2 + H_2O$.

The conduction characteristics of the systems $CeO_2-Y_2O_3$, CeO_2-CaO, CeO_2-MgO or CeO_2-SrO have also been studied and it was found that the oxide ion transference number decreases under low oxygen partial pressure.[45-47]

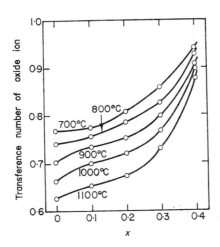

FIG. 18. Transference number of oxide ion of $(CeO_2)_{0.7-x}$ $(ThO_2)_x$ $(LaO_{1.5})_{0.3}$ as a function of x at various temperatures.

(f) ThO$_2$–CaO *System*

The conductivity of this system is relatively low even at $1000°C$[48] and p-type electronic conduction appears under high partial pressure of oxygen. Under partial pressure of oxygen lower than 10^{-5} atm, an oxide ion transference number of over 0·99 has been obtained in some concentration ranges of calcia though the value of its conductivity is too low to be used as the electrolyte for fuel cells.

The conduction characteristics of the systems ThO_2-BeO, ThO_2-MgO, ThO_2-SrO, ThO_2-BaO and $ThO_2-Y_2O_3$ have been studied, without finding a sufficiently high oxide ion conductor.[24] The conductivities of the above fluorite-type solid solutions are listed in Table II.

(g) *Perovskites*[49, 50, 51]

Oxides having perovskite type structure have been known as ferrimagnetic substances. By substituting the metallic ions in a perovskite lattice with other metallic ions at different valency, oxide ion vacancies appear and it then becomes an oxide ion conductor.

The electrical conductivity of various perovskite type oxides measured in air is shown in Fig. 19 in the form of Arrhenius plots. As is evident from this Figure, the conductivities of substitutional solid solutions are remarkably high compared to the corresponding pure substances, suggesting that the concentration of the charge carriers which are possibly oxide ion vacancies, cations, electrons and electron holes would be increased by replacing the native cations in the host crystals with foreign ions of lower valency.

TABLE II.   The conductivity of fluorite-type oxide solid solutions.

Oxide	Conductivity Ω^{-1} cm^{-1}		Note
	1000°C	800°C	
$(ZrO_2)_{0.89}$ $(CaO)_{0.11}$	4.5×10^{-2}	6×10^{-3}	
$(ZrO_2)_{0.85}$ $(CaO)_{0.15}$	2.5×10^{-2}	2×10^{-3}	
$(ZrO_2)_{0.91}$ $(Y_2O_3)_{0.09}$	9×10^{-2}	2×10^{-2}	
$(ZrO_2)_{0.91}$ $(Yb_2O_3)_{0.09}$	1.6×10^{-1}	3×10^{-2}	
$(ZrO_2)_{0.92}$ $(Yb_2O_3)_{0.08}$	—	6×10^{-2}	
$(ZrO_2)_{0.69}$ $(LaO_{1.5})_{0.33}$	1.5×10^{-3}	—	t_i:
$(ZrO_2)_{0.85}$ $(Sc_2O_3)_{0.15}$	1.3×10^{-1}	5×10^{-3}	Transference
$(ZrO_2)_{0.9}$ $(Nd_2O_3)_{0.1}$	6×10^{-3}	1.1×10^{-3}	number of ion
$(ZrO_2)_{0.9}$ $(Gd_2O_3)_{0.1}$	—	2×10^{-2}	
$(ThO_2)_{0.95}$ $(CaO)_{0.05}$	9×10^{-4}	1.2×10^{-4}	
$(ThO_2)_{0.85}$ $(MgO)_{0.15}$	1.0×10^{-3}	1.5×10^{-4}	
$(ThO_2)_{0.85}$ $(SrO)_{0.15}$	1.5×10^{-3}	4×10^{-4}	
$(ThO_2)_{0.75}$ $(LaO_{1.5})_{0.25}$	5×10^{-3}	—	
$(ThO_2)_{0.95}$ $(Y_2O_3)_{0.05}$	1.2×10^{-2}	1.9×10^{-3}	$t_i = 0.63$ at 1,000°C
$(CeO_2)_{0.8}$ $(LaO_{1.5})_{0.2}$	8×10^{-2}	2×10^{-2}	
$(CeO_2)_{0.8}$ $(YO_{1.5})_{0.2}$	1.2×10^{-1}	3×10^{-2}	$0.5 < t_i < 0.95$

Ion transference numbers are shown in Fig. 20, where cells I and II are represented as follows.

Cell I O_2(air), Pt | specimen | Pt, O_2(1 atm)

Cell II H_2(1 atm) + H_2O(18–32 mmHg), Pt | specimen | Pt, O_2(1 atm)

As is clearly shown in the figure, ion transference numbers are influenced by oxygen partial pressure. Though the conduction of the specimen under the

ondition of cell I is partly ionic, the electromotive force of cell II was stable
nd reproducible indicating that the electrodes behaved as reversible oxygen
lectrodes and the movable ions are oxide ions.

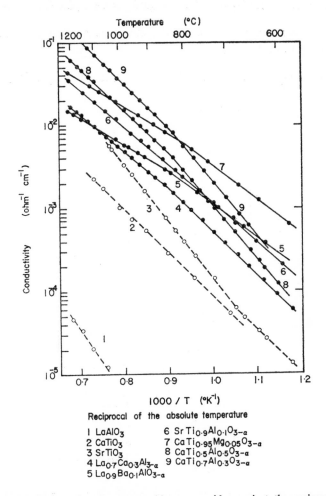

FIG. 19. Conductivities of various pervoskite type oxides against the reciprocal of the
absolute temperature in air.

Further, the conductivity depends on the partial pressure of surrounding
oxygen as shown in Fig. 21. This tendency is characteristic for p-type conductive oxides and shows that the conduction in this solid solution may be
due partly to the electron hole.

The perovskite type oxide, in this case, is a mixed conductor under high partial pressure of oxygen. For example, in the solid solution obtained by substituting La in $LaAlO_3$ with Ca, the following equilibrium is established,

$$AlO_{1.5} + (1 - x)LaO_{1.5} + xCaO + \tfrac{1}{4}xO_2 \rightleftarrows La_{1-x}Ca_xAlO_3 + xh \quad (36)$$

The ionic conduction in this solid solution is attributed to oxide ion vacancy formation, that is,

$$AlO_{1.5} + (1 - y) LaO_{1.5} + yCaO \rightleftarrows La_{1-y}Ca_yAlO_{3-\frac{1}{2}y} + \tfrac{1}{2}y \,\square_{O^{2-}}.$$

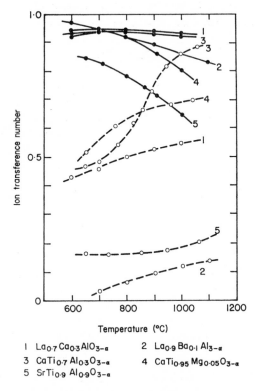

I $La_{0.7}Ca_{0.3}AlO_{3-\alpha}$ 2 $La_{0.9}Ba_{0.1}Al_{3-\alpha}$
3   $CaTi_{0.7}Al_{0.3}O_{3-\alpha}$ 4 $CaTi_{0.95}Mg_{0.05}O_{3-\alpha}$
5   $SrTi_{0.9}Al_{0.9}O_{3-\alpha}$

FIG. 20. Ion transference numbers of various perovskite type oxides measured by cell I (broken line) and by cell II (solid line) at various temperatures.

Almost the same ion transference number characteristics have been found in $CaTi_{1-x}Al_xO_{3-\alpha}$. In this case, the equilibrium of the defect in a crystal with oxygen atmosphere can be described as follows.

$$\tfrac{1}{4}O_2 + CaTi_{1-x}Al_xO_{3-\frac{1}{2}x} + (x/2) \,\square_{O^{2-}} \rightleftarrows CaTi_{1-x}Al_xO_3 + xh \quad (37)$$

Under ordinary oxygen atmosphere, an intermediate state will be presented, that is,

$$CaTi_{1-x}Al_xO_{3-\alpha} + \alpha\square_{O^{2-}} + (x - 2\alpha)\,h,$$

where α is the content of oxide ion vacancy per unit formula leading the mixed conduction by oxide ion vacancy and positive hole in the solid solution. Under the sufficiently low partial pressure of oxygen the equilibrium of eqn (37) is partial on the left hand side, thus making the specimen a highly ionic conductor. This situation is realized in solid electrolyte fuel cells. The electronic conduction in this case is probably due to the excess electron which may be produced according to the following equation,

$$O^{2-} \rightarrow \square_{O^{2-}} + \tfrac{1}{2}O_2 + 2e^- \tag{38}$$

As is clearly shown in Figs. 19 and 20, $CaTi_{0.7}Al_{0.3}O_{3-\alpha}$ and $CaTi_{0.95}Mg_{0.05}O_{3-\alpha}$ have relatively high oxide ion conductivities which are comparable to those of zirconias. Though $La_{0.7}Ca_{0.3}AlO_3$ showed relatively large ion transference numbers, the oxide ion conductivity was not so high and $BaZr_{0.9}Bi_{0.1}O_{3-\alpha}$, $BaCe_{0.9}Bi_{0.1}O_{3-\alpha}$ and $BaTh_{0.95}La_{0.05}O_{3-\alpha}$ show-

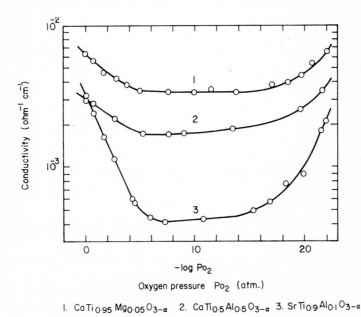

1. $CaTi_{0.95}Mg_{0.05}O_{3-\alpha}$ 2. $CaTi_{0.5}Al_{0.5}O_{3-\alpha}$ 3. $SrTi_{0.9}Al_{0.1}O_{3-\alpha}$

FIG. 21. Dependence of conductivity on oxygen pressure at 800°C.

ed higher ionic conductivities. But the latter three solid solutions were rather unstable and reproducible values of conductivity could not be obtained, as they reacted with the electrode materials.

III. Electrode Processes in Solid Oxide Electrolyte Systems

The essential reactions which occur at the electrodes of the solid oxide electrolyte fuel cell are represented in eqns (1) and (2). These reactions start at the place where the oxide electrolyte, electrode constructing material and oxygen gas are in contact with each other. The mechanism of this contact will be more complex than in the case when the electrolyte is an aqueous solution. The hysteresis phenomena which have often been observed when the current is passed through an electrode, composed of platinum powder baked on the oxide ion conducting electrolyte, will suggest a change in the above stated contact mechanism.

In order to minimize the polarization, that is to draw appreciable current, it is, of course, necessary to make the effective electrode contact area as large as possible. But as the microscopic contact area is difficult to estimate, comparative studies of polarization have been done with various kinds and pressures of gas at an apparently identical electrode type or with various electrode materials in apparently the same configuration.

It is of interest to consider the effect on the characteristics of the electrode of pretreatment by electric current. For example, Takahashi, Iwahara and Itoh[52] studied the cathodic characteristics of an oxygen electrode constructed with sintered platinum powder on a solid electrolyte $(ZrO_2)_{0.9}$ $(Y_2O_3)_{0.1}$ after preliminary treatment with a high current density of $800 \, mA/cm^2$ at $800°C$. With such preliminary treatment, the polarization of the electrode was decreased and this effect was found to be more pronounced in the case of cathodic treatment than anodic. The limiting cathodic current due to the diffusion limit of oxygen gas was apparent at the cathodic treated electrode. The impedance of the electrode was increased by anodic treatment, while it was decreased by cathodic treatment compared with that of the untreated electrode. From the dependence of impedance on frequency, they considered that the diffusion of oxygen gas was a rate-determining step of cathodic reaction at the cathodic treated electrode and that charge transfer polarization was negligibly small in all cases. The large depolarization effect of the cathodic treatment was deduced from considering that a two-phase reaction occurred at the boundary between the gas phase and the electrolyte surface where electronic and ionic conduction coexisted.

Hartung and Möbius[53] have found, by the measurement of the faradaic impedance, that it was changed by the atmosphere around the electrode. They also found that the transmission resistance, which was independent of the measuring frequency, was small compared with the diffusion resistance which depended on the frequency leading to the reversal of the slow discharge mechanism. This opinion has been partly supported by the measurement of the anodic polarization in a mixture of CO and CO_2 which shows that the rate of the diffusion of CO to the surface of the electrode plays an important part.[54]

On the other hand, Neuimin, Karpachev and Palguev[55] showed that for air electrodes negligible overvoltages occur, whereas for a $CO–CO_2$ mixture electrodes considerable overvoltages were produced under anodic current. Plots of $\log i$ against the overvoltage at 900°–1100°C displayed a linear Tafel line with the transfer coefficient $\alpha = 0.5$. That is, the charge transfer polarization has been shown to appear.

Perfilev and Palguev[56] also measured the effect of the current pretreatment and reported that a preliminary anodic treatment of the oxygen electrode made the characteristics of the electrode excellent and this has been attributed to the enlargement of the contact area between the electrolyte and the electrode material. By measuring the polarization characteristics of these pre-treated electrodes, it was found that the polarization potential showed a linear relation with current at low current densities and at relatively high current densities, Tafel relations were established by both anode and cathode performances. With the electrolyte containing ceria from the Tafel line, the slope b, the exchange current density i_0 and the activation energy (which is calculated from the temperature dependency of i_0) were obtained and are tabulated in Table III.[57] This treatment has been carried out on a stabilized zirconia electrolyte and it was found that a charge transfer process played an important role.[58]

TABLE III.   Slope b of the Tafel line

Electrolyte	Temperature	b (anodic) mV	b (cathodic) mV	i_0 mA	Activation Energy E kcal
$(CeO_2)_{0.85}(SrO)_{0.15}$	700–900	110–200	110–265	0.35–10	32
$(CeO_2)_{0.85}(La_2O_3)_{0.15}$	700–890	150–240	210–340	0.6–9	15
$(CeO_2)_{0.85}(CaO)_{0.15}$	650–880	105–200	300–500	0.1–2.8	41

Recent experiments on Pt, CO, $CO_2 \mid ZrO_2$–CaO made by Etsell and Flengas[59] have yielded a linear Tafel line which coincided with the result obtained by Neuimin et al. Binder et al.[60] measured thea nodic polarization of porous platinum electrodes on $(ZrO_2)_{0.85}$ $(CaO)_{0.15}$ using $H_2–H_2O$ and $CO–CO_2$ gas mixtures as fuels, and found that no appreciable polarization other than resistance polarization was observed. These divergent results have been often obtained with the same system and this suggests that the further investigations should be carried out on the polarization phenomena of the gas electrode on the solid electrolytes.

Möbius and Rohland[61] studied polarization on $(ZrO_2)_{0.85}$ $(CaO)_{0.15}$ at 800°C using oxygen and $CO–CO_2$ as gas phases. As shown in Fig. 22, the cell characteristics are widely changed by the electrode materials. However, the state of the electrodes is not always the same with regard to porosity, gas permeability, absorptivity and surface diffusion rate. In this respect, the fundamental study made by Kleitz,[62] who measured polarization in the temperature range 700–1200°C using several kinds of metal wire electrode pressed on the stabilized zirconia electrolyte serves as a good reference. Using O_2, $H_2–H_2O$ and $CO–CO_2$ as gas phases, the conclusion that the rate determining process is mass transport rather than electrode reaction kinetics was obtained which coincides with the results reported by Takahashi and Hartung though the applicability of these results to conventional porous electrodes may be doubtful. Takahashi, Ito and Iwahara[44, 63] measured the overpotentials of the electrodes of the cells using $(ZrO_2)_{0.85}$ $(CaO)_{0.15}$ and $(ZrO_2)_{0.9}$ $(Y_2O_3)_{0.1}$ as electrolytes, hydrogen as fuel and platinum powder as electrodes. They have found that at 1000°C

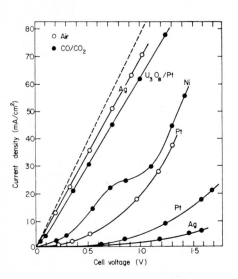

FIG. 22. Current-voltage curves for air and 50:50 $CO–CO_2$ gas electrodes on $(ZrO_2)_{0.85}$ $(CaO)_{0.15}$ at 800°C with various supporting electrode materials.

the polarization appeared merely at the anode which was consistent with the result of Neumin *et al.*, while at 800°C and 600°C, the polarizations were found to appear at both the cathode and anode. In the case of the cell, the electrolyte of which was $(CeO_2)_{0.6}$ $(LaO_{1.5})_{0.4}$, the polarization at the cathode was similar to that of the above stated cells while polarization at the anode was not observed. From considering that the difference of polarization by the different electrolytes was observed merely at the anode, the anodic polarization in the case of the stabilized zirconia electrolyte cell may occur at the interface of the electrolyte and the electrode, or in the electrolyte.

In stabilized zirconia electrolytes, moreover, blackening of the electrolyte, that is the decomposition of the electrolyte, has been noticed when strongly cathodic currents were passed. According to the report of Casselton,[64] with a ZrO_2–Y_2O_3 electrolyte and platinum electrode system, a critical current density at the cathode was found for the onset of blackening which was a linear function of the oxygen gas pressure. In the blackened zone, the conductivity increased which suggested the injection of electrons into the electrolyte from the electrode at above the critical current density. Some of the electrons are trapped in oxide ion vacancies giving colour centres. Sufficient oxide ion vacancies might be built up above the critical current density because of the rate of oxide ion transport from the cathode exceeds the oxygen gas supply.

Such being the case, the electrode reaction mechanism cannot be said to have been made clear compared with that in aqueous electrolytes. The main reasons seem to be the lack of knowledge regarding the contribution to polarizations of the inner structure of the solid oxide electrolyte. For example, the fact that the values of activation energy listed in Table III have similar values of activation energy of conduction, may point to the contribution of the oxide electrolyte to the electrode reaction processes.

In the case of the solid electrolyte, as no supporting electrolyte exists, a space charge is produced when the electrode is polarized, forming a distribution of the excess or the short conducting ion species at the interface of the solid electrolyte and the electrode. In order to clarify this situation, the the solid electrolyte and the electrode. In order clarify this situation, the potential distribution in the oxide electrolyte should be known. For example, from the measurement of the potential distribution in the solid solution of $(CeO_2)_{0.99}$ $(CaO)_{0.01}$, it has been found that when D.C. voltage was applied to the oxide electrolyte, the potential distribution was not linear through the electrolyte and it changed with time.[46] Though this sort of treatment may give some knowledge about the space charge, little work has been done and development in this field is left to the future.

It is certain, however, that an aim of the study of high temperature solid electrolyte fuel cells will be to reduce the above mentioned polarizations.

IV. Solid Electrolyte High Temperature Fuel Cells

A. General Aspects

The weakest point of the solid electrolyte fuel cell is that the conductivity of the solid electrolyte is relatively low compared to that of fused salts. For example, a mixed molten salt of $Na_2CO_3-K_2CO_3-Li_2CO_3$ has a conductivity of the order of above $10^{-1}\,\Omega^{-1}\,cm^{-1}$ even at about 600°C while the conductivities of $(ZrO_2)_{0.85}(CaO)_{0.15}$ and $(ZrO_2)_{0.91}(Y_2O_3)_{0.09}$ which have been used as the electrolyte for fuel cells are $2.5 \times 10^{-1}\,\Omega^{-1}\,cm^{-1}$ and $9 \times 10^{-1}\,\Omega^{-1}\,cm^{-1}$ respectively at 1000°C. These values mean that ohmic falls of 0.8 V and 0.2 V respectively will appear in cells with the electrolytes of $(ZrO_2)_{0.85}(CaO)_{0.15}$ and $(ZrO_2)_{0.91}(Y_2O_3)_{0.09}$ when a current density of $200\,mA/cm^2$ is required in cells with 1mm thick electrolyte. These ohmic falls will be serious for fuel cells as the open circuit voltage of a fuel cell is only about one volt or so. In addition, the operating temperature of 1000°C is a disadvantage in some respects and the lowering to 700–800°C of this temperature would be an advantage. One line of research would be to seek solid electrolytes having higher oxide ion conductivity and another to find methods of obtaining a thinner film of solid electrolyte. As an example of the first approach, some fluorite type electrolytes containing cerium dioxide[44, 65, 66] and certain perovskite structure electrolytes[49–51] have been proposed as electrolytes for solid electrolyte fuel cells. As an example of the second approach 0.001mm thick zirconia electrolyte has been deposited on metals by electron beam method,[67] 0.01mm thick electrolyte by thermal decomposition of $ZrCl_4$ and $CaCl_2$[68] and 0.1mm thickness by flame spray[69] or plasma jet method.[70]

B. Electrolytes

A plate type, bell-and-spigot type and thin layer electrolytes have been tested. In Fig. 23, the plate or disc type electrolyte developed by Brown and Boveri[71] is shown. The electrolyte is $(ZrO_2)_{0.91}(Y_2O_3)_{0.09}$ and by constructing the fuel cell with the 0.5 mm thick electrolyte, a power density of $250\,kW/m^3$ was obtained at 900°C. Though the fuel cell research at Brown and Boveri was discontinued after 1968, the disc type electrolyte has been used mainly in test or laboratory scale solid electrolyte fuel cells.

The bell-and-spigot type electrolyte battery[72] was developed by Westinghouse Electric Corporation in the form shown in Fig. 24. A schematic axial cross section is shown in Fig. 25, and the battery connected to a 100 W power system in Fig. 26.

The welding material of each bell-and-spigot was 18% Ni and 82% Au and it was heated at 1125°C for 20 minutes in a stream of hydrogen. The electrolyte material was $(ZrO_2)_{0.9}(Y_2O_3)_{0.1}$ and cell performance is shown

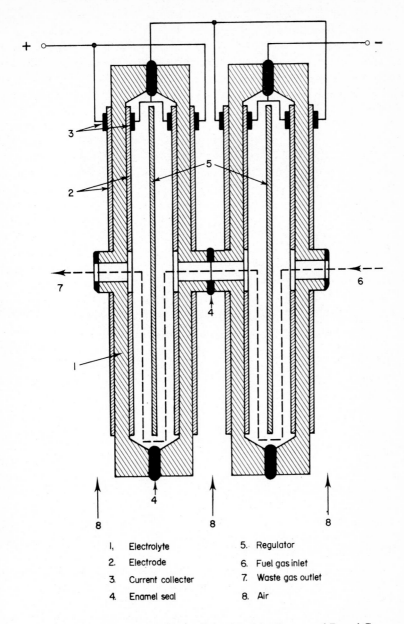

I,	Electrolyte	5.	Regulator
2.	Electrode	6.	Fuel gas inlet
3.	Current collecter	7.	Waste gas outlet
4.	Enamel seal	8.	Air

FIG. 23. The solid electrolyte fuel cell developed by Brown and Boveri Co.

in Fig. 27. Westinghouse Electric Corporation has discontinued development of this type of electrolyte in favour of thin film types. Thin film electrolyte has been obtained by atomizing a suspension of calcium zirconate and zirconium oxide in the ratio $(ZrO_2)_{0.85} (CaO)_{0.15}$ in *n*-butyl acetate, masking the parts of the interconnector $(CoCr_2O_4)$ before sintering at 1400°C to obtain 0.03–0.04 mm thick calcia stabilized zirconia.[73] The assembly of this thin electrolyte is schematically shown in Fig. 28.[74, 75] At Battelle Institut

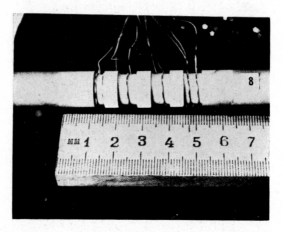

FIG. 24. The bell-and-spigot type electrolyte battery developed by Westinghouse Electric Corporation.

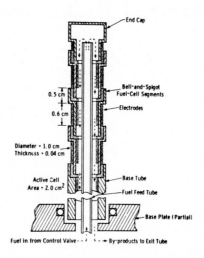

FIG. 25. Connection of the bell-and-spigot type solid electrolyte in series.

E.V. in Frankfurt/Main,[76] a fine powder of $(ZrO_2)_{0.91}$ $(Y_2O_3)_{0.09}$, the diameter of which was less than 50Å, was obtained from a mixture of $Zr(OC_3H_7)_4$ and $Y(OC_3H_7)_3$ which was sintered at 1400°C for 12 hr to obtain a dense disc of porosity less than 1%.

FIG. 26. Solid electrolyte 100W fuel cell power system.

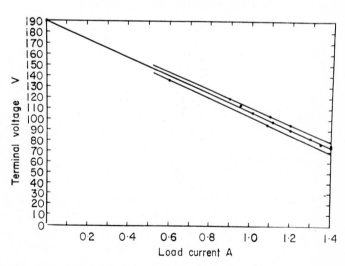

FIG. 27. Performance of the 100W cell constructed by Westinghouse Electric Corporation.

By these methods, fairly high discharge current densities have been obtained, but some of the cells have shown lower open-circuit voltages than their thermodynamically calculated values. This means, in most cases, that the electrolyte contains some electronic conduction so that a local cell is formed in which electronic current flows from cathode to anode in the electrolyte. In this case, Joule heat due to the electronic and ionic current of the local cell will be generated and this causes energy loss.[77]

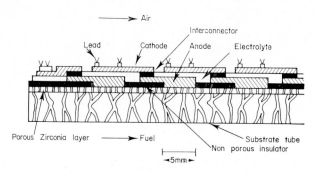

Fig. 28. Westinghouse thin film battery.

However, even when a purely ionic conductor is used, Joule heat is generated during discharge, and the energy loss may be unavoidable. Therefore, even if an electrolyte having some electronic conduction is used for fuel cells, if the absolute value of the ionic conductivity of the electrolyte is higher than that of an alternative purely ionic conductor, a higher output may be obtained.

Let the electrolyte show mixed conduction, i.e. oxide ion conduction and electronic conduction. When the fuel cell is constructed with such an electrolyte, electronic current flows through the electrolyte even at open circuit and the terminal voltage is somewhat lower than the theoretical value according to eqn (26). A typical relation between terminal voltage and current output in this case is shown in Fig. 29. In this figure, the polarizations other than those due to the ohmic resistance of the electrolyte are ignored for simplicity. The magnitude of the electronic current i_e at open circuit can be determined by extrapolating the line to E. When the cell is discharged through the outer circuit, the electronic current in the electrolyte decreases as the terminal voltage of the cell is lowered. i_e is derived from eqns (23), (24), (25) and (26) as

$$i_e = (1 - t_{O^{2-}})(\sigma E t_{O^{2-}} - i) \tag{39}$$

where i is the current output which is given by $i_{O^{2-}} - i_e$.

The total energy generated by this sort of fuel cell will be consumed by Joule heat due to $i_{O^{2-}}$ ($W_{O^{2-}}$), that due to i_e (W_e) and the power output W.

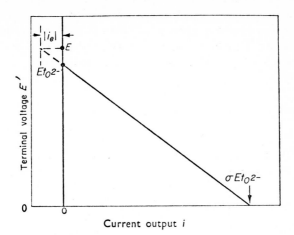

FIG. 29. Relation between terminal voltage and current output.

These quantities can be represented by

$$W_{O^{2-}} = (E - E') i_{O^{2-}} \tag{40}$$

$$W_e = E' i_e \tag{41}$$

$$W = E' i = t_{O^{2-}} Ei - (1/\sigma) i^2 \tag{42}$$

From these equations, the total energy W_0 is

$$W_0 = W_{O^{2-}} + W_e + W = E(i + i_e). \tag{43}$$

The energy efficiency η of the cell, that is, the ratio of power output to total energy is

$$\eta = W/W_0. \tag{44}$$

Then, from eqns (39), (42), (43) and (44), the relation between energy efficiency and current output can be given by the following equation,

$$\eta = \frac{\sigma E t_{O^{2-}} i - i^2}{\sigma E[t_{O^{2-}} (1 - t_{O^{2-}}) \sigma E + t_{O^{2-}} i]}. \tag{45}$$

Fig. 30 shows the relation between η and i for constant E and σ taking $t_{O^{2-}}$ as parameter indicating that the energy efficiency has a maximum at a certain current output. From this figure, it is evident that the efficiency of the cell, the electrolyte of which has some electronic conduction, is lower in the whole range of the current output than that of a cell with a purely ionic conductor for a constant value of total conductivity. However, if the total conductivity of the electrolyte is greater than that of the purely ionic conductor, the efficiency of the cell may be greater than that of the cell with the purely ionic conductor (above a certain current output). Fig. 31 shows η versus i for various σ indicating that the current output giving the maximum efficiency increases with increasing total conductivity, so that the power at maximum efficiency also increases and even when $t_{O^{2-}}$ is smaller than unity η can surpass the efficiency of the cell with $t_{O^{2-}} = 1$.

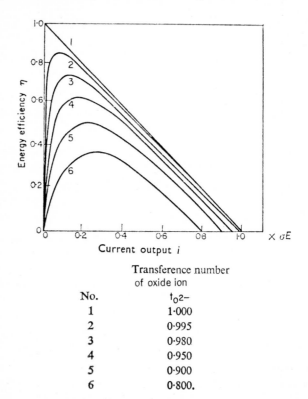

No.	Transference number of oxide ion $t_{O^{2-}}$
1	1·000
2	0·995
3	0·980
4	0·950
5	0·900
6	0·800.

FIG. 30. Relation between the energy efficiency and the current output.

Under the conditions shown in Fig. 30, the current giving maximum energy efficiency can be calculated from $\partial\eta/\partial i = 0$, that is,

$$i_{\eta_{max}} = \sigma E[\sqrt{(1 - t_{O^{2-}})} - (1 - t_{O^{2-}})] \tag{46}$$

ubstituting eqn (46) for i in eqn (45), the maximum energy efficiency η_{max} an be obtained as

$$\eta_{max} = \frac{1 - \sqrt{(1 - t_{O^{2-}})}}{1 + \sqrt{(1 - t_{O^{2-}})}}. \tag{47}$$

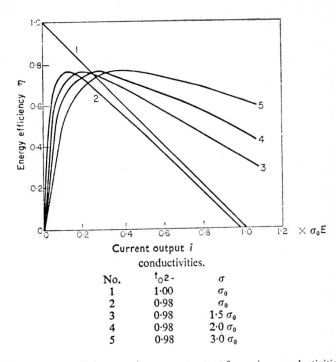

Current output i

conductivities.

No.	$t_{O^{2-}}$	σ
1	1·00	σ_0
2	0·98	σ_0
3	0·98	$1·5\,\sigma_0$
4	0·98	$2·0\,\sigma_0$
5	0·98	$3·0\,\sigma_0$

FIG. 31. Energy efficiency against current output for various conductivities.

Thus the maximum energy efficiency is dependent only on the oxide ion transference number and is independent of total conductivity. This is an important criterion for the selection of solid electrolytes for fuel cells. The relation of eqn (47) is illustrated in Fig. 32.

The power output at maximum energy efficiency can be obtained from eqns (42) and (46) as

$$W_{\eta_{max}} = \sigma E^2 \sqrt{(1 - t_{O^{2-}})}\,[1 - \sqrt{(1 - t_{O^{2-}})}]^2. \tag{48}$$

This equation is shown graphically in Fig. 33 indicating that a maximum value of $W_{\eta_{max}}$ is obtained at $t_{O^{2-}} = 8/9$. The relation between the power

output and the energy efficiency can be derived by eliminating i from eq
(44) and (45), as

$$W = (1/2)\,\sigma E^2\,\eta t_{O^2-}\{\sqrt{[(1-\eta)^2\,t_{O^2-}^2 - 4\eta(1-t_{O^2-})\,t_{O^2-}]}$$
$$+ (1-\eta t_{O^2-}) + (1-t_{O^2-})\}. \qquad (4$$

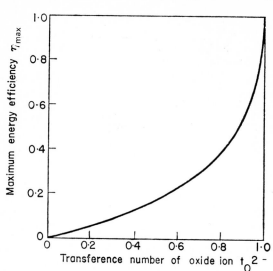

FIG. 32. Maximum energy efficiency versus t_{O^2-}.

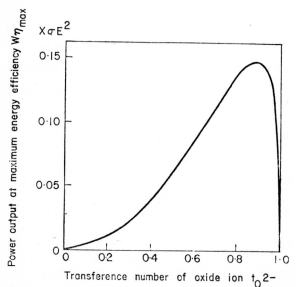

FIG. 33. Power output of maximum efficiency versus t_{O^2-}.

The relation between W and η for various t and for constant E and σ is shown in Fig. 34. In these curves, the arrows show the direction of the increase in current output.

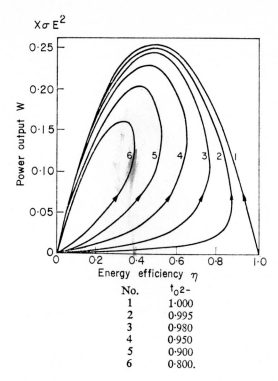

$$X\sigma E^2$$

FIG. 34. Relation between power output and $t_{O^{2-}}$ for constant conductivity.

No.	$t_{O^{2-}}$
1	1·000
2	0·995
3	0·980
4	0·950
5	0·900
6	0·800.

The power output at maximum energy efficiency does not coincide with the maximum power output, and Fig. 34 shows that higher power output can be obtained for a slight loss of efficiency, since the slope of W versus η curve is very steep in the vicinity of maximum efficiency.

As shown in eqn (49), the power output of the cell is a complicated function of η and $t_{O^{2-}}$, but proportional to σ. In Fig. 35, a σ–E relation is added to Fig. 34. Using this figure, the characteristics of various cells can be compared with respect to their energy efficiencies and power outputs, if the conductivities and oxide ion transference numbers are known. For example, consider two cells I and II, the electrolytes of which have total conductivities σ_I and σ_{II} ($\sigma_I < \sigma_{II}$) and oxide ion transference numbers t_I and t_{II} ($t_I > t_{II}$) respectively. When cells I and II are operated in the vicinity of $\eta = \eta_k$ (Fig. 35), cell II has a higher output power than cell I, i.e. the cell having the electrolyte with the smaller oxide ion transference number shows the highe output. In the vicinity of $\eta = \eta_l$ $(\eta_l > \eta_k)$, however, the power output o cell I is superior to that of cell II.

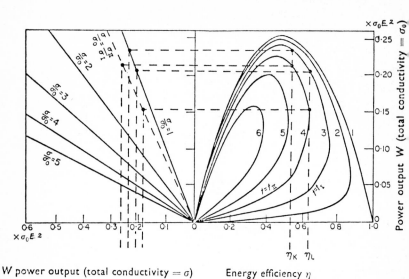

W power output (total conductivity = σ)   Energy efficiency η

FIG. 35. Relation between energy efficiency and power output with respect to conductivit and transference number of oxide ion.

TABLE IV.   Comparison of total conductivities of electrolytes with various t_{O^2-} a the same cell efficiencies

$t \backslash \eta$	0·35	0·40	0·45	0·50	0·55	0·60	0·65	0·70	0·75	0·80	0·85
1·000	1·0	1·0	1·0	1·0	1·0	1·0	1·0	1·0	1·0	1·0	1·0
0·995	1·1	1·1	1·1	1·1	1·1	1·1	1·1	1·1	1·1	1·1	1·5
0·990	1·1	1·1	1·1	1·1	1·1	1·1	1·1	1·1	1·2	1·4	
0·980	1·1	1·1	1·1	1·1	1·1	1·1	1·2	1·3	1·5		
0·960	1·1	1·1	1·1	1·1	1·2	1·2	1·5				
0·940	1·1	1·2	1·2	1·2	1·3	1·5					
0·920	1·2	1·2	1·2	1·3	1·5						
0·900	1·2	1·2	1·3	1·4							
0·850	1·3	1·4									
0·800	1·6										

Ratio of total conductivity of electrolyte to that of a purely ionic conductor $(t_{O^2-} = 1)$.

In Table IV, the ratio of total conductivity of the electrolyte to that of a purely ionic conductor giving the same efficiency is listed. For example, in order to get an efficiency of 0·60 using an electrolyte with oxide ion transference number 0·94, it is sufficient to use an electrolyte of conductivity 1·5 times greater than that of the purely ionic conductor.

C. Electrode Materials

The design and the selection of materials for electrodes and electrode leads are important problems in view of the operational requirements and working environment of the cell. The most significant material limitations are for the cathode and cathode lead. A wide variety of materials have been proposed for cathode materials.

1. Cathodes

(a) *Metals*. Because of the highly oxidizing environment at the cathode, only noble metals such as platinum,[22] palladium, gold and silver[78,79] have been tested for this purpose. Their use in practical fuel cell systems, however, has been limited in the cases of platinum, palladium and gold because of economics and because of volatilization in the case of silver.[80]

(b) *Oxides with Metallic Collectors*. An example of this type of electrode is porous stabilized zirconia applied over a platinum or palladium grid. This type of cathode has been estimated, however, to be impractical because of economics in the cases of both platinum or palladium grids.

(c) *Oxides*. The electrode made from a porous oxide having sufficiently high electronic conductivity is this class of cathode. Oxides proposed for cathode material so far are manganese oxide, copper oxide,[81] lithium doped nickel oxide,[82,84] strontium doped $LaCoO_3$,[85] tin doped indium oxide,[86] alumina or zirconia doped zinc oxide[87,88] and doped or undoped perovskites such as $LaCoO_3$ or $PrCoO_3$.[89,90] The electrical conductivities of these oxides are shown in Fig. 36 as a function of temperature. It is clear from this figure that none of these materials is without its limitations so far as electronic conductivity is concerned and none of these materials is fully satisfactory for use in high temperature solid electrolyte fuel cells due to adverse properties such as the volatilization of dopant, reaction with electrolyte, spalling and cracking during cooling cycles and unequal thermal expansion.[91]

2. Anodes

Various powdered metals such as platinum, nickel, iron, titanium, manganese, cobalt and copper,[92-94] metal oxides such as nickel oxide, chromic oxide, cobaltous oxide, ferric oxide, titanium oxide[91] and uranium oxide,[95]

cobalt–zirconia and nickel–zirconia cermets[96] and carbides such as zirconium carbide have been examined with a view to finding suitable anode
materials. Of these materials, nickel, manganese, titanium oxide and cermets

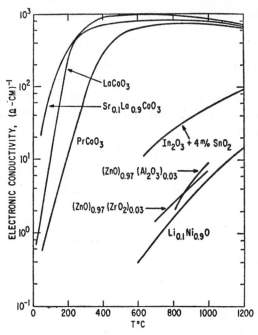

Fig. 36. Electrical conductivity as a function of temperature for various oxides proposed
for use in fuel cell cathodes.

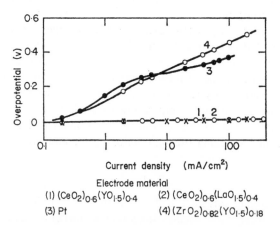

Electrode material

(1) $(CeO_2)_{0.6}(YO_{1.5})_{0.4}$ (2) $(CeO_2)_{0.6}(LaO_{1.5})_{0.4}$

(3) Pt   (4) $(ZrO_2)_{0.82}(YO_{1.5})_{0.18}$

Fig. 37. Anodic polarization characteristics of various electrode materials at 1000°C.

have been found to be desirable as regards to their conductivities and adherabilities. Further, the solid solutions of ceria and lanthana or yttria have been used as anode materials and a highly depolarizing effect has been observed as shown in Fig. 37.[91] This depolarizing effect may be attributed to the mixed conduction of oxide ion and electron in the solid solution of ceria which originates in the reduction of ceria in a reducing atmosphere. This means that these solid solutions serve not only as the anode but also as the electrolyte. In order to ascertain this point, the conductivities of $(CeO_2)_{0.6}(LaO_{1.5})_{0.4}$ and $(CeO_2)_{0.6}(YO_{1.5})_{0.4}$ have been measured in air and in anodic gas atmosphere to obtain the results shown in Fig. 38. This

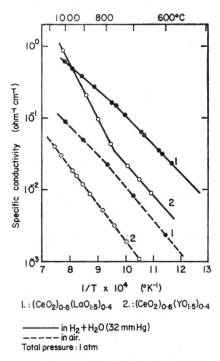

$$1.:(CeO_2)_{0.6}(LaO_{1.5})_{0.4} \quad 2.:(CeO_2)_{0.6}(YO_{1.5})_{0.4}$$

——— in $H_2 + H_2O$ (32 mmHg)
– – – – in air.
Total pressure : 1 atm

FIG. 38. Specific conductivity of the solid solutions containing ceria.

figure indicates that the conductivity in the anodic gas atmosphere is about ten times as high as that in air. This is considered to be due to electronic conduction which is produced by the partial reduction of ceria. Consequently, $(CeO_2)_{0.6}(LaO_{1.5})_{0.4}$ and $(CeO_2)_{0.6}(YO_{1.5})_{0.4}$ are mixed ionic and electronic conductors in an anodic gas atmosphere. In these conductors, the electronic contribution is dominant making the electrode reaction zone larger than when a pure electronic conductor is used as the electrode material (illustrated schematically in Fig. 39).

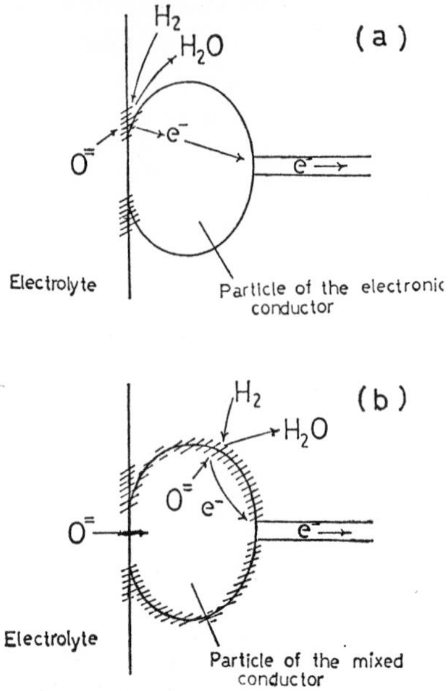

(a) in the case of the electronic conductor;
(b) in the case of the mixed conductor.
(The hatched parts indicate the reaction zone.)

FIG. 39. Model of the electrode reaction zone.

D. Cell Characteristics

Using $(ZrO_2)_{0.85}(CaO)_{0.15}$ as the electrolyte, Weissbart and Ruka[22] studied the current voltage relation of a cell at 800–1094°C, the fuel of this cell was hydrogen or methane and the electrode material was platinum. They obtained non-resistance polarization less than 10 mV for apparent current densities up to 20 mA/cm² at 810°C and 110 mA/cm² at 1094°C. A similar result was obtained by Neuimin et al.,[55] indicating that their cells showed mainly resistance polarization. Moreover, the cell characteristics obtained by the investigators of Westinghouse Electric Corporation[97] supported these results as shown in Figs. 40 and 41. Figure 40 shows the performance of the plate type cells and Fig. 41 illustrates that of the bell-and-spigot type electrolyte cell composed of three unit cells. These figures indicate that deviations from linear relations are observed at small current density ranges suggesting that the polarizations other than resistance may be found at least

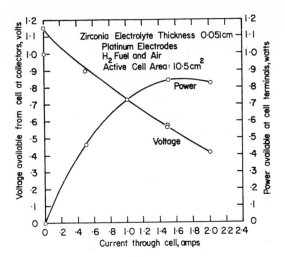

FIG. 40. Voltage and power output of Westinghouse solid electrolyte fuel cell at 1010°C.

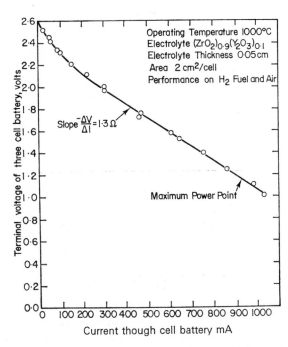

FIG. 41. Voltage-current characteristics of three-cell battery with bell-and-spigot joints.

in these ranges. The CeO_2–La_2O_3 electrolyte cell[44, 98] shows the performance as illustrated in Fig. 42[44] and indicates that polarization is only that of resistance at 1000°C when hydrogen is used as fuel. This advantage may offset the disadvantage of somwhat smaller open circuit voltage than in the case of stabilized zirconia electrolyte cells. In Fig. 43 the cell performances obtained using various kinds of fuel are shown. The relatively small open circuit voltages are attributed to the existence of the electronic conduction in the electrolyte.

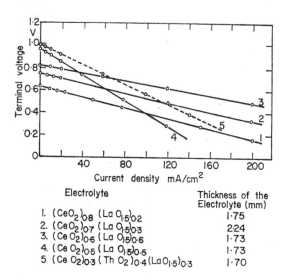

Electrolyte	Thickness of the Electrolyte (mm)
1. $(CeO_2)_{0.8}(LaO_{1.5})_{0.2}$	1·75
2. $(CeO_2)_{0.7}(LaO_{1.5})_{0.3}$	2·24
3. $(CeO_2)_{0.6}(LaO_{1.5})_{0.6}$	1·73
4. $(CeO_2)_{0.5}(LaO_{1.5})_{0.5}$	1·73
5. $(CeO_2)_{0.3}(ThO_2)_{0.4}(LaO_{1.5})_{0.3}$	1·70

FIG. 42. Characteristics of hydrogen-oxygen fuel cells at 1000°C.

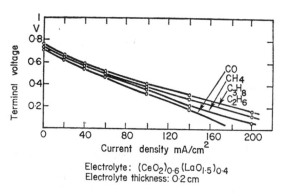

Electrolyte: $(CeO_2)_{0.6}(LaO_{1.5})_{0.4}$
Electrolyte thickness: 0·2 cm

FIG. 43. Cell performances with various fuels at 1000°C.

In Table V, the maximum power densities of cells under the following conditions are listed to show that the cell with $(CeO_2)_{0.6}$ $(LaO_{1.5})_{0.4}$ electrolyte has the best characteristics.

electrolyte thickness : 1 mm

apparent electrode area : 1 cm^2

fuel : H_2, oxidant : O_2

TABLE V.   Maximum Output Power.

Electrolyte	Maximum Output Power (W)	Current and Voltage at Maximum Output	
		Current Density (A)	Terminal Voltage (V)
$(CeO_2)_{0.8}$ $(LaO_{1.5})_{0.2}$	0·07	0·23	0·30
$(CeO_2)_{0.7}$ $(LaO_{1.5})_{0.3}$	0·14	0·39	0·36
$(CeO_2)_{0.6}$ $(LaO_{1.5})_{0.4}$	0·17	0·36	0·47
$(CeO_2)_{0.5}$ $(LaO_{1.5})_{0.5}$	0·07	0·14	0·475
$(CeO_2)_{0.3}$ $(ThO_2)_{0.4}$ $(LaO_{1.5})_{0.3}$	0·09	0·17	0·48
$(ZrO_2)_{0.85}$ $(CaO)_{0.15}$†	0·05	0·09	0·57
$(Zr_2)_{0.85}$ $(CaO)_{0.15}$‡	0·04	0·09	0·50
$(ZrO_2)_{0.9}$ $(Y_2O_3)_{0.1}$‡	0·08	0·23	0·38

† reduced from reference 22
‡ reduced from reference 72

In Fig. 44,[99] the performance of 1·39 mm thick $(ZrO_2)_{0.92}$ $(Yb_2O_3)_{0.08}$ electrolyte cell at 800°C is shown. The electrode materials of this cell are nickel for the anode and silver for the cathode. The cell characteristics are excellent though the high cost of Yb_2O_3 is a problem for practical use.

In Figs. 45 and 46,[88] the performances of cells with metal oxide electrodes are illustrated showing that these oxide electrodes can improve the discharge characteristics of high temperature solid electrolyte fuel cells. The same improvement of cell performance has been found with a cell using the perovskite type electrolyte. Typical cell performances are shown in Figs 47 and 48.[100]

E. Fossil Fuel Solid Electrolyte Fuel Cells

Fossil fuels are used for solid electrolyte fuel cells in the form of gases such as gaseous hydrocarbons, or reformed or partially oxidized gases of hydrocarbons, petroleum or coal. As carbonaceous gas fuels easily form carbon

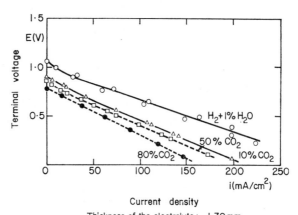

Current density
Thickness of the electrolyte : 1·39mm

FIG. 44. Performance of the $(ZrO_2)_{0.92}(Yb_2O_3)_{0.08}$ electrolyte cell.

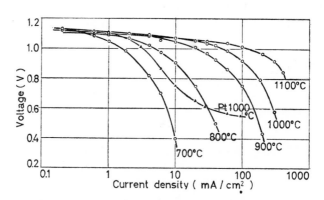

Anode gas : H_2 (1 atm., containing 12.7 mm Hg H_2O)
Cathode gas : O_2 (1 atm.)
—○— : $(TiOx)$-H_2/$(ZrO_2)_{0.91}(Y_2O_3)_{0.09}$/
    O_2-$(ZnO)_{0.97}(Al_2O_3)_{0.03}$
—▲— : (Pt)-H_2/$(ZrO_2)_{0.91}(Y_2O_3)_{0.09}$/$O_2$-$(Pt)$

FIG. 45. Cell characteristics of the cell provided for metal oxide electrode (Resistance polarization free).

deposits by pyrolysis at high temperatures, reformed hydrogen rich gases are usually used as fuels. For example, in the case of propane, the reactions of

$$C_3H_8 \rightarrow C_3H_6 + H_2 \text{ (pyrolysis)}$$

$$C_3H_6 + 3H_2O \rightarrow 3CO + 6H_2 \text{ (reforming)}$$

$$CO + H_2O \rightarrow CO_2 + H_2 \text{ (water gas shift)}$$

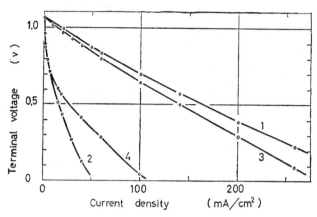

Anode material :
1) $(CeO_2)_{0.6}(YO_{1.5})_{0.4}$ 2) Pt
3) $(CeO_2)_{0.6}(LaO_{1.5})_{0.4}$ 4) Pt
Electrolyte : $(ZrO_2)_{0.82}(YO_{1.5})_{0.18}$.   Thickness : 1.5 mm

Fig. 46. Cell performance of the hydrogen cell at 1000°C.

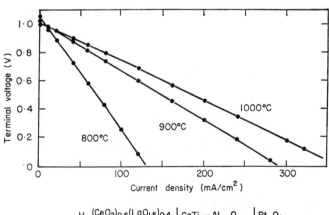

$$H_2, \frac{(CeO_2)_{0.6}(LaO_{1.5})_{0.4}}{Pt} \mid CaTi_{0.7}Al_{0.3}O_{3-\alpha} \mid Pt, O_2$$

Electrolyte thickness ; 0.35 mm

Fig. 47. Performance of the perovskite type electrolyte cell at various temperatures.

1040   Takehiko Takahashi

Several types of fuel cell system provided with reformers are in the stage of development. One of them is an inner reforming process which means that the electrode materials of the anode serve for also reforming catalysts, thus the anode chamber of the cell being employed also serves as the reformer. The other process is an outer reforming process which means that the reformer is provided separately from the fuel cell.

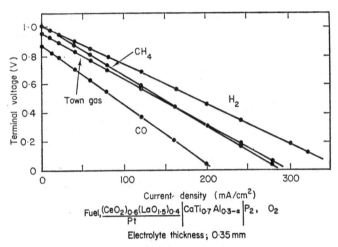

FIG. 48. Cell characteristics with the perovskite type electrolyte at 100°C using various fuels.

In Fig. 49[60] an example of the outer reforming process is schematically drawn and its performance is shown in Fig. 50.[60] In a fuel cell which is operated with reforming gas, the waste heat of the cell which is generated by

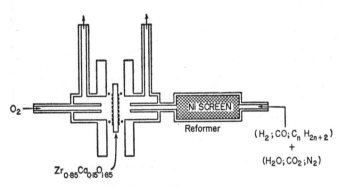

FIG. 49. Schematic diagram of an example of the outer reforming solid electrolyte fuel cell system.

passing the current through the electrolyte must, in any case, make up for the heat required to continue the reforming reaction. For example, at 600°C, the endothermic heat required for the reforming of propane,

$$C_3H_8 + 3H_2O \rightarrow 3CO + 7H_2$$

is 127,060 cal, while the exothermic heat of the cell reaction,

$$3CO + 7H_2 + 5O_2 \rightarrow 3CO_2 + 7H_2O$$

is 615,650 cal, the former being about 20% of the latter. This means that the heat required for the reforming reaction is supplied by the heat generated from the cell if the fuel cell is operated under the thermal efficiency below 80%.[101]

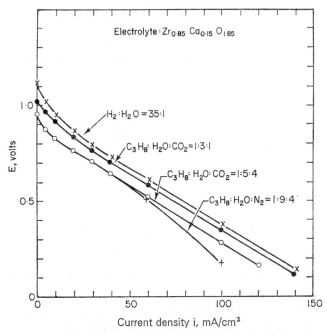

FIG. 50. Characteristics of the outer reforming solid electrolyte fuel cell at 1000°C.

The object of developing solid electrolyte high temperature fuel cells is to use them for on-site power generating stations or large scale central station powerplants. The former is represented by the TARGET (Team to Advance Research for Gas Energy Transformation) project[102] which has been sponsored by 32 companies in the United States. The object of the

TARGET project is a power plant which supplies the power to serve a family dwelling, town house groupings, schools, hospitals and other commercial establishments using natural gas fuel cell systems. These systems may include either partial oxidation or steam reforming process, but as is mentioned above, a decrease in the available chemical energy occurs in either case. Figure 51 and $52^{(103)}$ show the flow diagrams of these processes in which q denotes the heat loss. The cost of the electricity obtainable by steam reforming system is evaluated as in Table VI.$^{(103)}$

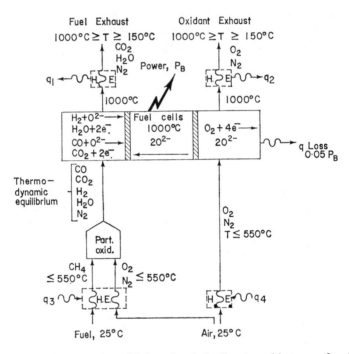

FIG. 51. Flow diagram for solid electrolyte fuel cell system with steam reforming.

In this table, the various regions in the United States have been taken to be characterized by averages of certain cities as follows:

East: average of Boston and New York
Central: average Cleveland, Chicago, Detroit and Pittsburgh
West: Los Angeles
Southwest: average of Kansas City, Denver, Houston and El Paso

As this evaluation is based on the assumption that the cell is constructed with the calcia stabilized zirconia electrolyte and the platinum electrodes, the

TABLE VI.

Estimated cost of electricity via solid-oxide fuel cells with steam reforming

	East	Central	West	South-west
		Electricity cost,† mills/kWh		

Residential (59·5% eff., 0·05 load factor, 17% cap. charge rate, $173/kW)

1960
	East	Central	West	South-west
Gas at well-head	0·86	0·86	0·86	0·86
Transmission and distribution	4·21	3·31	3·11	2·79
Generation	67·6	67·6	67·6	67·6
Total	72·67	71·77	71·57	71·25

1980
Gas at well-head	1·86	1·86	1·86	18·6
Transmission and distribution	7·77	6·16	5·77	5·17
Generation	67·6	67·6	67·6	67·6
Total	77·23	75·62	73·23	74·63

2000
Gas at well-head	2·87	2·87	2·87	2·87
Transmission and distribution	11·47	9·03	8·48	7·59
Generation	67·6	67·6	67·6	67·6
Total	81·94	79·50	78·95	78·06

Industrial (53% eff., 0·8 load factor, 17% cap, charge rate, $312/kW)

1960
Gas at well-head	0·96	0·96	0·96	0·96
Transmission and distribution	2·24	1·25	1·02	0·64
Generation	7·55	7·55	7·55	7·55
Total	10·75	9·76	9·53	9·15

1980
Gas at well-head	2·10	2·10	2·10	2·10
Transmission and distribution	4·11	2·30	1·88	1·21
Generation	7·55	7·55	7·55	7·55
Total	13·76	11·95	11·53	10·86

2000
Gas at well-head	3.20	3.20	3.20	3.20
Transmission and distribution	6.14	3.38	2.77	1.77
Generation	7.55	7.55	7.55	7.55
Total	16·89	14·13	13·52	12·52

† Assumes very substantial improvements in technology of solid-oxide fuel cells.

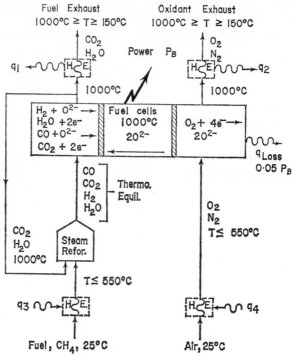

FIG. 52. Flow diagram for solid electrolyte cell system with partial oxidation.

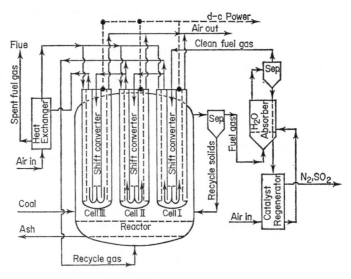

FIG. 53. Fuel cell reactor vessel.

isted cost may be made somewhat lower if the electrode material is converted to cheaper materials instead of platinum.

Solid electrolyte fuel cells for central power generation have been developed by the investigators of Westinghouse Electric Corporation.[74, 75, 104] As stated previously, calcia doped zirconia thin electrolyte, tin doped indium oxide cathode and cobalt or nickel cermat anodes have been used for cell materials. The construction of the cells which Westinghouse is producing is shown in Fig. 28. The reactor which will be made up from the unit cells and the schematic diagrams of the fuel cell power plant are illustrated in Figs. 53 and 54,[105] respectively. Figure 54 shows the power plant using coal as fuel, but it will be possible to use natural gas or petroleum if a gasification process such as reforming or partial oxidation is added to the system.

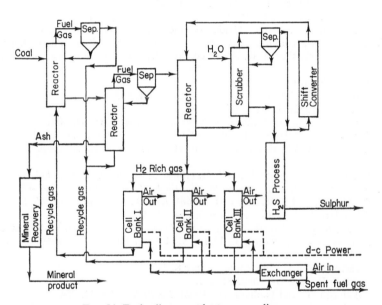

Fig. 54. Fuel cell power plant process diagram.

The advantages of a fuel cell power plant are the possibility of obtaining high efficiency, no cooling water requirement, air polution minimized, compactness and efficiency and cell costs per unit of power relatively insensitive to the power capacity.

Westinghouse Electric Corporation has estimated the installed cost of 1000 MW coal burning fuel cell power plant at $107,900,000. The plant output is to be set at 800 V direct current and 1250 kA which is to be converted to 345 kV alternating current. The estimated total energy cost was 2·8 mils/kWh when the load factor is 0·9 and the cost of coal is $0·18 pe M.B.T.U.[68, 74]

Though there are many problems which remain to be solved, Westing house's development schedule for solid electrolyte fuel cell power plants i 20 MW operation by 1973 and 100 MW by 1974[104].

Further, the application of solid electrolyte fuel cells to traction powe sources may be possible if the operating temperature of the cell could b lowered to 600–800°C.

References

1. Nernst, W. (1900). *Z. Elektrochem.* **6,** 41.
2. Baur, E. and Preis, H. (1937). *Ibid.* **43,** 727.
3. Ruff, O. and Ebert, F. (1929). *Z. anorg. Chem.* **180,** 19 .
4. Ryshkewitch, E. (1960). "Oxide Ceramics", p.353. Academic Press, New Yorl
5. Levin, E. M., Robbins, C. R. and McMurdie, H. F. (1964). "Phase Diagram for Ceramists". (M. K. Reser, ed.) p. 105, Amer. Ceram. Soc., Columbus Ohio.
6. *Ibid.,* (1964) p.140.
7. Garvie, R. C. (1968). *J. Amer. Ceram. Soc.* **51,** 553.
8. Hund, F. (1952). *Z. phys. Chem.* **199,** 142.
9. Tien, T. Y. and Subbarao, E. C. (1963). *J. Chem. Phys.* **39,** 1041.
10. Dietzel, A. and Tober, H. (1953). *Ber. Dtsch. Keram. Ges.* **30,** 47, 71.
11. Duwez, P., Odell, F. and Brown, F. H. Jr. (1952). *J. Amer. Ceram. Soc.* **35** 107.
12. Johansen, H. A. and Cleary, J. G. (1964). *J. Electrochem. Soc.* **111,** 100.
13. Strickler, D. W. and Carlson, W. G. (1964). *J. Amer. Ceram. Soc.* **47,** 122.
14. Baukal, W. and Scheidegger, R. (1968). *Ber. Dtsch. Keram. Soc.* **45,** 610.
15. Schmalzried, H. (1962). *Z. Electrochem.* **66,** 572.
16. Carter, R. E. and Roth, W. L. (1963). G. E. Res, Rept. No. 63–RL–3479M
17. Kingery, W. D., Pappis, J., Doty, M. E. and Hill, D. C. (1959). *J. Amer Ceram. Soc.* **42,** 393.
18. Lefèvre, J. (1963). *Ann. Chim. France* **8,** 117.
19. Bratton, R. J. (1969). *Ibid.* **52,** 213.
20. Takahashi, T. and Iwahara, H. (1966). *Denki Kagaku,* **34,** 254.
21. Kiukkola, K. and Wagner, C. (1961). *J. Electrochem. Soc.* **104,** 379.
22. Weissbart, J. and Ruka, R. (1963). "Fuel Cells" (G. J. Young, ed.) Vol. 2 p. 37, Reinhold Pulb. Corp, New York.
23. Palguev, S. F. and Neuimin, A. D. (1961). *Electrochem. Molten Solid Electrolyt* **1,** 90.
24. Volchenkova, Z. S., and Palguev, S. F. (1964). *Ibid.* **2,** 53.
25. Subbarao, E. C. and Sutter, P. H. (1963). *J. Amer. Ceram. Soc.* **48,** 433.
26. Steele, B. C. H. and Alcock, C. B. (1965). *Trans. Metal Soc. A.I.M.E.* **233** 1359.
27. Tubandt, C., Reinhold, H. and Liebold, G. (1931). *Z. anorg. allg. Chem.* **197** 225.
28. Hebb, M. (1952). *J. Chem. Phys.* **20,** 185.

29. Wagner, C. (1956). *Proc. C.I.T.C.E.* **7**, 361; (1956). *Z. Elektrochem.* **60**, 4.
30. Odell, F. and Brown, F. H. Jr. (1952). *J. Amer. Ceram. Soc.* **35**, 107.
31. Trombe, F. and Foex, M. (1953). *Compt. Rend.* **236**, 1783.
32. Volchenkova, Z. S. and Palguev, S. F. (1961). *Electrochem. Molten Solid Electrolyte* **1**, 97.
33. Dixon, J. M. *et al.* (1963). *J. Electrochem. Soc.* **110**, 276.
34. Subbarao, E. C. and Sutter, P. H. (1964). *J. Phys. Chem. Solid* **25**, 148.
35. Kröger, F. A. (1966). *J. Amer. Ceram. Soc.* **49**, 215.
36. Takahashi, T. and Suzuki, Y. (1967). *Proc. J. Intl. Etude Piles Combust.* **2**, 378.
37. Smoot, T. W. and King, D. F. (1963). *I.E.E.E. Trans. Aerospace A.S.I.* p.1192.
38. Strickler, D. W. and Carlson, W. G. (1965). *J. Amer. Ceram. Soc.* **48**, 286.
39. Baukal, W. (1969). *Electrochim. Acta* **14**, 1071.
40. Tannenberger, H., Schachner, H. and Kovacs, P. (1965). *Proc. J. Intl. Etude Piles Combust.* **1-III**, 19.
41. Levin, E. M., Robbins, C. R. and McMurdie, H. F. (1964). "Phase Diagrams for Ceramists" (M. K. Reser, ed.) p. 138, Amer. Ceram. Soc., Columbus, Ohio.
42. Robert, G., Forestier, M. and Deportes, C. (1969). *Proc. J. Intl. Etude Piles Combust.* **3**, 97.
43. Takahashi, T., Suzuki, Y. and Yoshida, T. (1970). *Denki Kagaku* **38**, 51.
44. Takahashi, T., Ito, K. and Iwahara, H. (1965). *Proc. J. Intl. Piles Combust.* **1-III**, 42.
45. Holverson, E. L. and Kevane, C. J. (1966). *J. Chem. Phys.* **44**, 3692.
46. Blumenthal, R. N. and Pinz, B. A. (1967). *J. Appl. Phys.* **38**, 2376.
47. Yushina, L. D. and Palguev, S. F. (1964). *Electrochem. Molten Solid Electrolyte* **2**, 74.
48. Steele, B. C. H. and Alcock, C. B. (1965). *Trans. Metal. Soc. A.I.M.E.* **233**, 1359.
49. Takahashi, T. and Iwahara, H. (1967). *Denki Kagaku* **35**, 433.
50. Takahashi, T., Iwahara, H. and Ichimura, T. (1969). *Denki Kagaku* **37**, 857.
51. Takahashi, T., Iwahara, H., Ichimura, T. and Aoyama, H. (1969). *Annual Rept. Asahi Glass Foundation Contrib. Ind. Techn.* **15**, 237.
52. Takahashi, T., Iwahara, H. and Itoh, I. (1970). *Denki Kagaku* **38**, 288.
53. Hartung, R. and Möbius, H. H. (1967). *Z. Chem. (Leipzig)* **8**, 325.
54. Karpachev, S. V., Filayev, A. T. and Palguev, S. F. (1964). *Electrochim. Acta* **9**, 1681.
55. Neuimin, A. D., Karpachev, S. V. and Palguev, S. F. (1961). *Proc. Acad. Sci. U.S.S.R. Phys. Chem. Sect.* **141**, 875.
56. Perfilev, M. V. and Palguev, S. F. (1966). *Electrochem. Molten Solid Electrolyte* **3**, 105.
57. Perfilev, M. V. and Palguev, S. F. (1967). *Ibid.* **4**, 153.
58. Filayev, A. T., Karpachev, S. V. and Palguev, S. F. (1967). *Ibid.* **4**, 161.
59. Etsell, T. H, and Flengas, S. N. (1970). *Ext. Abst. No.* 329 *Electrochem. Soc. Los. Angeles Meeting,* May .
60. Binder, H., Köhling, A., Krupp, H., Richter, K. and Sanstede, G. (1963). *Electrochim. Acta.* **8**, 781.
61. Möbius, H. H. and Rohland, B. (1966). *Z. Chem. (Leipzig)* **4**, 158.
62. Kleitz, M. (1968). Thesis, University Grenoble .
63. Takahashi, T., Ito, K. and Iwahara, H. (1965). *Denki Kagaku* **34**, 205.

64. Casselton, R. E. W. (1968). "Electromotive Force Measurements in High Temperatrue Systems". (C. B. Alcock, ed) p. 151, Elsevier, New York.
65. Croatto, V. and Mayer, A. (1943). *Gazz Chim. Ital.* **73**, 199.
66. Neuimin, A. D. and Palguev, S. F. (1964). *Electrochem. Molton Solid Electrolyte* **2**, 79.
67. Shultz, E. B. Jr., Vorres, K. S., Marianowski, L. G. and Linden, H. R. (1963). "Fuel Cells" (G. J. Young, ed). Vol. 2 p. 24. Reinhold Publ. Corp. New York.
68. White, H. F. (1966). "Project Fuel Cell", O.C.R. Rept. No. 17, Contract No. 01–0001–500.
69. Bliton, J. L., Rechter, H. L. and Harada, Y. (1963). *Ceram. Bull.* **42**, 6.
70. Tannenberger. H, (1966). Inst. Battelle Genève, private communication.
71. Antonsen, O., Baukal, W. and Fischer, W. (1966). *Brown Boveri Rev.* **53**, 21.
72. Archer, D. H. *et al.* (1964); *Proc. Ann. Power Sources Conf.* **18**, 36; (1965), "Fuel Cell System" p. 332, 343, Am. Chem. Soc., Washington D. C.; (1965). "Hydrocarbon Fuel Cell" (B. S. Baker, ed.) p. 51 Academic Press, New York and London.
73. Maskalick, N. J. and Sun, C. C. (1970). Ext. Abst. No. 14, Electrochem. Soc. Atlantic City Meeting, Oct.
74. White, H. F. (1968). I.E.E.E. Trans. Power Apparatus System, PAS–87, 1956.
75. Sverdrup, E. F. *et al.* (1970). "1970 Final Report Project Fuel Cell Research and Development Report No. 57". U.S. Government Printing Office, Washington D.C.
76. Baukal, W. (1970). Battelle Inst, E. V. Frankfurt/Main, private communication.
77. Takahashi, T., Ito, K. and Iwahara, H. (1967). *Electrochim. Acta.* **12**, 21.
78. Schachner, H. and Tannenberger, H. (1965). *Comp. Rend.* **III**, 49.
79. Tannenberger, H. and Siegert, H. (1969). "Fuel Cell System-II" p. 281 Am. Chem. Soc. Washington D.C.
80. Honig, R, E. (1962). *RCA Rev.* **23**, 567.
81. Ranny, M. W. (1969). "Fuel Cells, Recent Developments" p. 99 Noyes Development Corp., N. J.
82. Verwey, E. J. W., Haaijman, P. W., Romeijin, F. C. and van Oosterhout, G. W. (1950). *Philips Res. Repts.* **5**, 173.
83. Heikes, R. R. and Johnston, W. D. (1957). *J. Chem. Phys.* **26**, 582.
84. van Houten, S. (1960). *J. Phys. Chem. Solids* **17**, 7.
85. Button, D. D. and Archer, D. H. (1966). 68th Ann. Meeting Am. Ceram. Soc. Washington D.C.
86. Sverdrup, E. F., Archer, D. H. and Glasser, A. D. (1969). "Fuel Cell System-II", p. 301 Am. Chem. Soc. Washington D.C.
87. Takahashi, T., Suzuki, Y., Ito, K. and Hasegawa, H. (1967) *Denki Kagaku* **35**, 201.
88. Takahaski, T., Iwahara, H. and Suzuki, Y. (1969). *Proc. J. Intl. Etude Piles Combust.* **3**, 113.
89. Heikes, R. R., Miller, R. C. and Mazelsky, R. (1964). *Physica* **30**, 1600.
90. Goodenough, J. B., and Raccah P. M. (1963). *J. Appl. Phys.* **36**, 1031.
91. Tedman. C. S. Jr., Spacil, H. S. and Mitoff, S. P. (1969). G. E. Rept. No. 69-c–056.
92. Takahashi, T., Suzuki, Y., Ito, K. and Yamanaka, H. (1968). *Denki Kagaku* **36**, 345.

93. Eysel, H. H. (1970). Ext, Abst. No. 36, Electrochem, Soc. Atlantic City Meeting, Oct.
94. Clark, D. P, and Meredith, R. E. (1967). *Electrochem. Techn.* **5,** 9.
95. Rohland, B. and Möbius, H. H. (1968). *Naturwiss.* **55,** 227.
96. Isenberg, A. O. (1970). Ext, Abst. No. 13, Electrochem. Soc. Atlantic City Meeting Oct.
97. Archer, D. H., Alles, J. J., English, W. A., Elikan, L., Sverdrup, E. F. and Zahranik, R. L. (1965). "Fuel Cell Systems". p. 332. Am. Chem. Soc., Washington, D.C.
98. Singman, D. (1966). *J. Electrochem. Soc.* **113,** 502.
99. Schachner, H. and Tannenberger, H. (1965). *Proc. J. Intl. Etude Piles Combust,* **1-III,** 49.
100. Takahashi, T. and Iwahara, H. (1971). *Energy Conv.* **11,** 105.
101. Smith, J. G. (1966). "An Introduction to Fuel Cells", (K. R. Williams, ed.) p. 183, Elsevier, London.
102. Burlingame, M. V. (1969). "Fuel Cell System-II" (B. S. Baker, ed.). p. 377, Am. Chem. Soc. Washington D. C.
103. Boll, R. H. and Bhada, R. K. (1968). *Energy Conv.* **8,** 3.
104. Archer, D. H. (1968). *Mech. Eng.* **90,** 42.
105. Cochran, N. P. (1969). "Fuel Cell System-II" (B. S. Baker, ed.). p. 383, Am. Chem. Soc. Washington D. C.

25. PHOTO-ELECTROCHEMICAL PHENOMENA

A. C. H. van Peski

Dalco, Inc., Soestduinen, Netherlands

J. Schoonman

Laboratorium voor Kernfysica en Vaste Stof,
State University, Utrecht, Netherlands

and

D. A. de Vooys

Van 't Hoff Laboratorium,
State University, Utrecht, Netherlands

Introduction

Photochemistry, the study of the chemical reaction between light quanta and matter, has many aspects. Life on earth is only made possible by the existence of the oxygen cycle (photosynthesis). In modern organic synthesis photochemistry can be a valuable tool, e.g. in the commercial production of vitamin D2. Modern communication systems are unimaginable without photography and photopolymeric printing plates.

In the framework of this book this chapter will be limited to three topics in which photons give rise to an electrical current in a solid, thus resulting in a chemical conversion. The three topics are: (i) The latent-image formation and photolysis of silver halides (by A. C. H. van Peski); (ii) The photochemical decomposition of lead chloride and lead bromide (by J. Schoonman); (iii) The electrode photoeffect of metals in an electrolyte solution (by D. A. de Vooys).

I. The Latent-Image Formation and Photolysis of Silver Halides

A. C. H. van PESKI

A large number of inorganic and organic compounds are light-sensitive, but only a few have found practical application. Although other systems such as electrophotography and photopolymerization are slowly gaining in importance, the most important class of light-sensitive salts are still the silver halides. A good review of important photographic processes not based on silver halides was written by the late Dr. Kosar (1965). The study of the photolysis of silver halides falls into two categories; one in which after a relative short exposure with actinic light the latent image is processed in a specific reduction solution ("developer solution"), and one in which the exposure is so long that a visible density appears (print-out effect). The former is observed in photographic emulsions which consist of very small crystals suspended in a protective colloid such as gelatin. During the development the effect of the light reaction is amplified some 10^5 to 10^9 times (Matejec, 1968a). It has been shown that these two categories are closely related. Both the latent image and the print-out density are formed according to the reaction:

$$AgX \xrightarrow{\text{photon}} Ag + \tfrac{1}{2}X_2, \text{ in which } X = Cl, Br \text{ or } I.$$

In many respects the growth of the print-out silver behaves as though it were a later stage of the latent image formation in a photographic grain.

A. The Mechanism of the Photochemical Reduction of Silver Halides

In the development of the theory of the photochemical changes in silver halides one of the most important events was the publication of a new theory of latent-image formation based on the concepts of solid state chemistry in 1938 by R. W. Gurney and N. F. Mott.

In the subsequent fifteen years this Gurney–Mott theory was the base of almost all work that was done on the photochemistry of the silver halides. In the early fifties J. W. Mitchell modified this theory strongly to bring it into agreement with his findings. However, his modifications were criticized by some groups of workers, who preferred the Gurney–Mott theory as a better starting-point for theoretical researches. Both theories with the recent additions will be discussed here. The effect of space-charge layers on latent-image formation will be reviewed too.

1. The Gurney–Mott Theory

According to the theory of Gurney and Mott (Mott and Gurney, 1964) the photochemical formation of silver from silver halides occurs in two stages, namely an electronic and an ionic process:

1a. The absorption of a photon by a crystal releases an electron from the full valence band into the empty conduction band:

$$Br_{Br}^{-x} \xrightarrow{\text{photon}} e' + Br^{\cdot}\dagger$$

This photoelectron has a high mobility.

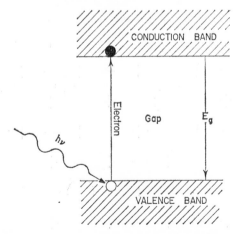

Fig. I–1. Absorption of a photon by silver halides. E_g = band gap energy = 2·5 eV for AgBr, 3·0 eV for AgCl and 2·8 eV for AgI at room temperature (Peterson, 1964).

† A modified defect notation of Kröger (1964) is used. The sign of the charge (′, · or ^{x}) indicates the real space charge in the ionic lattice.

1b. The photoelectron is trapped at some point of the crystal, which thus effectively becomes negatively charged.

2. An interstitial silver ion moves up and combines with the trapped electron, thus forming a silver atom.

Interstitial silver ions:

The occurance of interstitial silver ions is the consequence of the existence of a certain type of imperfection in the silver halide crystal due to a thermal disorder. In this vacancy defect, the Frenkel defect, an ion is transferred from a lattice site to an interstitial position, a position not normally occupied by an ion. The formation of a Frenkel defect can be formulated as:

$$Ag_i^{+\cdot} + V'_{Ag} \rightleftharpoons Ag_{Ag}^{+x} + V_i^x.$$

According to the "law of mass action" the equilibrium constant for the above reaction can be expressed as (Vink, 1964):

$$K = \frac{[Ag_{Ag}^{+x}][V_i^x]}{[Ag_i^{+\cdot}][V'_{Ag}]} \quad \dagger$$

This equation can be simplified by the entirely justifiable assumption that changes in the concentration of Ag_{Ag}^{+x} and V_i^x do not appreciably alter their total concentration. Thus:

$$\frac{1}{[Ag_i^{+\cdot}][V'_{Ag}]} = K'_F = K_F'^0 \exp[-H_F/kT],$$

where H_F is the enthalpy of the Frenkel disorder reaction. The equilibrium

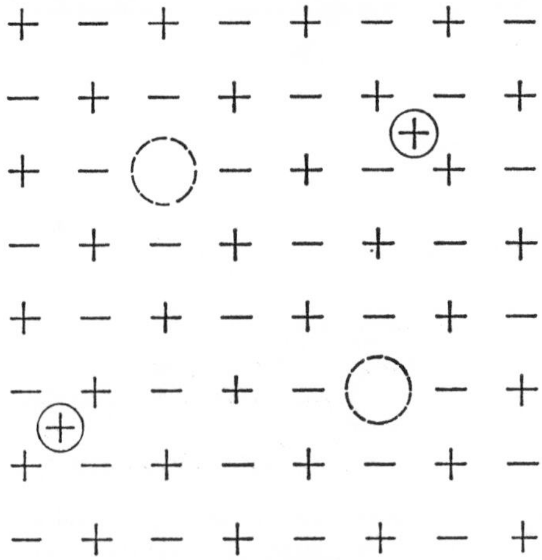

FIG. I–2. Frenkel defect.

† Concentrations are indicated by square brackets.

constant $K_F'^{-1}$ is given by:

$$K_F'^{-1} = [\text{Ag}_i^+\cdot][V_{\text{Ag}}'] = 9 \times 10^2 \exp\left[-1\cdot06\,\text{eV}/kT\right](\text{mol fract})^2 \text{ for AgBr and}$$
$$= 65\exp\left[-1\cdot25\,\text{eV}/kT\right](\text{mol fract})^2 \text{ for AgCl}$$

(Müller, 1965a, b; van der Meulen and Kröger, 1970).

At room-temperature the density of Frenkel pairs due to the intrinsic Frenkel disorder is around $10^{+14}\,\text{cm}^{-3}$ in silver bromide single crystals and $10^{+12}\,\text{cm}^{-3}$ in silver chloride (Brown, 1967; Matejec, 1968b). The room-temperature ionic conductivity of photographic silver bromide grains is more than 100 times the value for single crystals, probably due to surface effects (Hamilton and Brady, 1959).

The diffusion of the interstitial silver ions in silver bromide single crystals occurs according to an interstitial jumping mechanism, in which a silver ion belonging to the lattice is replaced by an interstitial ion and is pushed in turn into another interstitial site. A direct jump from one interstitial site to the next is energetically unfavourable (Friauf, 1957; Steiger et al., 1966).

Trapping of the photoelectron:

The photons absorbed by a crystal of silver halide are incident at random over the grain. However, by studying the crystals microscopically it can be seen that the silver in the grain is found in a relatively small number of specks. So there must be a mechanism by which a photon absorbed at a given point can precipitate silver on a speck located at a considerable distance. In order to explain this "concentration principle" Gurney and Mott assumed that the photoelectron, after wandering around for a certain time (may be temporarily trapped at one or more places in the crystal before the final capture) will eventually find a particle of metal and be captured by it. Hamilton (1966) suggests that the initial electron-trapping centres are interstitial silver ions which have previously become attached to twin boundaries.

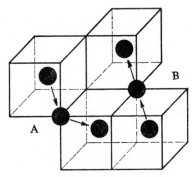

FIG. I–3. Showing the interstitial jumping mechanism: A: noncollinear interstitial jump. B: collinear interstitial jump.

In a photographic emulsion the trapping centres for photoelectrons are particles of silver metal or silver sulphide formed at the surface of the crystal during the chemical ripening process (one of the steps in the manufacture of a photographic emulsion).

Due to the capture of an electron the speck has a negative charge and attracts an interstitial silver ion that is discharged at that speck. This enlarged speck can again trap a photoelectron, etc. So the function of sensitivity specks, which are too small to act as a latent image, is to concentrate the silver atoms formed by the light. A negative charged speck can not only attract interstitial silver ions, but also positive holes (Br· or h·) produced by the removal of the photoelectron from the valence band to the conduction band. The result of this process would be a re-formation of silver halide.

To prevent this reaction, as well as the recombination of photoelectrons with the positive holes from which they came, Gurney and Mott considered it fundamental to their theory that the speck discharges more quickly by attracting interstitial silver ions than by attracting photoholes and that the chance of recombination is small. Having no *a priori* knowledge of the mobility of the holes, their assumptions were that this mobility is very much smaller than that of the photoelectrons and of the interstitial silver ions, and that after some time the positive holes diffuse to the surface forming halogen atoms which can escape from the crystal. The positive holes can be trapped in the crystal by the negatively charged vacancies V_{Ag}', thus forming a neutral complex $[V_{Ag} . h]^x$ which might diffuse to the surface as a whole.

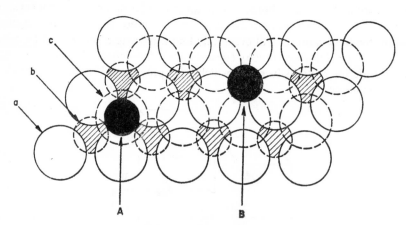

Fig. I-4. Entry of Ag^+-ion on (111) face.
A: additional Ag^+-ion in normal position.
B: additional Ag^+-ion in twin positions.
a: upper layer Br^--ions.
b: second layer Ag^+-ions.
c: third layer Br^--ions.

Hamilton and Brady (1964) disagreed with the original assumption of Gurney and Mott that the hole mobility must be less than that of the silver ions in interlattice positions in order to prevent the neutralization of the negative charge on the silver by the positive holes. Although hole mobility has been shown to be much higher than that of silver ions, they stated that it is actually the product of mobilities and concentrations which determines the relative probabilities of the arrival of photoholes and interstitial silver ions. Since the concentration of mobile silver ions is so high, the successful formation of silver atoms competes favourably with recombination.

The migration of the photoholes on illumination is proved experimentally by Lieser *et al.* (1966). Pellets of silver iodide were labelled on one side by isotope exchange using ^{131}I. Illumination of the non-labelled side with ultraviolet light set free ^{131}I$_2$ from the labelled side, which was away from the light. The number of iodine atoms liberated is directly proportional to the time of illumination. In photographic emulsion grains the motion of holes is demonstrated by Hamilton *et al.* (1956). By electron-microscope examination of exposed silver bromide crystals they found a diffuse cloud with a relatively high electron-scattering power symmetrically located around the grain. However, if an electric field is applied to this system, the cloud density becomes concentrated at only one side of the crystal. This cloud has been interpreted as being bromine atoms, molecules or bromo-gelatin complexes.

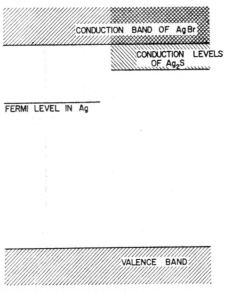

FIG. I–5. Energy levels in AgBr, Ag$_2$S and metallic Ag.

Malinowski (1970) recently published a review paper in which he accepted the Gurney–Mott theory as a working hypothesis for the formation of latent-image silver in view of the experimental results now available. The merit of the Sofia School indeed is the contribution which they have provided in the theory of the role of the photoholes.

In the ionic stage of the latent-image formation the trapped electrons are neutralized by interstitial silver ions according to the theory of Gurney and Mott. The temporary trapping of the photoelectrons occurs at levels of about 0·03–0·04 eV, the terminal trapping at surface sites (or at crystal defects). The temporary trapping of the photoholes occurs at much deeper levels—about 0·4 eV, the terminal trapping as $[V_{Ag} . h]^x$-complexes (0·6 eV). This last value is very close to that of latent-image silver (0·7 eV).

Malinowski suggests that the difference in the depth of traps for both charge carriers may well be the basis of the concentration principle of the silver halides. Because of the very small depth of the temporary traps for the photoelectrons, the frequency of thermal re-ejection should be very high. Combined with their comparatively large drift mobility (75 cm²/V sec), this makes it highly probable that the electron neutralization process occurs at a single, specifically active site on the surface. The much deeper levels and the lower drift mobility of the photoholes (1 cm²/V sec) are the cause of a neutralization practically in the immediate vicinity of their origin. Once the hole is trapped as the $[V_{Ag} . h]^x$-complex, this neutral complex diffuses to the

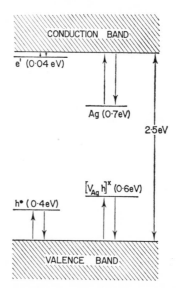

Fig. I-6. Band scheme of AgBr with the trapping levels for photogenerated electrons and holes.

surface (diffusion coefficient of about 3×10^{-7} cm^2/s), where it dissociates into bromine atoms according to:

$$[V_{Ag} \cdot h]^x \rightarrow V_{Ag}' + h^{\cdot} \rightarrow Br \rightarrow \tfrac{1}{2}Br_2.$$

The bromine atoms may react with sensitizers at the surface (bromine acceptors). The concentration gradient of hole complexes established in this way, is able to pump them out of the volume of the crystal. Platikanowa and Malinowski (1966) proved that the role of gelatin as an effective halogen acceptor is not as big as has sometimes been supposed.

Herschel Effect

If an exposed but undeveloped photographic emulsion is exposed to red or infrared light during a sufficiently long time, the density after developing is less than it would have been without exposure to the inactinic light. An explanation for this phenomenon, known as the Herschel effect, can be given by the Gurney–Mott theory (Mott and Gurney, 1964).

The red light is capable to remove electrons from the energy level of the latent-image into the conduction band of the silver halide:

$$Ag \rightarrow Ag^{\cdot} + e'.$$

This electron can be trapped at the same speck, but also at another speck. This redistribution of latent-image silver may result in specks that are too small to act as a latent-image speck. On the other hand, a loss of electrons by recombination, etc. is possible.

As a result of the escape of the electron, the silver speck becomes positively charged. This speck can attract electrons, but it can also be neutralized by other negative particles, viz. silver ion vacancies V_{Ag}'. The overall effect is that in some of the grains of the emulsion the latent-image has been partly destroyed or rendered unsusceptible to the developer solution.

2. The Theory of Mitchell

Around 1950 it became clear that the internal photographic sensitivity was determined by the presence of crystal imperfections. Therefore Mitchell introduced two new hypotheses concerning primitive sensitivity:

(i) The primitive internal sensitivity of microcrystals in emulsions is due to the presence of edge dislocations.

(ii) The first stage in the formation of the latent image is the combination of an interstitial silver ion and an electron with a silver ion at a kink site (Fig. I–7) or a jog (Fig. I–8).

He postulated that the photoelectron is not trapped unless the depth of the positive potential energy well (the silver ion at the kink or jog) is well increased by the presence of a mobile silver ion in the immediate neighbourhood.

The band schemes of silver chloride and silver bromide are given in Fig. I-9 (surface effects are ignored).

It is plausible that a small particle of silver represents a level of higher energy than a big particle of metallic silver with respect to the trapping of an electron.

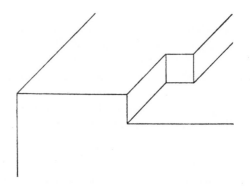

Fig. I-7. Kink defect

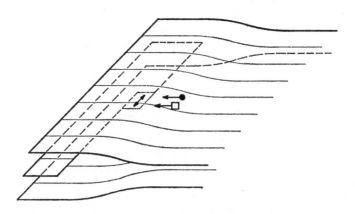

Fig. I-8. A jog in an edge dislocation. An interstitial atom (●) can be adsorbed at the jog, thus causing it to shift one distance to the left. The arrival of a vacancy (☐) at the site of the jog results in a displacement of the jog by one atomic spacing along the dislocation axis to the right.

The theory of Mitchell is most conveniently discussed in terms of the energy level scheme (Mitchell, 1957a). (see Fig. I–10).

(I) Energy levels within the crystal:

AB = the conduction band

CD = the full valence band

BC = the forbidden band

(II) Localized energy levels at the surface of the crystals:

BB_s = range of unoccupied levels associated with surface silver ions

C_sC = range of occupied levels associated with surface halide ions.

(III) Localized levels associated with "latent sub-image specks" Ag_2 adsorbed at sites adjacent to surface silver ions:

a = acceptor levels which provide very shallow traps for conduction electrons

b = highest occupied levels which provide traps for positive holes.

(IV) Localized levels associated with positively charged "latent pre-image specks" $Ag_2{}^{\cdot}$ adsorbed at sites adjacent to surface silver ions:

a = acceptor levels which furnish traps for conduction electrons

b = highest occupied levels which are unable to trap positive holes.

These specks have an extremely short lifetime and dissociate with the loss of a silver ion.

(V) Localized levels associated with small uncharged latent image specks adsorbed at sites adjacent to surface silver ions:

a = acceptor levels which furnish shallow traps for conduction electrons

b = highest occupied levels which provide traps for positive holes.

(VI) Localized levels associated with small positively charged latent image specks $Ag_n{}^{\cdot}$ adsorbed at sites adjacent to surface silver ions:

a = acceptor levels which provide effective traps for conduction electrons

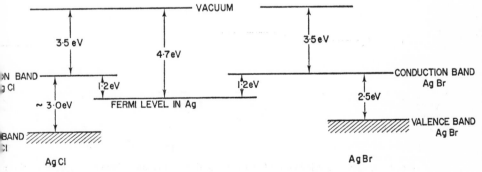

Fig. I–9. Energy band schemes of AgCl and AgBr.

b = highest occupied levels which do not provide traps for positive holes.

The ionization energy of a halide ion at the surface is smaller than of one within the crystal. The surface halide ions thus introduce a group of localized levels above C in the forbidden band (CC_s in Fig. I–10II). The energy released when an electron combines with a surface silver ion is greater than when it combines with one within the crystal. The surface silver ions thus introduce a group of localized levels below B in the forbidden band (BB_s in Fig. I–10II).

The levels in the group BB_s are not well able to trap electrons at room temperature. Due to the thermal agitation the electrons have only a short lifetime in these traps.

Occupied levels in the group CC_s provide deeper traps for positive holes than do levels in the group BB_s for electrons. Therefore it is assumed by Mitchell that the positive holes are trapped before the electrons during the initial stage of the latent-image formation. The trapping of a hole results from the transference of an electron from a localized level in the group CC_s to the unoccupied level in the valence band. As a result of this, an (Ag^+X^-)-group at a kink site becomes an $(Ag^+X)\cdot$-group. (see Fig. I–11 and Fig. I–12).

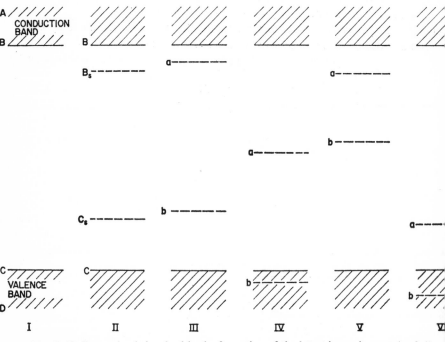

Fig. I–10. Energy levels involved in the formation of the latent-image in crystals of silver bromide.

It now becomes necessary to explain why the trapping of a hole is not immediately followed by the trapping of an electron at the same site. This possibility can be excluded only if the $(Ag^+X)^{\cdot}$ group dissociates, because for example the silver ion, the charge of which is equivalent to the charge of the hole, jumps into an interstitial position and diffuses away within the lifetime of a conduction electron. As a result of this an X_2^- molecule is formed at the surface. An interstitial silver ion cannot trap an electron at room temperature.

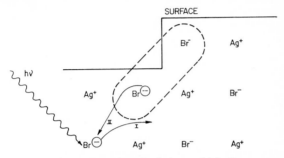

FIG. I–11. Trapping of a hole at a kink site.

A trap for conduction electrons is formed by the shallow positive potential energy well (consisting of a surface silver ion) which is deepened by the presence of an interstitial silver ion in the immediate neighbourhood. The

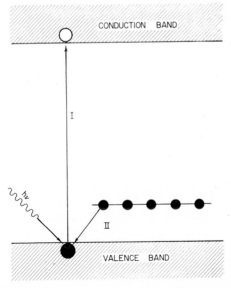

FIG. I–12. Energy band scheme of the trapping of a hole at a kink site.

electron is then drawn into the space between the surface silver ion and the interstitial silver ion and combines with the surface silver ion. The local negative charge which would otherwise arise is neutralized by the interstitial silver ion. Thus an unstable latent pre-image speck is formed. By the timely trapping of a second electron by the Ag_2· a latent sub-image speck is formed (otherwise dissociation). The lifetimes of latent pre-image specks are of the order of seconds whereas those of surface latent sub-image specks are of the order of days.

The latent sub-image speck, which does not provide a stable trap for a photoelectron, is converted into a latent image speck by first adsorbing an interstitial silver ion followed by the neutralization by another conduction photoelectron

$$(Ag_2 + Ag_i^+\cdot) + e' \rightleftharpoons Ag_3\cdot + e' \rightleftharpoons Ag_3.$$

3. The Combined Mitchell–Gurney–Mott Theory of Chibisov and Broun

According to Chibisov and Broun (Chibisov and Broun, 1966; Broun and Chibisov, 1965) the process of photolysis is initiated by elementary processes in which the simplest particles of silver of a molecular–colloidal dispersity are formed with the direct participation of the mobile silver ions in the defects. Luminescence studies showed that the primary centres effectively localize the positive holes but are not electron traps. However, the secondary, pre-colloidal centres serve as centres for trapping the electrons. Hence they concluded that the growth in size of the molecular centres takes place in accordance with the Mitchell mechanism until it reaches the critical state in which it acquires electron-acceptor properties. From that moment on the further growth takes place in accordance with the Gurney–Mott mechanism:

$$Ag_0^+\cdot + e' \rightarrow Ag_1 \rightarrow Ag_2 \dots Ag_n$$
$$(Ag_n + Ag_i^+\cdot) + e' \rightarrow Ag_{n+1}\cdot + e' \rightarrow Ag_{n+1}$$
$$\left.\right\} \text{Mitchell}$$
$$(Ag_m + e') + Ag_i^+\cdot \rightarrow Ag_m'\cdot + Ag_i^+\cdot \rightarrow Ag_{m+1} \quad (m > n) \text{ Gurney–Mott.}$$

B. Nature of Sensitivity Centres

Introduction

As mentioned before, sensitivity centres are formed during one of the stages in the manufacture of the photographic emulsion, viz. the chemical or after-ripening. Sensitivity centres may consist of silver (reduction sensitization), silver sulphide (sulphur sensitization), or gold complexes (gold sensitization). Reduction sensitization produces effective traps for photoholes, whereas the centres formed by sulphur sensitization and gold sensitization act as traps for photoelectrons. Both types of sensitization result in a greater probability of the formation of a stable latent-image speck.

1. Reduction Sensitization

During the chemical reduction ripening process silver specks are formed. The dimensions of these specks are below that of a stable latent-image centre. The behaviour of these small specks is exactly the same as that of bromine acceptors (Spencer et al., 1967; Moisar, 1970). Therefore it is concluded that reduction sensitivity centres act as photohole acceptors. In this way the recombination probability is lowered and the efficiency of the photochemical reaction enhanced.

A stable latent-image centre consists of at least 4–5 atoms of silver. The reduction sensitivity speck therefore has to consist of about 2–4 atoms of silver. If it is true that the reduction sensitivity specks are identical with the so-called sub-image specks, and these specks indeed are photohole traps, then the question arises in which way a latent-image speck is formed in a crystal without a sulphur- or gold-sensitivity speck. The growth according to the scheme:

$$(e' + Ag_i^{+\cdot}) \rightarrow Ag + (e' + Ag_i^{+\cdot}) \rightarrow Ag_2 + (e' + Ag_i^{+\cdot}) \rightarrow Ag_3 + \ldots$$

implies that the photoelectrons are captured by the sub-image specks Ag_2, Ag_3, etc. The results obtained by Spencer and Moisar indicate that these silver specks are hole traps, so this scheme seems to be incorrect. An alternative way is proposed by Moisar. Due to statistical fluctuations in the concentration of the photoelectrons a "supersaturation" may be possible, resulting in a silver speck of a sufficient size and with electron-trapping properties. From this moment on a "normal" growth of the image speck is possible.

2. Gold Sensitization

In 1936 Koslowsky discovered in his laboratory of the Agfa Filmfabrik in Wolfen (Germany) the possibility of enhancing the sensitivity of a photographic emulsion by gold sensitization (see Koslowsky, 1951). In this type of sensitization gold salts are added to the emulsion preferably at the start of the chemical ripening. Commonly used gold salts are sodium bis(thiosulphato)-aurate(I) ($Na_3[Au(S_2O_3)_2]$) and ammonium bis(thiocyanato)aurate(I) ($NH_4[Au(SCN)_2]$). According to Faelens (1968) the mechanism of the gold sensitization with sodium bis(thiosulphato)aurate(I) is based on the formation of an Ag_3AuS_2 speck. He assumes that a reaction takes place at the silver bromide surface with the $Au(S_2O_3)_2^{3-}$-ion according to the following reaction:

$$3AgBr + Au(S_2O_3)_2^{3-} + 2H_2O \rightarrow Ag_3AuS_2 + 2H_2SO_4 + 3Br^-.$$

This gold ion cannot occupy a substitutional site in the silver bromide lattice, but must form an interstitial. The Ag_3AuS_2 speck is an effective electron trap.

3. Sulphur Sensitization

A physical model for sulphur sensitization, based on solid state physics, was presented recently by Fatuzzo and Coppo (1970). With this model it is possible to explain the known chemical and photographic effects of sulphur-sensitization, such as increase of grain sensitivity, decrease of the low-intensity reciprocity failure, enhancement of the high-intensity reciprocity failure, increase of surface speed at the expense of internal speed, and so on.

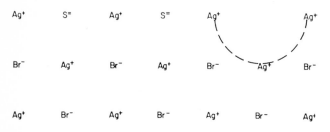

FIG. I–13. The Ag_3AuS_2 sensitivity speck in the silver bromide lattice with $Au(S_2O_3)_2^{3-}$ in a gold sensitized emulsion.

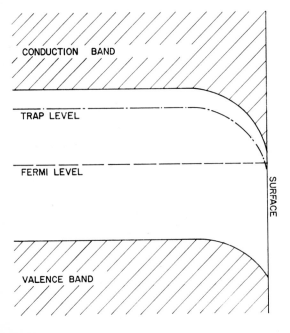

FIG. I–14. The downward bending of the energy band (crystal surface in contact with silver sulphide).

It has been demonstrated that a surface double layer occurs in silver halides. A discussion of this phenomenon will be given in the next subsection. Based on the existence of such a double layer Fatuzzo and Coppo developed their new model for sulphur sensitization. At the surface of the silver bromide crystal a bending of the energy bands takes place due to the charge present in the double layer; bending downward if silver sulphide is adsorbed onto the crystal surface (Fig. I–14), bending upward if no silver sulphide is

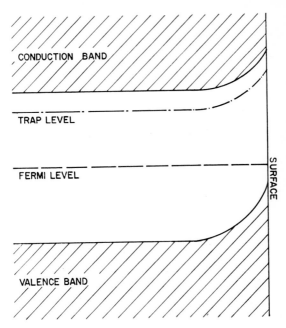

FIG. I-15. The upward bending of the energy band (crystal surface not in contact with silver sulphide).

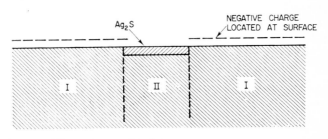

FIG. I–16. Silver sulphide at the crystal surface.
    I: area with excess interstitial silver ions $Ag_i^{+\cdot}$.
    II: area with excess silver ion vacancies V_{Ag}'.

present (Figs. I–15,16). Thus, if silver sulphide is adsorbed onto the crystal surface, an electric field results which attracts the electrons towards the surface. Here the depth of the electron traps is larger than it is in the bulk of the crystal. The trapped lifetime now will be long enough to allow the "ionic step" to take place. If, on the other hand, no silver sulphide is present, the trapped lifetime of photoelectrons at the surface is smaller than in the bulk of the crystal as a result of the repulsing force of the surface and of the possibility for the electrons to tunnel to the conduction band. With this model the experimental results on sulphur sensitization in literature can be explained.

4. Surface Space Charge

Surfaces of ionic crystals can have a charged surface with a region of space charge of the opposite sign adjacent to the surface. The cause of this is that vacancies and interstitials can only be created in pairs within the bulk of a perfect crystal. At the surface they can be created separately. The free energy of formation of the silver ion vacancy V_{Ag}' and of the interstitial silver ion $Ag_i^{+\cdot}$ differs. Kliewer (1966) succeeded in separating the Frenkel pair formation free energy F^F into a free energy of the formation of the vacancy F^V and of the interstitial F^I:

$$F^F = F^V + F^I.$$

For AgCl the value of F^F is determined by Abbink and Martin (1966): $F^F = 1.44 \, eV - 9.4 \, kT$. Splitting this value into F^V and F^I results in $F^V = 0.47 \, eV - 7.7 \, kT$ and $F^I = 0.97 \, eV - 1.7 \, kT$ (Kliewer, 1966). Slifkin *et al.* (1967) reported for these formation energies the values $U^V = 0.58 \, eV$ and $U^I = 0.86 \, eV$. For AgBr U^V and U^I are $0.60 \, eV$ and $0.32 \, eV$ respectively (Trautweiler, 1968). For silver chloride the isoelectric point is above room temperature, if the excess divalent metal impurity exceeds approximately 1 part in 10^7. Thus the surface of a silver chloride crystal of photographic purity must have a negative charge in consequence of an accumulation of interstitial silver ions near the surface. The large electric field of the order of 10^4 to 10^5 V/cm in the space charge region should be rather effective in splitting the photogenerated electron–hole pair and in drawing the photo-hole quickly to the surface thus reducing the probability of recombination. The excess of negative charge at the surface reduces the number of possible electron trapping sites. This helps to concentrate the photolytic silver in one or a few surface sites. Indeed the effects of adsorbed ions at the surface of emulsion grains have to be considered, because it may modify the internal space charge. Summarising it can be said that the effects of the surface charge must play an important role in the latent-image formation.

References

Abbink, H. C. and Martin Jr., D. S. (1966). *J. Phys. Chem. Solids* **27**, 205–215.

Broun, Z. L. and Chibisov, K. V. (1965). *Z. wiss. Phot.* **59**, 52–58.

Brown, F. C. (1967). "The Physics of Solids: Ionic Crystals, Lattice Vibrations, and Imperfections". p. 312. W. A. Benjamin, Inc., New York.

Chibisov, K. V. and Broun, Z. L. (1966). *J. Phot. Sci.* **14**, 159–163.

Faelens, P. A. (1968). *Phot. Korr.* **104**, 137–146.

Fatuzzo, E. and Coppo, S. (1970). Paper presented at the International Congress of Photographic Science, Moscow.

Friauf, R. J. (1967). *Phys. Rev.* **105**, 843–848.

Hamilton, J. F. (1966). Paper presented at 1966 Colloquium on the Photographic Interaction between Radiation and Matter (Soc. Phot. Sci. Eng.), Washington D. C.

Hamilton, J. F. and Brady, L. E. (1959). *J. appl. Phys.* **30**, 1893–1913.

Hamilton, J. F. and Brady, L. E. (1964). *Phot. Sci. Eng.* **8**, 189–196.

Hamilton, J. F., Hamm, F. A. and Brady, L. E. (1956). *J. appl. Phys.* **27**, 874–885.

Kliewer, K. L. (1966). *J. Phys. Chem. Solids* **27**, 705–717.

Kosar, J. (1965). "Light Sensitive Systems: Chemistry and Application of Nonsilver Halide Photographic Processes". John Wiley & Sons, Inc., New York.

Koslowsky, R. (1951). *Z. wiss. Phot.* **46**, 65–72.

Kröger, F. A. (1964). "The Chemistry of Imperfect Crystals". Chapter 7. North-Holland Publishing Company, Amsterdam.

Lieser, K. H., Lakatos, E. and Hoffmann, P. (1966). *Ber. Bunsenges. Physik. Chem.* **70**, 176–180.

Malinowski, J. (1970). *Phot. Sci. Eng.* **14**, 112–121.

Matejec, R. (1968a). *In* "Die Grundlagen der Photographischen Prozesse mit Silberhalogeniden" (H. Fieser, G. Haase and E. Klein, eds.) Band 2, p. 751. Akademische Verlagsgesellschaft, Frankfurt am Main.

Matejec, R. (1968b). *In* "Die Grundlagen der Photographischen Prozesse mit Silberhalogeniden" (H. Frieser, G. Haase and E. Klein, eds.) Band 1, p. 93. Akademische Verlagsgesellschaft, Frankfurt am Main.

Mitchell, J. W. (1957a). *Rept. Progr. Phys.* **20**, 433–515.

Mitchell, J. W. (1957b). *Phot. Korr.*, 1. Sonderheft, 1–35.

Mitchell, J. W. (1962). *J. Phys. Chem.* **66**, 2359–2367.

Moisar, E. (1970). *Phot. Korr.* **106**, 149–160.

Mott, N. F. and Gurney, R. W. (1964). "Electronic Processes in Ionic Crystals". Chapter 7. Dover Publications, Inc., New York.

Müller, P. (1965a). *Phys. Stat. Sol.* **9**, K193–K197.

Müller, P. (1965b). *Phys. Stat. Sol.* **12**, 775–794.

Peterson, C. W. (1964). Thesis, Cornell University. p. 12.

Platikanowa, W. and Malinowski, J. (1966). *Phys. Stat. Sol.* **14**, 205–214.

Slifkin, L., McGowan, W., Fukai, A. and Kim, J. S. (1967). *Phot. Sci. Eng.* **11**, 79–81.

Spencer, H. E., Brady, L. E. and Hamilton, J. F. (1967). *J. Opt. Soc. Am.* **57**, 1020–1024.

Steiger, R., Boustany, K. and Boissonnas, Ch. G. (1966). *Helv. Chim. Acta* **49**, 787–791.

Trautweiler, F. (1968). *Phot. Sci. Eng.* **12**, 98–101.

van der Meulen, Y. J. and Kröger, F. A. (1970). *J. Electrochem. Soc.* **117**, 69–76.

Vink, H. J. (1964). *Chem. Weekblad* **45**, 601–611.

II. The Photochemical Decomposition of
Lead Chloride and Lead Bromide

J. SCHOONMAN

A. Introduction

The decomposition of silver halides into silver and halogen under the action of light has been a separate field of research for a considerable time (Part I).

For the silver halides a mechanism for the photochemical decomposition has been proposed by Mees and James (1966) based upon the Gurney–Mott principle. Mott and Gurney (1964) established assumptions also applicable to other ionic conductors. In general the photochemical decomposition consists of an ionic and electronic part:

(a) as regards the ionic conductivity, cations are the mobile species,

(b) optically produced electrons are much more mobile with respect to the photoholes,

(c) silver ions trap photoelectrons, and surface halide ions are converted to atoms by trapping photoholes, followed by the desorption of these atoms.

Both lead chloride and lead bromide darken after irradiation at room temperature with light in the region of the fundamental absorption of the compounds. Although this century began with this knowledge (Schmid, 1866; Wells, 1893; Norris, 1895; Renz, 1921), extensive studies of the photo-chemical decomposition of both halides started in the preceding decade (Verwey, 1966; Kaldor and Somorjai, 1966; Tennant, 1966; Spencer and Darlak, 1968; Pierrard, 1969). The availability of large, pure, single crystals has been influential.

In order to understand the interaction between photons and a solid in photo-chemical studies, knowledge of the defect state of that solid is of great importance. For lead chloride and lead bromide measurements of transport phenomena and optical properties together with magnetic resonance experiments have given insight into the nature of ionic and electronic defects.

The crystal structure of lead chloride and lead bromide is orthorhombic with space group $D_{2h}^{16}(P_{nma})$. X-ray diffraction measurements (Bräkken, 1932; Nieuwenkamp, 1932) have shown that $PbCl_2$ and $PbBr_2$ crystals exhibit a coordination structure formed by a disturbed hexagonal packing of the halogen ions, between which the lead ions are placed in the interstices. The lead ions are surrounded by nine halogen ions at different distances; the halogen ions are alternately surrounded by four and five lead ions.

The compounds of interest show an ionic character and have a bandgap of about 4 eV (Fesefeldt, 1930) as is shown in Fig. II–1.

de Vries and van Santen (1963, 1965) concluded from their experiments on pure and doped lead chloride crystals that chloride ion vacancies are the mobile species in lead chloride. Schoonman and Verwey (1967, 1968) concluded, from similar conductivity experiments on lead bromide crystals, that also in lead bromide, bromide ion vacancies carry the electrical current.

Photoconductivity experiments by Verwey and Westerink (1967, 1969) revealed that photoholes are the mobile charge carriers in lead chloride and lead bromide.

In spite of the striking differences with the aforementioned assumptions (a) and (b), the results of the photo-chemical decomposition of both lead halides seem to be identical with the "print-out" effect in the silver halides (West and Saunders, 1959; Moser et al., 1959).

It is the purpose of the next sections to elucidate in more detail the defect chemistry and the mechanism of the photo-chemical decomposition of lead chloride and lead bromide.

B. Ionic Conductivity

The ionic conductivity of lead chloride and lead bromide is increased by incorporating monovalent cations at lead ion sites, or divalent anions at halogen ion sites. The incorporation of trivalent cations at lead ion sites decreases the ionic conductivity with respect to the undoped crystals (de Vries, 1965; Schoonman and Verwey, 1967, 1968). Some results of the conductivity experiments on lead bromide are presented in Fig. II–2 as examples.

Tubandt et al. (Tubandt, 1921, 1932; Tubandt and Eggert, 1920; Tubandt et al., 1931) concluded from transport measurements that the electrical current in both lead halides is carried exclusively by the halide ions, requiring activation heats for the migration of halide ion vacancies and the thermal

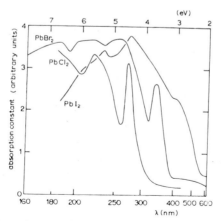

FIG. II–1. Optical absorption (arbitrary units) at room temperature as a function of wave length of thin layers of lead chloride, lead bromide and lead iodide (Fesefeldt, 1930).

formation of point defects. Therefore it is not necessary to consider the lattice defects in the lead ion sub-lattice as charge carriers. If ions might occur in interstitial sites in lead chloride it may be expected that such ions occur only in the $(100)_{\frac{1}{4}}$ and $(100)_{\frac{3}{4}}$ planes, where in the neighbourhood of the gliding mirror planes at $(001)_{\frac{1}{4}}$ and $(001)_{\frac{3}{4}}$ (Bräkken, 1932) chloride ions at 3·70 Å from lead ions have left enough space for ions with a radius of at most $0·7 \pm 0·1$ Å. In the neighbourhood of the gliding

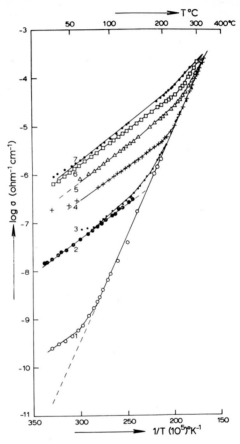

FIG. II.2. The electrical conductivity, σ, of pure and doped lead bromide, plotted as log σ versus $1/T$.

    1 $PbBr_2$–$BiBr_3$, 0·019   mole %
    2 $PbBr_2$–$BiBr_3$, 0·0008 mole %
    3 $PbBr_2$ (nominally pure)
    4 $PbBr_2$–$AgBr$, 0·002   mole %
    5 $PbBr_2$–$TlBr$, 0·010   mole %
    6 $PbBr_2$–$TlBr$, 0·020   mole %
    7 $PbBr_2$–$TlBr$, 0·027   mole %

mirror planes in lead bromide the bromide ions have left enough space for ions with a radius of atmost 0·94 Å (Nieuwenkamp, 1932). The Pauling radii of chloride, bromide and lead ions are 1·81, 1·95 and 1·21 Å, respectively, so the occurrence of interstitial halide and lead ions may be disregarded.

Anion and cation vacancies are to be considered the only point defects in lead chloride and lead bromide. According to a Schottky mechanism their thermal generation is given by:

$$0 \rightarrow V_{Pb^{2+}} + 2V_{X^-} \qquad \qquad \text{(II–1)}$$

where $V_{Pb^{2+}}$ and $V_{X^-}(X^- = Cl^-, Br^-)$ denote a missing lead ion at a lead ion site and a missing halide ion at a halide ion site, respectively, and 0 denotes the perfect lattice.

When the thermal generation of the ionic point defects is given by eqn (II–1), the incorporation of the monovalent cations Me^+ can be described by the following lattice reaction:

$$MeX \rightarrow Me^+_{Pb^{2+}} + V_{X^-} + X^- \qquad \qquad \text{(II–2)}$$

followed in principle by a shift of the Schottky equilibrium. $Me^+_{Pb^{2+}}$ denotes a monovalent cation at a lead ion site. Incorporation of trivalent metal ions requires the lattice reaction:

$$2MeX_3 \rightarrow 2Me^{3+}_{Pb^{2+}} + V_{Pb^{2+}} + 6X^-. \qquad \qquad \text{(II–3)}$$

So in both anionic conductors the lattice reaction (II–2) increases, and the lattice reaction (II–3) decreases the number of anion vacancies and therefore the ionic conductivity with respect to the conductivity of the pure crystals.

The total electroneutrality condition upon doping with monovalent cations, divalent anions (A^{2-}), or trivalent cations according to eqns (II–1), (II–2) and (II–3) is then given by:

$$[V_{X^-}] + [Me^{3+}_{Pb^{2+}}] = 2[V_{Pb^{2+}}] + [Me^+_{Pb^{2+}}] + [A^{2-}_{X^-}] \qquad \text{(II–4)}$$

where square brackets denote concentrations. The activation heat for the migration of the halide ion vacancies, $U(V_{X^-})$, and the heat E, required for the formation of one set of vacancies are listed in Table I.

The mobility of the chloride ion vacancies as a function of temperature (Hoshino et al., 1969) is described by:

$$\mu(T) = (2\cdot6 \times 10^{-1}/T)\exp(-0\cdot20 \pm 0\cdot03\,\text{eV}/kT)\,\text{cm}^2\,\text{Volt}^{-1}\,\text{sec}^{-1} \quad \text{(II–5)}$$

whereas the temperature dependence of the mobility of the bromide ion vacancies (Schoonman and Verwey, 1968) is given by:

$$\mu(T) = (10^2/T)\exp(-0\cdot29 \pm 0\cdot04\,\text{eV}/kT)\,\text{cm}^2\,\text{Volt}^{-1}\,\text{sec}^{-1}: \quad \text{(II–6)}$$

TABLE I. Heat effects in lead chloride and lead bromide: Activation heat for vacancy migration, $U(V_{X^-})$, and the heat required for intrinsic point defect formation, E.

	de Vries (1965)	Smakula (1965)	Schoonman et al. (1967, 1968)	Hoshino et al. (1969)
$U(V_{Cl^-})$	0.30 ± 0.02 eV			0.20 ± 0.03 eV
$U(V_{Br^-})$		0.28 eV	0.29 ± 0.04 eV	
$E(PbCl_2)$	1.6 ± 0.1 eV			1.5 ± 0.2 eV
$E(PbBr_2)$			1.5 ± 0.1 eV	

C. Identification of the Photolysis Products

1. X-ray Fluorescence Analysis

Both lead chloride and lead bromide darken after irradiation at room temperature with radiation in the region of the fundamental absorption of the crystals. The photochemical induced change is located near the surface. The identification of the dark coloration resulted from a particularly suitable analytical method: X-ray fluorescence (Verwey, 1967).

In the context of this chapter it is not meaningful to give a detailed description of the method. We will briefly mention some experimental circumstances for lead bromide, because for this substance some quantitative results will be presented.

After cleavage perpendicular to the c-axis pieces of lead bromide crystals were irradiated with a high pressure mercury lamp (HPK 125). For the determination of the change of ratio of the halogen-to-lead content the crystals were excited by the radiation from an X-ray tube with chromium anode operated at 30 kV and 10 mA. The analytical lines (X-ray fluorescence with a characteristic wave length) K_α and $L_{\beta 1}$ of bromine and lead, respectively, were detected by a scintillation counter in combination with a LiF crystal. After an exposure of one hour to ultra-violet irradiation ($\lambda_i = 250$ nm) the bromine-to-lead ratio showed a decrease of 1 per cent with respect to an unirradiated crystal used as a standard. All fluorescent radiation is emitted by a surface layer with thickness of about 20μ. This decrease in the halogen-to-lead ratio can be explained by assuming that the photolysis is described by an overall reaction.

$$PbBr_2 \xrightarrow[250 \text{ nm}]{h\nu} Pb(s) + Br_2(g). \qquad (II\text{-}7$$

No visible change occurred in the crystals owing to the X-ray irradiation during the analysis. It must be pointed out that with this experimental technique Verwey (1967) obtained the same results with lead chloride crystals.

X-ray diffraction of blackened lead chloride showed in addition to the crystal pattern three small peaks which coincided with the three principal reflections of lead metal (Tennant, 1966).

2. Optical Microscopy

Microscopic investigation of blackened lead chloride and lead bromide crystals showed that the lead was formed in discrete portions of the crystal. Treatment with nitric acid showed that the blackening had penetrated to some depth within the crystal, since the black colour remained unchanged. The maximum depth observed was $3\,\mu$. Lead bromide showed a strong orientation dependence which can be seen in Fig. II–3, and which was absent in lead chloride. The photolysis products lie into the direction of the a-axis, as was concluded from X-ray diffraction (Verwey and Schoonman, 1967). From ionic conductivity experiments Smakula (1965) found no differences in activation heats for the bromide ion vacancy migration when it is measured along the three principal axes. So the anisotropical distribution of the photodecomposition products cannot be correlated with the isotropical anion vacancy migration.

In lead halide crystals a surface damage may occur attributable to photo-enhanced evaporation due to the fact that substance is completely evaporated from the surface (Verwey, 1967).

3. Oxidation of the Crystal Ambient

The electron–hole pair created in either lead chloride or lead bromide causes, as was shown, an electronic reaction: the reduction of a lead ion to a lead atom. Vacuum increases the rate of the photochemical decomposition of both crystals compared to irradiation in air. Photochemical decomposition in alcohol vapour in turn shows a still higher rate for lead chloride. Alcohol for example readily reacts with chlorine gas giving monochloroacetal C_2H_3Cl $(OC_2H_5)_2$ as the first stable product. As a possible explanation we may assume that during photolysis of the crystals, halogen atoms are easily formed by the capture of a photohole at a halide ion at the surface. The holes are more easily transported to the surface if the concentration of adsorbed halogen atoms is low, which occurs if in the case of lead chloride the chlorine atoms readily react with alcohol molecules giving monochloroacetal.

Another possibility is the direct oxidation of alcohol according to the reaction

$$C_2H_5OH + h^+ \rightarrow C_2H_5O + H^+ \qquad\qquad (II\text{-}8)$$

followed by the diffusion of hydrogen ions into the crystal or the desorption of hydrochloric acid.

Spencer and Darlak (1968) investigated the oxidation of leucocrystalviolet (LCV) to crystal violet (CV) during the photochemical decomposition of lead bromide. LCV-containing lead bromide dispersions in polyvinylalcohol-water solutions were used. Since microcrystals have a large surface-to-volume

FIG. II.3. Surface of a cleaved lead bromide crystal irradiated by a high pressure mercury lamp (Philips HPK 125) for 30 min through a grid of an electron microscope. A small part of the shadow of this grid can be seen where the crystal was not irradiated. Optical microscope: Reichert Zetopan with interference contrast technique according to Nomarski.

atio, they are particularly suited for study in solid state reactions which occur at the surfaces. The afore-mentioned oxidation can be followed spectrophotometrically, since LCV is colourless in aqueous solution and CV has an absorption maximum at 592 nm (molar extinction coefficient 9×10^4 $M^{-1} cm^{-1}$).

$$\left((CH_3)_2N - \langle O \rangle\right)_2 CH - \langle O \rangle - N(CH_3)_2 + Br_2 \rightleftarrows H^+ + 2Br^-$$

$$+ \left((CH_3)_2N - \langle O \rangle\right)_2 C = \langle \rangle = N^+(CH_3)_2 \qquad (II\text{-}9)$$

As is the case in the oxidation of alcohol vapour it is not possible to conclude unambiguously whether or not one bromine molecule, two bromine atoms or two photoholes must be considered as the reacting species, since they can all be used in balancing reaction (II-9) in which as an example one bromine molecule is used.

D. Charge and Mass Transport

1. Lead Bromide in a Bromine Atmosphere

As an introduction to the transport and trapping processes of photogenerated electrons and holes during irradiation of lead chloride and lead bromide at liquid nitrogen temperature and the mass transport at room temperature the ionic conductivity data (Section B) will be extended with the results of bromine induced conduction. Moreover the reaction of bromine with irradiated lead bromide crystals will be discussed.

If molecules of bromine can dissolve and diffuse as such into the lead bromide lattice the electrical conductivity would remain unchanged. If the reaction of bromine with the crystal is connected with the formation of charge carriers, the spatial migration of which is not directly related to the diffusion of bromide ions the electrical conductivity will change. Indeed it is found that the conductivity of lead bromide brought into contact with a bromine atmosphere increases (Schoonman, 1970). This excess conductivity varies as the square-root of the bromine pressure in isothermal experiments. In isobaric experiments the excess conductivity of crystals doped with mono-valent cations at lead ion sites exceeds the excess conductivity in the pure crystals indicating that bromide ion vacancies play an important role as can be concluded from lattice reaction (II-2).

Bromine vapour causes bleaching of lead bromide crystals in which lead colloids are present in a surface layer. This is illustrated in Fig. II–4.†

X-ray fluorescence analysis before and after the bromine treatment of irradiated crystals showed that after bleaching the bromine-to-lead ratio exceeds the value of this ratio measured in an untreated crystal. Lead bromide exposed to bromine vapour exhibits excess conductivities attributable to holes, since structure considerations preclude the diffusion of molecules of bromine. The chemical reaction for bromine incorporation in the crystal is represented by formulation (II–10)

$$\tfrac{1}{2}Br_2 + V_{Br^-} \rightleftharpoons Br^-_{\text{lattice}} + h^+. \tag{II–10}$$

The bromine induced bleaching is then described by

$$Pb^\circ + 2h^+ \rightarrow Pb^{2+}. \tag{II–11}$$

In this model it is assumed that in the temperature range 25–200°C (Schoonman, 1970) the mobility, μ_V, of the bromide ion vacancies is small compared to the mobility, μ_p, of the holes. At room temperature for the ratio μ_V/μ_p the value 10^{-2} was measured.

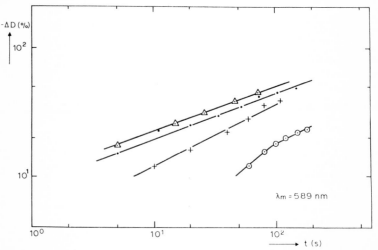

Fig. II–4. Bromine induced bleaching of irradiated lead bromide crystals. The decrease in the optical density is plotted versus the time during which the crystal was in a bromine atmosphere (185 mmHg). The crystals were irradiated before bleaching with the unfiltered radiation of a high pressure mercury lamp (HPK 125): ○ 4 min, + 2 min, · 2 min, △ 4½ min.

† These results can only be used as an illustration, since no serious efforts have been made to obtain reproducible quantities of lead colloids in the crystal during ultra-violet irradiation.

2. E.S.R. Experiments

Irradiation of lead halides at liquid nitrogen temperature gave reaction products of electrons and holes with ionic defects, detectable by electron spin resonance (Arends and Verwey, 1967). The electron spin resonance spectra of single crystals of pure and doped lead chloride and lead bromide, irradiated at liquid nitrogen temperature, consist of a single absorption line (Fig. II–5) with a g-value of 2·0006 for lead chloride and 2·0009 for lead bromide crystals. The deviation from the free electron value is negative. The position of the resonance line shows no observable dependence on the orientation of either halides with respect to the static magnetic field. The line width depends upon the irradiation time: longer irradiation times produce broader E.S.R. signals with the same g-value. In lead bromide crystals annealed in a bromine atmosphere the signal intensity almost vanishes (Section D–1). On warming the samples the centres undergo thermal bleaching. The signal is constant to 120° K but at higher temperatures the intensity drops abruptly. It appears to be very important that thermal bleaching only affects the intensity of the lines but not their shape.

These facts together with the observation that the signal intensity as a function of the microwave power incident on the sample indicate a homogeneously broadening of the line, lead to the conclusion that this E.S.R. signal is due to conduction electrons in colloidal lead particles.†

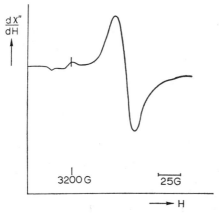

Fig. II–5. The derivative of the paramagnetic absorption, $d\chi''/dH$, as a function of H for a lead chloride crystal measured at 80°K: irradiation time 15 min.

† The photochemical change at about 100°K was observable in optical absorption upon warming the crystals. At room temperature when lead particles have a considerable size it was impossible to obtain an electron spin resonance signal from conduction electrons in the clearly visible metallic lead particles.

After irradiation of freshly prepared pure and doped lead bromide crystals a characteristic group of seven lines with centre at $g = 1.9998$ mostly super-imposed upon the aforementioned "lead" signal is observed: Fig. II–6. Patankar and Schneider (1966) have observed the same set of seven resonance lines in γ-irradiated lead chloride and lead iodide after thermal bleaching. The spectrum also was completely independent of the crystal orientation with respect to the static magnetic field. The group of seven lines could be ascribed to Pb^+ ions. After storage of the irradiated crystal for several hours at liquid nitrogen temperature the group of three lines marked with an asterisk have disappeared. This group is due to a cluster of three Pb^+ ions. The remaining group of four lines is due to pairs of Pb^+ ions (Arends and Verwey, 1967). Electron capture by a Pb^{2+} ion increases the ionic radius from 1·21 Å to about 1·45 Å leading to a distortion at sites where a Pb^+ ion is formed. It is known that upon exposure at room temperature and at high intensities large quantities of matter are removed from the surface. It is therefore quite reasonable to assume that the surface layer of the ultra-violet irradiated halides behaves like a polycrystalline layer.

Theoretical considerations concerning the line shapes of magnetic resonance lines indicate that pairs and triplets of spins $S = \frac{1}{2}$ may be present in a polycrystalline layer. From the experiments it appears that the pairs are more stable than the triplets.

Until now the experiments have confirmed the existence of two stable "electron" centres Pb^+ and $Pb°$. One might suggest two possible reactions leading to the formation of a lead atom, $Pb°$.

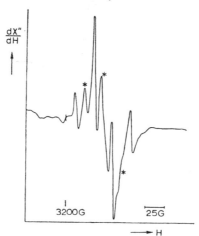

FIG. II–6. The derivative of the paramagnetic absorption, $d\chi''/dH$, as a function of H for a silver-doped lead bromide crystal measured at 80°K: irradiation time 3 min. The lines marked with an asterisk have disappeared after storing the irradiated crystal for five hours at liquid nitrogen temperature.

$$Pb^+ + e^- \rightarrow Pb° \qquad\qquad (II\text{--}12)$$

$$2Pb^+ \rightarrow Pb^{2+} + Pb°. \qquad\qquad (II\text{--}13)$$

The fact that lead atoms are formed out of Pb^+ ions *after* the ultra-violet irradiation (disappearance of the three lines marked with an asterisk in Fig. II–6) favours reaction (II–13).

Kaldor and Somorjai (1966) gave a discussion in terms of colour centres of the type $(V_X\text{-}·e^-)$ to describe the formation of photolytic lead. The creation of a Pb^+ ion may indeed occur via a Pooley–Hersh mechanism, i.e. the creation of a halide ion vacancy which traps an electron. The E.S.R. results indicate the absence of this colour centre. Possibly this centre is not stable at liquid nitrogen temperature, giving Pb^+ upon stabilization, that means, the photoelectron is located at one of the neighbouring Pb^{2+} ions. Perhaps the very initial centre has escaped observation, since the $(V_X\text{-}·e^-)$ centre might be stable at liquid helium temperature.

Another line with a g-value larger than the free electron value ($\Delta g = + 0·0280$) has been observed too. This resonance signal is due to a hole trapped at a lead ion vacancy: $(V_{Pb^{2+}}·h^+)$. The signal becomes very intense in lead bromide crystals annealed in bromine vapour. From reactions (II–1) and (II–10) it is clear that in these crystals larger concentrations of cation vacancies are present with respect to untreated crystals. Moreover bromine induced holes are present.

The axial symmetry of the centre is in agreement with the experimental fact that the angular dependence of the signal can be described by the resonance condition

$$hv = (g^2_{\parallel} \cos^2 \theta + g^2_{\perp} \sin^2 \theta)^{1/2} \beta H$$

when the crystal is rotated around the b-axis (Fig. II–7). Here, v is the microwave frequency and θ the angle between the a-axis and the magnetic

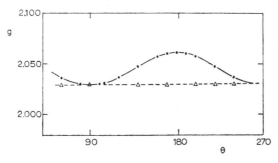

FIG. II–7. The angular dependence of the g-values of the hole centre: $(V_{Pb^{2+}} · h^+)$. —·— denotes the rotation about the b-axis. $-\triangle-$ denotes the rotation about the a-axis.

field. The numerical values of $g_{\parallel}$ and $g_{\perp}$ are 2·061 and 2·029, respectively. If the crystal is rotated around the a-axis the signal can be described by $h\nu = g_{\perp}\beta H$.

3. Photoconductivity Experiments

Lead chloride and lead bromide show some interesting photoconductive properties. The results have contributed to a better understanding of the mechanism underlying the photochemical decomposition of these crystals. All the experiments must be carried out at liquid nitrogen temperature because at room temperature both halides show a large ionic conductivity. This gives space-charge-polarization effects in the crystals after application of a d.c.-voltage. At liquid nitrogen temperature the ionic conductivity is frozen in.

For a detailed discussion of the photoconductive properties we refer to the work of Verwey and Westerink (1967, 1969). It is the aim of this section to show that the behaviour of the optically produced electrons differs from the general rule (b): optically produced electrons are much more mobile with respect to the photoholes, mentioned in the introduction.

Illumination with ultra-violet radiation through one electrode insulated from the crystal gives information about the sign of the dominant mobile charge carriers. If for example the photoholes have a higher mobility with respect to photoelectrons they will diffuse more easily into the crystal and a photodiffusion voltage (Dember voltage) will develop in such a way that the illuminated surface becomes negative with respect to the non-illuminated surface. In both halides the illuminated surfaces became actually negative, indicating that the holes might have a higher mobility than the electrons.

The Dember field, E_D, is given by Tabak and Warter (1966).

$$E_D = \frac{kT}{e} \frac{(r_h - r_e)}{(r_h + r_e)} \frac{dp/dx}{p} \tag{II–14}$$

with k = Boltzmann constant,
 T = Temperature (°K),
 e = elementary charge,
 r_h = hole range per unit field,
 r_e = electron range per unit field,
 p = concentration of the holes.

Integrating of this formula gives the Dember voltage, V_D

$$V_D = \frac{kT}{e} \frac{(r_h - r_e)}{(r_h + r_e)} \ln (p_i/p_d) \tag{II–15}$$

in which p_i denotes the concentration of the holes at the illuminated side, and p_d the concentration at the dark side. The Dember voltage, V_D, in dependence of the radiation intensity in lead bromide at about 100° K is presented in two different ways in Figs II–8 and II–9. To understand these results we write V_D as

$$V_D(:) \ln (1 + \Delta p/p_d) \qquad (II\text{--}16)$$

where Δp is the conentration of the photo-created holes at the surface. With the assumption that Δp is proportional to F^β, in which β is a constant, eqn (II–16) leads for high intensities to $V_D(:) \ln F$ (Fig. II–8) and for low intensities to $V_D(:) F^\beta$ (Fig. II–9).

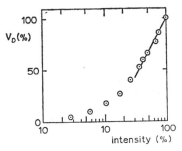

Fig. II–8. The intensity dependence of the Dember voltage, V_D, in lead bromide at 100°K. Illumination with unfiltered radiation of a high pressure mercury lamp (HPK 125). The radiation was attenuated by neutral density filters.

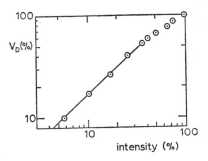

Fig. II–9. The intensity dependence of the Dember voltage, V_D, in lead bromide at 100°K. The results presented in Fig. II.8 are plotted here in a log–log curve.

If the photoconductivity is measured with insulating electrodes the current saturates at the voltage where the electron (hole) range ω ("Schubweg") has the same order of magnitude as the electrode distance. As a matter of fact in this case the displaced charge as a function of voltage is measured. A characteristic experimental result for a lead bromide crystal measured at about 100°K is presented in Fig. II–10. The response was greater if the insulated electrode at the illuminated side of the crystal was positive with respect to the electrode at the non-illuminated side. The displaced charge in this forward polarity condition was about 10 to 20 times the charge in the reversed condition both in lead chloride and lead bromide. This is better evidence that in these crystals the holes have a higher mobility than the electrons. The displaced charge, Q, as a function of voltage, V, is given by Hecht's formula

$$Q = Q_0(\omega/l)(1 - e^{-l/\omega}) \tag{II-17}$$

where l = thickness of the crystal,
    ω = $\mu\tau V/l$ range of the holes (Schubweg),
    μ = mobility of the holes,
    τ = life time of the holes.

The quantity Q_0 represents the charge of the holes created near the surface and, therefore, is linearly correlated to the number of photons absorbed and the quantum efficiency (next section). For a detailed discussion of formula II–17, we refer to the literature (Mott and Gurney, 1964; van Heyningen and Brown, 1958).

From the best fit of Hecht's formula to the measured points in Fig. II–10 the hole range in lead bromide was calculated: $r_h = 1.2 \times 10^{-4}\,\mathrm{cm^2/V}$. The estimated values for the hole ranges per unit field in lead chloride, lead bromide and the corresponding silver halides are shown in Table II.

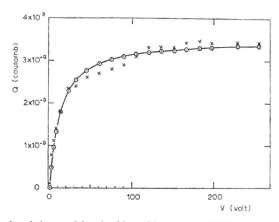

FIG. II–10. Displaced charge, Q in a lead bromide crystal at 100°K measured with electrodes insulated from the crystal. The thickness of the crystal was 0·05 cm. Crosses are measured points, the illuminated electrode being positive with respect to the non-illuminated one. Circles are results of a computer calculation of the best fit curve of Hecht's formula to the measured points. The curve is

$$Q = 34 \times 10^{-10}\,\frac{V}{21\cdot5}\,(1 - e^{-21\cdot5/V})$$

where V denotes applied voltage.

TABLE II.   Ranges per unit field $(\mu\tau)$ of the dominant charge carriers.

Substance	Charge carrier	Range (cm^2/V)	References
AgCl	e	$1{\cdot}05 \times 10^{-3}$	van Heyningen et al. (1958)
AgBr	e	$1{\cdot}03 \times 10^{-4}$	Burnham et al. (1960)
PbCl$_2$	h	4×10^{-4}	Verwey et al. (1969)
PbBr$_2$	h	$1{\cdot}2 \times 10^{-4}$	Verwey et al. (1969)

It is clear that the ranges per unit field of the dominant mobile carriers in the lead halides and the silver halides mentioned above have the same order of magnitude at liquid nitrogen temperature. In lead chloride and lead bromide the electron ranges per unit field, r_e, are 5 to 10 per cent of r_h (Verwey, 1967).

Preliminary experiments by the Haynes–Shockley technique have indicated that the photoholes are much more mobile with respect to the photoelectrons at room temperature (Verwey, 1967). Moreover, electronic conduction in lead chloride and lead bromide (Wagner and Wagner, 1957) indicates $\mu_p > \mu_e$. Therefore we conclude that in the mechanism of the photodecomposition photoholes play a more important role than photoelectrons.

4. Kinetic Measurements

In addition to charge transport there is a mass transport due to ionic and atomic motion. This happens in lead halides upon irradiation at a temperature near room temperature. As a result of atomic motion lead particles are formed inside the crystal near the surface. At the same time halogen leaves the surface. To get more insight into the mechanism of the formation of the lead colloids and the halogen molecules some kinetic measurements are discussed now.

The lead chloride and lead bromide crystals were irradiated in vacuum. As was mentioned before vacuum is a very good halogen accepter, since irradiation in vacuum gives a much higher darkening compared with irradiation in air. A typical result for a lead bromide crystal is presented in Fig. II–11. The increase in optical density, ΔD, at the wave length $\lambda = 589$ nm was measured as a function of the irradiation time. The increase in the optical density varied as the square root of the irradiation time: i.e. $\Delta D \sim \sqrt{t}$. Verwey (1970) pointed out that only for radiation intensities of about 10^{18} photons sec^{-1} cm^{-2} the $\sqrt{t}$ dependence can be measured in lead bromide. A diffusion step seems rate limiting.

Let us consider the diffusion of lead and halogen atoms away from the trapping centre at which the charge transfer has occurred. The diffusion of these atoms depends exponentially on temperature, whereas charge trapping is independent or only weakly dependent on temperature. Therefore if atomic diffusion is the rate-limiting step during the photodecomposition the precipitation rate of lead should be a function of temperature. Experiments have confirmed that the photochemical decomposition is a weak function of temperature in the range $-77°C$ to $+115°C$ (Kaldor and Somorjai, 1966; Verwey, 1967, 1970), so atomic diffusion can be the rate-limiting step.

If it is assumed that photoholes diffuse to the surface, instead of halogen atoms, and that halide ion vacancies diffuse into the crystal we can understand among other things the influence of impurities on the decomposition rate as will be discussed later on. The diffusion of these species through a layer formed inside the lead bromide crystals at some particular intensity (10^{18} photons $\text{sec}^{-1}\text{cm}^{-2}$) can be rate-limiting.† It seems difficult to correlate the square root dependence with a particular step in the photolysis process.

The intensity (F) dependence of the photolysis is a strong function of the irradiation time (Fig. II–12), i.e. $\Delta D \sim F^n$. For lead bromide it was found that $n = 1$ at irradiation times shorter than 1 second. This indicates the existence of a one-electron process, i.e. the trapping of an electron by a Pb^{2+} ion at a suitable site. For lead chloride photolysis n lies between 0.93 and 0.17

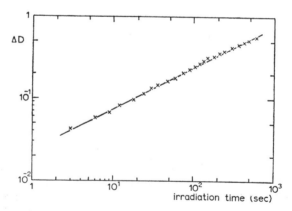

Fig. II–11. Change in the optical density, ΔD, of a lead bromide crystal measured parallel to the c-axis, as a function of irradiation time. Wave-length of the measuring light $\lambda = 589$ nm.

† J. F. Verwey: private communication.

(Verwey 1967, 1970). Kaldor and Somorjai (1966) found for lead chloride in the high intensity region even $n = 3.2$. At low intensities photodecomposition occurs with rather high quantum efficiency, η. The calculated values for lead production in lead chloride ranges from 0·06 to 0·14 Pb atoms per absorbed photon. For lead bromide η is 0·25 Pb atom per photon. Thus in lead bromide lead production accounted for one half of the electrons during the initial stage of the photolysis.

At high radiation intensities, many lead atoms precipitate as a result of the absorption of a single photon at the lead chloride surface: $n = 3.2$. Kaldor and Somorjai (1966) suggested that at high radiation levels the

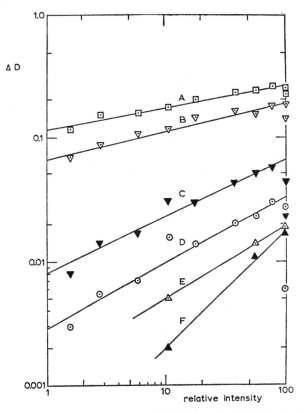

FIG. II–12. Optical density change ΔD ($\lambda = 589$ nm) upon irradiating lead chloride (A–D) and lead bromide (E, F) crystals with different radiation intensities. Shown is ΔD as a function of the relative radiation intensity (percentage transmission of the neutral density filters used) for several total irradiation times t,

Curve A $t = 100$ sec	D $t = 5$ sec
B $t = 50$ sec	E $t = 1$ sec
C $t = 10$ sec	F $t = 0.5$ sec

surface density of the chloride ion vacancies becomes so great that the crystal lattice in the surface region collapses and a superlinear growth of lead aggregates results. As a matter of fact there is a significant difference not only in results but also in experimental circumstances if we compare the studies of Kaldor and Somorjai with the studies of Verwey. Verwey (1970) measured the change in the light transmitted by the crystals and found that the curves $\Delta D \sim F^n$ tend to decrease at high intensities. Kaldor and Somorjai (1966) measured the light reflected from the surfaces of evaporated lead chloride layers and found at high intensities the value 3·2 for n. If photo-enhanced evaporation occurs a strong change in the reflected light intensity will take place, but in optical absorption the change in the optical density ΔD, due to this photo-evaporation is less than ΔD due to the equivalent amount of separated lead. The transmission measurements are more reliable, since at high radiation intensities photoenhanced evaporation actually occurs.

Not only suitable crystal ambients (e.g. alcohol vapour) can increase the rate of the photodecomposition, but impurities though not very pronounced can influence this rate too. For example the photolysis of pure and doped lead bromide crystals is shown in Fig. II–13. An enhanced photolysis was observed in lead chloride doped with Tl^+ (Verwey, 1967). From ionic conductivity experiments on both halides we know that the substitution of monovalent cations enhances the anion vacancy concentration and substitution of trivalent cations increases the cation vacancy concentration. We conclude from Fig. II–13 that increasing the concentration of anion vacancies

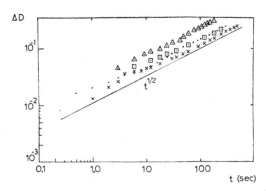

FIG. II–13. Photolysis of pure and doped lead bromide crystals. The figure shows ΔD ($\lambda = 589$ nm) as a function of irradiation time in vacuum. Wave length of the ultra-violet radiation $\lambda = 250$ nm.

• undoped $PbBr_2$
△ $PbBr_2$–CuBr, about 0·002 mole %
▯ $PbBr_2$–TlBr,   0·001 mole %
× $PbBr_2$–BiBr_3, about 0·02 mole %

leads to an enhanced rate of photolysis. When we consider the Schottky equilibrium, described by eqn (II–1), with equilibrium constant, K_s, given by

$$K_s = [V_{Pb^{2+}}][V_{X^-}]^2 \qquad (II-18)$$

we see that an increase in the concentration of V_{X^-} decreases the concentration of $V_{Pb^{2+}}$. From the E.S.R. experiments we know the occurrence of $(V_{Pb^{2+}} \cdot h^+)$ centres, i.e. an electronhole, h^+, trapped at a cation vacancy, $V_{Pb^{2+}}$. This centre is stable at liquid nitrogen temperature, but unstable at room temperature. This indicates that $V_{Pb^{2+}}$ is an effective recombination centre for holes and therefore reduces the life-time of the holes. During photolysis halogen is formed at the surface of the crystal. It is believed that a photohole diffuses to the surface and is trapped at a surface halide ion. So $V_{Pb^{2+}}$, acting as a shallow trap for holes, can decrease the rate of photolysis. Incorporation of monovalent cations decreases the cation vacancy concentration. This leads to an enhanced rate of photolysis.

E. Reaction Scheme

From the preceding sections it is clear that the photochemical decomposition of lead chloride and lead bromide at room temperature consists of an ionic and an electronic part:

(a) as regards the ionic conductivity, anions are the mobile species,

(b) optically produced electronholes are much more mobile with respect to the photoelectrons,

(c) lead ions trap photoelectrons, giving lead atoms, which aggregate under annihilation of halide ion vacancies. Surface halide ions are converted to atoms by trapping photoholes, followed by the desorption of these atoms.

The absence of halogen desorption and lattice defect migration after the creation of electrons and holes will only occur at very low temperature. After prolonged irradiation or at high radiation intensities substance is completely evaporated from the surface.

As mentioned in the introduction (Section A) the results of the photochemical decomposition of both lead halides seem to be identical with the "print-out" effect in the silver halides, but the assumptions made by Mott and Gurney (1964) concerning ionic conductivity and charge carrier mobility are not applicable to lead chloride and lead bromide, ultimately leading to a different mechanism.

The photochemical decomposition at room temperature can be described by the next reaction steps.

An absorbed photon creates an electronhole pair

$$0 \rightarrow h^+ + e^-. \qquad (II-19)$$

The electronhole, h^+, diffuses to the surface and is trapped at a surface halogen ion X_s^-: a neutral halogen atom X_s° is formed

$$X_s^- + h^+ \rightarrow X_s^\circ. \tag{II–20}$$

Probably two X_s° give a halogen molecule X_2, which desorbs from the surface

$$2X_s^\circ \rightarrow X_2 + 2(V_{X^-})_s. \tag{II–21}$$

The anion vacancies formed at the surface $(V_{X^-})_s$ diffuse into the crystal

$$(V_{X^-})_s \rightarrow V_{X^-}. \tag{II–22}$$

The anion vacancy V_{X^-} may trap an electron e^- giving an F-centre structure $(V_{X^-} e^-)$. This centre is not stable.

It probably produces a Pb^+ ion according to

$$Pb^{2+} + (V_{X^-} \cdot e^-) \rightarrow (V_{X^-} \cdot Pb^+). \tag{II–23}$$

This $(V_{X^-} \cdot Pb^+)$ or Pb^+ (eqn II–13) aggregates giving lead particles Pb_n and lead ions Pb^{2+}. Also annihilation of V_{X^-} occurs to supply space for the lead particles since the lead atom has a larger radius than the lead ion Pb^{2+}. The reaction steps (II–20) and (II–21) can be compared with the proposed model for lead bromide in a bromine atmosphere presented in Section (D.1).

References

Arends, J. and Verwey, J. F. (1967). *Phys. Stat. Sol.* **23**, 137.

Bräkken, H. (1932). *Z. Krist.* **83**, 222,

Burnham, W. C., Brown, F. C. and Knox, R. S. (1960). *Phys. Rev.* **119**, 1560.

Fesefeldt, H. (1930). *Z. Physik* **64**, 741.

Heyningen, R. S. van, and Brown, F. C. (1958). *Phys. Rev.* **111**, 462.

Hoshino, H., Yamazaki, M., Nakamura, Y. and Shimoji, M. (1969). *J. Phys. Soc. Japan* **26**, 1422.

Kaldor, A. and Somorjai, G. A. (1966). *J. Phys. Chem.* **70**, 3538.

Mees, C. E. K. and James, T. H. (1966). "Theory of the Photographic Process." New York.

Moser, F., Nail, N. R. and Urbach, F. (1959). *J. Phys. Chem. Solids* **9**, 217.

Mott, N. F. and Gurney, R. W. (1964). "Electronic Processes in Ionic Crystals." Dover Publications Inc., New York.

Nieuwenkamp, W. (1932). Ph.D. Thesis, University of Utrecht.

Norris, R. S. (1895). *Am. Chem. J.* **17**, 1895.

Patankar, A. V. and Schneider, E. E. (1966). *J. Phys. Chem. Solids* **27**, 575.

Pierrard, J. M. (1969). Ph.D. Thesis, University of Washington.

Renz, C. (1921). *Z. Anorg. Chem.* **116**, 62.

Schmid, W. (1866). *Ann. Phys. Pogg.* **127**, 493.

Schoonman, J. (1970). *J. Solid State Chem.* **2**, 31.

Schoonman, J. and Verwey, J. F. (1968). *Physica* **39**, 244.

Smakula, A. (1965). M.I.T. Technical Report No. 6.

Spencer, H. E. and Darlak, J. O. (1968). *J. Phys. Chem.* **72**, 2384.
Tabak, M. D. and Warter, P. J. (1966) *Phys. Rev.* **148**, 982.
Tennant, W. C. (1966). *J. Phys. Chem.* **70**, 3523.
Tubandt, C. (1921). *Z. Anorg. Chem.* **115**, 105.
Tubandt, C. (1932). Handbuch der Expermental Physik **12**, (1), Akademischer Verlag Leipzig.
Tubandt, C. and Eggert, S. (1920). *Z. Anorg. Chem.* **110**, 196.
Tubandt, C., Reinhold, H. and Liebold, G. (1931). *Z. Anorg. Chem.* **197**, 225.
Verwey, J. F. (1966). *J. Phys. Chem. Solids* **27**, 468.
Verwey, J. F. (1967). Ph.D. Thesis, University of Utrecht.
Verwey, J. F. (1970) *J. Phys. Chem. Solids* **31**, 163.
Verwey, J. F. and Schoonman, J. (1967). *Physica* **35**. 386.
Verwey, J. F. and Westerink, N. G. (1969). *Physica* **42**, 293.
Vries, K. J. de. (1965). Ph.D. Thesis, University of Utrecht.
Vries, K. J. de, and Santen, J. H. van. (1963). *Physica* **29**, 482.
Wagner, J. B. and Wagner, C. (1957). *J. Electrochem. Soc.* **104**, 509.
Wells, H. L. (1893). *Am. J. Sci.* **45**, 134.
West, W. and Saunders, V. I. (1959). *J. Phys. Chem.* **63**, 45.

III. The Electrode Photoeffect of Metals in an Electrolyte Solution

D. A. DE VOOYS

In Part I the formation of the latent image is discussed. The disappearance of the latent image due to suitable radiation is mentioned (the Herschel effect). In this aspect the electrode photoeffect is interesting (Barchevsky and Gurevich, 1970), because the electrode photoeffect includes the behaviour of a metal electrode in a solvent, when light falls on the electrode.

In general, when metal electrodes, in a non absorbing electrolyte, are irradiated with ultra-violet or visible light, a photocurrent is generated. The magnitude of this photocurrent depends, among other things, on: the wavelength of the light, the concentration of ions or molecules that scavenge the electrons generated by light, the potential applied to the electrode, and the products of the scavenger reaction which can occur at the interface.

Several theoretical models have been put forward to explain the photocurrent phenomenon.

The proposed models should account for the following experimental observations (Barker *et al.*, 1966; Bomchill *et al.*, 1970; Delahay and Srinivasan, 1966; Korshunov *et al.*, 1968; de Levie and Kreuser, 1969; Pleskov and Rotenberg, 1968, 1969; Rotenberg *et al.*, 1968; Sharma *et al.*, 1968): (a) the relationship between the photocurrent and the frequency of the light used is, to the best of our present knowledge, given by:

$$i_p \sim (\omega - k')^\alpha \tag{III-1}$$

where i_p = the observed photocurrent,

k', α = constants,

$\dfrac{\omega}{2\pi}$ = the frequency of the light used.

(b) the current depends on the applied electrode potential according the relationship:

$$i_p \sim (k'' - \phi)^\alpha \tag{III-2}$$

where k'' = constant,

ϕ = potential applied to the electrode.

(c) the relationship between the current and the scavenger concentration, c_s. This relationship can be described by:

at low c_s: $i_p \sim c_s^{\frac{1}{2}}$ (III-3)

at high c_s: $i_p = i_{ps} =$ constant (III-4)

where i_{ps} = the photocurrent at high scavenger concentrations. This is often called the limiting or saturation value.

To account for these experimental observations, a number of theoretical models have been proposed. In general no model has been developed to account for the most complicated behaviour or for the complete set of observations. For the simplified case of monochromatic incident light and for negative photocurrents, models have been proposed to include as variables the frequency of the light, the applied potential and the scavenger concentration. One model also includes possible subsequent reactions of the products of the scavenger reaction. For the more complicated case of polychromatic light, so far the models only include as variables the scavenger concentration and the ionic strength of the indifferent electrolyte. Some models include negative and positive photocurrents. Consider now some of the details of the proposed models and the evidence used to corroborate them.

A. The Model of Barker *et al.*

Barker *et al.* (1966) proposed a three step model: (a) electron emission, (b) solvation and (c) reaction between the solvated electrons and a scavenger. The model applies when the incident light is monochromatic and for negative photocurrents. The details are:

(a) photoemission is the first step. The magnitude of the photocurrent depends on the frequency of the light used and on the applied potential on the electrode. They assumed this dependency to be that for photoemission in vacuum. Fowler's law (1931) relates the photocurrent, i_p, to the radiation frequency. This relationship is:

$$i_p \sim (\omega - \omega_0(\phi))^2 \tag{III-5}$$

where $\omega_0(\phi)$ = frequency of the red boundary. This is the minimum frequency at which photoemission occurs and is a function of the applied potential.

For $\omega_0(\phi) > \omega$ no photoemission occurs. During the polarization of the electrode, the red boundary shifts. This shift is given by (Pleskov and Rotenberg, 1969):

$$\hbar\omega_0(\phi) = \hbar\omega_0(0) + e\phi \tag{III-6}$$

or

$$\omega_0(\phi) = \omega_0(0) + \frac{e\phi}{\hbar} \tag{III-7}$$

where $\hbar$ = Planck's constant divided by 2π, e = the charge of an electron.

When the charge on the metal is zero, $\phi = 0$, by definition. Substitution of equation (III-7) in (III-5) gives:

$$i_p \sim \left(\omega - \omega_0(0) - \frac{e\phi}{\hbar}\right)^2 \tag{III-8}$$

or

$$i_p \sim (\hbar\omega - \hbar\omega_0(0) - e\phi)^2. \tag{III-9}$$

One should expect from (III-6) that the coefficient of proportionality between the change in the red boundary and the corresponding change in the potential to be unity. That is, that the following ratio holds:

$$\frac{\Delta\hbar\omega_0(\phi)}{\Delta e\phi} = 1. \tag{III-10}$$

This was experimentally confirmed by Barker et al. (1966) and Korshunev et al. (1969). However Heyrovsky (1966a, b) found the ratio was 0·5, and de Levie and Kreuser (1969), 1·26 ± 0·04.

One must keep in mind that formula (III-10) was derived only on the basis of an emission model, and other effects, such as the diffusion of an electron and the influence of the double layer on the electron, are not included. This model so far can be expected to apply at high scavenger concentrations because, for these scavenger conditions, the electron is directly captured so that no diffusion occurs and no electrons reach the electrode. Furthermore, at high ionic concentrations the thickness of the double layer is small relative to the distance of emission.

(b) Next, they assumed that some of the emitted electrons travel some distance through the solution and become solvated, and some electrons are retrapped by the electrode.

(c) After solvation, the electrons react with a scavenger, such as hydrogen ion, nitrous oxide, or dissolved oxygen. At high scavenger concentrations, the electrons are captured immediately and no diffusion occurs. At higher scavenger concentrations the reaction rate is limited by the number of electrons. Hence the photocurrent becomes independent of scavenger concentration. In general for lower scavenger concentrations, the electron can diffuse before scavenging.

The diffusion can be described by Fick's second law for the x-direction:

$$\frac{\partial c}{\partial t} = D \frac{\partial^2 c}{\partial x^2} - k_0 c_s c + \Phi(x, t),$$ (III–11)

where c = electron concentration,
   t = time,
   D = diffusivity of a solvated electron,
   x = distance from the electrode,
   k_0 = rate constant for the scavenging of solvated electrons,
   Φ = the source function, which describes the production of solvated electrons.

Barker et al. assumed the diffusion was undirectional. They postulated that all the solvation occurs in a plane at distance x_0 from the surface of the electrode, and from this plane the solvated electrons move only by diffusion, not influenced by the diffuse double layer. They assumed further that no scavenging occurs for values of $0 < x < x_0$. Based on these assumptions Barker et al. obtained the result:

$$\frac{i_p}{i_{ps}} = \frac{Q x_0}{1 + Q x_0}$$ (III–12)

where

$$Q = \left(\frac{k_0 c_s}{D} \right)^{\frac{1}{2}}$$ (III–13)

They did not use the formalism of eqn (III–11). This model predicted the observed relationship between the current and the scavenger concentrations for low values of c_s, in the limit when the denominator in eqn (III–12) approaches unity (eqn III–3) and for high scavenger concentrations (eqn III–4) when $Q x_0 \gg 1$.

However, later work by Gurevich and Rotenberg (1968) showed that the combination of a source function consistent with the model of Barker *et al.* is given by:

$$\Phi(x, t) = \frac{i_{ps}\delta(x - x_0)}{e} \cdot \exp{(i\omega t)} \qquad \text{(III–14)}$$

together with the elimination of the assumption that no scavenging occurs for $0 < x < x_0$ does not yield eqn (III–12). To get eqn (III–12) another source function should be used, given by:

$$\Phi(x, t) = \frac{i_{ps}}{ex_0} \cdot \exp{\frac{-x}{x_0}} \cdot \exp{(i\omega t)}. \qquad \text{(III–15)}$$

Barker *et al.* calculated from (III–12) values of x_0 of 34, 41 and 48 Å at potential $-1\cdot2$, $-1\cdot2$ and $-1\cdot4$ volts respectively. The potentials were measured against SCE, with the hydrogen ion as scavenger. These values will be compared with those of Bomchill later on.

The merit of Barker *et al.*'s model is, that it includes for the first time the assumption that electrons travel some distance through the solution before solvation takes place.

B. The Model of Gurevich *et al.*

Gurevich *et al.* (1967) modified the model of Barker *et al.* Their model applied to monochromatic light and to negative photocurrents. They assumed the following steps:

(a) the electron photoemission proper, i.e. the transition of electrons absorbing light quanta through the interface.

(b) the retardation of electrons in the liquid due to their transmission of energy to solvent molecules (the retardation usually occurs at distances of ten and hundreds of angstrom units from the surface; i.e. beyond the reach of surface forces).

(c) the solvation of "heat" electrons and subsequent physicochemical transformations (diffusion, capture by acceptors and "traps", chemical reaction in the body, on the electrode and so forth).

Generally speaking, isolation of step (b) and (c) is to some extent conditional; for example, retardation may be accompanied by chemical reaction.

Consider now some of the details of their modifications of Barker *et al.*'s model:

(a) the first step: emission. They eliminated the assumption that the photoemission was represented by emission in vacuum. They derived, from quantum mechanics, that the frequency dependence should be:

$$i_p \sim (\omega - \omega_0(\phi))^{5/2}. \qquad \text{(III–16)}$$

Therefore the slope of log i_p versus log $\Delta\omega$ should be 2·5 rather than 2·0. Several authors (de Levie and Kreuser, 1969; Pleskov and Rotenberg, 1969) found 0·4 fit the experimental data better than 0·5. To prove this model Pleskov and Rotenberg (1969) experimented with solutions of low ionic strength and with different amalgams as electrode.

For solutions with low ionic strength one would expect that the diffuse double layer may play a role in the photoemission because the electrons interact with the electric field of the double layer. Pleskov and Rotenberg made, therefore, a correction for ϕ : $\phi - \psi_1 \cdot (\psi_1 =$ the potential in the plane of closest approach). Eqn (III–16) was rewritten in the form:

$$i_p \sim (\hbar\omega - \hbar\omega_0 - e(\phi - \psi_1))^{5/2}. \qquad \text{(III–17)}$$

When they applied this correction, they got more linear plots of their log i_p versus $\frac{5}{2}$ log $(\hbar\Delta\omega - e(\phi - \psi_1))$ curves, but this was not found by de Levie and Kreuser (1969).

Now consider the effect of different amalgams. The electrochemical potential of an electron does not depend on the metal nature at the metal-electrolyte interface (Frumkin, 1965; Parsons, 1964). This is valid for any process of electron transfer—including photoemission—that occurs across the metal–electrolyte interface. This statement was confirmed by Pleskov and Rotenberg. They found the same threshold potential with different amalgams.

Bomchill et al. (1970) pointed out, that this experimental evidence may be insufficient to prove the appropriateness of the model because the scavenger concentration used in the corroborating experiments of Pleskov and Rotenberg was too low to yield the required saturation photocurrent. Thus the diffusion processes occurring after emission can be important.

(b) and (c): retardation and subsequent physicochemical transformations of the electrons. Gurevich and Rotenberg (1968) calculated the influence of diffusion, assuming that there is no interaction between the diffuse double layer and the electrons, and that the products of the scavenger reaction are not discharged at the electrode. For the electron concentration and the source function they used:

$$c = c(x) \exp(i\omega t) \qquad \text{(III–18)}$$

and

$$\Phi = f(x) \exp(i\omega t). \qquad \text{(III–19)}$$

For the diffusion eqn (III–11) they assumed the following limiting condition:

$$D\left(\frac{\partial c}{\partial x}\right) = \kappa c(0) \qquad \text{(III–20)}$$

where κ = electron capture by the surface, and

$$c(\infty) = 0. \qquad \text{(III–21)}$$

For low scavenger concentration they calculated that $i_p \sim c_s^{\frac{1}{2}}$. This result was always obtained at sufficiently small values of Q, without any assumption regarding the form of the source function.

They defined a new quantity, θ, as the phaseshift between the measured photocurrent and the limiting photocurrent using modulated illumination. Gurevich et al. made the following predictions based on their model:

(a) The character of the dependence of $\tan\theta$ on the frequency of the light source to evaluate the rate of solvated electron scavenging by the metal surface, κ.

(b) The rate constant of scavenging, k_0, is easily determined by measuring $\tan\theta$. For this purpose, it is sufficient to measure this quantity at $k_0 c_s > \omega$. Thus, given $k_0 c_s$ and i_{ps}, the measurement of i_p can serve as the basis for determining the average electron solvation path in the solution.

(c) Measurements made under modulated illumination are extremely convenient for the investigation of the residual photocurrent (at $c_s = 0$).

(d) By investigating the photocurrent at variable c_s and ω one can expect to deduce the form of the source function, $\phi(x, t)$.

Gurevich et al. have also calculated the extra photocurrent, which arises from the reaction of the products of the scavenger reaction at the electrode.

Although the model of Gurevich is an improvement on that of Barker because it eliminates the assumption that the emission of electrons is in a vacuum and includes a better description of the diffusion, more experiments are needed to prove that the model fits the experimental data better.

C. The Model of Bomchill et al.

Bomchill et al. (1970) considered the condition for polychromatic light and negative photocurrents. They calculated the influence of the scavenger concentration on the magnitude of the photocurrent with a model different from that of either of the preceding workers. They assumed a uniform, initial distribution of the electrons between the electrode (or some distance very close to it, where they define the origin of their coordinate system $x' = 0$) and a plane determined by the maximum frequency of the incident light $x' = \delta$, where the electrons are deposited at a constant rate. They assumed the source function to be :

$$\Phi = k(\delta)(H(x') - H(x' - \delta)) \tag{III-22}$$

where $k(\delta)$ = constant rate of deposition of the solvated electrons
between $x' = 0$ and $x' = \delta$.

$H(x' - \delta)$ = the Heaviside function; this is a double step function of
value one f or x' between 0 and δ, and zero elsewhere.

The origin of the coordinate system, $x' = 0$, is not necessarily the metal
surface, but rather the plane from which the electron transfer reaction takes
place. Thus, x' should be substituted for x in eqn (III–11). They assumed
steady state. The boundary conditions were:

$$c(0) = 0 \tag{III-23}$$

and

$$c(\infty) = 0. \tag{III-24}$$

Their results were:

$$\frac{i_p}{i_{ps}} = 1 - \frac{1}{\delta Q}\,(1 - \exp\,(-\delta Q)). \tag{III-25}$$

This result fitted their experimental data well. A comparison of the constant
in this equation for low Q with that of Barker *et al.* is worthwhile. That is,
for low Q:

Barker:

$$\frac{i_p}{i_{ps}} = Q x_0$$

and Bomchill:

$$\frac{i_p}{i_{ps}} = \frac{Q\delta}{2}. \tag{III-27}$$

A comparison of these equations suggests that:

$$x = \frac{\delta}{2}. \tag{III-28}$$

This conclusion is reasonable since Barker assumed a plane at distance x_0
and Bomchill a region of $x' = 0$ to $x' = \delta$, so the average distance to the
electrode is about $\delta/2$.

Bomchill *et al.* found that $\delta = 50\text{Å}$ for the experiments for different
potentials: -0.6 to -1.0 volts versus 1 N calomel electrode. The hydrogen
ion was the scavenger. This value of δ is of the same order of magnitude as
calculated by Barker *et al.*

The model of Bomchill *et al.* agrees with the experimental data for the
more complicated case of polychromatic light for the range of variables
considered.

D. The Model of Korshunov

Korshunov *et al.* investigated the photoemission current from metals in solution with single flashes of monochromatic and polychromatic light. In this case one must integrate eqn (III–17) to obtain the total current. They used the equation:

$$V = \frac{\exp(-t/RC)}{C} \int_0^t \exp\left(\frac{t'}{RC}\right) i_p(t')\mathrm{d}t'$$

where V = the recorded potential signal,
  C = the capacitance of the double layer,
  R = the standard resistance connected in series with the cell and the polarizing battery,
  t = time,
  i_p = the photocurrent.

For negative potentials, the results were predicted from eqns (III–10) and (III–17). In the neighbourhood of the potential of the zero point of charge ($\phi = 0$) they found the behaviour of the photocurrent different from what is expected from eqn (III–17). They found a change in the sign of the photocurrent on each side of the potential of the z.p.c. The same effect was measured by de Levie and Kreuser (1969). This effect cannot be explained by the model of either Barker or Gurevich. Korshunov *et al.* assumed that the lightflash heats up the solution near the electrode and this would explain the observed alternation of the photocurrent. However, this suggested cause cannot account for the observations of Levin and Delahay (1970). They found no significant decay in their potential–time curves with radiation without scavenger. Such a decay would be expected if a momentary heating effect is the cause of this behaviour. However they postulated another possible cause of the heating effect: the heat of the scavenger-solvated electron reaction. The heating effect should be independent of the nature of the electrode. Indeed similar effects are observed with AgI (van Peski *et al.*, 1969). More investigations of the heating effect are necessary.

E. The Model of Heyrovsky

The model of Heyrovsky accounts for either positive and negative photocurrents. According to Heyrovsky, the observed photocurrent is due to the formation of a complex with charge transport on the surface. Molecules or ions from the solution close to the surface of the electrode may create a higher or lower electron concentration there. When light falls on the electrode this complex may be destroyed, and the molecule or ion may leave the surface. The charge enters the electrode, and this is what is measured as photocurrent. Depending on the properties of the molecule or ion complexes formed, the photocurrent may either be positive or negative.

Heyrovsky gives few equations to test his model.

F. Conclusion

In conclusion for monochromatic conditions and for negative photocurrents the model of Gurevich *et al.* is an improvement over the model of Barker *et al.* The model of Gurevich is attractive, for it gives a method to calculate the source function, a difficulty that was discerned by Bomchill *et al.* However, the experimental confirmation of this model is not convincing. More data are needed especially considering the variety of experimental results obtained for the shift of the red boundary. Furthermore, experimental evidence concerning the effect of the scavenger concentration needs to be collected for conditions consistent with the model. More experiments are necessary to decide if the model of Gurevich is useful without modifications. The models, which explain the positive and negative photocurrents are highly speculative and need more confirmation.

References

Baker, G. C., Gardner, A. W. and Sammon, D. C. (1966). *J. Electrochem. Soc.* **113**, 1182–1197.

Barshevsky, B. U. and Gurevich, Y. Y. (1970). Paper presented at the International Congress of Photographic Science, Moscow.

Bomchill, G., Schiffrin, D. J. and D'Alessio, T. J. (1970). *J. Electroanal. Chem.* **25**, 107–119.

Brodskii, A. M. and Gurevich, Y. Y. (1968). *Zh. Eksp. Teor. Fiz.* **54**, 213–227.

Delahay, P. and Srinivasan, V. S. (1966). *J. Phys. Chem.* **70**, 420–426.

de Levie, R. and Kreuser, J. C. (1969). *J. Electroanal. Chem.* **21**, 221–236.

Fowler, R. H. (1931). *Phys. Rev.* **38**, 35

Frumkin, A. N. (1965). *J. Electroanal. Chem.* **9**, 173.

Gurevich, Y. Y. and Rotenberg, Z. A. (1968). *Elektrokhimiya* **4(5)**, 529–533.

Gurevich, Y. Y., Brodskii, A. M. and Levich, V. G. (1967). *Elektrokhimiya* **3(11)**, 1302–1310.

Heyrovsky, M. (1965). *Nature* **206 (4991)**, 1356–1357.

Heyrovsky, M. (1966a). Thesis, Cambridge University.

Heyrovsky, M. (1966b). *Nature,* **209 (5029)**, 708.

Heyrovsky, M. (1967a). *Proc. Roy. Soc. A.* **301**, 411–431.

Heyrovsky, M. (1967b). *Z. physik. Chem. Neue Folge,* **52**, 1–7.

Korshunov, L. I., Zolotovitskii and Benderskii (1968). *Electrokhimiya* **4 (5)**, 499–503.

Levin, I. and Delahay, P. (1970). *J. Electroanal. Chem.* **24**, App. 17–21.

Parsons, R. (1964). *Surface Sci.* **2**, 418.

Pleskov, Y. V. and Rotenberg, Z. A. (1969). *J. Electroanal. Chem.* **20**, 1–12.

Rotenberg, Z. A. and Gurevich, Y. Y. (1968). *Elektrokhimiya* **4 (8)**, 984–987.

Rotenberg, Z. A. and Pleskov, Y. V. (1968). *Electrokhimiya* **4 (7)**, 826–827.
Rotenberg, Z. A., Pleskov, Y. V. and Lakomov, V. I. (1968). *Electrokhimiya* **4(8)**, 1022.
Sharma, V. P., Delahay, P., Susbielles, G. G. and Tessari, G. (1968). *J. Electroanal. Chem.* **16**, 285–294.
van Peski, A. C. H., de Vooys, D. A. and Engel, D. J. C. (1969). *Nature* **223**, **(5202)**, 177.

APPENDIX

TABLES OF IONIC CONDUCTIVITY AND DIFFUSION IN IONIC CRYSTALS

Robert J. Friauf

Department of Physics and Astronomy, University of Kansas, Lawrence, Kansas, U.S.A.

The following material is reprinted with permission from the American Institute of Physics Handbook, D. E. Gray, Editor, 3rd Edition, McGraw–Hill Book Co., New York, 1972.

9f–1. Ionic Conductivity and Diffusion. These phenomena are ascribed to the presence of ionic defects—vacancies where ions are missing from normally occupied positions and ions in interstitial positions in the structure. *Schottky defects* are combinations of cation and anion vacancies, as in the alkali halides and alkaline earth oxides. *Frenkel defects* are combinations of vacancies and interstitial ions, for cations as in the silver halides, or for anions as in the alkaline earth halides. At high temperatures the defects exist in thermodynamic equilibrium in the crystal; for Schottky defects in MX crystals, for example, the concentration or mole fraction increases with temperature according to (ref. 23)

$$x = x_0 \exp\left(-\tfrac{1}{2}h_f/kT\right), \qquad x_0 = \exp\left(\tfrac{1}{2}s_f/k\right), \qquad (9f–1)$$

where h_f and s_f are the enthalpy and entropy of formation of a pair of defects. At lower temperatures the mole fraction is usually controlled by the presence of aliovalent impurities.

The random jumping of a defect gives rise to a *microscopic diffusion coefficient* for the defect of

$$d = d_0 \exp\left(-\Delta h/kT\right), \qquad d_0 = \tfrac{1}{6}\nu a^2 \exp\left(\Delta s/k\right), \qquad (9f–2)$$

where v is an attempt frequency, a is the jump distance, and Δh and Δs are the activation enthalpy and entropy for the jump. (The factor $\frac{1}{6}$ is appropriate for a cubic lattice.) In an electric field there is also a drift *mobility*

$$\mu = \mu_0 \exp\left(-\Delta h/kT\right), \qquad \mu_0 = (q/kT)\,d_0. \qquad (9f\text{-}3)$$

Here μ_0 has been obtained from d_0 with the *microscopic Einstein relation*

$$d/\mu = kT/q; \qquad (9f\text{-}4)$$

the conversion factor is $k/e = 0.862 \times 10^{-4}$ volt/°K with d in cm^2/sec and μ in cm^2/volt–sec. Eqns (9f–1) to (9f–3) are used to express the observed conductivity and diffusion coefficients in the following sections.

Ionic crystals covered in these tables include halides, simple inorganic radicals (such as nitrates and azides), binary oxides, and other chalcogenides (sulphides, selenides, and tellurides). Excluded from consideration are III–V compounds, ternary oxides (such as spinels and perovskites), and glasses and zeolites. Conductivity and self-diffusion coefficients are given for pure crystals only, but some information from experiments on doped crystals is contained in Table 9f–2. The effect of high pressure on conductivity and data for mixed electronic and ionic conductors are also presented. Space limitations prevent any consideration of the extensive recent literature on dielectric and anelastic relaxation, thermoelectric phenomena, and effects of radiation and plastic deformation on conductivity and diffusion. Similarly the diffusion of all foreign ions is excluded because of the proliferation of results. Many of the excluded topics are discussed in some of the books and review articles given in the general references.

9f–2. Conductivity for Ionic Conductors. The conductivity can be determined by passage of direct current through the sample if sufficient precautions are taken. More recently, however, most measurements have been made with current pulses of the order 10^{-2} to 10^{-3} sec duration or alternating currents at frequencies of 1 to 10 kHz in order to avoid large polarisation effects at the electrodes.

In most cases a plot of log σ versus $1/T$ is approximately a straight line, at least for a limited temperature range, allowing an empirical representation of the data as

$$\sigma = \sigma_0 \exp\left(-W/kT\right). \qquad (9f\text{-}5)$$

The parameters σ_0 and W are listed in Table 9f–1. The conductivity at the melting temperature has been calculated from eqn (9f–5) if it is not given in the references.

The values for σ_0 and W are not always so accurate as the number of significant figures would indicate. With good single crystal or polycrystalline samples of high purity a careful worker can reproduce results within a few per cent, but data from different laboratories may differ by 5 to 10 per cent, and discrepancies of 50 per cent are not uncommon. Hence W may be reliable to a few per cent in favourable circumstances or to perhaps 10 per cent in less favourable cases, and a discrepancy of 50 per cent in σ_0, which is very sensitive to the choice of W, is not surprising. For this reason several representative sets of data, if available, have been given for each substance.

9f–3. Concentration and Mobility of Defects in Ionic Crystals. The conductivity of a crystal containing several types of defects is

$$\sigma = N\Sigma_j q_j x_j \mu_j, \tag{9f–6}$$

where N is the number of molecules per unit volume of the perfect crystal and q_j is the magnitude of the charge of the jth defect. If only one type of defect makes an appreciable contribution to the conductivity, the use of eqns (9f–1) and (9f–3) gives the observed form of eqn (9f–5). In the *intrinsic* region for temperatures near the melting point $W_{\text{intr}} = \frac{1}{2}h_f + \Delta h$ and $\sigma_0 = Nqx_0\mu_0$, and in the *extrinsic* region for lower temperatures $W_{\text{extr}} = \Delta h$ and $\sigma_0 = Nqc\mu_0$ since x is maintained constant at the impurity concentration c. This simple explanation corresponds to the frequent observation of two different temperature ranges with different slopes in the plot of log σ versus $1/T$, especially for the initial observations on a substance, and the two slopes are often combined to obtain h_f and Δh from the expressions for W_{intr} and W_{extr}. This is presumably the extent of the analysis when only activation enthalpies are given in Table 9f–2. Recent work has shown, however, that such an analysis is at best only tentative because of contributions of other types of ions, association and precipitation of impurities, and overlapping of the different temperature regions.

In early work many transport number determinations were made by electrolysis in order to identify the ions carrying the current. When only one type of ion contributes to the conductivity, these experiments have, in fact, verified Faraday's laws of mass transport to an accuracy of 1%. When several types of ions, or both ions and electrons, however, make appreciable contributions, such experiments have not given very reliable results, presumably because of experimental difficulties at the electrodes and at the interfaces between the several samples involved. Hence only a handful of these experiments have been reported in the last ten years, and no separate table of results is provided.

In a few recent investigations of alkali halides an attempt has been made to separate cation and anion contributions to the conductivity by fitting a sum of two terms of the form of eqn (9f–5) to the observed total conductivity, as indicated by eqn (9f–6), and some results are given in this form in Table 9f–1. Often measurements of tracer diffusion coefficients allow evaluation of ionic transport numbers, but even these may not be completely unambiguous if vacancy pairs contribute noticeably to diffusion (ref. Ne1). The most reliable results are obtained from analysis of measurements on crystals intentionally doped with aliovalent impurities, with due account taken of mass action laws, association of charged defects and impurities, and long range Debye–Hückel interactions (ref. Be1). Most of the results in Table 9f–2 have been obtained in this way.

The temperature dependence of x and μ is given by eqns (9f–1) and (9f–3). It should be observed, however, that μ_0 contains a factor $1/T$, which is also carried over into σ_0. For this reason d_0 is listed rather than μ_0 in Table 9f–2; the conversion is obtained immediately from eqn (9f–4). When the factor of $1/T$ is not explicitly removed from μ_0 or σ_0, the apparent activation energy is smaller than the correct value by kT, which is of the order of 0·05 to 0·15 eV for temperatures from 300° to 1500°C.

9f–4. Effect of Pressure on Conductivity. When the effect of high pressure is taken into consideration, eqns (9f–1) and (9f–3) are modified to give

$$x = x_0 \exp\left(-\frac{h_f + Pv_f}{2kT}\right),$$

$$\text{(9f–7)}$$

$$\mu = \mu_0 \exp\left(-\frac{\Delta h + P\Delta v}{kT}\right),$$

where v_f is the change in volume of the crystal when a pair of Schottky defects is formed and Δv is the activation volume when a defect moves from one position to another. If only one type of defect contributes appreciably to the conductivity, the pressure dependence of the conductivity is given by

$$\sigma = \sigma_0 \exp\left(-P\Delta V/kT\right), \qquad \text{(9f–8)}$$

where $\Delta V_{\text{intr}} = \frac{1}{2}v_f + \Delta v$, for instance, in the intrinsic range.

The pressure dependence of the original data is expressed by a *pressure coefficient*

$$\alpha = -\left(\partial \ln \sigma/\partial P\right)_T. \qquad \text{(9f–9)}$$

The corresponding *free volume* from eqn (9f–8) is

$$\Delta V = RT\alpha = 82 \cdot 0 \times T \times \alpha,$$

with ΔV in cm^3/mole and α in atm^{-1}. Values of pressure coefficients and free volumes are given for a number of substances in Table 9f–3.

9f–5. Mixed Electronic and Ionic Conductors. Many ionic crystals have an appreciable electronic conductivity in addition to their ionic conductivity. Exclusive ionic conductivity occurs for nearly all halides (the cuprous halides being the only noteworthy exception) and for crystals with simple inorganic radicals. Beryllia also has mainly ionic conductivity, but the other alkaline earth oxides show progressively larger amounts of electronic conductivity, especially at higher temperatures. The only other predominantly ionic conductors are crystals with the fluorite structure such as calcia stabilized zirconia and even sodium sulphide, perhaps some rare earth type trioxides such as scandia and neodymia, and a new class of complex sulphides typified by Ag_3SI. Appreciable, but not exclusive, ionic conductivity is displayed by the cuprous halides, some simple metal oxides such as alumina and tetragonal zirconia, and most rare earth oxides such as ceria and dysprosia. Traces of ionic conductivity (a few per cent) are present in the copper and silver chalcogenides. Electronic conductivity (by electrons or holes) is dominant in transition metal oxides such as Cr_2O_3, and in all other divalent chalcogenides such as ZnO and PbS.

It should be clear that a fairly complicated situation exists when both electronic and ionic defects are present to an appreciable extent in a crystal. The treatment of the various interactions (refs 21, 22) shows that the defect structure may be profoundly influenced by the atmosphere surrounding the crystal or by deviations from stoichiometry of the crystal. Hence conductivity results are practically meaningless unless these conditions are specified, and similar remarks apply to diffusion. Fortunately much more attention has been devoted in recent years to control and measurement of the environment, and this information is provided where pertinent in Tables 9f–5 and 9f–6 in one of three ways: saturation of one constituent by contact with the metal or high vapour pressure of a volatile component, measurement of the oxygen partial pressure, or determination of the deviation from stoichiometry.

Several experimental techniques have been used to distinguish between electronic and ionic conductivity (ref. 56). (1) The earliest was direct determination of mass transport by electrolysis, but this has often been unreliable (ref. He1) and is seldom used at present. (2) Polarization effects are often observed, namely, the a.c. conductivity at moderately high frequencies like 100 kHz is considerably less than the d.c. conductivity. The simplest assumption is that the ac value is due to the electronic conductivity only whereas the d.c. value represents the total conductivity (ref. Ve3). Despite the appeal of this interpretation the results are not usually unambiguous, and much clarification is needed to make this method reliable (ref. Mc1). (3) If the potential drop between the electrodes is kept below the decomposition voltage of the sample, it may be assumed that an ionic current cannot flow to the electrode, and the remaining current is then ascribed to electronic conductivity (refs Wa5, Wa6). This method appears to be fairly reliable in some cases, but note must be taken of the range of chemical potentials occurring in such experiments. (4) If the conductivity is completely ionic, an e.m.f. that can be calculated from thermodynamic data should be established when the ends of the sample are at different chemical potentials (ref. Wa2), and this has been amply verified for calcia stabilized zirconia, for instance. If some electronic conductivity is also present, part of the e.m.f. is effectively shorted out, and hence the reduction of the observed e.m.f. below the thermodynamic value gives an indication of the amount of electronic transport (ref. Sc6). This method is the most commonly used, especially for oxides and appears to give a reliable estimate of the average transport number if care is taken to establish a well-defined chemical potential at each end of the sample and to ensure thermodynamic equilibrium. An unfortunate aspect of this method is that it does not distinguish between electrons and holes for the electronic part of the conductivity or between different types of ions for the ionic conductivity, but often other information is available. (5) The amount of ionic conductivity can be calculated from tracer diffusion coefficients with the Einstein relation if the charge on the defect and the correlation factor are known (see Section 9f–7). Since the last two items require a rather detailed knowledge of the diffusion mechanism, this approach is most often useful to establish an order of magnitude, especially when the ionic conductivity is very much smaller than the electronic part.

Table 9f–4 gives in most cases the *total conductivity*, which can often be determined more accurately than the transport numbers. Table 9f–5 gives the *ionic transport number*, which is defined as the fraction of the total current carried by ions. The two tables should be used together to obtain an estimate of the magnitude and nature of the conductivity for a particular substance. Substances have been listed only when there is some information about the ionic part of the conductivity; thus the numerous articles dealing solely with semiconducting behaviour in ionic crystals such as ZnO and CdS are not included.

9f–6. Diffusion. The tracer diffusion coefficient for an ion which can diffuse by means of several types of defects is

$$D_T = \Sigma_j f_j x_j d_j, \tag{9f–10}$$

where f_j is the correlation factor (see Section 9f–7) and x_j and d_j are given by eqns (9f–1) and (9f–2). When only a single mechanism is important the temperature dependence is given by

$$D_T = D_0 \exp(-W/kT), \tag{9f–11}$$

and this form is usually used to represent experimental results. Empirically determined values of D_0 and W are given in Table 9f–6.

In the intrinsic region the parameters in eqn (9f–11) are given by

$$D_0 = \tfrac{1}{6}va^2 x_0 f \exp \frac{\tfrac{1}{2}s_f + \Delta s}{k} \quad \text{and} \quad W = \tfrac{1}{2}h_f + \Delta h. \tag{9f–12}$$

Theoretical estimates indicate that W should be several electron volts, as observed, and that $\tfrac{1}{2}s_f + \Delta s$ should be at most a few entropy units, leading to a value of $D_0 \sim 10^{-3}$ to $10\ \text{cm}^2/\text{sec}$. When an appreciably different value of D_0 is obtained empirically, it is usually an indication that some disturbing influence, such as impurities or grain-boundary diffusion, has domination over the assumed thermodynamic equilibrium for volume diffusion.

Since the temperature dependence is the same for all types of defects, indirect methods must be used to distinguish a particular type of defect; this has been done with considerable success in many instances, as indicated in Table 9f–6. Some of these methods are (1) determination of the influence of aliovalent impurities in doped crystals, (2) study of correlation effects as described in Section 9f–7, and (3) observation of the effect of varying the stoichiometry or ambient pressure of one of the constituents of the crystal.

Among experimental methods for measuring diffusion coefficients with radioactive or isotopic tracers, *sectioning* is the most direct and reliable. *Surface counting, gaseous exchange,* and *solution exchange* are more sensitive but sometimes less reliable. Other methods of detection involve changes in *optical absorption, X-ray emission,* and *semiconducting properties* or observation of *additive coloration* or *electrotransport*. The line width in *nuclear magnetic resonance* allows a determination of the temperature dependence and an estimate of the magnitude of diffusion for stable nuclei. The rate of *oxidation* and *sintering* processes can also be used to evaluate diffusion coefficients when the process is sufficiently well understood.

The remarks concerning the accuracy of the results for ionic conductivity apply here with even more need for caution. For most halides pure single crystals are available, the melting points are not excessively high, and the influence of the surrounding atmosphere is often unimportant (ref. 50); hence in favourable cases an accuracy approaching that for the conductivity may be realised. For the usually semiconducting and often refractory chalcogenides, however, the situation is much less favourable. The high melting temperatures and difficulties of obtaining pure materials suggest that very few intrinsic properties have yet been observed for these substances (ref. 54). Furthermore the influence of grain boundaries is just beginning to be investigated, and yet a number of measurements have been made on sintered or pressed powder samples with porosities up to 5 or 10 per cent. Finally the defect structure is strongly influenced by any excess or deficit of the constituents, as discussed in Section 9f–5. The data in Table 9f–5 may nonetheless be useful both as a survey of existing experimental efforts and as a stimulus to better understanding.

9f–7. Correlation Effects in Diffusion. Both the ionic conductivity and diffusion of a charged defect are caused by the jumping of the defect through the crystal, and the connection of these two phenomena is given by the microscopic Einstein relation in eqn (9f–4). If a single type of defect is responsible for all of the observed conductivity and diffusion, eqns (9f–6) and (9f–10) may be combined (without the correlation factor) to give a *macroscopic Einstein relation* that defines $D_{\text{conductivity}}$.

$$D_{\text{conductivity}} = (kT/Nq^2)\sigma. \tag{9f–13}$$

In many cases this relationship is at least approximately satisfied, but there are four ways in which deviations may occur. (1) There may be another contribution to the conductivity, such as an electronic part or another type of ionic defect. (2) There may be neutral complexes of defects, such as vacancy pairs in the alkali halides, which contribute to the diffusion but not to the conductivity. (3) In the diffusion of tracers there are correlations in the random-walk motion of tracer atoms that lead to correlation factors, as first described by Bardeen and Herring (ref. Ba2). (4) In the case of interstitialcy mechanisms there are also different displacements for the tracer atom and for the charge of the defect. This displacement effect is usually included with the genuine correlation effects to give an overall correlation factor for interstitialcy mechanisms.

The *experimental correlation factor* is defined by

$$f = D_{\text{tracer}}/D_{\text{conductivity}}.$$

Theoretical correlation factors may be calculated by considering the geometry of the diffusion mechanism and of the lattice (see refs. below). Comparison of experimental and theoretical values will then often point to a particular mechanism for diffusion. Experimental and theoretical correlation factors are presented in Table 9f–7.

Guide to references on theoretical correlation factors.

General treatment: 24, Ba2, Co2, Co3, Ho10
Vacancy mechanisms: Ba2, Co2, Fr1, Sc8
Interstitial and interstitialcy mechanisms: Co3, Fr2, Mc2
Vacancy pairs and impurity complexes: Co2, Ho10, Le1, Li1
Anisotropic lattices: Co2, Gh1, Hu1, Hu2, Md1, Mu3
Disordered lattices: Ri2, Yo3
Diffusion by nuclear magnetic resonance: Ei2, St5
Isotope effects: Le2, Th1

TABLE 9f-1.—Conductivity for Ionic Conductors

[The conductivity is given as $\sigma = \sigma_0 \exp(-W/kT)$.]

Substance	Form	T_m (°C)	$\sigma(T_m)$ $(\text{ohm-cm})^{-1}$	T Range (°C)	σ_0 $(\text{ohm-cm})^{-1}$	W (eV)	Specific Ref.	Other Ref.
Alkali Halides								
LiH	sc	688	4×10^{-2}	480–630	4×10^{7}	1·72	Pr1	
LiD	sc			240–480	1	0·53	Pr1	
LiF	sc	842	$2{\cdot}4 \times 10^{-3}$	480–630	1×10^{7}	1·72	Pr1	Be7, Be8,
				540–720	$6 \times 10^{9}/T$	2·07	Ja6	Ha19, Le5,
				340–540	$5 \times 10/T$	0·70	Ja6	St7
	sc		$1{\cdot}5 \times 10^{-3}$	560–750	$1{\cdot}6 \times 10^{9}/T$	1·99	Ba7	
				330–560	$4{\cdot}5 \times 10^{10}/T$	1·65	Ba7	
LiCl	sc	606	9×10^{-3}	480–570	$2{\cdot}5 \times 10^{6}$	1·47	Ha19	Le5
	pc		$1{\cdot}8 \times 10^{-3}$	400–550	$2{\cdot}5 \times 10^{5}$	1·42	Gi1	
				30–350	$1{\cdot}2$	0·59	Gi1	
LiBr	sc	550	$1{\cdot}8 \times 10^{-2}$	440–540	$1{\cdot}4 \times 10^{6}$	1·29	Ha19	
	pc		$1{\cdot}4 \times 10^{-2}$	350–500	$4{\cdot}2 \times 10^{5}$	1·22	Gi1	
				30–300	$3{\cdot}3$	0·56	Gi1	
LiI	sc			160–360	8×10^{-2}	0·43	Al3	
	sc	452	5×10^{-2}	340–420	$9{\cdot}6 \times 10^{5}$	1·05	Ha19	
	pc		7×10^{-2}	250–350	$1{\cdot}8 \times 10^{5}$	0·92	Gi1	
				30–150	$1{\cdot}4 \times 10^{-1}$	0·36	Gi1	
NaF	pc	992	3×10^{-3}	330–980	$1{\cdot}3 \times 10^{3}$	1·42	Ph1	Le5

Crystal	Type	T	D_0	Temp. range	Prefactor	E	Ref.	Refs.
NaCl	sc	800	$1{\cdot}0 \times 10^{-3}$	520–740	$4{\cdot}7 \times 10^{8}/T$	$1{\cdot}86^{c}$	Fu3	Bi2, Br5, Do6, Dr2, Et1, Ja4, Ka4, Ko1, La4, Ma6, Ue1
				520–740	$1{\cdot}2 \times 10^{9}/T$	$2{\cdot}07^{a}$	Fu3	
			$1{\cdot}2 \times 10^{-3}$	720–800	$2{\cdot}4 \times 10^{10}/T$	$2{\cdot}19$	Ne1	
				550–650	$9{\cdot}2 \times 10^{8}/T$	$1{\cdot}92$	Ne1	
				275–425	$3 \quad /T$	$0{\cdot}65$	Ne1	
NaBr	sc	755	$1{\cdot}2 \times 10^{-3}$	450–700	$2{\cdot}1 \times 10^{8}/T$	$1{\cdot}68$	Ma6	Le5, Ph1, Sc2
				300–450	$3{\cdot}5 \times 10^{2}/T$	$0{\cdot}84$	Ma6	
	sc		$2{\cdot}1 \times 10^{-3}$	610–730	$2{\cdot}3 \times 10^{8}/T$	$1{\cdot}64$	Ho8	
				490–570	$3 \times 10^{2}/T$	$0{\cdot}80$	Ho8	
NaI	pc	661	$2{\cdot}1 \times 10^{-3}$	350–600	$8{\cdot}1 \times 10^{3}$	$1{\cdot}23$	Ph1	Le5
				170–350	6×10^{-2}	$0{\cdot}60$	Ph1	
KF	sc	846	6×10^{-4}	660–790	2×10^{7}	$2{\cdot}34$	Ka1	Le5
				400–500	4×10^{-1}	$1{\cdot}02$	Ka1	
KCl	sc	768	$2{\cdot}4 \times 10^{-4}$	570–750	$4{\cdot}1 \times 10^{6}/T$	$1{\cdot}66^{c}$	Fu4	As3, Be2, Bi4, Gr4, He4, Le5, Me2, Ph2, Pe1, Wa4
				570–750	$5{\cdot}6 \times 10^{10}/T$	$2{\cdot}36^{a}$	Fu4	
	sc		$1{\cdot}4 \times 10^{-4}$	340–640	$2{\cdot}3 \times 10^{8}/T$	$1{\cdot}90$	Mi4	
	sc		$4{\cdot}5 \times 10^{-5}$	480–680	$5{\cdot}9 \times 10^{7}/T$	$1{\cdot}88$	Al4	
KBr	sc	728	$1{\cdot}3 \times 10^{-4}$	300–700	$3{\cdot}1 \times 10^{8}/T$	$1{\cdot}91^{c}$	Da9	Gr4, Le5
				300–700	$7{\cdot}1 \times 10^{9}/T$	$2{\cdot}21^{a}$	Da9	
	sc		$2{\cdot}1 \times 10^{-4}$	440–680	$1{\cdot}1 \times 10^{6}$	$1{\cdot}93$	Ro4	Pe1, Ph2
	sc		$1{\cdot}0 \times 10^{-4}$	560–680	$7{\cdot}9 \times 10^{5}$	$1{\cdot}97$	Ho9	
KI	sc	680	$1{\cdot}0 \times 10^{-3}$	430–600	$1{\cdot}4 \times 10^{7}/T$	$1{\cdot}69^{c}$	Pe1	Bi4, Ec1, He4, Le5, Ph2
				430–600	$1{\cdot}6 \times 10^{12}/T$	$2{\cdot}31^{a}$	Pe1	
	sc		$2{\cdot}1 \times 10^{-4}$	450–650	$1{\cdot}6 \times 10^{6}$	$1{\cdot}87$	Ka1	Ph2

TABLE 9f-1.—Conductivity for Ionic Conductors (*Continued*)

Substance	Form	T_m (°C)	$\sigma(T_m)$ (ohm-cm)$^{-1}$	T Range (°C)	σ_0 (ohm-cm)$^{-1}$	W (eV)	Specific Ref.	Other Ref.
RbCl	sc	717	$1\cdot3 \times 10^{-5}$	550–700	$3\cdot6 \times 10^6/T$	$1\cdot58^c$	Fu3	Le5, Pi1
				550–700	$8\cdot8 \times 10^{11}/T$	$2\cdot55^a$	Fu3	
RbBr	sc	681	$3\cdot4 \times 10^{-5}$		$1\cdot8 \times 10^6$	$2\cdot03$	Le5	
CsF	pc	684	$1\cdot1 \times 10^{-3}$	550–660	$1\cdot6 \times 10^5$	$1\cdot55$	Ha6	
				330–550	2	$0\cdot85$	Ha6	
CsCl(α)	sc	636	$6\cdot1 \times 10^{-5}$	480–610	$1\cdot0 \times 10^8/T$	$1\cdot67$	Ar3	Ar2
	pc		7×10^{-5}	470–580	1	$0\cdot95$	Ha6	
CsCl(β)	sc	tr. 469		250–480	$8\cdot0 \times 10^7/T$	$1\cdot33$	Ar3	Ha6, Ha9,
	sc			150–460	$1\cdot0 \times 10^2$	$1\cdot05$	Mo7	He4, Ho6
CsBr	sc	636	$2\cdot6 \times 10^{-3}$	475–590	$2\cdot5 \times 10^5$	$1\cdot44$	Ly1	
				300–475	$2\cdot5 \times 10^4$	$1\cdot28$	Ly1	
	pc		4×10^{-4}	340–620	1×10^3	$1\cdot15$	Ha6	
CsI	sc	621	$1\cdot9 \times 10^{-3}$	480–595	$2\cdot2 \times 10^5$	$1\cdot43$	Ly1	Be10, Ec1,
				300–480	$1\cdot4 \times 10^4$	$1\cdot25$	Ly1	Ha6
	sc		$2\cdot1 \times 10^{-3}$	300–550	$1\cdot1 \times 10^5$	$1\cdot37$	Ho7	
Other Monovalent Halides								
NH_4Cl	sc			40–170	$4\cdot4 \times 10^5$	$1\cdot15$	He4	
NH_4Br	pp			70–150	$2\cdot6$	$0\cdot83$	He4	

Substance	Type	T (°C)/tr	σ	Temp. range	Pre-exp.	E	Ref.	References
NH$_4$I	pp							
AgCl	sc	455	3.5×10^{-2}	0–130	1.9×10^7	1.23	He4	Ab2, Co4, Sh6, Wa9
				160–380	$3.9 \times 10^7/T$	0.99[c]	Ab1	
	sc		4.6×10^{-2}	160–380	$5.0 \times 10^6/T$	0.78[i]	Ab1	
				50–250	$2.8 \times 10^7/T$	0.97[c]	Mu1	
				50–250	$7.1 \times 10^6/T$	0.78[i]	Mu1	
AgBr	sc	422	1.3×10^{-1}	225–400	2.4×10^5	0.93	Eb1	Ku3, Lu1, Mi1, Te1, Wa6, Wa10
	sc		1.3×10^{-1}	20–180	$6.3 \times 10^7/T$	0.87[c]	Mu1	
				20–180	$4.9 \times 10^6/T$	0.68[i]	Mu1	
	sc		7.0×10^{-1}	345–410	1.1×10^8	1.13	Fr2	
				250–345	1.4×10^6	0.89	Fr2	
				140–250	2.1×10^4	0.70	Fr2	
AgI(α)	pc	555	2.7	220–530	5.5	0.051	Kv5	Li3
AgI(β)	pc		2.7	145–555	9.2	0.064	Bi3	Li3, Mr1
	sc[c]	tr.146		85–145	6×10^7	0.97[x]	La2	
				15–85	5×10^2	0.61	La2	
	sc[a]			90–145	3×10^7	0.93	La2	
				20–90	1.3	0.40	La2	
TlCl	sc	427	6.6×10^{-3}	375–425	$8.8 \times 10^6/T$	0.87	Fr4	Ha10, Le5, Ph2, Sa1
				325–380	$2.3 \times 10^6/T$	0.80	Fr4	
				235–330	$1.15 \times 10^6/T$	0.76	Fr4	
	sc			200–350	$8.6 \times 10^5/T$	0.75	Ja8	
TlBr	pc	458	5.7×10^{-3}	295–420	3.8×10^4	1.10	He3	Le5, Mo6, Sa1
			1.0×10^{-3}	175–295	4.0×10^2	0.80	He3	
TlI(α)	pc		2.0×10^{-3}	150–400	2.2×10^2	0.73	Ph2	
TlI(β)	pc	438		163–400	4.2×10	0.63	Ph2	Sa1
	pc	tr.163	1.4×10^{-3}	90–163	2.5×10^{-3}	0.41	Ph2	

TABLE 9f-1.—Conductivity for Ionic Conductors (Continued)

Substance	Form	T_m (°C)	$\sigma(T_m)$ (ohm-cm)$^{-1}$	T Range (°C)	σ_0 (ohm-cm)$^{-1}$	W (eV)	Specific Ref.	Other Ref.
Mixed Halides								
Na_2CdCl_4	pc			230–350	$2\cdot8 \times 10^3$	0·86	Ja7	
$KHF_2(\beta)$	pc	239		195–225	$1\cdot1 \times 10^5$	0·91	Da6	
KAg_4I_5	pc	253	$3\cdot4 \times 10^{-1}$	40–250	$3\cdot1 \times 10^3/T$	0·13	Br2	Ow1
K_2BaCl_4	pc	662		500–635	$1\cdot6 \times 10^6$	1·54	Kr2	Sc10
K_2BaBr_4	pc			430–600	$7\cdot1 \times 10^4$	1·33	Ja7	
$RbAg_4I_5$	pc	228	$8\cdot6 \times 10^{-1}$	20–220	$1\cdot1 \times 10^4/T$	0·14	Br3	Ow1
$(NH_4)_2 SnCl_6$	sc			20–180	$4\cdot0 \times 10^{-8}$	0·30	He4	
NH_4PF_6	pp			20–160	$7\cdot7 \times 10^3$	1·03	He4	
Cu_2HgI_4 (α)	pp			67– 80	$3\cdot3 \times 10^3$	0·59	Su2	Ja2
Cu_2HgI_4 (β)	pp	tr.67		15– 66	$2\cdot2 \times 10^{-1}$	0·39	Su2	
Ag_2HgI_4 (α)	pp			50– 80	$7\cdot5 \times 10^3$	0·44	Su2	Ne2
Ag_2HgI_4 (α)	pp			50– 84	$1\cdot8 \times 10^6$	0·61	We3	
Ag_2HgI_4 (β)	pp	tr.50		10– 50	$1\cdot4 \times 10^6$	0·71	Su2	Ma5
	pp			22– 35	$9\cdot9 \times 10^4$	0·65	Ne2	Ne2
Polyvalent Halides								
CaF_2	sc	1418	2·9	620–980	$1\cdot1 \times 10^{10}/T$	2·13	Ba6	Ar4, As1, Ch2, Ni1, So3
	sc		2·7	560–1000	$5\cdot6 \times 10^9/T$	2·04	Ur1	
	sc			200–560	3×10^{-1}	0·80	Ur1	

Compound	Type	T	σ	T range (°C)	σ	E (eV)	Ref	Ref
SrF$_2$	sc	1190	$1{\cdot}6 \times 10^{-1}$	700–1010	$5{\cdot}6 \times 10^{9}/T$	2·14	Ba6	Bo3, Cr1, Ni1
	sc		5×10^{-2}	510–800	$2{\cdot}6 \times 10^{4}$	1·65	Ar4	
SrCl$_2$	sc	873	$5{\cdot}0 \times 10^{-2}$	50–510	2×10^{-1}	0·9	Ar4	
	sc		2	380–660	$1{\cdot}8 \times 10^{7}/T$	1·25	Ba6	
				650–780	$3 \times 10^{17}/T$	3·2	Ho5	
BaF$_2$	sc	1287	2·2	400–500	$3 \times 10/T$	Ho5	Ba6	Ba6
				620–900	$6{\cdot}8 \times 10^{8}/T$	1·64	Ba8	
				300–620	$7{\cdot}2 \times 10^{7}/T$	1·49	Ba8	
				50–800	$1{\cdot}3 \times 10^{3}$	1·24	Ar4	
BaCl$_2$	pp	950	$1{\cdot}3 \times 10^{-1}$	310–760	$3{\cdot}5 \times 10^{-2}$	0·43	Ja3	
BaBr$_2$	pc	847	6×10^{-4}	390–750	$1{\cdot}0 \times 10^{-1}$	0·41	Ja7	
CdCl$_2$	pc	568	$1{\cdot}4 \times 10^{-3}$	260–520	$1{\cdot}6 \times 10^{5}$	1·03	Ja7	Bi6
HgI$_2$ (α)	pc		$1{\cdot}1 \times 10^{-1}$	127–150	$1{\cdot}1 \times 10^{7}$	0·43	Ja2	
HgI$_2$ (β)	pc	tr.126		92–125	$3{\cdot}8 \times 10^{14}$	0·96	Ja2	
PbCl$_2$	sc	500	$4{\cdot}0 \times 10^{-3}$	340–440	$2{\cdot}5 \times 10^{3}$	0·89	De8	Sc9
				200–340	$3{\cdot}5 \times 10^{-2}$	0·30	De8	
	pc			175–350	8·7	0·40	Si1	
PbBr$_2$	sc	373	$2{\cdot}2 \times 10^{-4}$	230–330	$4{\cdot}9 \times 10^{4}/T$	0·71	Ve1	
				60–140	$4{\cdot}8\ /T$	0·36	Ve1	
PbI$_2$	sc	402	$4{\cdot}0 \times 10^{-4}$	210–270	2×10^{5}	1·30	Da8	
				170–210	2×10^{-1}	0·71	Da8	
AlCl$_3$	sc^c		$1{\cdot}2 \times 10^{-5}$	270–400	$2{\cdot}1 \times 10^{4}$	1·24	Se7	
	sc^a		9×10^{-5}	180–370	6×10^{-2}	0·38	Se7	
	pc	189	$2{\cdot}4 \times 10^{-6}$	150–189	$3{\cdot}8 \times 10^{13}$	1·76	Bi5	
	pc		$1{\cdot}1 \times 10^{-5}$	160–189	2×10^{4}	0·85	Se11	
GaCl$_3$	pc	78	$1{\cdot}5 \times 10^{-6}$	60–70	$2{\cdot}5 \times 10^{12}$	1·21	Gr3	
GaBr$_3$	pc	122	$1{\cdot}5 \times 10^{-6}$	80–100	$3{\cdot}4 \times 10$	0·52	Gr3	

TABLE 9f-1.—Conductivity for Ionic Conductors (*Continued*)

Substance	Form	T_m (°C)	$\sigma(T_m)$ (ohm–cm)$^{-1}$	T Range (°C)	σ_0 (ohm–cm)$^{-1}$	W (eV)	Specific Ref.	Other Ref.
GaI$_3$	pc	211	6×10^{-5}	190–205	$6 \cdot 7 \times 10^{18}$	2·23	Gr3	Kv1, Kv3
LaF$_3$	sc	1490		160–560	$3 \quad /T$	0·084	Sh3	Ja1
				20– 80	$2 \quad \times 10^5/T$	0·46	Sh3	
Simple Inorganic Radicals								
Li$_2$SO$_4$	pc	860	3·0	575–800	$1 \cdot 2 \times 10^2$	0·36	Kv2	
LiN$_3$	pp			172–280	$4 \cdot 3 \times 10^7$	1·87	Ga1	
				90–170	$5 \cdot 3 \times 10^{-2}$	0·53	Ga1	
NaNO$_2$	sc^c			163–190	$1 \cdot 4 \times 10^5$	1·15	As2	
		tr.163		130–160	$2 \cdot 5 \times 10^{12}$	1·85	As2	
	sc^f			163–200	$1 \cdot 4 \times 10$	0·64	So1	
		tr.163		40–163	$1 \cdot 0 \times 10$	0·68	So1	
NaNO$_3$	sc^c	306	5×10^{-6}	240–300	$4 \cdot 0 \times 10^{13}$	2·17	Ra2	Bi7
				20–230	$1 \cdot 3 \times 10$	0·94	Ra2	
	sc^c		5×10^{-5}	250–280	$3 \quad \times 10^{22}$	3·08	Ma7	
				40–240	$3 \cdot 6 \times 10$	0·87	Ma7	
NaBrO$_3$	sc	dec315		182–298	$1 \cdot 6 \times 10^{-3}$	1·1	Ra1	
				50–162	$9 \quad \times 10^{-6}$	0·46	Ra1	
Na$_2$C$_2$	pp	dec797		190–270	$1 \cdot 3 \times 10^4$	0·97	An3	

Compound	State	mp/tr	σ	ref	Range	value	E	ref
NaN_3	sc^c			Ja1	270–330	$2\cdot5\times10^{7}$	1·82	To2
NaN_3	sc^c				100–270	$1\cdot0\times10^{-2}$	0·87	To2
NaN_3	sc^c				230–400	$1\cdot0\times10^{-1}$	0·87	To2
NaN_3	sc^c				130–230	$1\cdot0\times10^{-4}$	0·56	To2
NaN_3	sc^a				130–165	$1\cdot0\times10^{2}$	0·80	As2
NaN_3	sc^a				90–130	2·3	0·80	As2
KNO_3	sc^c	334, tr.130	$1\cdot0\times10^{-4}$	Cl	320–630	$2\cdot8\times10^{3}$	1·34	Lo2
KNO_3					160–320	$3\cdot4\times10^{-1}$	0·93	Lo2
KCN	sc	634	3×10^{-6}		120–175	$3\cdot1\times10^{18}$	2·13	Lo1
KCN					20–120	$4\cdot4\times10^{7}$	1·30	Lo1
$KCNS$	sc	175		Pl1	40–160	$2\cdot0\times10^{7}$	1·36	To1
KN_3	pp				120–230	4×10^{4}	1·30	Ja1
$CsNO_3$	pc	404	$2\cdot0\times10^{-3}$		270–390	$1\cdot3\times10^{5}$	1·05	Bi8
$CsNO_3$					220–270	2×10^{2}	0·74	Bi8
$CsNO_3$	sc	tr.154			160–210	6×10^{-5}	0·20	So2
$CsNO_3$					35–130	8	0·72	So2
NH_4NO_3	pp	tr.125			125–143	$3\cdot8\times10^{4}$	0·68	Br6
NH_4NO_3		tr.84			48–124	$3\cdot1\times10^{6}$	0·96	Br6
NH_4NO_3					48–84	$4\cdot8\times10$	0·75	Br6
NH_4ClO_4	pp				50–110	$3\cdot5\times10^{5}$	1·15	Zi2
$CuSO_4$	pp	dec770	6×10^{-4}	Ja3	500–570	$2\cdot8\times10^{7}$	2·21	Ja3
$AgNO_3$	pc	208	5×10^{-1}		206–208			Ce2
Ag_2SO_3	pc	dec100			164–207	1×10^{12}	1·46	Da7
$Ag_2SO_4\ (\alpha)$	pp	656	6×10^{-2}		40–80	5×10^{2}	0·80	Za1
$Ag_2SO_4\ (\beta)$	pp	tr.425			430–600	$3\cdot6\times10^{2}$	0·70	Ha12
$Ag_2SO_4\ (\beta)$					200–410	$8\cdot0\times10^{-1}$	0·42	Ha12

TABLE 9f-1.—Conductivity for Ionic Conductors (*Continued*)

Substance	Form	T_m (°C)	$\sigma(T_m)$ (ohm–cm)$^{-1}$	T Range (°C)	σ_0 (ohm–cm)$^{-1}$	W (eV)	Specific Ref.	Other Ref.
$Ag_2C_2O_4$	pp			40–100	$8 \cdot 3 \times 10^{-2}$	0·70	Bo2	
	pp				$1 \cdot 0 \times 10^{-6}$	0·46	Fi1	
AgN_3	pp			180–250	2	0·46	Gr2	Ba10
	pp			25–110	$3 \cdot 3 \times 10^{4}$	0·82	Za2	
$Ca(N_3)_2$	pp			80–110	4·8	0·94	To1	
	pp			20–100	$2 \cdot 8 \times 10^{-10}$	0·23	Ja1	
$Sr(N_3)_2$	pp			30–110	$2 \cdot 0 \times 10^{-11}$	0·22	Ja1	
$Ba(N_3)_2$	sc^d			80–130	3×10^{-6}	0·56	To3	Ja1
	sc^e			70–110	2×10^{-7}	0·41	To3	
$ZnSO_4$	pp	761	$1 \cdot 4 \times 10^{-6}$	504–761	2·4	1·28	Ja3	
$PbSO_4$	pp	1080	8×10^{-4}	625–810	$5 \cdot 1 \times 10^{2}$	1·56	Ja3	

Activation Energy Notes

[a] Anion vacancy contribution

[c] Cation vacancy contribution

[i] Interstitial cation contribution

$\times\ \sigma = \sigma_p \exp(-W/kT)\,[1 + \exp(-W/kT)]^{-1}$

Crystal Directions

sc^a ∥ a-axis
sc^c ∥ c-axis
sc^d ⊥ (100) plane
sc^e ⊥ (001) plane
sc^f ∥ ferroelec. axis

Form of Sample

sc　single crystal
pc　polycrystalline
pp　pressed powder

TABLE 9f-2.—Concentration and Mobility of Defects in Ionic Crystals.

Substance	T Range (°C)	x_0	h (eV)	Defect	d_0 (cm²/sec)	Δh (eV)	Defect	Method	Specific Ref.	Other Ref.
Alkali Halides										
LiH	240– 630		2·38	Schottky		0·53	V'_{Li}	con	Pr1	Be9
LiF	400– 700	$9·2 \times 10$	2·42	Schottky	$1·6 \times 10^{-1}$	0·73	V'_{Li}	con	Ba7	Ei1
	460– 840	$5·0 \times 10^2$	2·68	Schottky	$5·2 \times 10^{-5} \times T$	0·65	V'_{Li}	con	Ha19	Ja6
	500– 800					1·0	V'_{F}	nmr	St7	St4
LiCl	310– 570	$1·7 \times 10^3$	2·12	Schottky	$2·8 \times 10^{-5} \times T$	0·41	V'_{Li}	con	Ha19	
LiBr	270– 540	$8·1 \times 10^2$	1·80	Schottky	$3·9 \times 10^{-5} \times T$	0·39	V'_{Li}	con	Ha19	
	160– 360		2·0	Schottky		0·43	V'_{Li}	nmr	Al3	
LiI	180– 420	$5·0 \times 10^2$	1·34	Schottky	$5·6 \times 10^{-5} \times T$	0·38	V'_{Li}	con	Ha19	
NaF	250– 650		3·0	Schottky	$\geqq 2·0 \times 10^{-3}$	0·52	V'_{Na}	nmr	Pe2	Ma2
NaCl	250– 720	$2·2 \times 10$	2·12	Schottky	$4·1 \times 10^{-1}$	0·80	V'_{Na}	con	Dr1	Bi2, Ch3,
	250– 790	3·4	2·17	Schottky	$3·7 \times 10^{-1}$	0·66	V'_{Na}	con	Al5	Et1, Ja5,
	250– 790				$2·0 \times 10^2$	1·15	V'_{Cl}	con	Al5	Ka4, Ki3,
	400– 550	8	0·95	V'_{Na}				dis	Da4	Pi1, St10
	630– 760	$3·0 \times 10^4$	1·30	$(V_{Na}V_{Cl})^{\times}$	$1·7 \times 10^{-2}$	1·25	Cl into $(V_{Na}V_{Cl})^{\times}$	die	Ec3	
NaBr	490– 730	8·3	1·68	Schottky	$7·8 \times 10^{-1}$	0·80	V'_{Na}	con	Ho8	Ma6, Sc2
KF	400– 790		2·64	Schottky		1·02	V'_{K}	con	Ka1	
KCl	400– 700	$3·5 \times 10$	2·22	Schottky	$4·8 \times 10^{-1}$	0·84	V'_{K}	con	Dr1	Al4, Bi4,

TABLE 9f-2.—Concentration and Mobility of Defects in Ionic Crystals (*Continued*)

Substance	T Range (°C)	x_0	h (eV)	Defect	d_0 (cm²/sec)	Δh (eV)	Defect	Method	Specific Ref.	Other Ref.
	270– 640	$1{\cdot}5 \times 10$	2·26	Schottky	$1{\cdot}1 \times 10^{-1}$	0·71	$V_K^\cdot$	con	Be1	Ch3, Ka1,
	270– 640				9·1	1·04	V_{Cl}'	con	Be1	Ma12, Pe1
	560– 760	$4{\cdot}4 \times 10$	2·31	Schottky	1·1	0·95	V_{Cl}'	dif	Fu2	Wa4
	250– 350	~ 1	0·7	$V_K^\cdot$				dis	Ka2	Ka3
	640– 770	$4{\cdot}5 \times 10^5$	1·34	$(V_K V_{Cl})^\times$	$1{\cdot}1 \times 10^{-3}$	1·04	Cl into $(V_K V_{Cl})^\times$	die	Sa2	Ec2
KBr	440– 680	$1{\cdot}4 \times 10^2$	2·53	Schottky	$9{\cdot}1 \times 10^{-2}$	0·66	$V_K^\cdot$	con	Ro4	Ch3, Ka1,
	440– 680				$7{\cdot}4 \times 10^{-1}$	0·87	V_{Br}'	con	Ro4	Ma12, Pe1
	410– 680	$7{\cdot}7 \times 10$	2·40	Schottky	$7{\cdot}9 \times 10^{-1}$	0·83	$V_K^\cdot$	con	Ho9	
	600– 730	$5{\cdot}5 \times 10^5$	1·41	$(V_K V_{Br})^\times$	$2{\cdot}3 \times 10^{-2}$	1·19	Br into $(V_K V_{Br})^\times$	die	Ec3	
KI	350– 560		1·59	Schottky		1·21	$V_K^\cdot$	con	Ec1	Ch3, Pe1
RbI	230– 260		2·00	Schottky				con	Ch3	
CsCl(α)	480– 610		2·1	Schottky		0·7	V_{Cl}'	con	Ar2	Ar2, Mo7
CsCl(β)	280– 460		1·86	Schottky		0·6	V_{Cs}'	dif	Ha9	
	280– 460					0·34	V_{Cl}'	dif	Ha9	
CsBr	320– 550		2·0	Schottky		0·58	V_{Cs}'	cdf	Ly1	
	320– 550					0·27	V_{Br}'	cdf	Ly1	
CsI	300– 500		2·14	Schottky		0·67	V_{Cs}'	dif	Ho7	Be10, Ly1
	300– 500					0·3	$V_i^\cdot$	con	Ho7	

Other Halides

Compound	Temp. range	A	Enthalpy (mechanism)	σ_0 / D_0	Energy	Defect		Ref.	Ref.
AgCl	160–380	1.6×10^2	1.44 Ag Frenkel	5.7×10^{-3}	0.27	V_{Ag}'	con	Ab1	Ab2, Eb1
	160–380			7.2×10^{-4}	0.055	Ag_i	con	Ab1	Mu1
	300–440			2.1×10^{-4}	0.008	collin. instlcy.	dif	We1	
	300–440			7.5×10^{-4}	0.13	noncoll. instlcy.	dif	We1	
	20–60		0.82 $(V_{Ag}V_{Cl})^{\times}$		1.0	Cl into $(V_{Ag}V_{Cl})^{\times}$	con	La6	
AgBr	200–300	5.3×10^2	1.13 Ag Frenkel	1.5×10^{-2}	0.30	V_{Ag}'	con	Te1	Ku3, Mu1
	200–300			1.2×10^{-2}	0.17	Ag_i	con	Te1	
	200–330			2.5×10^{-4}	0.058	collin. instlcy.	dif	We1	Fr2
	200–330			3.0×10^{-2}	0.27	noncoll. instlcy.	dif	We1	
AgI(α)	300–400	4×10^{-2}	1.62 Schottky		0.057	Ag_i	con	Kr1	
	150–250		0 Ag disorder		0.14	Ag_i	con	Li3	Mr1
AgI(β)	20–140		0.69 Ag Frenkel				con	Li3	
TlCl	200–330	5.5×10	1.36 Schottky	2.2×10^{-2}	0.44	V_{Tl}'	spc	Ja8	Ch8, Fr4
	280–400			9.9×10^{-4}	0.105	V_{Cl}	spc	Ja8	
CaF$_2$	640–920	6.0×10^2	2.81 F Frenkel	1.2×10^3	1.64	F_i'	con	Ur1	
	250–650			$9 \times 10^{-5} \times T$	0.60	$V_F^{·}$	con	Ur1	
SrF$_2$	230–980	6×10^2	2.4 F Frenkel		0.9	$V_F^{·}$	con	Ba6	Bo3, Cr1
	370–1010		2.3 F Frenkel		1.0	$V_F^{·}$ or F_i'	con	Ba6	
SrCl$_2$	180–660	1×10^3	1.8 Cl Frenkel	4×10^{-2}	0.4	V_{Cl}	con	Ba6	
	350–650	2.1×10		5.5×10^{-2}	0.56	$V_F^{·}$	con	Ba8	
BaF$_2$	150–500		1.86 F Frenkel	3.5×10^{-1}	0.79	F_i'	con	Ba9	Ba6, Ma2, Ni1

TABLE 9f-2.—Concentration and Mobility of Defects in Ionic Crystals (*Continued*)

Substance	T Range (°C)	x_0	h (eV)	Defect	d_0 (cm²/sec)	Δh (eV)	Defect	Method	Specific Ref.	Other Ref.
PbCl$_2$	160– 310	$4{\cdot}6 \times 10^{-1}$	1·16	Schottky	$1{\cdot}4 \times 10^{-6} \times T$	0·60	V''_{Pb}	con	Sc9	De8
	160– 310			$V''_{Pb} + 2V'_{Cl}$	$2{\cdot}6 \times 10^{-7} \times T$	0·12	V'_{Cl}	con	Sc9	
PbBr$_2$	60– 330		1·44	$V''_{Pb} + 2V'_{Br}$		0·29	V'_{Br}	con	Ve1	
LaCl$_3$	20– 440	$5{\cdot}8 \times 10$	0·28	$V^x_{La} + 3V^x_F$		0·42	V^x_F	thx	Sh3	
Simple Inorganic Radicals										
NaNO$_3$	120– 300		2·46	Na Frenkel		0·94	$Na_i^{\cdot}$	con	Ra2	
NaN$_3$	100– 330		1·90	Schottky		0·87	V'_{Na}	con	To2	
KNO$_3$	130– 330	7×10^{-1}	1·04	K Frenkel	$8 \times 10^{-5} \times T$	0·40	$V^{\cdot}_K$	con	Cl1	
CsNO$_3$	220– 390		0·62			0·74		con	Bi8	
Ag$_2$C$_2$O$_4$	50– 110			Ag Frenkel	$7 \times 10^{-7} \times T$	0·82	V'_{Ag}	con	Bo2	
Oxides										
SrO	1040–1340		2·0	$\tfrac{1}{2}O_{2(g)} \to O'_i + h^{\cdot}$		0·6	O'_i	P1	Co8	
	790–1100		3·0	$V^{\cdot\cdot}_O + O'_i$		0·6	O'_i	P2	Co8	
ZnO	900–1025	$1{\cdot}2 \times 10^2$	2·78	Zn Frenkel	$3{\cdot}3 \times 10^{-5} \times T$	1·78	Zn'_i	dif	Se2	
CdO	630– 850		6·5	$2V^{\cdot\cdot}_O + Cd^{\cdot\cdot}_{Cd} + 2Cd^{\cdot}_{Cd} + O^x_O + O_{2(g)}$		2·7	$V^{\cdot\cdot}_O$	dif / P3	Ha17	
Al$_2$O$_3$	1670–1900		6·8	$2V'''_{Al} + 3V^{\cdot\cdot}_O$		2·2	V'''_{Al}	dif / dif	Oi1 / Pa1	

Compound	T range		Defect reaction					Defect species	Method	Ref
TiO_2	1000–1500	9·6	$Ti_i^{\cdots} + 3e' + O_{2(g)}$						sem	Bl2
$Nb_2O_5(\alpha)$	800–1160	2·8	$V_O^{\cdot\cdot} + e' + \tfrac{1}{2}O_{2(g)}$				1·2	$V_O^{\cdot\cdot}$	con	Ch4 Ch6
Cr_2O_3	1040–1550	3·9	$V_{Cr}''' + Cr_i^{\cdots}$				1·3	V_{Cr}''' or $Cr_i^{\cdots}$	cdf	Ha1
UO_2	320– 850	3·1	$V_O^{\cdot\cdot} + O_i''$				1·3	O_i'' instcy	dif	Be4
UO_2	100– 800						2·3	V_U'''' or $U_i^{\cdots\cdot}$	con	Na1

Sulphides and Tellurides

Compound	T range		Defect reaction					Defect species	Method	Ref
Na_2S	350– 800	1·77	Na Frenkel				0·76	V_{Na}' or $Na_i^{\cdot}$	cdf	Mo2
	350– 800						1·17	$V_S^{\cdot\cdot}$	cdf	Mo2
$Cu_2S(\beta)$	130– 210	8×10^{-4}	0	V_{Cu}'	$3\cdot6 \times 10^{-3} \times T$		0·26	$V_{Cu}^{\cdot}$	sem	Yo1
$CdTe$	700–1000	$1\cdot7 \times 10^{-4}$	$1·04$	$V_{Cd}' + Cd_i^{\cdot}$					sem	De5

Method

cdf	Ionic conductivity and diffusion
con	Ionic conductivity in doped samples
die	Dielectric relaxation
dif	Diffusion with tracers
dis	Charged dislocations
nmr	Nuclear magnetic resonance
sem	Semiconducting properties
the	Ionic thermoelectric effect
thx	Thermal expansion
spc	Space charge polarization

P1	$P_{O_2} = 10^{-3}$ to 1 atm
P2	$P_{O_2} = 10^{-12}$ to 10^{-5} atm
P3	$P_{O_2} = 10^{-5}$ to 1 atm

TABLE 9f-3.—Effect of Pressure on Conductivity.

(1) Pressure coefficient $\alpha = -(\partial \ln \sigma/\partial \sigma)_T$.

Substance	Form	T Range (°C)	P Range (kiloatm)	$\alpha_0 \times 10^4$ (atm^{-1})	Comments	Specific Ref.	Other Ref.
NaCl	sc	600–700	0– 5	3·7	intrinsic	Bi2	
	sc	220–510	0–10	1·33	extrinsic	Pi1	
KCl	sc	550–700	0– 5	5·1	intrinsic	Bi4	Ta12
	sc	220–490	0–10	1·46	extrinsic	Pi1	
KI	sc	400–630	0– 5	5·3	intrinsic	Bi4	
RbCl	sc	300	0– 5	4·0	extrinsic	Pi1	
CuBr(β)	pc	380	0– 0·15	0·29	Cu electrodes	Bi1	
CuBr(γ)	pc	250–380	0– 0·15	1·2	Cu electrodes	Bi1	Ne3
AgCl	sc	144–336	0– 8	1·3 to 2·6	intrinsic	Ab2	Sh7
	sc	200–350	0– 1·5	3·6 to 4·5	Ag diffusion	Mu6	
AgBr	sc	202–406	0– 2	1·2 to 2·1	intrinsic	Ku3	Sc3
	pc	250–350	0– 0·15	3·3 to 3·9		Wa1	
AgI(α)	pc	191	0– 0·15	0·37		Wa1	Li3
		400	0– 0·15	0·15		Wa1	
AgI(β)	pc	110–130	0– 0·03	−1·9 to −12		Li3	Pa8,
	pc	110–135	0– 0·15	−2·2 to −2·8		Wa1	Ri4,
	pp	90	0– 2·0	−7·1	Ag diffusion	Mu5	Sc7
	pc	20	3– 7	5·8 to 5·4	High P phases	Ne3	

Substance		High P phases	P Range (kiloatm)	Δv motion (cm³/mole)	Specific Ref.	Other Ref.
Ag₂HgI₄(α)	pp	65– 85	0– 4	$-2\cdot1$	We3	Bi2,
KNO₃	pp	65– 85	4– 8	5·0 to 4·3	We3	Sh8
ZrO₂	pp	200–320	0– 0·5	2·0 to 2·8	Cl2	Ta12
(monocl.)	pp	600–800	10–30	$-1\cdot1$	Wh1	
		1000	0– 5	$-3\cdot9$	Wh1	

(2) Free volumes.

Substance	T Range (°C)	P Range (kiloatm)	v formation (cm³/mole)	Defect	Δv motion (cm³/mole)	Defect	Specific Ref.	Other Ref.
NaCl	400–700	0– 5	43	Schottky	9·5	V'_{Na}	Bi4	Bi2,
	220–510	0–10			7·7	V'_{Na}	Pi1	Sh8
KCl	400–700	0– 5	67	Schottky	10	V'_{K}	Bi4	Ta12
	220–490	0–10			7·0	V'_{K}	Pi1	
	685, 745	7–17			11·7	V'_{Cl}	Ra3	
AgCl	30–350	0– 8	16·7	Ag Frenkel	4·7	V'_{Ag}	Ab2	
	30–350	0– 8			3·2	Ag_i	Ab2	
AgBr	200–290	0– 8	16	Ag Frenkel	7·4	V'_{Ag}	Ku3	
	200–290	0– 8			2·6	Ag_i	Ku3	
KNO₃	350–410	0– 5	43	Schottky	20	V'_{Br}	Ta10	
	200–330	0– 0·5	8 to 11	K Frenkel	4 to 9	V'_{K}	Cl2	

TABLE 9f-4.—Conductivity for Mixed Conductors.

[Total conductivity $\sigma = \sigma_0 \exp(-W/kT)$.]

Substance	Form	T_m (°C)	T Range (°C)	σ_0 (ohm–cm)$^{-1}$	W (eV)	Environment	Specific Ref.	Other Ref.
Halides								
CuCl	sc	426	100– 400	6.3×10^6	1.06	Cu	Hs1	Tu2, Wa7
CuBr(α)	pc	491	470– 491	6.5	0.039	Cu	Bi3	
CuBr(β)	pc	tr.470	379– 450	2.0×10	0.21	Cu	Bi1	Wa7
CuBr(γ)	pc	tr.379	230– 379	2.8×10^{10}	1.47	Cu	Bi1	Wa3, Wa7
CuI(α)	pc	602	402– 440	8	0.20	Cu	Bi1	Wa7
CuI(β)	pc	tr.402	370– 400	2×10^7	1.09	Cu	Bi1	Wa7
CuI(γ)	pc	tr.369	330– 369	1.8×10^{10}	1.52	Cu	Bi1	Wa7, We4
Ag$_2$HgI$_4$(β)	pp	tr. 50	27– 48	7.5×10^8	0.89	Ag	We3	
Oxides						P_{O_2}(atm)		
BeO	pc		1100–1300	$1.6 \times 10^5/T$	2.52	air	De3	Br1, Pr3
			600–1100	$2.8 \times 10^{-2}/T$	0.69	air	De3	
	sp		1380–1600	$4.7 \times 10^6/T$	3.12	air	Cl4	
MgO	sc		930–1500	1.3[i]	2.00[i]	O$_2$, air	Mi3	Bu1, Da5
	sc		770–1300	1.8×10^3	2.7	10^{-8}	Mi2	
	sc		400– 750	1.3×10^3	2.8	10^{-7}	Le6	

Material	Form	m.p. (°C)	Temp. range (°C)	σ	E (eV)	P_{O_2} (atm)	Ref.	Ref.
CaO	sc	2550	1000–1300	$2 \cdot 9 \times 10$	$1 \cdot 73$	$\sim 10^{-4}$	Gu2	Pa2
	sc		770–1150	8×10^{7}	$3 \cdot 5$	vacuum	Su3	Su3
SrO	sp		1040–1340	$2 \cdot 8 \times 10^{2}$	$1 \cdot 6$	1	Co8	
				$\sigma \propto P_{O_2}^{1/4}$		10^{-3} to 1	Co8	
BaO	sc		790–1100	$2 \cdot 0 \times 10^{2}$	$2 \cdot 1$	10^{-12} to 10^{-5}	Co8	
Al_2O_3	sc, pc		440–730	4×10^{-2}	$0 \cdot 5$	vacuum	Do2	Ch1, Da3, Fl1, Ha7, Ha8, Pe4
	sc		750–1320	8×10^{-3}	$1 \cdot 9$	air	Ma9	
			1625–1725	$5 \cdot 1 \times 10^{9}$	$5 \cdot 5$	10^{-5}	Pa5	
			1300–1600	$9 \cdot 1 \times 10$	$2 \cdot 62$	10^{-5}	Pa5	
Sc_2O_3	pp		800–1300	8×10^{2}	$1 \cdot 70$	air	No1	No1
Y_2O_3	sp		1200–1600		$1 \cdot 94$	10^{-1}	Ta8	No1
				$\sigma \propto P_{O_2}^{3/16}$		10^{-1} to 10^{-7}	Ta8	
La_2O_3 (hex)	pp	2315	400–700	$4 \cdot 6 \times 10^{6}$	$0 \cdot 67$	air	Me1	
(cub)		tr.590	350–580	$1 \cdot 5 \times 10^{10}$	$1 \cdot 05$	air	Me1	
ZrO_2 (tetr)	sc		1500–1780	$2 \cdot 1 \times 10^{6}$	$3 \cdot 3$	air	An2	Mc1, Ve2
			1150–1340	$6 \cdot 7 \times 10^{-2}$	$0 \cdot 74$	air	An2	
$ZrO_2:CaO(15\%)$	sp		100–1100	$1 \cdot 4 \times 10^{2}$	$1 \cdot 13$	10^{-9}	Ve3	Di1, Jo1, Ki1, St9
$ZrO_2:YO_{1.5}(20\%)$	sp		700–1350	$1 \cdot 8 \times 10^{2}$	$0 \cdot 83$	air	St9	
HfO_2	sp		1000–1600	$9 \cdot 3 \times 10$	$1 \cdot 45$	1	Ro1	No1
				$\sigma \propto P_{O_2}^{0.19}$		1 to 10^{-4}	Ro1	
$HfO_2:CaO(12\%)$	sp		800–2000	$1 \cdot 8 \times 10^{3}$	$1 \cdot 43$	max. cond.	Jo1	
$HfO_2:YO_{1.5}(16\%)$	sp		900–1600	8×10^{2}	$1 \cdot 12$	max. cond.	Be11	
CeO_2	sc		130–1000	$1 \cdot 5 \times 10^{3}$	$1 \cdot 28$	air	Vi1	
ThO_2	sp	3050	800–1100	$1 \cdot 1 \times 10$	$1 \cdot 41$	10^{-10} to 10^{-22}	La3	Da1
$ThO_2:YO_{1.5}(15\%)$	sp		800–1100	$3 \cdot 1 \times 10$	$0 \cdot 92$	10^{-5} to 10^{-22}	La3	

TABLE 9f-4.—Conductivity for Mixed Conductors (Continued)

Substance	Form	T_m (°C)	T Range (°C)	σ_0 (ohm–cm)$^{-1}$	W (eV)	Environment	Specific Ref.	Other Ref.
Pr₂O₃ (hex)	pp		720– 850		0·86	dry H₂	Me1	No1
(cub)	pp	tr.780	700– 780		0·84	dry H₂	Me1	No1
Nd₂O₃ (hex)	pp		400– 700		0·75	air	Me1	No1
(cub)	pp	tr.550	400– 540		1·26	air	Me1	
Sm₂O₃	pp		800–1300		1·17	air	No1	
Eu₂O₃	pp		800–1300		1·24	air	No1	
Gd₂O₃	pp		800–1300		1·36	air	No1	
Tb₄O₇	pp		800– 900		0·40	air	No1	
Dy₂O₃	pp		800–1300		1·39	air	No1	
Er₂O₃	pp		800–1300		1·40	air	No1	
Yb₂O₃	pp		800–1300		1·53	air	No1	

Sulphides, Selenides, and Tellurides

Substance	Form	T_m (°C)	T Range (°C)	σ_0 (ohm–cm)$^{-1}$	W (eV)	Environment	Specific Ref.	Other Ref.
Na₂S	sp	1169	520– 800	$3·4 \times 10^7$	1·64	H₂	Mo2	
			350– 520	$8·0 \times 10$	0·75	H₂	Mo2	
Cu₂S(β)		tr.470	110– 470	$9 \times 10^4/T$	0·24		Mi5	Ha16, We2,
	pc		400	$\sigma_i = 2·4$		Pt, Cu	Yo2	Yo1
Cu₂Se	pc		580– 750	3×10^i	$0·17^i$	$Cu_{1·96}$ Se	Ce1	
Ag₂S(α)	pc	835	180– 300	$2·9 \times 10^4/T^i$	$0·11^i$		Ok1	He1

	Form				Conductivity			
$Ag_2S(\beta)$		tr.179	$1\cdot3 \times 10^8$	130–160	0·71	Ag	Mi6	Ri1
			$1\cdot4 \times 10^6$	130–170	0·69	S	Mi6	
Ag_3SBr	sp		8^i	100–250	$0\cdot24^i$	Ag	Re4	
$Ag_3SI(\alpha)$	sp	700	2	235–400	0·04	Ag	Re4	Ta1
$Ag_3SI(\beta)$	sp	tr.235	2·9	0–80	0·14	Ag	Ta4	Ta5
$Ag_2Se(\alpha)$	pc		$1\cdot7 \times 10^4/T^i$	130–300	$0\cdot10^i$		Ok1	Mi7
$Ag_2Se(\beta)$	pc	tr.133	$3\cdot4 \times 10^i$	100–130	$0\cdot39^i$	Ag	Mi9	
$Ag_2Te(\alpha)$			9×10^4	165–225	0·31		Mi8	
$Ag_2Te(\beta)$	pc		6×10^4	80–140	0·56		Mi9	Ta6

^{i}Ionic portion of conductivity

Form of Sample

sc single crystal
pc polycrystalline
pp pressed powder
sp sintered powder

TABLE 9f-5.—Transport Numbers for Mixed Conductors

Substance	T (°C)	Transport Numbers $t_c = 1 - t_e$	Form	Environment	Specific Ref.	Other Ref.
Halides						
CuCl	150	[1]0·008 [2]0·04	[1]pc [2]pc	Cu Cu, Cu$_2$O	[1]Ma3 [2]Tu2	Wa7
	250	0·027 0·85				
	350	0·51 0·99				
CuBr(α)	470– 491	1·00	pc	Cu	Tu3	
CuBr(β)	390– 445	1·00	pc	Cu	Tu3	Wa7
CuBr(γ)	100	[1]0·004 [2]0·01 [3]0·005	[1]pc	Cu	[1]Ma3	Wa7
	200	0·042 0·10 0·032	[2]pc	Cu, Cu$_2$O	[2]Tu3	
	300	0·26 0·87 0·36	[3]pc	Cu, Cu$_2$O	[3]Ki4	
	390	0·25 1·00 0·65				
CuI(α)	402– 500	1·00	pc	Cu	Tu1	Wa7
CuI(β)	375– 400	0·99	pc	Cu	Tu1	Wa7
CuI(γ)	200	[1]0·003 [2]0·00	[1]pc	Cu, Cu$_2$O	[1]Ma3	Wa7, We4
	300	0·004 0·25	[2]pc	Cu	[2]Tu1	
Ag$_2$HgI$_4$(α)	50– 84	0·97	pp	Ag, Pt	We3	
Ag$_2$HgI$_4$(β)	27– 48	0·4	pp	Ag, Pt	We3	
Ag$_2$SO$_4$(α)	602	$t_e = 3 \times 10^{-3}$		P_{O_2} = 10 to 700 Torr	Ha2	
	446	$t_h = 1 \times 10^{-4}$		P_{O_2} = 0·1 to 700 Torr	Ha12	

TABLE 9f-5.—Transport Numbers for Mixed Conductors

Substance	T (°C)	Transport Numbers		Form	Environment	Specific Ref.	Other Ref.
		$t_c = 1 - t_e$			P_{O_2}(atm)		
Oxides							
Cu₂O	800–1000	(2 to 5) × 10⁻⁴		pp	Cu	Gu1	
BeO	1000–1200	[2]1·00	[3]0·80 to 0·96	[1]sp	10 to 10⁻³	[1]Cl3	Sc4
			1·00				
MgO	1200–1300	[1]0·94		[2]pp	O₂, air	[2]Pa2	
	1400–1700	1·00		[3]pp	O₂, CO — CO₂	[3]Pa2	
	900	[1]1·0	[3]0·57	[1]sc	O₂, air	[1]Mi3	
	1100	0·9	0·91	[2]sc	10⁻⁴	[2]Sc6	
	1300	0·5	0·63	[3]pp	O₂, CO — CO₂	[3]Pa2	
CaO	900	[2]0·87	[3]0·02	[1]pp	O₂, air	[1]Pa2	
	1100	[2]0·57	0·02	[2]pp	O₂, CO — CO₂	[2]Pa2	
	1300	0·91	0·02	[3]sc	~10⁻⁴	[3]Gu2	
		0·63					
SrO	1000	[1]0·03	[2]0·68	[1]pp	O₂, air	[1]Pa2	
	1150	0·08	0·89	[2]pp	O₂, CO — CO₂	[2]Pa2	
BaO	200– 700	0·0		sc	vacuum	Do2	
		$t_i = 1 - t_e - t_h$			P_{O_2}(atm)		
Al₂O₃	800–1300	0·9 to 0·0		sc	O₂, air	Ma9	
Sc₂O₃	1100–1300	1·0 to 0·0		sp	1 to 10⁻¹³	Sc6	
	800–1000	1·0		sp	10⁻¹², 10⁻¹⁷	Sc6	

TABLE 9f-5.—Transport Numbers for Mixed Conductors (*Continued*)

Substance	T (°C)	Transport Numbers	Form	Environment	Specific Ref.	Other Ref.
Y$_2$O$_3$	700–800	0·3 to 0·15	sp	10^{-15}	Ta8	
	1200–1600	0·00	sp	10^{-1} to 10^{-17}	Ta8	
	825	⩾0·5	sp	6×10^{-5} to 3×10^{-22}	Ta11	
La$_2$O$_3$	400–1000	~1	pp	air	Mel	
		$t_a = 1 - t_e - t_h$		P_{O_2}(atm)		
ZrO$_2$(tetr)	1300–1600	0·9 to 0·4	sp	10^{-5} to 10^{-11}	Ve2	
	1140–1340	0·6 to 0·2	sp	10^{-11} to 10^{-14}	Mc1	
ZrO$_2$ (monocl)	990	$4·5 \times 10^{-3}$	sc	0·4	Ma1	Ta7
ZrO$_2$:CaO (15%)	1000	⩾0·99	sp	1 to 10^{-15}	St1	Ve3
	800	[1]$t_h = 10^{-2·6}$ to $10^{-3·3}$ [2]$t_e = 10^{-6·6}$ to $10^{-4·9}$	[1]sp	1 to 10^{-5}	Pa6	
	1000	$t_h = 10^{-1·7}$ to $10^{-3·1}$ $t_e = 10^{-5·0}$ to $10^{-2·7}$	[2]sp	10^{-10} to 10^{-20}	Pa6	
ZrO$_2$:YO$_{1.5}$ (20%)	900–1100	$t_e \leqslant 0·01$	sp	air or H$_2$	Br4	Sm1, Tr1
HfO$_2$	1000	0·71 to 0·94	sp	10^{-6} to 10^{-15}	Ro1	St1
	1000	0·01 to 0·03		1 to 10^{-3}	Ro1	
	1400	0·06 to 0·13		1 to 10^{-3}	Ro1	
	1000–1500	0·00	sp	1 to 10^{-18}	Ta9	

Compound	Temp	Value	Method	P_{O_2} (atm) / cond.	Ref	Ref
$HfO_2 : YO_{1.5}$ (16%)	800–1050	1·0	sp	10^{-10} to 10^{-15}	Be11	Vo1
$Nb_2O_5(\alpha)$	1000–1200	0·03	sp	1 to 10^{-5}	El1	
Cr_2O_3	980–1550	$t_c = 1 - t_e - t_h$; $5\cdot7 \times 10^{-6}$ to $1\cdot2 \times 10^{-3}$	sp	N_2	Ha1	
MnO	1010	1×10^{-5}	sp	Mn	Bo1	
FeO	720–1020	$1\cdot4 \times 10^{-5}$ to $2\cdot6 \times 10^{-4}$	pc	$Fe_{0.887}O$	De6	Ha11
CeO_2	450–1300	$t_a = 1 - t_e - t_h$; $0 < t_a < 1$	sp	air	Ho4	No2, Vi2
$CeO_2 : LaO_{1.5}$ (30%)	1000–1100	1·00	sp	air	Ne4	Ta3
ThO_2	1000	0·06 to 0·93	sp	1 to 10^{-9}	La3	Da1, St1
ThO_2	1000	1·00 to 0·96	sp	10^{-12} to 10^{-21}	La3	
$ThO_2 : YO_{1.5}$ (15%)	1000	$\geqslant 0\cdot99$	sp	10^{-5} to 10^{-24}	St1	La3, Su1
	800	$[1]t_a = 1 - t_e - t_h$ $[2]t_e = 10^{-5\cdot7}$; $[1]t_h = 10^{-2\cdot7}$ to $10^{-3\cdot2}$	[1]sp	1 to 10^{-10}	Pa6	Wi1
	1000	$t_h = 10^{-4\cdot7}$ to $10^{-2\cdot7}$; $t_e = 10^{-4\cdot4}$	[2]sp	10^{-5} to 10^{-25}	Pa6	
UO_2	900–1100	4×10^{-6} to $1\cdot2 \times 10^{-4}$	sp	$UO_{2\cdot0003}$	Do3	Ii1
Nd_2O_3	800–1000	$t_i = 1 - t_e - t_h$; 1·00	pp	$10^{-12}, 10^{-17}$	Sc6	
Sm_2O_3	800–1000	1·00	pp	$10^{-12}, 10^{-17}$	Sc6	
Gd_2O_3	800–1100	$0 < t_i < 1$			Ra4	
Dy_2O_3	800–1100	$0 < t_i < 1$			Ra4	
Yb_2O_3	800–1000	1·00	pp	$10^{-12}, 10^{-17}$	Sc6	

TABLE 9f-5.—Transport Numbers for Mixed Conductors (*Continued*)

Substance	T (°C)	Transport Numbers $t_c = 1 - t_e$	Form	Environment	Specific Ref.	Other Ref.
Sulphides, Selenides, and Tellurides						
Na$_2$S	350– 800	1·00	sp		Mo2	
Cu$_2$S	134– 207	0·05 to 0·15			Yo1	Hi2, Ku4,
	400	0·86	pc	Pt, Cu	Yo2	Wa8, We2
Cu$_2$Se	580– 750	0·06 to 0·11	pc	Cu$_{1·96}$Se	Ce1	
Cu$_2$Te	335– 410	$t_c = 1 - t_e$ 1×10^{-4}			Re3	
Ag$_2$S(α)	200	$t_c = 10^{-2}$ to 10^{-3} $t_a = 10^{-8}$			Wa2	Bu3
Ag$_2$S(β)	130– 160	0·011		Ag	Mi6	He1
	130– 170	0·59		S	Mi6	
Ag$_3$SBr	100– 250	0·98 to 0·93	sp	Ag	Re4	
Ag$_3$SI(α)	300	0·95	sp	Ag	Re4	Ta1, Ta2
Ag$_3$SI(β)	100– 200	0·995 to 0·986	sp	Ag	Re4	Ta4
Ag$_2$Se(α)	220	2×10^{-3}		Ag	Mi7	
Ag$_2$Se(β)	124	1×10^{-7}	pc	Ag	Mi9	Bu2
Ag$_2$Te(α)	270	0·010	pc	Ag	Yo2	Mi8
Ag$_2$Te(β)	100	4×10^{-7}		Ag	Mi9	Ta6
ZnSe	200– 400	0·08 to 0·13	sc	Fe − FeO, Cu − Cu$_2$O	Ki4	

Form of Sample

sc single crystal
pc polycrystalline
pp pressed powder
sp sintered powder

Transport Numbers

t_i Ionic
t_c Cation
t_a Anion
t_e Electron
t_h Hole

TABLE 9f-6.—Diffusion in Ionic Crystals.

[The diffusion coefficient is given as $D = D_0 \exp(-W/kT)$.]

Substance	Form	Isotope	T Range (°C)	D_0 (cm²/sec)	W (eV)	Defect	Method	Environment	Comments	Specific Ref.	Other Ref.
Alkali Halides											
LiF	sc	^{7}Li	560–770	2.3	1.81	V'_{Li}	nmr			Ei1	Ma2, St6
			360–560	4.5×10^{-7}	0.71	V'_{Li}	nmr			Ei1	
		^{19}F	600–790	6.1×10	2.2	V'_{F}	nmr			Ei1	St7
NaF	pw	^{23}Na	550–650	1.6×10	2.0	V'_{Na}	nmr			Pe2	Ma2
			250–400	2.3×10^{-7}	0.52	V'_{Na}	nmr			Pe2	
NaCl	sc	^{22}Na	600–720	2.9×10	1.97	V'_{Na}	sct			Be5	Ba4, Do6,
		^{22}Na	670–770	1.8×10	2.10	V'_{Na}	sct		single vac.	Ne1	Ei3, La5,
			670–770	1.1×10^3	2.35	$(V_{Na} V_{Cl})^{\times}$	sct		vacancy pair	Ne1	Ma6
		^{36}Cl	520–745	2.2	2.07	V'_{Cl}	sct		single vac.	La4	Ba5, La5
			520–745	9.9×10^2	2.50	$(V_{Na} V_{Cl})^{\times}$	sct		vacancy pair	Fu3	
NaBr	sc	^{24}Na	425–700	6.7×10^{-1}	1.53	V'_{Na}	sct			Ma6	
	sc	^{82}Br	450–690	5.0×10^{-2}	2.02	V'_{Br}	sct			Sc2	Do1
KF	pp	^{42}K	580–840	2	1.78	V'_{K}	sct			La5	

Compound		tracer	T range	value	E (eV) + defect	mechanism	method	ref.	ref.
KCl	sc	^{42}K	450– 750	4×10^{-2}	$1{\cdot}48\ V'_K$		sct	La5	Ar6, Wi3
	sc	^{36}Cl	560– 760	$3{\cdot}6 \times 10$	$2{\cdot}10\ V^{\bullet}_{Cl}$	single vacancy	sct	Fu2	Ba3, La5,
			560– 760	$8{\cdot}6 \times 10^{3}$	$2{\cdot}65\ (V_K V_{Cl})^{\times}$	vacancy pair	sct	Fu2	Ra3
KBr	sc	^{42}K	470– 730	1×10^{-2}	$1{\cdot}26\ V'_K$		sct	La5	Do1, La5
	sc	^{82}Br	400– 700	3×10^{4}	$2{\cdot}61\ V^{\bullet}_{Br}$		gsx	Da10	
KI	sc	^{42}K	430– 690	1×10^{-5}	$0{\cdot}64\ V'_K$		sct	No6	La5
	sc	^{131}I	430– 690	$1{\cdot}2 \times 10^{-3}$	$1{\cdot}12\ V'_i$		sct	No6	La5
RbCl	sc	^{86}Rb	600– 760	$3{\cdot}3 \times 10$	$1{\cdot}99\ V'_{Rb}$		sct	Ar1	Ma4
CsF	pc	^{137}Cs	480– 640	$3{\cdot}1$	$1{\cdot}67\ \{V'_{Cs}\}$		sct	La5	
CsCl(α)	pp	^{137}Cs	465– 620	1×10^{-1}	$1{\cdot}39\ \{V'_{Cs}\}$		sct	La5	
	pp	^{36}Cl	465– 620	7×10^{-1}	$1{\cdot}56\ \{V^{\bullet}_{Cl}\}$		sct	La5	
CsCl(β)	sc	^{137}Cs	280– 460	$2{\cdot}4 \times 10$	$1{\cdot}53\ V'_{Cs}$		sfc	Ha9	La5
	sc	^{36}Cl	280– 460	$1{\cdot}5$	$1{\cdot}27\ V^{\bullet}_{Cl}$		sfc	Ha9	Ho6, La5
CsBr	sc	^{134}Cs	320– 550	$1{\cdot}5 \times 10$	$1{\cdot}54\ V'_{Cs}$		sct	Ly1	
	sc	^{82}Br	415– 530	$3{\cdot}9$	$1{\cdot}42\ V^{\bullet}_{Br}$		sct	Ly1	
CsI	sc	^{134}Cs	320– 550	$1{\cdot}4 \times 10$	$1{\cdot}53\ V'_{Cs}$		sct	Ly1	Ho7, K15
	sc	^{131}I	410– 540	$2{\cdot}1$	$1{\cdot}37\ V^{\bullet}_i$		sct	Ly1	K15

Other Monovalent Halides

Compound		tracer	T range	value	E (eV) + defect	mechanism	method	ref.	ref.
CuI(α,β)	pp	^{131}I	370– 500	$7{\cdot}5$	$1{\cdot}48\ \{V'_i\}$	instlcy.	sfc	No3	
CuI(γ)	pp	^{131}I	350– 370	$5{\cdot}0 \times 10^{4}$	$1{\cdot}96\ \{V'_i\}$	instlcy.	sfc	No3	
AgCl	sc	^{110}Ag	380– 440	$6{\cdot}5 \times 10^{2}$	$1{\cdot}13\ V'_{Ag} + Ag^{\bullet}_i$		sct	We1	Co4, Re1
			300– 380	$1{\cdot}8$	$0{\cdot}92\ V'_{Ag} + Ag^{\bullet}_i$		sct	We1	
	sc	^{36}Cl	300– 450	$8{\cdot}5 \times 10$	$1{\cdot}57\ V^{\bullet}_{Cl}$		sct	La1	Co4, No5

TABLE 9f-6.—Diffusion in Ionic Crystals (Continued)

Substance	Form	Isotope	T Range (°C)	D_0 (cm²/sec)	W (eV)	Defect	Method	Environment	Comments	Specific Ref.	Other Ref.
AgBr	sc	^{110}Ag	345–410	6.8×10^2	1·10	$V'_{Ag} + Ag_i$	sct		instlcy.	Fr2	Mi1, Mu4,
			250–345	5.7×10	0·97	$V'_{Ag} + Ag_i$	sct		instlcy.	Fr2	Se6, St2,
			140–250	1·3	0·79	$V'_{Ag} + Ag_i$	sct		instlcy.	Fr2	St3
	sc	^{82}Br	370–415	9.4×10^8	2·53	$V^{\cdot}_{Br}$	sct			Ta10	
			332–370	3.0×10^4	1·93	$V^{\cdot}_{Br}$	sct			Ta10	
AgI(α)	pp	^{110}Ag	145–220	1.6×10^{-4}	0·097	V'_{Ag}	sct		cation disorder	Jo3	Ba1, Zi1
AgI(β)	pp	^{131}I	145–540	4.4×10^{-4}	0·70	$\{V'_I\}$	sfc			No4	Jo3, Jo4
	pp	^{110}Ag	20–145	3.2×10^{-1}	0·62	$\{V'_{Ag} + Ag_i^{\cdot}\}$	slx			Jo3	Mu5, Zi1
	sc	^{131}I	80–140	4.7×10^{-10}	0·29	$\{V'_I\}$	sct		$\parallel a$-axis	La2	
	sc	^{131}I	50–140	$D = (8.2 \text{ to } 3.4) \times 10^{-4}$		$\{V'_I\}$	sct		$\parallel c$-axis	La2	
AgI(γ)	pp	^{110}Ag	20–147	5×10^{-4}	0·37	$\{V'_{Ag} + Ag_i^{\cdot}\}$	slx			Jo3	
Ag$_2$HgI$_4$(α)	pc	^{111}Ag	40–110	1·8	0·47	V'_{Ag}	sfc		cation disorder	Zi1	
		^{203}Hg	50–160	5×10^{-3}	0·89	V''_{Hg}	sfc		cation disorder	Zi1	
TlCl	sc	^{204}Tl	290–390	6.2×10^{-1}	1·10	V'_{Tl}	sct			Fr4	
	sc	^{36}Cl	270–420	3.1×10^{-2}	0·77	$V^{\cdot}_{Cl}$	sct			Fr4	Fr3

Polyvalent Halides

Compound		Isotope	T range	Value	ΔH / defect	Method	Notes	Refs
CaF_2	sc	^{45}Ca	800–1250	1.3×10^2	$3.75\ \{V''_{Ca}\}$	sct		Ma11 Sh9
CaF_2	sc	^{18}F	670–950	1.9×10	$1.91\ V^{\cdot}_{F}+F'_{i}$	sfc		Ma10
		^{18}F	360–670	1.1×10^{-4}	$0.91\ V^{\cdot}_{F}$	sfc		Ma10
$CaCl_2$	sc	^{90}Sr	800	$D=3 \times 10^{-10}$		sfc	(see next table)	Ho5
		^{36}Cl	700	$D=3 \times 10^{-7}$	$V^{\cdot}_{Cl}$ or Cl'_{i} $\{V''_{Ca}\}$	sfc		Ho5
$PbCl_2$	pp	^{212}Pb	180–270	1.2×10^2	$1.65\ \{V''_{Pb}\}$	sfc		He5
PbI_2	sc	^{212}Pb	260–320	2.0×10^6	$1.37\ \{V''_{Pb}\}$	sfc		Se7 He5
LaF_3	sc	^{19}F	0–230	7.1×10^{-4}	$0.50\ \{V^{\cdot}_{F}\}$	nmr		Lu3 Le3

Simple Inorganic Radicals

Compound		Isotope	T range	Value	ΔH / defect	Method	Notes	Refs
Li_2SO_4	pc	^{6}Li	640–790	1.9×10^{-3}	$0.34\ V'_{Li}$	sct	cation disorder	Kv4 Ku5, Lu2
$LiOH$	sc	^{6}Li	25	$D=7.3 \times 10^{-11}$				Ku5
$Ag_2SO_4(\alpha)$	sp	^{111}Ag	430–600	2.5	$1.15\ \{V'_{Ag}+Ag^{\cdot}_{i}\}$	sfc		Jo2 Li4
$Ag_2SO_4(\beta)$	sp	^{111}Ag	100–430	6.7×10^{-5}	$0.58\ \{V'_{Ag}+Ag^{\cdot}_{i}\}$	sfc		Jo2 Ja3
$CaCO_3$	sc	^{13}C	606–848	4.5×10^{-4}	2.51	gsx		Ha13

Monovalent Oxides

Compound		Isotope	T range	Value	ΔH / defect	Method	P_{O_2}(atm)	Refs
Cu_2O	pc	^{64}Cu	800–1050	4.4×10^{-2}	$1.57\ V'_{Cu}$	sct	1.3×10^{-3}	Mo3 Ca4, Sh4
	sc	^{18}O	1020–1120	5.4×10^{-4}	$1.70\ O''_{i}$	gsx	1.3×10^{-1} $D \propto P_{O_2}^{1/2}$	Eb2 Mo4, Sl1
Ag_2O	pp	^{110}Ag	20–160	5.4×10^{-8}	$0.30\ \{V'_{Ag}\}$	sfc		Ro7

Divalent Oxides

Compound		Isotope	T range	Value	ΔH / defect	Method	P_{O_2}(atm)	Refs
BeO	sc	^{7}Be	1720–1960	1.2×10^{-6}	$1.56\ V''_{Be}$	sct	Ar	Au2 Au1, Au3
			1500–1760	1.3×10^{-3}	$2.78\ V''_{Be}$	sct	Ar	Au2

TABLE 9f-6.—Diffusion in Ionic Crystals (*Continued*)

Substance	Form	Isotope	T Range (°C)	D_0 (cm²/sec)	W (eV)	Defect	Method	Environment	Comments	Specific Ref.	Other Ref.
	pc	^7Be	1100–1800	3.2×10^{-3}	2.73	V''_{Be}	sct	vacuum		De3	De2
	sp	^7Be	1500–2130	5.9×10^{-5}	2.12	V''_{Be}	sct		$D_c/D_a = 1.3$	Co7	Au2
	sc	^{18}O	1560–1730	3.0×10^{-5}	2.97		gsx	4.1×10^{-1}		Ho1	
MgO	sc	^{28}Mg	1400–1600	2.5×10^{-1}	3.42		sct	air		Li13	Hi3, Sc4
	sc	^{18}O	1000–1150	4.3×10^{-5}	3.56	O''_i	gsx	$(1-120)\times10^{-3}$		Ro6	Oi2
	sc	^{18}O	750–1000	4.8×10^{-14}	1.31	O'_i	gsx	$(1-120)\times10^{-3}$		Ro6	
	pc	^{18}O	1650	$D<10^{-14}$			ard	1.6×10^{-1}		Ho2	
CaO	sc	^{45}Ca	1000–1400	8.8×10^{-8}	1.50	$\{V''_{Ca}\}$	sct			Gu2	Li6
BaO	sc	^{140}Ba	1080–1230	1×10^{29}	11	$\{Ba_i^\times\}$	sct	1.3×10^{-4}		Re2	Be12, De1
				1×10^{31}	12	$\{V''_{Ba}\}$	sct			Re2	
			330–1080	1×10^{-9}	0.44	$\{Ba_i^\times\}$	sct			Re2	
				3×10^{-10}	0.3	$\{V''_{Ba}\}$	sct			Re2	
	sc	O	800–1300	2.5×10^{3}	2.8	$V_O^\times$	adc	Ba(g)		Sp2	
ZnO	pw	^{65}Zn	720–840	4×10^{-1}	3.32	$Zn_i^{\cdot\cdot}$	gsx	Zn(g)	$D\propto P_{Zn}^{0.65}$	Se5	Le4, Li5, Pa4, Ro2,
	sc	^{65}Zn	940–1025	5.0	3.25		sfc			Mu2	Ro3, Sel,
	sc	^{65}Zn	850–940	3.0×10^{-9}	0.87		sfc			Mu2	Se2, Se4,
	sc	^{65}Zn	1000–1250	1.3×10^{-5}	1.9	$\{Zn_i^{\cdot\cdot}\}$	sct	O_2		Mo5	Sp1
CdO	sc	^{18}O	1100–1300	6.5×10^{11}	7.15	$\{disloc\}$	gsx	$1(\text{to }10^{-1})$	$D\propto P_{O_2}^{1/2}$	Mo5	Ha14, Ha1
	sc	^{18}O	630–855	3.8×10^{6}	3.99	$V_{\ddot{O}}$	gsx	$1(\text{to }10^{-3})$	$D\propto P_{O_2}^{-1/5}$	Ha17	Si4
SnO₂	sp	^{119}Sn	980–1380	1×10^{6}	5.14		sfc	air		Li10	

Compound	Type	Isotope	Temp range (°C)	D_0	Q	Defect	Technique	Atmosphere	P_{O_2}(atm)	Notes	Refs
PbO(α)	pp	^{210}Pb	600–680	4×10^9	3.5		sfc	air			Da2, Li8
	pw	^{18}O	500–650	5.4×10^{-5}	0.93		gsx	O$_2$			Th3
PbO(β)	pc	^{212}Pb	200–460	1.6×10^{-11}	0.56		sfc	air			Li9
Polyvalent Oxides									P_{O_2}(atm)		
Al$_2$O$_3$	sp	^{26}Al	1670–1905	2.8×10	4.95	V'''_{Al}	sct	air			Pa1
	sc	^{18}O	1500–1780	1.9×10^3	6.6	$V_O^{\cdot\cdot}$	gsx		2.0×10^{-2}		Oi1, Co1, He6
			1200–1620	6.3×10^{-8}	2.5	$V_O^{\cdot\cdot}$	gsx		2.0×10^{-2}		Oi1
In$_2$O$_3$	pf	In	308–407	7.8×10^{-3}	1.35	$In_i^{\cdot\cdot\cdot}$	oxy		5×10^{-4}	on InSb	Ro5
Bi$_2$O$_3$	pp	^{210}Bi	720–780	4.5×10^{-1}	2.00	V'''_{Bi}	sfc	air			Pa3
			600–700	4.3×10^{-6}	0.90	V'''_{Bi}	sfc	air			Pa3
Y$_2$O$_3$	sp	^{91}Y	1400–1800	2.4×10^{-4}	1.90		sct	vacuum			Be6
	sp	O	1000–1500	7.2	2.54		oxy	air			Wi2
TiO$_2$	sc	^{18}O	710–1300	2.0×10^{-3}	2.60	$V_O^{\cdot\cdot}$	gsx		$(2\text{–}6) \times 10^{-1}$	$D_\perp/D_{//} = 1.6$	Ha18, Ha16
ZrO$_2$	sc	^{18}O	800–1000	9.7×10^{-3}	2.43	$V_O^{\cdot\cdot}$	gsx		4×10^{-1}		Ma1, De4, Do4
	pf	O	300–390	9×10^{-4}	1.24	$V_O^{\cdot\cdot}$	oxy		1		Sm2
ZrO$_2$:CaO(16%)	pc	^{95}Zr	1700–2150	3.5×10^{-2}	4.01	V''''_{Zr}	sct	H$_2$			Rh1, Mo1
ZrO$_2$:CaO(14%)	sc	^{45}Ca	1700–2100	4.4×10^{-1}	4.35	$V_{Ca}^{\cdot\cdot}$	sct	H$_2$			Rh1, Mo1
	sc	^{18}O	780–1100	1.8×10^{-2}	1.35	$V_O^{\cdot\cdot}$	sct	air			Si3, Ha2, Sm1
Nb$_2$O$_5$(α)	sc	^{18}O	850–1200	1.2×10^{-2}	2.14	$V_O^{\cdot\cdot}$	gsx		$1(\text{to }10^{-2})$	$D \propto P_{O_2}^{-\frac{1}{4}}$	Ch6, Ch4, Do5,
	sc	^{18}O	850–900	$D_{//}/D_\perp = 60, 190$		O$_2$	ard		$10^{-17},\ 10^{-12}$	[010]axis	Sh1, Sh2
Nb$_2$O$_5$(γ)	pf	O	540–840	1.0	1.85	$\{V_O^{\cdot\cdot}\}$	oxy		1 to 10^{-3}		Sh2
Transition Metal Oxides									P_{O_2}(atm)		
Cr$_2$O$_3$	sp	^{51}Cr	1040–1550	1.4×10^{-1}	2.64	V'''_{Cr} or $Cr_i^{\cdot\cdot\cdot}$	sct	N$_2$			Ha1, Fe1, Li11
	sc	^{51}Cr	1300	$D \propto (P_{H_2O}/P_{H_2})^{0.4}$		V'''_{Cr}	sct		$P_{H_2O}/P_{H_2} = 2$ to 18		Wa1
	pc	^{18}O	1100–1450	1.6×10	4.38		gsx		1.6×10^{-1}		Ha3, Ha5

TABLE 9f-6.—Diffusion in Ionic Crystals (*Continued*)

Substance	Form	Isotope	T Range (°C)	D_0 (cm²/sec)	W (eV)	Defect	Method	Environment	Comments	Specific Ref.	Other Ref.
MnO	sp	^{54}Mn	900–1150	7.4×10^{-7}	0.79	V''_{Mn}	sct	10^{-12} (to 10^{-14})	$D\propto P_O^{1/6}$	Bo1	
FeO	pc	^{59}Fe	700–1120	1.1×10^{-2}	1.31	V''_{Fe}	sct	$Fe_{0.917}O$		De6	De7, He2,
	pc	^{55}Fe	700–1000	1.4×10^{-2}	1.31	V''_{Fe}	sct	$Fe_{0.921}O$		Ca2	Hi1
Fe$_2$O$_3$(α)	sp	^{59}Fe	950–1050	1.3×10^{6}	4.35		sct	air		Iz2	Li7
	pc	^{18}O	900–1250	2.0	3.38	$\{V''_{\ddot{o}}\}$	gsx	1.6×10^{-1}		Ha4	Ki2
Fe$_3$O$_4$	pc	^{55}Fe	750–1000	5.2	2.38	V''_{Fe}	sfc	$Fe_{2.993}O_4$		Hi1	Iz1
	sc	^{55}Fe	850–1075	6×10^{5}	3.64		sct	Ar		Ki1	
	sp	^{59}Fe	1115	$D\propto(P_{CO_2}/P_{CO})^{0.8}$		V''_{Fe}	sfc	$P_{CO_2}/P_{CO}=10$ to $10^{3.5}$		Sc5	
CoO	sc	^{18}O	300–550	3.2×10^{-14}	0.74		gsx	$H_2O(g)$		Ca5	Ca3, Pr2
	pc	^{60}Co	1010–1340	2.2×10^{-3}	1.50	V''_{Co}	sct	1	$D\propto P_{O_2}^{0.3}$	Ca2	
	sc	^{18}O	1150–1500	9×10	4.2		gsx	2.1×10^{-1}		Ch5	Ho3, Th2
Co$_3$O$_4$	pc	^{18}O	830–860	2.4×10^{22}	7.6		gsx			Th2	
NiO	sc	^{63}Ni	1000–1470	1.8×10^{-3}	1.98	V''_{Ni}	sct	air		Ch7	K12
	sc	^{63}Ni	1000–1400	4.4×10^{-4}	1.92	V''_{Ni}	sfc	air		Sh5	Li12
	sc	^{18}O	1100–1500	6.2×10^{-4}	2.49	$\{O''_i\}$	gsx	7×10^{-2}		Ok2	
Rare Earth Oxides								P_{O_2}(atm)			
Pr$_2$O$_3$	pw	O	700–990	4.5×10^{-2}	1.82	$V_{\ddot{o}}$	oxy	vacuum		Ku2	

Compound		Isotope	Temp range	D	E	Defect	Code	Conditions		Ref	
Nd_2O_3	pc	^{18}O	700–1000	1.3×10^{-4}	1·34		gsx	4×10^{-2} to 4×10^{-1}		St8	
Sm_2O_3	pc	^{18}O	700–1000	6.0×10^{-6}	0·93		gsx	4×10^{-2} to 4×10^{-1}		St8	
Er_2O_3	pw	O	850–1250	1·2	2·07		oxy	air		Wi2	
UO_2	sc	^{233}U	1450–1700	4×10^{-7}	3·04	$\{V_U'''\}$	sec	H_2		All	Be4, Li2,
	sp	^{235}U	1500	$D = 1.6 \times 10^{-11} \times y^{1.9}$			UO_{2+y} sfc		$y = 0.007$ to 0.17	Ma8	Li14, Ya1
	pw	^{18}O	550–850	1.2×10^3	2·83	O_i''	gsx	$UO_{2.002}$	instlcy.	Be4	Ar7, Be3,
			320–500	2.1×10^{-3}	1·29	O_i'	gsx	$UO_{2.063}$	instlcy.	Be4	Do3, Th4

Monovalent Chalcogenides

Compound		Isotope	Temp range	D	E	Defect	Code	Conditions	Ref	
Na_2S	sp	^{22}Na	520–700	8.3×10^2	1·66		sfc	H_2	Mo2	
		^{35}S	400–520	1.6×10^{-3}	0·77		sfc	H_2	Mo2	
	sp	^{35}S	420–800	3.8×10^{-3}	1·77		sfc	H_2	Mo2	
Cu_2S		^{64}Cu	140–450	2×10^{-4}	<0·2			no effect	Pa7	We2
Cu_2Se	pc	Cu	580–750	9×10^{-4}	0·27	Cu disorder	smc	$Cu_{1.96}Se$	Cel	Ok1
$Ag_2S(\alpha)$	pc	^{110}Ag	200–400	2.8×10^{-4}	0·15	Ag disorder	sct	no effect	Al2	
	pc	^{35}S	650–1000	2.4×10^{-4}	1·04	$V_S\ddot{}$	sct		Is1	Ja3, Sh4
$Ag_2S(\beta)$	pc	^{110}Ag	95–175	6×10^{-3}	0·45	$\{Ag_i\}$	sct		Al2	
	pw	^{110}Ag	25–70	9.3×10^{-3}	0·40		slx		Pe3	
	pw	^{35}S	120–141	2.4×10^{-1}	1·07		slx		Pe3	
$AgSbS_2$	pc	Ag	400	$D = 4 \times 10^{-7}$			elt	S_2	Ri3	
		Sb	400	$D = 3 \times 10^{-11}$			oxy	S_2	Ri3	
Ag_2Se	pc	^{110}Ag	150–280	2.1×10^{-4}	0·12	Ag disorder	sct	no effect	Ok1	

TABLE 9f-6.—Diffusion in Ionic Crystals (*Continued*)

Substance	Form	Isotope	T Range (°C)	D_0 (cm²/sec)	W (eV)	Defect	Method	Environment	Comments	Specific Ref.	Other Ref.
Tl₂Se	pc	^{204}Tl	150– 300	1.2×10^{-3}	0.61		sct	vacuum		Ak1	
		^{75}Se	150– 300	2.2×10^{-5}	0.58		sct	vacuum		Ak1	
Divalent Chalcogenides								P(atm)			
ZnS	sc	^{65}Zn	1030–1075	1×10^{16}	6.5	$\{V''_{Zn}\}$	gsx	$P_{Zn}=1$		Se3	
			940–1030	1.5×10^4	3.25	$\{Zn_i^{··}\}$	gsx	$P_{Zn}=1$		Se3	
		^{35}S	700– 890	2.9×10^4	3.4		sct	$P_S=5 \times 10^{-1}$		Bl1	
		^{35}S	740–1100	8×10^{-5}	2.2		ard	$P_S=5 \times 10^{-1}$		Go1	
ZnSe	sc	^{75}Se	1000–1150	2.3×10^{-1}	2.7	$Se_i^{\times}$	sct	P_{Se_2} max		Wo2	
ZnTe	sc	^{65}Zn	780– 950	1.4×10	2.69	$V_{Zn}^{\times}$ or $Zn_i^{\times}$	sct	no effect		Re5	
		^{123}Te	780– 950	1.9×10^4	3.78	$Te_i^{\times}$	sct	Te(g)	or $(V_{Zn}V_{Te})^{\times}$	Re5	
CdS	sc	^{115}Cd	700–1130	3	2.0	$V_{Cd}^{\times}$	sct	Cd sat.		Wo1	
CdSe	sc	^{75}Se	700–1000	1.3×10^6	4.43	$Se_i^{\times}$	sct	$P_{Cd}=10^{-1}$ (to 10^{-7})	$D \propto P_{Cd}^{-1}$	Wo2	
	sc	^{75}Se	700–1000	2.6×10^{-3}	1.55	$Se_i^{\times}$	sct	P_{Se_2} max		Wo2	
CdTe	sc	^{109}Cd	660– 920	3.3×10^2	2.67	$V_{Cd}^{\times} + Cd_i^{\times}$	sct	Cd sat.		Bo6	
			640– 850	1.6×10	2.44	$V_{Cd}^{\times} + Cd_i^{\times}$	sct	Te sat.		Bo6	
	sc	^{123}Te	660– 900	8.5×10^{-7}	1.42	$Te_i^{\times}$	sct	Cd sat.		Bo6	
		^{123}Te	510– 780	1.7×10^{-4}	1.38	$Te_i^{\times}$	sct	Te sat.		Bo6	Wo2

Compound	Form	Tracer	Temp. range	D_0	Method	Condition	Defect	E	Remarks	References
PbS	sc	^{210}Pb	500–800	$8\cdot6\times10$	sct	stoich.	$V_{Pb}+Pb_i$	$1\cdot5$		Si2, Se9 (Am, Sc,)
	sc		500–800	$2\cdot6\times10^{-5}$	sct	Pb 10^{18}cm^{-3}	$Pb_i^\times$	$1\cdot4$		Si2
	sc		500–800	$5\cdot5\times10^{-7}$	sct	S excess	$V_{Pb}^\times$	$1\cdot0$		Si2
	sc	^{35}S	500–750	$6\cdot8\times10^{-5}$	sct	stoich.		$1\cdot38$		Se10
			500–750	$1\cdot9\times10^{-6}$	sct	Pb 10^{18}cm^{-3}	$\{(V_{Pb}V_S)\}^\times$	$1\cdot16$		Se10
			500–750	$4\cdot6\times10^{-5}$	sct	S 10^{18}cm^{-3}	$\{S_i^\times\}$	$1\cdot2$		Se10
PbSe	sc	^{210}Pb	400–800	$5\cdot0\times10^{-6}$	sct	vacuum	$V_{Pb}^\times+Pb_i^\times$	$0\cdot83$		Se8, Bo5
	sc	^{75}Se	650–800	$2\cdot1\times10^{-5}$	sct	vacuum	$\{V_{Se}^\times\}$	$1\cdot2$		Bo5
PbTe	sc	^{210}Pb	310–820	3×10^{-2}	sct		$\{Pb_i^\times\}$	$1\cdot51$		Ca1, Bo4, Go2
	sc	^{127}Te	500–800	$2\cdot7\times10^{-6}$	sct	vacuum	$\{V_{Te}^\times\}$	$0\cdot75$		Bo5, Go2
BiSe	pc	^{75}Se	100–200	$2\cdot5\times10^{-9}$	sct	vacuum		$0\cdot67$		Ku1
Bi$_2$Se$_3$	pc	^{75}Se	100–200	$8\cdot5\times10^{-9}$	sct	vacuum		$2\cdot17$		Ku1
FeS	sc	Fe	350–700	$3\cdot2\times10^{-2}$		Fe$_{0\cdot85}$S	$V_{Fe}^\times$	$0\cdot97$		Co6, Co5, Me3
	sc	S	900–1060	1×10^{33}		Fe$_{0\cdot85}$S		$10\cdot4$		Co6, Co5
CoTe(γ)	pc	^{60}Co	400–800	$D=1\cdot5\times10^{-7}$	sct			D at 805°		Ar5
	pc	^{125}Te	400–800	$D=1\cdot6\times10^{-9}$	sct			D at 410°		Ar5
NiS	sc	^{63}Ni	725–880	$1\cdot1\times10^{-2}$	sct	Ni$_{0\cdot97}$S	$V_{Ni}^\times+Ni_i^\times$	$1\cdot11$	‖ c-axis	Fu1, KI3, KI4
	sc		725–880	$8\cdot5\times10^{-3}$	sct	Ni$_{0\cdot97}$S	$V_{Ni}^\times+Ni_i^\times$	$1\cdot11$	⊥ c-axis	KI3
	sc	^{35}S	800–880	$2\cdot5\times10^{2}$	sct	Ni$_{0\cdot97}$S	$V_S^\times$	$2\cdot84$	‖ c-axis	KI3, KI4
	sc		800–880	$2\cdot2\times10^{6}$	sct	Ni$_{0\cdot97}$S	$V_S^\times$	$3\cdot79$	⊥ c-axis	KI3

Form of Sample

- sc single crystal
- pc polycrystalline
- pf polycrystalline film
- pw powder
- pp pressed powder
- sp sintered powder

Defect

- V_{Ba}'' Ba ion vacancy
- $V_{Zn}^\times$ Zn vacancy, neutral
- $V_{Br}^\cdot$ Br ion vacancy
- $(V_{Na}V_{Cl})^\times$ Vacancy pair, neutral
- $Ag_i^\cdot$ Interstitial Ag ion
- $Te_i^\times$ Interstitial Te atom
- $\{\ \}$ Tentative assignment

Method

- sct sectioning
- sfc surface counting
- gsx gaseous exchange
- slx solution exchange

- adc additive coloration
- ard autoradiography
- elt electrotransport
- nmr nuclear magnetic resonance
- oxy oxidation or weight loss
- smc semiconducting properties

TABLE 9f-7.—Correlation Effects in Diffusion.

[The correlation factor is $f = D_{tracer}/D_{conductivity}$.]

Substance	Isotope	T Range (°C)	f_{expt}	Specific Ref.	Other Ref.	Defect and Lattice	f_{theory}	Reference
NaCl	^{22}Na + ^{36}Cl	580- 680	0·9 - 1·0	Do6	Ma6	$V'_{Na} + V^{·}_{Cl}$, fcc	0·781	Ba2, Co2
NaCl	^{22}Na + ^{36}Cl	640- 790	0·85- 1·00	Ne1			0·781	,,
NaBr	^{24}Na	450- 700	1·0	Ma6	Wi3	V'_{Na} , fcc	0·781	,,
KCl	^{42}K	500- 700	1·0	As3		V'_{K} , fcc	0·781	,,
CsCl(β)	^{137}Cs + ^{36}Cl	280- 460	1·4 - 1·5	Ha9		$V'_{Cs} + V^{·}_{Cl}$, sc	0·653	,,
CsBr	^{134}Cs + ^{82}Br	330- 530	0·83- 0·88	Ly1		$V'_{Cs} + V^{·}_{Br}$, sc	0·653	,,
CsI	^{134}Cs + ^{131}I	330- 530	0·68- 0·86	Ly1		$V'_{Cs} + V^{·}_{I}$, sc	0·653	,,
AgCl	^{110}Ag	300- 440	0·48- 0·54	We1	Co4, Mu6	$V'_{Ag} + Ag^{·}_{i}$, fcc	0·33 -0·78	Co3, Mc2
AgCl: Cd	^{110}Ag	130- 230	0·78- 0·74	Gr1	Co4, Lu4	V'_{Ag} , fcc	0·781	Ba2, Co2
AgBr	^{110}Ag	140- 410	0·47- 0·65	Fr2	Mi1, Mu4	$V'_{Ag} + Ag^{·}_{i}$, fcc	0·33 -0·78	Co3, Fr2
AgBr: Cd	^{110}Ag	100- 200	0·80	Mi1	St2	V'_{Ag} fcc	0·781	Ba2, Co2
AgI(β)	^{110}Ag	90- 110	6 -11	Mu5		{Ag ring}		
AgI	^{110}Ag	90- 110	2 - 8	Mu5		High P phase		
TlCl	^{204}Tl + ^{36}Cl	290- 390	0·75- 0·70	Fr4		$V'_{Tl} + V^{·}_{Cl}$, sc	0·653	Ba2, Co2
CaCl$_2$	^{36}Cl	650- 700	0·6	Ho5		$V^{·}_{Cl}$ or Cl'_{i}, sc		
		200- 500	2·0 - 2·6	Ho5				
Li$_2$SO$_4$	^{6}Li	600- 800	0·9 - 0·7	Kv4	Kv2	Li disorder		
BeO	^{7}Be	1100-1250	0·4 - 0·8	De3		V''_{Be} , hcp	0·783	Co2, Mu3
	^{7}Be	1550	0·8	Cl5				
ZrO$_2$: CaO(15%)	^{18}O	780-1100	0·5 - 0·8	Si3	Ki1	$V^{··}_{O}$, sc	0·653	Ba2, Co2
Na$_2$S	^{22}Na	400- 700	1·0	Mo2		V'_{Na} , sc	0·653	,,
Ag$_2$S(α)	^{110}Ag	180- 280	0·26- 0·30	Ok1		Ag disorder	0·3	Yo3
Ag$_2$S(α)	^{110}Ag	200- 400	0·27- 0·38	Ri1		Ag disorder	0·5	Ri2
Ag$_2$Se(α)	^{110}Ag	140- 280	0·33- 0·40	Ok1		Ag disorder	0·3 - 0·5	Ri2, Yo3

References

References for Tables 9f-1 to 9f-7

The literature has been surveyed to about the middle of 1968. Most of the references cover only the period 1958 to 1968; some earlier articles have been included, especially when more recent work is not available. The *specific references* in the tables provide the most complete or reliable information, but the *other references* also contain either appreciable data or extensive discussion.

COLLECTIONS OF DATA AND BIBLIOGRAPHIES

Halides: 9
Oxides: 1, 2, 4, 5, 7
General: 3, 6, 8

1. Berard, M. F. *Diffusion in Ceramic Systems; A Selected Bibliography.* (Ames Lab., Iowa State Univ., 1962).
2. Cumming, P. A. and Harrop, P. J. *U.K. Atom. Energy Authority, Res. Group, Bibliog.* AERE–BIB **143**, 22 pp. (1965).
3. *Diffusion Data.* (Diffusion Information Centre, Cleveland, Ohio).
4. Dragoo, A. L. *J. Res. Nat. Bur. Stand.* **72A**, 157 (1967).
5. Harrop, P. J. *J. Mater. Sci.* **3**, 206 (1968).
6. Landolt-Boernstein. *Zahlenwerte und Funktionen aus Physik, Chemie, Astronomie, Geophysik, und Technik.* Springer-Verlag, Berlin, 6th ed. Band II, Teil 6 (1959).
7. O'Keffe, M. In *Sintering and Related Phenomena.* G. C. Kuczynski, N. A. Horton, C. F. Gibbon (eds.) (Gordon and Breach, New York, 1967). p. 57.
8. Touloukian, Y. S. (ed.). *Thermophysical Properties Research Literature—A Retrieval Guide.* 2nd ed. (Plenum Press, New York, 1967).
9. Sueptitz, P. and Teltow, J. *Phys. Status Solidi* **23**, 9 (1967).

BOOKS AND MONOGRAPHS

10. Adda, Y. and Philibert, J. *La Diffusion dans les Solides.* (Presses Univ. France, Paris, 1966).
11. Boltaks, B. I. *Diffusion in Semiconductors.* (Infosearch Ltd., London, 1961).
12. Franklin, A. D. (ed.). *Calculation of the Properties of Vacancies and Interstitials.* Nat. Bur. Stand. Misc. Publ. **287**, (1966).
13. Girifalco, L. A. *Atomic Migration in Solids.* (Blaisdell, New York, 1964).
14. Gruber, B. (ed.). *Theory of Crystal Defects.* (Academic Press, New York, 1966).
15. Hasiguti, R. R. (ed.). *Lattice Defects and Their Interactions.* (Gordon and Breach, New York, 1967).
16. Hauffe, K. *Oxydation von Metallen und Metallegierungen.* (Springer-Verlag, Berlin, 1956); *Oxidation of Metals.* (Plenum Press, New York, 1965).

1149

17. Howard, R. E. and Lidiard, A. B. Matter Transport in Solids. *Rept. Prog. Phys.* **27**, 161 (1964).
18. Inokuchi, H. *Electrical Conduction in Solids.* (Routledge and Kegan Paul, London, 1965).
19. Jost, W. *Diffusion in Solids, Liquids, and Gases.* (Academic Press, New York, 1952).
20. Kingery, W. D. (ed.). *Kinetics of High Temperature Processes.* (Tech. Press of MIT, Cambridge, Mass., 1959).
21. Kroeger, F. A. and Vink, H. J. Relations Between the Concentrations of Imperfections in Crystalline Solids, *Solid State Physics* **3**, 310 (1956).
22. Kroeger, F. A. *The Chemistry of Imperfect Crystals.* (North Holland Publ. Co., Amsterdam, 1964).
23. Lidiard, A. B. *Ionic Conductivity.* Handbuch der Physik, Fluegge S. (ed.). (Springer-Verlag, Berlin, 1957), Vol XX, p. 246.
24. Manning, J. R. *Diffusion Kinetics for Atoms in Crystals.* (Van Nostrand, Princeton, N.J., 1968).
25. Murin, A. N. and Lur'e, B. G. *Diffuziya Mechenykh Atomov i Provodimost v Ionnykh Kristallakh* (Diffusion of Labelled Atoms and Conductivity in Ionic Crystals). (Izd. Leningradsk. Univ., Leningrad, 1967).
26. Pick, H. *Struktur von Stoerstellen in Alkalihalogenidkristallen.* (Springer-Verlag, Berlin, 1966).
27. Schmalzried, H. Point Defects in Ternary Ionic Crystals. *Progr. Solid State Chem.* **2**, 265 (1965).
28. Shewmon, P. G. *Diffusion in Solids.* (McGraw–Hill, New York, 1963).
29. Smith, A. C., Janak, J. F. and Adler, R. B. *Electric Conduction in Solids.* (McGraw–Hill, New York, 1967).
30. Stasiw, O. *Elektronen-und Ionenprozesse in Ionenkristallen.* (Springer-Verlag, Berlin, 1959).
31. Van Bueren, H. G. *Imperfections in Crystals.* 2nd ed., (North-Holland Publ. Co., Amsterdam, 1961).
32. Wachtman, J. B. Jr. and Franklin, A. D. (eds.). *Mass Transport in Oxides,* Nat. Bur. Stand. Spec. Publ. **296**, (1968).

REVIEW ARTICLES

Conduction processes, ionic and electronic: 53, 56, 61, 63, 64, 68
Correlation and isotope effects: 35, 42, 43, 44
Dielectric and anelastic relaxation: 33, 46, 47
Diffusion, general discussion: 54, 55, 58, 59, 60, 62, 66, 67
Diffusion and ionic conductivity: 36, 38, 48, 50, 57, 69
Diffusion of divalent ions and inert gases: 37, 39, 45, 48
Point defects, formation and general properties: 34, 40, 49, 51, 52
Pressure effects on ionic conductivity: 41, 65

Halides
33. Cole, R. H. *Prog. in Dielectrics* **3**, 1 (1961).
34. Curien, H. *J. Phys. (Paris)* **24**, 543 (1963).
35. Friauf, R. J. *J. Appl. Phys.* **33**, 494 (1962).
36. Friauf, R. J. *J. Phys. Chem.* **66**, 2380 (1962).
37. Friauf, R. J. *J. Phys. Chem. Solids* **30**, 429 (1969).
38. Haven, Y. *Proc. Brit. Ceram. Soc.* **1**, 93 (1964).

39. Kelly, R. and Jech, C. *Proc. Brit. Ceram. Soc.* **9**, 243 (1967).
40. Lawson, A. W. *J. Appl. Phys.* **33**, 466 (1962).
41. Lazarus, D. *Progr. Very High Pressure Research.* Proc. Intern. Conf., Bolton Landing, Lake George, N.Y., 46 (1961).
42. LeClaire, A. D. *High Temp. Technol.* Proc. Int. Symp., Pacific Grove, Calif., 255 (1964).
43. Lidiard, A. B. *Proc. Intern. Symp. Reactivity Solids.* 4th., Amsterdam, 52 (1961).
44. Lidiard, A. B. *Interaction Radiation Solids,* Proc. Intern. Summer School, Mol, Belg., 804 (1964).
45. Matzke, H. *Can. J. Phys.* **46**, 621 (1968).
46. Meakins, R. J. *Progr. Dielectrics* **3**, 151, (1961).
47. Nowick, A. S. *Advan. Phys.* **16**, 1 (1967).
48. Seitz, F. *Interaction Radiation Solids,* Proc. Intern. Summer School, Mol, Belg., 362 (1964).
49. Shlichta, P. J. *Geol. Soc. Amer.* Spec. Pap. **88**, 597 (1968).
50. Slifkin, L. In *Mass Transport in Oxides.* J. B. Wachtman, Jr., and A. D. Franklin, (eds.). Nat. Bur. Stand. Spec. Publ. **296**, 1 (1968).
51. Smakula, A. *Mol. Designing Mater. Devices,* **69**, (1965).
52. Smoluchowski, R. *Interaction Radiation Solids,* Proc. Intern, Summer School, Mol, Belg., 378 (1964).

Oxides
53. Anthony, A. M. *Journees Intern. Combust. Conversion Energie,* Paris, 719 (1964).
54. Birchenall, C. E. In *Mass Transport in Oxides.* J. B. Wachtman, Jr. and A. D. Franklin (eds.). Nat. Bur. Stand. Spec. Publ. **296**, 119 (1968).
55. Garbunov, N. S. and Izvekov, V. I. *Uspekhi Fiz. Nauk* **72**, 273 (1960).
56. Heyne, L. In *Mass. Transport in Oxides.* J. B. Wachtman, Jr. and A. D. Franklin (eds.). Nat. Bur. Stand. Spec. Publ. **296**, 149 (1968).
57. Hirano, K. *Yogyo Kyokai Shi* **74**, 215 (1966).
58. Kingery, W. D. *Kinet. High-Temp. Processes.* Conf., Dedham, Mass., 37 (1959).
59. Lindner, R. *Inst. Intern. Chim. Colvay,* 10e Conseil chim., Brussels, 459 (1956).
60. Lindner, R. *Proc. U. N. Intern. Conf. Peaceful Uses At. Energy.* 2nd, Geneva, **20**, 116 (1958).
61. Mitoff, S. P. *Prog. Ceramic. Soc.* **4**, 217 (1966).
62. Moore, W. J. *Radioisotopes Sci. Research,* Proc. Intern. Conf. **1**, 528 (1957).
63. Volger, J. *Progr. Semicond.* **4**, 205 (1960).

General
64. Krogh-Moe, J. *Selected Topics High Temp. Chem.* **79**, (1966).
65. Lacam, A. and Lallemand, M. *J. Phys. (Paris)* **25**, 402 (1964).
66. Philibert, J. *J. Phys. (Paris)* **24**, 417 (1963).
67. Philibert, J. *Silicates Ind.* **28**, 449 (1963).
68. Wagner, C. *Proc. 7th Meeting Intern. Comm. Electrochem. Thermodynam. and Kinetics,* 361 (1957).
69. Wagner, C. *Mol. Designing Mater. Devices,* 122 (1965).

ARTICLES
Ab1.   Abbink, H. C. and Martin, D. A. *J. Phys. Chem. Solids* **27**, 205 (1966).
Ab2.   Abey, A. E. and Tomizuka, C. T. *J. Phys. Chem. Solids* **27**, 1149 (1966).

Ak1. Akhundov, G. A. and Abdullaev, G. B. *Sov. Phys. Dokl.* **3**, 390 (1958).
All.   Alcock, C. B., Hawkins, R. J., Hills, A. W. and McNamara, P. *Thermodynamics, Proc. Symp.*, Vienna **2**, 57 (1966).
Al2.   Allen, R. L. and Moore, W. J. *J. Phys. Chem.* **63**, 223 (1959).
Al3.   Allen, R. R. and Weber, M. J. *J. Chem. Phys.* **38**, 2970 (1963).
Al4.   Allnatt, A. R. and Jacobs, P. W. M. *Trans. Faraday Soc.* **58**, 116 (1962).
Al5.   Allnatt, A. R. and Pantelis, P. *Solid State Commun.* **6**, 309 (1968).
An1.   Anderson, J. S. and Richards, J. R. *J. Chem. Soc.* **1946**, 537.
An2.   Anthony, A. M., Guillot, A. and Nicolau, P. *C. R. Acad. Sci. (Paris)* **B262**, 896 (1966).
An3.   Antropoff, A. V. and Muller, J. Fr. *Z. Anorg. Allg. Chem.* **204**, 305 (1932).
Ar1.   Arai, G. and Mullen, J. G. *Phys. Rev.* **143**, 663 (1966).
Ar2.   Arends, J. and Nijboer, H. *Sol. State Commun.* **5**, 163 (1967).
Ar3.   Arends, J. and Nijboer, H. *Phys. Stat. Solidi* **26**, 537 (1968).
Ar4.   Arkhangelskaya, V. A., Mikheev, B. G., Nikitinskaya, T. I. and Tyutin, M. S. *Sov. Phys.-Solid State* **9**, 539 (1967).
Ar5.   Arkharov, V. I., Klotsman, S. M., Timofeev, A. N. and Trakhtenberg, I. Sh. *Phys. Met. Metall.* **14**, No. 1, 62 (1962).
Ar6.   Arnikar, H. J. and Chemla, M. *C.R. Acad. Sci. (Paris)* **242**, 2132 (1956).
Ar7.   Aronson, S., Roof, R. B. Jr. and Belle. J. *J. Chem. Phys.* **27**, 137 (1957).
As1.   Asadi, P. *Phys. Stat. Solidi* **20**, K55, K59 (1967).
As2.   Asao, Y., Yoshida, I., Ando, R. and Sawada, S. *J. Phys. Soc. Japan* **17**, 442 (1962).
As3.   Aschner, J. F. *Phys. Rev.* **94**, 771 (1954).
Au1.   Austerman, S. B. In *Kinetics of High Temp. Processes,* W. D. Kingery (ed.) (Technology Press of MIT, Cambridge, Mass., 1959). p. 66.
Au2.   Austerman, S. B. *J. Nucl. Mater.* **14**, 248 (1964).
Au3.   Austerman, S. B. and Wagner, J. W. *J. Am. Ceram. Soc.* **49**, 94 (1966).
Ba1.   Baranovskii, V. L., Lure, B. G. and Murin, A. N. *Dokl. Akad. Nauk SSSR* **105**, 1188 (1955).
Ba2.   Bardeen, J. and Herring, C. In *Atom Movements,* J. H. Holloman (ed.). (Amer. Soc. for Metals, Cleveland, 1951) p. 87; also In *Imperfections in Nearly Perfect Crystals,* W. Shockley (ed.). (Wiley, New York, 1952), p. 261.
Ba3.   Barr, L. W., Hoodless, I. M., Morrison, J. A. and Rudham, R. *Trans. Faraday Soc.* **56**, 679 (1960).
Ba4.   Barr, L. W. and LeClaire, A. D. *Proc. Brit. Ceram. Soc.* **1**, 109 (1964).
Ba5.   Barr, L. W., Morrison, J. A. and Schroeder, P. A. *J. Appl. Phys.* **36**, 624 (1965).
Ba6.   Barsis, E. and Taylor, A. *J. Chem. Phys.* **45**, 1154 (1966).
Ba7.   Barsis, E., Lilley, E. and Taylor, A. *Proc. Brit. Ceram. Soc.* **9**, 203 (1967).
Ba8.   Barsis, E. and Taylor, A. *J. Chem. Phys.* **48**, 4357 (1968).
Ba9.   Barsis, E. and Taylor, A. *J. Chem. Phys.* **48**, 4362 (1968).
Ba10. Bartlett, B. E., Tompkins, F. C. and Young, D. A. *Proc. Roy. Soc. (London)* **A246**, 206 (1958).
Be1.   Beaumont, J. H. and Jacobs, P. W. M. *J. Chem. Phys.* **45**, 1496 (1966).
Be2.   Beaumont, J. H. and Jacobs, P. W. M. *Phys. Stat. Solidi* **17**, K45 (1966).
Be3.   Belle, J. and Auskern, A. B. In *Kinetics of High Temp. Processes,* W. D. Kingery (ed.). (Tech. Press of MIT, Cambridge, Mass., 1959). p. 44; also A. B. Auskern and J. Belle, *J. Chem. Phys.* **28**, 171 (1958).
Be4.   Belle, J., Auskern, A. B., Bostrom, W. A. and Susko, F. S. *Proc. Intern. Symp. Reactivity Solids,* 4th., Amsterdam, 452 (1961).

Be5.   Bénière, F. and Chemla, F. *C. R. Acad. Sci. (Paris)* **C266,** 660 (1968).
Be6.   Berard, M. F. and Wilder, D. R. *J. Appl. Phys.* **34,** 2318 (1963).
Be7.   Bergé, P. *Bull. Soc. Franc. Mineral et Crist.* **83,** 57 (1960).
Be8.   Bergé, P., Benveniste, M., Blanc, G. and Dubois, M. *C.R. Acad. Sci. (Paris)* **258,** 5839 (1964).
Be9.   Bergé, P., Gago, C., Blanc, G., Adam-Benveniste, M. and Dubois, M. *J. Phys. Radium* **27,** 295 (1966).
Be10. Besson, H., Chauvy, D. and Rossel, J. *Helv. Phys. Acta* **35,** 211 (1962).
Be11. Besson, J., Deportes, C. and Robert, G. *C.R. Acad. Sci. (Paris)* **C262,** 527 (1966).
Be12. Bever, R. S. *J. Appl. Phys.* **24,** 1008 (1953).
Bi1.   Biermann, W. and Oel, H. J. *Z. Physik. Chem. (Frankfurt)* **17,** 163 (1958).
Bi2.   Biermann, W. *Z. Physik. Chem. (Frankfurt)* **25,** 90 (1960).
Bi3.   Biermann, W. and Jost, W. *Z. Physik. Chem. (Frankfurt)* **25,** 139 (1960).
Bi4.   Biermann, W. *Z. Physik. Chem. (Frankfurt)* **25,** 253 (1960).
Bi5.   Biltz, W. and Voigt, A. *Z. Anorg. Allgem. Chem.* **126,** 39 (1923).
Bi6.   Biltz, W. and Klemm, W. *Z. Physik. Chem.* **110,** 318 (1924).
Bi7.   Bizouard, M. and Cerisier, P. *C.R. Acad. Sci. (Paris)* **B262,** 1 (1966).
Bi8.   Bizouard, M., Cerisier, P. and Pantaloni, J. *C.R. Acad. Sci. (Paris)* **C264,** 144 (1967).
Bl1.   Blount, G. H., Marlor, G. A. and Bube, R. H. *J. Appl. Phys.* **38,** 3795 (1967).
Bl2.   Blumenthal, R. N., Baukus, J. and Hirthe, W. M. *J. Electrochem. Soc.* **114,** 172 (1967).
Bo1.   Bocquet, J. P., Kawahara, M. and Lacombe, P. *C.R. Acad. Sci. (Paris)* **C265,** 1318 (1967).
Bo2.   Boldyrev, V. V., Zakharov, Yu. A., Lykhin, V. M. and Votinova, L. A. *Kinetics and Catalysis* **4,** 587 (1963).
Bo3.   Bollmann, W., Goerlich, P., Karras, H. and Mothes, H. in ref. 9.
Bo4.   Boltaks, B. I. and Mokhov, Yu. N. *Sov. Phys.—Tech. Phys.* **1,** 2366 (1957).
Bo5.   Boltaks, B. I. and Mokhov, Yu. N. *Sov. Phys.—Tech. Phys.* **3,** 974 (1958).
Bo6.   Borsenberger, P. M. and Stevenson, D. A. *J. Phys. Chem. Solids* **29,** 1277 (1968).
Br1.   Bradhurst, D. H. and de Bruin, H. J. *J. Nucl. Mater.* **24,** 261 (1967).
Br2.   Bradley, J. N. and Greene, P. D. *Trans. Faraday Soc.* **62,** 2069 (1966).
Br3.   Bradley, J. N. and Greene, P. D. *Trans. Faraday Soc.* **63,** 424 (1967).
Br4.   Bray, D. T. and Merten, U. *J. Electrochem. Soc.* **111,** 447 (1964).
Br5.   Brown, N. and Hoodless, I. M. *J. Phys. Chem. Solids* **28,** 2297 (1967).
Br6.   Brown, R. N. and McLaren, A. C. *Proc. Roy. Soc.* **266,** 329 (1962).
Bu1.   Budnikov, P. P. and Yanovskii, V. K. *J. Appl. Chem. USSR* **37,** 1249 (1964).
Bu2.   Busch, G. and Junod, P. *Helv. Phys. Acta* **30,** 470 (1957).
Bu3.   Busch, G. and Junod, P. *Helv. Phys. Acta* **31,** 567 (1958).
Ca1.   Card, F. E. Thesis, Syracuse Univ. (1957).
Ca2.   Carter, R. E. and Richardson, F. D. *Trans. Met. Soc. AIME* **200,** 1244 (1954).
Ca3.   Carter, R. E., Richardson, F. D., Wagner, C. *Trans. Met. Soc. AIME* **203,** 336 (1955).
Ca4.   Castellan, G. W. and Moore, W. J. *J. Chem. Phys.* **17,** 41 (1949).
Ca5.   Castle, J. E. and Surman, P. L. *J. Phys. Chem.* **71,** 4255 (1967).
Ce1.   Celustka, B. and Ogorelec, Z. *J. Phys. Chem. Solids* **27,** 957 (1966).
Ce2.   Cerisier, P. and Bizouard, M. *C.R. Acad. Sci. (Paris)* **261,** 5100 (1965).
Ch1.   Champion, J. A. *Brit. J. Appl. Phys.* **15,** 633 (1964).
Ch2.   Champion, J. A. *Brit. J. Appl. Phys.* **16,** 805 (1965).
Ch3.   Chang, R. *Proc. Brit. Ceram. Soc.* **9,** 193 (1967).

Ch4. Chen, W. K. and Swalin, R. A. *J. Phys. Chem. Solids* **27**, 57 (1966).
Ch5. Chen, W. K. and Jackson, R. A. *Am. Ceram. Soc. Bull.* **46**, 357 (1967).
Ch6. Chen, W. K. and Jackson, R. A. *J. Chem. Phys.* **47**, 1144 (1967).
Ch7. Choi, J. S. and Moore, W. J. *J. Phys. Chem.* **66**, 1308 (1962).
Ch8. Christy, R. W. and Dobbs, H. S. *J. Chem. Phys.* **46**, 722 (1967).
Cl1. Cleaver, B. *Z. Physik. Chem. (Frankfurt)* **45**, 346 (1965).
Cl2. Cleaver, B. *Z. Physik. Chem. (Frankfurt)* **45**, 359 (1965).
Cl3. Cline, C. F., Carlberg, J. and Newkirk, W. *J. Am. Ceram. Soc.* **50**, 55 (1967).
Cl4. Cline, C. F., Newkirk, H. W. and Vandervoort, R. R. *J. Am. Ceram. Soc.* **50**, 221 (1967).
Cl5. Cline, C. F., Newkirk, H. W., Condit, R. H. and Hashimoto, Y. In *Mass Transport in Oxides,* J. B. Wachtman, Jr., and A. D. Franklin (eds.). Nat. Bur. Stand. Spec. Publ. **296**, 177 (1968).
Co1. Coble, R. L. *J. Am. Ceram. Soc.* **41**, 55 (1958).
Co2. Compaan, K. and Haven, Y. *Trans. Faraday Soc.* **52**, 786 (1956).
Co3. Compaan, K. and Haven, Y. *Trans. Faraday Soc.* **54**, 1498 (1958).
Co4. Compton, W. D. *Phys. Rev.* **101**, 1209 (1956); Compton, W. D. and Maurer, R. J. *J. Phys. Chem. Solids* **1**, 191 (1956).
Co5. Condit, R. H. and Birchenall, C. E. *U.S. Dept. Comm.,* Office Tech. Serv., P.B. Rept. 147, 772 (1960).
Co6. Condit, R. H. Thesis, Princeton Univ. (1961).
Co7. Condit, R. H. and Hashimoto, Y. *J. Am. Ceram. Soc.* **50**, 425 (1967).
Co8. Copeland, W. D. and Swalin, R. A. *J. Phys. Chem. Solids* **29**, 313 (1968).
Cr1. Croatto, V. and Bruno, M. *Gazz. Chim. Ital.* **78**, 95 (1948).
Da1. Danforth, W. E. and Bodine, J. H. *J. Franklin Inst.* **260**, 467 (1955).
Da2. Dasgupta, A. K., Sitharamarao, D. N., Palkar, G. D. *Nature* **207**, 628 (1965).
Da3. Dasgupta, S. *Brit J. Appl. Phys.* **17**, 267 (1966).
Da4. Davidge, R. W. *Phys. Stat. Solidi* **3**, 1851 (1963).
Da5. Davies, M. O. *J. Chem. Phys.* **38**, 2047 (1963).
Da6. Davis, M. L. and Westrum, E. F. Jr. *J. Phys. Chem.* **65**, 338 (1961).
Da7. Davis, W. J., Rogers, S. E. and Ubbelohde, A. R. *Proc. Roy. Soc.* **220A**, 14 (1953).
Da8. Dawood, R. I. and Forty, A. J. *Phil. Mag.* **7**, 1633 (1962).
Da9. Dawson, D. K. and Barr, L. W. *Phys. Rev. Lett.* **19**, 844 (1967).
Da10. Dawson, D. K. and Barr, L. W. *Proc. Brit. Ceram. Soc.* **9**, 171 (1967).
De1. Debiesse, J. and Neyret, G. *Vide* **6**, 1098 (1951).
De2. DeBruin, H. J. and Watson, G. M. *J. Nucl. Mater.* **14**, 239 (1964).
De3. DeBruin, H. J., Watson, G. M. and Blood, C. M. *J. Appl. Phys.* **37**, 4543 (1966).
De4. Debuigne, J. and Lehr, P. *C. R. Acad. Sci. (Paris)* **256**, 1113 (1963).
De5. de Nobel, D. *Philips Res. Rept.* **14**, 361, 430 (1959).
De6. Desmarescaux, P., Bocquet, J. P. and Lacombe, P. *Bull. Soc. Chim. France* **1965**, 1106.
De7. Desmarescaux, P. *Publ. Sci. Tech. Min. Air (Fr.)* **434** (1967).
De8. de Vries, K. J. and van Santen, J. F. *Physica* **29**, 482 (1963).
Di1. Dixon, J. M., Lagrange, L. D., Merten, U., Miller, C. F. and Porter, J. T. *J. Electrochem. Soc.* **110**, 276 (1963).
Do1. Dobrovinska, O. R., Solunskii, V. I. and Shakhova, A. G. *Ukr. Fiz. Zh.* **12**, 868 (1967).
Do2. Dolloff, R. T. *J. Appl. Phys.* **27**, 1418 (1956).
Do3. Dornelas, W. and Lacombe, P. *C.R. Acad. Sci. (Paris)* **C265**, 359 (1967).

Do4. Douglas, D. L. *Corrosion Reactor Mater.*, Proc. Conf., Salzburg, Austria **2**, 224 (1962).
Do5. Douglas, D. L. *Corrosion Reactor Mater.*, Proc. Conf., Salzburg, Austria **2**, 233 (1962).
Do6. Downing, H. L. Jr., and Friauf, R. J. *J. Phys. Chem. Solids* **31**, 845 (1970).
Dr1. Dreyfus, R. W. and Nowick, A. S. *J. Appl. Phys.* **33**, 473 (1962).
Dr2. Dreyfus, R. W. and Nowick, A. S. *Phys. Rev.* **126**, 1367 (1962).
Eb1. Ebert, I. and Teltow, J. *Ann. Physik* **15**, 268 (1955).
Eb2. Ebisuzaki, Y. Thesis, Indiana Univ. (1963).
Ec1. Ecklin, D., Nadler, C., Rossel, J. *Helv. Phys. Acta* **37**, 692 (1964).
Ec2. Economu, N, A. *Phys. Rev.* **135**, A1020 (1964).
Ec3. Economu, N, A. and Sastry, P. V. *Phys. Stat. Solidi* **6**, 135 (1964).
Ei1. Eisenstadt, M. *Phys. Rev.* **132**, 630 (1963).
Ei2. Eisenstadt, M. and Redfield, A. G. *Phys. Rev.* **132**, 635 (1963).
Ei3. Eisenstadt, M. *Phys. Rev.* **133**, A191 (1964).
El1. Elo, R., Swalin, R. A. and Chen, W. K. *J. Phys. Chem. Solids* **28**, 1625 (1967).
Et1. Etzel, H. W. and Maurer, R. J. *J. Chem. Phys.* **18**, 1003 (1950).
Fe1. Fedorchenko, I. M. and Ermolovich, Yu. B. *Ukrain. Khim. Zh.* **26**, 429 (1960).
Fi1. Finch, A., Jacobs, P. W. and Tompkins, F. *J. Chem. Soc.* **1954**, 2053.
Fl1. Floris, J. V. *J. Am. Ceram. Soc.* **43**, 262 (1960).
Fr1. Franklin, A. D. *J. Res. Nat. Bur. Standards* (Phys. and Chem.) **69A**, 301 (1965).
Fr2. Friauf, R. *J. Phys. Rev.* **105**, 843 (1957).
Fr3. Friauf, R. J. *J. Phys. Chem. Solids* **18**, 203 (1961).
Fr4. Friauf, R. J. *Z. Naturforsch.* **26a**, 1210 (1971).
Fu1. Fueki, K., Oguri, Y. and Mukaibo, T. *Bull. Chem. Soc. Japan* **41**, 569 (1968).
Fu2. Fuller, R. G. *Phys. Rev.* **142**, 524 (1966).
Fu3. Fuller, R. G. and Reilly, M. H. *Phys. Rev. Lett.* **19**, 113 (1967).
Fu4. Fuller, R. G., Reilly, M. H., Marquardt, C. L. and Wells, J. C. Jr. *Phys. Rev. Lett.* **20**, 662 (1968).
Ga1. Gallais, F. and Masdupuy, E. *C.R. Acad. Sci.* (*Paris*) **227**, 635 (1948).
Gh1. Ghate, P. B. *Phys. Rev.* **133**, A1167 (1964).
Gi1. Ginnings, D. C. and Phipps, T. E. *J. Am. Chem. Soc.* **52**, 1340 (1930).
Go1. Gobrecht, H., Nelkowski, H., Baars, J. W. and Weigt, M. *Solid State Commun.* **5**, 777 (1967).
Go2. Gomez, M. P. Thesis, Stanford Univ. (1965).
Gr1. Gracey, J. P. and Friauf, R. J. *J. Phys. Chem. Solids* **30**, 421 (1969).
Gr2. Greener, E. H., Fehr, G. A. and Hirthe, W. M. *J. Chem. Phys.* **38**, 133 (1963).
Gr3. Greenwood, N. N. and Worrall, I. J. *J. Inorg. and Nuclear Chem.* **3**, 357 (1957).
Gr4. Gruendig, H. *Z. Phys.* **158**, 577 (1960).
Gu1. Gundermann, J., Hauffe, K. and Wagner, C. *Z. Physik. Chem.* **B37**, 148 (1937); Gundermann, J. and Wagner, C. *Ibid.*, 155, 157 (1937).
Gu2. Gupta, Y. P. and Weirick, L. J. *J. Phys. Chem. Solids* **28**, 811 (1967).
Ha1. Hagel, W. C. and Seybolt, A. U. *J. Electrochem. Soc.* **108**, 1146 (1961).
Ha2. Hagel, W. C. *J. Electrochem. Soc.* **110**, 63C (1963).
Ha3. Hagel, W. C. *J. Am. Ceram. Soc.* **48**, 70 (1965).
Ha4. Hagel, W. C. *Trans. Met. Soc. AIME* **236**, 179 (1966).

Ha5. Hagel, W. C., Jorgensen, P. J. and Tomalin, D. S. *J. Am. Ceram. Soc.* **49,** 23 (1966).

Ha6. Harpur, W. W., Moss, R. L. and Ubberlohde, A. R. *Proc. Roy. Soc.* **A232,** 196 (1955).

Ha7. Harrop, P. J. and Creamer, R. H. *Brit. J. Appl. Phys.* **14,** 335 (1963).

Ha8. Harrop, P. J. *Brit. J. Appl. Phys.* **16,** 729 (1965).

Ha9. Harvey, P. J. and Hoodless, I. M. *Phil. Mag.* **16,** 543 (1967).

Ha10. Hauffe, K. and Griessbach-Vierk, A. L. *Z. Electrochem.* **57,** 248 (1953).

Ha11. Hauffe, K. and Pfeiffer, H. *Z. Metallkunde* **44,** 27 (1953).

Ha12. Hauffe, K. and Hoeffgen, D. *Z. Physik. Chem. (Frankfurt)* **49,** 94 (1966).

Ha13. Haul, R. A. W. and Stein, L. H. *Trans. Faraday Soc.* **51,** 1280 (1955).

Ha14. Haul, R. and Just, D. *Naturwiss.* **45,** 435 (1958).

Ha15. Haul, R., Just, D. and Duembgen, G. *Proc. Intern. Symp. Reactivity Solids,* 4th., Amsterdam, 65 (1961).

Ha16. Haul, R. and Duembgen, G. *Z. Electrochem.* **66,** 636 (1962).

Ha17. Haul, R. and Just, D. *J. Appl. Phys.* **33,** 487 (1962).

Ha18. Haul, R. and Duembgen, G. *J. Phys. Chem. Solids* **26,** 1 (1965).

Ha19. Haven, Y. *Rec. Trav. Chim. Pays-Bas* **69,** 1471 (1950).

He1. Hebb, M. H. *J. Chem. Phys.* **20,** 185 (1952).

He2. Hembree, P. L. Thesis, Northwestern Univ. (1967).

He3. Hermann, P. *Z. Physik. Chem. (Leipzig)* **227,** 338 (1964).

He4. Herrington, T. M. and Staveley, L. A. K. *J. Phys. Chem. Solids* **25,** 921 (1964).

He5. Hevesy, G. and Seith, W. *Z. Physik.* **56,** 790 (1929).

He6. Hewson, C. W. and Kingery, W. D. *J. Am. Ceram. Soc.* **50,** 218 (1967).

Hi1. Himmel, L., Mehl, R. F. and Birchenall, C. E. *Trans. Met. Soc. AIME* **197,** 827 (1953).

Hi2. Hirahara, E. *J. Phys. Soc. Japan* **6,** 422 (1951).

Hi3. Hirashima, M. *J. Phys. Soc. Japan* **10,** 1055 (1955).

Ho1. Holt, J. B. *J. Nucl. Mater.* **11,** 107 (1964).

Ho2. Holt, J. B. and Condit, R. H. *Mater. Sci. Res.* **3,** 13 (1966).

Ho3. Holt, J. B. *Proc. Brit. Ceram. Soc.* **9,** 157 (1967).

Ho4. Holverson, E. L. and Kevane, C. J. *J. Chem. Phys.* **44,** 3692 (1966).

Ho5. Hood, G. M. and Morrison, J. A. *J. Appl. Phys.* **38,** 4796 (1967).

Ho6. Hoodless, I. M. and Morrison, J. A. *J. Phys. Chem.* **66,** 557 (1962).

Ho7. Hoodless, I. M. and McNicol, B. D. *Phil. Mag.* **17,** 1223 (1968).

Ho8. Hoshino, H. and Shimoji, M. *J. Phys. Chem. Solids* **28,** 1169 (1967).

Ho9. Hoshino, H. and Shimoji, M. *J. Phys. Chem. Solids* **29,** 1431 (1968).

Ho10. Howard, R. E. *Phys. Rev.* **144,** 650 (1966).

Hs1. Hsueh, Y. W. and Christy, R. W. *J. Chem. Phys.* **39,** 3519 (1963).

Hu1. Huntington, H. B. and Ghate, P. B. *Phys. Rev. Lett.* **8,** 421 (1962).

Hu2. Huntington, H. B., Ghate, P. B. and Rosolowski, J. H. *J. Appl. Phys.* **35,** 3027 (1964).

Ii1. Iida, S. *Jap. J. Appl. Phys.* **6,** 77 (1967).

Is1. Ishiguro, M., Oda, F. and Fujino, T. *Mem. Inst. Sci. Ind. Research, Osaka Univ.* **10,** 1 (1953).

Iz1. Izvekov, V. I. *Inzhener. Fiz. Zh., Akad. Nauk. Belorus SSR* **1,** 64 (1958).

Iz2. Izvekov, V. I., Garbunov, N. S. and Babad-Zakhryakin, A. A. *Phys. Met. Metall.* **14,** 30 (1962).

Ja1. Jacobs, P. W. M. and Tompkins, F. C. *J. Chem. Phys.* **23,** 1445 (1955).

Ja2. Jaffray, J. J. *recherches centre natl. recherche sci.*, *Lab Bellevue* (*Paris*) **39**, 125 (1957).
Ja3. Jagitsch, R. *Trans. Chalmers Univ. Tech. Gothenburg*, No. 11, 1 (1942).
Ja4. Jain, S. C. and Dahake, S. L. *Phys. Letters* (*Netherlands*) **3**, 308 (1963).
Ja5. Jain, S. C. and Dahake, S. L. *Indian J. Pure Appl. Phys.* **2**, 71 (1964).
Ja6. Jain, S. C. and Sootha, G. D. *Phys. Stat. Solidi* **22**, 505 (1967).
Ja7. Jander, W. Z. *Anorg. Allgem. Chem.* **199**, 306 (1931).
Ja8. Jackson, B. J. H. and Young, D. A. *Trans. Faraday Soc.* **63**, 2246 (1967).
Jo1. Johansen, H. A. and Cleary, J. G. *J. Electrochem. Soc.* **111**, 100 (1964).
Jo2. Johansson, G. and Lindner, R. *Acta Chem. Scand.* **4**, 782 (1950).
Jo3. Jordan, P. and Pochon, M. *Helv. Phys. Acta.* **30**, 33 (1957).
Jo4. Jost, W. and Noelting, J. *Z. Physik. Chem.* (*Frankfurt*) **7**, 383, (1956); Jost, W. and Oel, H. J. *Disc. Faraday Soc.* **23**, 137 (1957).
Ka1. Kalbitzer, S. *Z. Naturforsch.* **17a**, 1071 (1962).
Ka2. Kanzaki, H., Kido, K. and Ninomiya, T. *J. Appl. Phys.* **33**, 482 (1962).
Ka3. Kanzaki, H., Kido, K. and Ohzora, S. *J. Phys. Soc. Japan* **18**, Suppl. 3, 115 (1963).
Ka4. Kanzaki, H., Kido, K., Tamura, S. and Oki, S. *J. Phys. Soc. Japan* **20**, 2305 (1965).
Ki1. Kingery, W. D., Pappis, J., Doty, M. E. and Hill, D. C. *J. Am. Ceram. Soc.* **42**, 393 (1959).
Ki2. Kingery, W. D., Hill, D. C. and Nelson, R. P. *J. Am. Ceram. Soc.* **43**, 473 (1960).
Ki3. Kirk, D. L. and Pratt, P. L. *Proc. Brit. Ceram. Soc.* **9**, 215 (1967).
Ki4. Kirovskaya, I. A., Maidanovskaya, L. G. and Zhukova, V. D. *Inorganic Materials* **3**, 260 (1967).
Kl1. Klotsman, S. M., Timofeev, A. N. and Trakhtenberg, I. Sh. *Phys. Met. Metall.* **10**, No. 5, 93 (1960).
Kl2. Klotsman, S. M., Timofeev, A. N. and Trakhtenberg, I. Sh. *Phys. Met. Metall.* **14**, No. 3, 91 (1962).
Kl3. Klotsman, S. M., Timofeev, A. N. and Trakhtenberg, I. Sh. *Phys. Met. Metall.* **16**, No. 5, 92 (1963).
Kl4. Klotsman, S. M., Timofeev, A. N. and Trakhtenberg, I. Sh. *Phys. Met. Metall.* **17**, No. 1, 119 (1964).
Kl5. Klotsman, S. M., Polikarkova, I. P., Timofeev, A. N. and Trakhtenberg, I. Sh. *Sov. Phys.-Solid State* **9**, 1956 (1967).
Ko1. Kobayashi, K. and Tomiki, T. *J. Phys. Soc. Japan* **15**, 1982 (1960).
Kr1. Kröger, F. A. *J. Phys. Chem. Solids* **26**, 901 (1965).
Kr2. Krogh-Moe, J., Vikan, M. and Krohn, C. *Acta Chem. Scand.* **21**, 309 (1967).
Ku1. Kuliev, A. A. and Abdullaev, G. B. *Sov. Phys.-Solid State* **1**, 545 (1959).
Ku2. Kuntz, U. E. and Eyring, L. In *Kinetics of High Temp. Processes*, W. D. Kingery (ed.) (Technology Press of MIT, Cambridge, Mass., 1959), p. 50.
Ku3. Kurnick, S. *J. Chem. Phys.* **20**, 218 (1952).
Ku4. Kushida, T. *J. Sci. Hiroshima Univ.*, Ser. A14, 147 (1950).
Ku5. Kuznets, E. D. and Yakimenko, L. M. *J. Appl. Chem. USSR* **40**, 754 (1967).
Kv1. Kvist, A. and Lunden, A. *Z. Naturforsch.* **19a**, 1058 (1964).
Kv2. Kvist, A. and Lunden, A. *Z. Naturforsch.* **20a**, 235 (1965).
Kv3. Kvist, A. *Z. Naturforsch.* **21a**, 487 (1966).
Kv4. Kvist, A. and Trolle, U. *Z. Naturforsch.* **22a**, 213 (1967).
Kv5. Kvist, A. and Josefson, A. M. *Z. Naturforsch.* **23a**, 625 (1968).
La1. Lakatos, E. and Lieser, K. H. *Z. Physik. Chem.* (*Frankfurt*) **48**, 213 (1966).
La2. Lakatos, E. and Lieser, K. H. *Z. Physik. Chem.* (*Frankfurt*) **48**, 228 (1966).

La3. Lasker, M. F. and Rapp, R. A. *Z. Physik. Chem. (Frankfurt)* **49**, 198 (1966).
La4. Laurance, N. *Phys. Rev.* **120**, 57 (1960).
La5. Laurent, J. F. and Bénard, J. *J. Phys. Chem. Solids* **3**, 7 (1957); **7**, 218 (1958); Laurent, J. F. *Ann. Chim. (Paris)* **3**, 712 (1958); Bénard, J. and Laurent, J. F. *Radioisotopes Sci. Research, Proc. Intern. Conf.* **1**, 577 (1957).
La6. Layer, H., Miller, M. G. and Slifkin, L. *J. Appl. Phys.* **33**, 478 (1962); Layer H. and Slifkin, L. *J. Phys. Chem.* **66**, 2396 (1962); Kabler, M. N., Layer, H., Miller, M. G. and Slifkin, L. *Mater. Sci. Res.* **1**, 82 (1963).
Le1. LeClaire, A. D. and Lidiard, A. B. *Phil. Mag.* **1**, 518 (1956).
Le2. LeClaire, A. D. *Phil. Mag.* **14**, 1271 (1966).
Le3. Lee, K. and Sher, A. *Phys. Rev. Lett.* **14**, 1027 (1965).
Le4. Lee, V. J. and Parravano, G. *J. Appl. Phys.* **30**, 1735 (1959).
Le5. Lehfeldt, W. *Z. Physik* **85**, 717 (1933).
Le6. Lewis, T. J. and Wright, A. J. *Brit. J. Appl. Phys.* **1**, 441 (1968).
Li1. Lidiard, A. B. *Phil. Mag.* **46**, 815, 1218 (1955).
Li2. Lidiard, A. B. *J. Nucl. Mater.* **19**, 106 (1966).
Li3. Lieser, K. H. *Z. Physik. Chem. (Frankfurt)* **9**, 302, 308 (1956).
Li4. Lindner, R. *Z. Elektrochem.* **54**, 430 (1950).
Li5. Lindner, R., Campbell, D. and Akerstroem, A. *Acta Chem. Scand.* **6**, 457 (1952).
Li6. Lindner, R., Austruemdal, St. and Akerstroem, A. *Acta Chem. Scand.* **6**, 468 (1952).
Li7. Lindner, R., Verduch, A. G. and Akerstroem, A. *Ark. Kemi* **4**, 381 (1952).
Li8. Lindner, R. *Ark. Kemi* **4**, 385 (1952).
Li9. Lindner, R. and Terem, H. N. *Ark. Kemi* **7**, 273 (1954).
Li10. Lindner, R. and Enqvist, O. *Ark. Kemi* **9**, 471 (1956).
Li11. Lindner, R. and Akerstroem, A. *Z. Physik. Chem. (Frankfurt)* **6**, 162 (1956).
Li12. Lindner, R. and Akerstroem, A. *Disc. Faraday Soc.* **23**, 133 (1957).
Li13. Lindner, R. and Parfitt, G. D. *J. Chem. Phys.* **26**, 182 (1957).
Li14. Lindner, R. and Schmitz, F. *Z. Naturforsch.* **16a**, 1373 (1961).
Lo1. Lomelin, J. M. and Neubert, T. J. *J. Phys. Chem.* **67**, 1115 (1963).
Lo2. Lothian, T. A. and Neubert, T. J. *J. Chem. Phys.* **47**, 3092 (1967).
Lu1. Luckey, G. and West, W. *J. Chem. Phys.* **24**, 879 (1956).
Lu2. Lunden, A. *Z. Naturforsch.* **17a**, 142 (1962).
Lu3. Lundin, A. G., Gabuda, S. P. and Lifshits, A. I. *Sov. Phys.-Solid State* **9**, 273 (1967).
Lu4. Lure, B. G., Marin, A. N. and Murin, I. V. *Sov. Phys.-Solid State* **8**, 2991 (1967).
Ly1. Lynch, D. W. *Phys. Rev.* **118**, 468 (1960).
Ma1. Madeyski, A. and Smeltzer, W. W. *Mater. Res. Bull.* **3**, 369 (1968).
Ma2. Mahendroo, P. P. and Nolle, A. W. *Phys. Rev.* **126**, 125 (1962).
Ma3. Maidanovskaya, L. G., Kirovskaya, I. A. and Lobanova, G. L. *Inorganic Materials* **3**, 839 (1967).
Ma4. Makarov, L. L., Lure, B. G. and Malyshev, V. N. *Sov. Phys.-Solid State* **2**, 79 (1960).
Ma5. Manakin, B. A., Voloshchenko, I. A. and Kolesnikov, V. A. *Russ. J. Phys. Chem.* **41**, 1861 (1967).
Ma6. Mapother, D., Crooks, H. N. and Maurer, R. J. *J. Chem. Phys.* **18**, 1231 (1950).
Ma7. Mariani, E., Eckstein, J. and Rubinova, E. *Czech. J. Phys.* **17**, 552 (1967).
Ma8. Marin, J. F., Michaud, H. and Contamin, P. *C.R. Acad. Sci. (Paris)* **C264**, 1633 (1967).

Ma9. Matsumura, T. *Can. J. Phys.* **44**, 1685 (1966).
Ma10. Matzke, H. *J. Nucl. Mater.* **11**, 344 (1964).
Ma11. Matzke, H. and Lindner, R. *Z. Naturforsch.* **19a**, 1178 (1964).
Ma12. Maycock, J. N. *J. Appl. Phys.* **35**, 1512 (1964).
Mc1. McClaine, L. A. and Coppel, C. P. *J. Electrochem. Soc.* **113**, 80 (1966).
Mc2. McCombie, C. W. and Lidiard, A. B. *Phys. Rev.* **101**, 1210 (1956).
Md1. Mdivani, O. M. *Tr. Tbilis. Gos. Univ.* **103**, 151 (1965).
Me1. Mehrotra, P. N., Chandrasekhar, G. V., Rao, C. N. R. and Subbaro, E. C. *Trans. Faraday Soc.* **62**, 3586 (1966).
Me2. Melik-Gaikazyan, I. Ya., Roshchina, L. I., Ignateva, M. I. and Kurakina, E. P. *Izv. Vyssh. Ucheb. Zaved., Fiz.* **10**, 141 (1967).
Me3. Meussner, R. A. and Birchenall, C. E. *Corrosion* **13**, 677 (1957).
Mi1. Miller, A. S. and Maurer, R. J. *J. Phys. Chem. Solids* **4**, 196 (1958).
Mi2. Mitoff, S. P. *J. Chem. Phys.* **31**, 1261 (1959).
Mi3. Mitoff, S. P. *J. Chem. Phys.* **36**, 1383 (1962).
Mi4. Miyata, T., Sano, R. and Tomiki, T. *J. Phys. Soc. Japan* **20**, 638 (1965).
Mi5. Miyatani, S. Y. and Suzuki, Y. *J. Phys. Soc. Japan* **8**, 680 (1953).
Mi6. Miyatani, S. *J. Phys. Soc. Japan* **10**, 786 (1955).
Mi7. Miyatani, S. *J. Phys. Soc. Japan* **13**, 317 (1958).
Mi8. Miyatani, S. *J. Phys. Soc. Japan* **13**, 341 (1958).
Mi9. Miyatani, S. *J. Phys. Soc. Japan* **14**, 996 (1959).
Mo1. Möbius, H. H., Witzmann, H. and Gerlach, D. *Z. Chem.* **4**, 154 (1964).
Mo2. Möbius, H. H., Witzmann, H. and Hartung, R. *Z. Physik. Chem. (Leipzig)* **227**, 40 (1964).
Mo3. Moore, W. J. and Selikson, B. *J. Chem. Phys.* **19**, 1539 (1951); **20**, 927 (1952).
Mo4. Moore, W. J., Ebisuzaki, Y. and Sluss, J. A. *J. Phys. Chem.* **62**, 1438 (1958).
Mo5. Moore, W. J. and Williams, E. L. *Disc. Faraday Soc.* **28**, 86 (1959).
Mo6. Morkel, A. and Schmalzried, H. *J. Chem. Phys.* **36**, 3101 (1962).
Mo7. Morlin, Z. *Acta Phys. Acad. Sci. Hungar.* **21**, 137 (1966).
Mr1. Mrgudich, J. N. *J. Electrochem. Soc.* **107**, 475 (1960).
Mu1. Mueller, P. *Phys. Stat. Solidi* **9**, K193 (1965); **12**, 775 (1965).
Mu2. Muennich, F. *Naturwiss.* **42**, 340 (1955).
Mu3. Mullen, J. G. *Phys. Rev.* **124**, 1723 (1961); *Phys. Rev. Lett.* **9**, 383 (1962).
Mu4. Murin, A. N., Lure, B. G. and Lebedev, N. A. *Sov. Phys.-Solid State* **2**, 2324 (1960).
Mu5. Murin, A. N., Lure, B. G. and Tarlakov, Yu. P. *Sov. Phys.-Solid State* **3**, 2395 (1961).
Mu6. Murin, A. N., Lure, B. G. and Murin, I. V. *Sov. Phys.-Solid State* **9**, 1840 (1968).
Na1. Nagels, P., van Lierde, W., de Batist, R., Denayer, M., de Jonghe, L. and Gevers, R. *Thermodynamics, Proc. Symp., Vienna* **2**, 311 (1966).
Ne1. Nelson, V. C. and Friauf, R. J. *J. Phys. Chem. Solids.* **31**, 825 (1970).
Ne2. Neubert, T. J. and Nichols, G. M. *J. Am. Chem. Soc.* **80**, 2619 (1958).
Ne3. Neuhaus, A. and Hinze, E. *Ber. Bunsenges. Physik. Chem.* **70**, 1073 (1966).
Ne4. Neuimin, A. D. and Palguev, S. F. *Dokl. Phys. Chem.* **143**, 315 (1962).
Ni1. Nikitinskaya, T. I., Suntsov, E. V. and Tyutin, M. S. *Sov. Phys.-Solid State* **9**, 1656 (1967).
No1. Noddack, W. and Walch, H. *Z. Elektrochem.* **63**, 269 (1959).
No2. Noddack, W. and Walch, H. *Z. Physik. Chem. (Leipzig)* **211**, 194 (1959).
No3. Noelting, J. *Z. Physik. Chem. (Frankfurt)* **19**, 118 (1959).
No4. Noelting, J. *Z. Physik. Chem. (Frankfurt)* **32**, 154 (1962).

No5. Noelting, J. *Z. Physik. Chem.* (*Frankfurt*) **38**, 154 (1963).
No6. Noyer, F. and Laurent, J. F. *C. R. Acad. Sci.* (*Paris*) **242**, 3068 (1956).
Oi1. Oishi, Y. and Kingery, W. D. *J. Chem. Phys.* **33**, 480 (1960).
Oi2. Oishi, Y. and Kingery, W. D. *J. Chem. Phys.* **33**, 905 (1960).
Ok1. Okazaki, H. *J. Phys. Soc. Japan* **23**, 355 (1967).
Ok2. O'Keffe, M. and Moore, W. J. *J. Phys. Chem.* **65**, 1438, 2277 (1961).
Ow1. Owens, B. B. and Argue, G. R. *Science* **157**, 308 (1967).
Pa1. Paladino, A. E. and Kingery, W. D. *J. Chem. Phys.* **37**, 957 (1962).
Pa2. Palguev, S. F. and Neuimin, A. D. *Sov. Phys.-Solid State* **4**, 629 (1962).
Pa3. Palkar, G. D., Sitharamararo, D. N. and Dasgupta, A. K. *Trans. Faraday Soc.* **59**, 2634 (1963).
Pa4. Panasyuk, G. P., Danchevskaya, M. N. and Kobozev, N. I. *Russ. J. Phys. Chem.* **41**, 354 (1967).
Pa5. Pappis, J. and Kingery, W. D. *J. Am. Ceram. Soc.* **44**, 459 (1961).
Pa6. Patterson, J. W., Bogren, E. C. and Rapp, R. A. *J. Electrochem. Soc.* **114**, 752 (1967).
Pa7. Pavyluchenko, M. M., Pokrovskii, I. I. and Tikhonov, A. S. *Dokl. Akad. Nauk. Belorussk. SSR* **9**, 235 (1965).
Pa8. Payne, R. T. and Lawson, A. W. *J. Chem. Phys.* **34**, 2201 (1961).
Pe1. Pershits, Ya. N. and Pavlov, E. V. *Sov. Phys.-Solid State* **10**, 1125 (1968).
Pe2. Persyn, G. A. and Nolle, A. W. *Phys. Rev.* **140**, A1610 (1965).
Pe3. Peschanski, D. *J. Chim. Phys.* **47**, 933 (1950).
Pe4. Peters, D. W., Feinstein, L. and Peltzer, C. *J. Chem. Phys.* **42**, 2345 (1965).
Ph1. Phipps, T. E., Lansing, W. D. and Cooke, T. G. *J. Am. Chem. Soc.* **48**, 112 (1926).
Ph2. Phipps, T. E. and Partridge, E. G. *J. Am. Chem. Soc.* **51**, 1331 (1929).
Pi1. Pierce, C. B. *Phys. Rev.* **123**, 744 (1961).
Pl1. Plester, D. W., Rogers, S. E. and Ubbelohde, A. R. *Proc. Roy. Soc.* **235**, 469 (1956).
Pr1. Pretzel, F. E., Vier, D. T., Szklarz, E. G. and Lewis, W. B. *USAEC LA2463* (1960).
Pr2. Price, J. B. and Wagner, J. B. Jr., *Z. Physik. Chem.* (*Frankfurt*) **49**, 257 (1966).
Pr3. Pryor, A. W. *J. Nucl. Mater.* **14**, 258 (1964).
Ra1. Ramasastry, C. and Murti, Y. V. G. S. *J. Phys. Chem. Solids* **24**, 1384 (1963).
Ra2. Ramasastry, C. and Murti, Y. V. G. S. *Proc. Roy. Soc.* **A305**, 441 (1968).
Ra3. Rapoport, E. Thesis, Univ. of Maryland (1964).
Ra4. Rapp, R. A. *USAEC COO*-1440–3 (1967).
Re1. Reade, R. F. and Martin, D. S. *J. Appl. Phys.* **31**, 1965 (1960).
Re2. Redington, R. W. *Phys. Rev.* **87**, 1066 (1952).
Re3. Reinhold, H. and Braueninger, H. *Z. Physik. Chem.* **B41**, 397 (1938).
Re4. Reuter, B. and Hardel, K. *Ber. Bunsenges. Physik. Chem.* **70**, 82 (1966).
Re5. Reynolds, R. A. Thesis, Stanford (1966).
Rh1. Rhodes, W. H. and Carter, R. E. *J. Am. Ceram. Soc.* **49**, 244 (1966).
Ri1. Rickert, H. *Z. Physik. Chem.* (*Frankfurt*) **23**, 355 (1960).
Ri2. Rickert, H. *Z. Physik Chem.* (*Frankfurt*) **24**, 418 (1960).
Ri3. Rickert, H. and Wagner, C. *Z. Elecktrochem.* **64**, 793 (1960).
Ri4. Riggleman, B. M. and Drickamer, H. G. *J. Chem. Phys.* **38**, 2721 (1963).
Ro1. Robert, G., Deportes, C. and Besson, J. *J. Chim. Phys.* **64**, 1275 (1967).
Ro2. Roberts, J. P. and Wheeler, C. *Phil. Mag.* **2**, 708 (1957).
Ro3. Roberts, J. P. and Wheeler, C. *Trans. Faraday Soc.* **56**, 570 (1960).

Ro4. Rolfe, J. *Can. J. Phys.* **42**, 2195 (1964).
Ro5. Rosenberg, A. J. and Lavine, M. C. *J. Phys. Chem.* **64**, 1135, 1143 (1960).
Ro6. Rovner, L. H. Thesis, Cornell Univ. (1966).
Ro7. Rozenblyum, N. D., Bubyreva, N. C., Bukhareva, V. I. and Kazakevich, G. Z. *Russ. J. Phys. Chem.* **40**, 1324 (1966).
Sa1. Samara, G. A. *Phys. Rev.* **165**, 959 (1968).
Sa2. Sastry, P. V. and Srinivasan, T. V. *Phys. Rev.* **132**, 2445 (1963).
Sc1. Scanlon, W. W. and Brebrick, R. F. *Physica* **20**, 1090 (1954).
Sc2. Schamp, H. W. and Katz, E. *Phys. Rev.* **94**, 828 (1954).
Sc3. Schmalzried, H. *Z. Physik. Chem. (Frankfurt)* **22**, 199 (1959).
Sc4. Schmalzried, H. *J. Chem. Phys.* **33**, 940 (1960).
Sc5. Schmalzried, H. *Z. Physik. Chem. (Frankfurt)* **31**, 184 (1962).
Sc6. Schmalzried, H. *Z. Physik. Chem. (Frankfurt)* **38**, 87 (1963).
Sc7. Schock, R. N. and Katz, S. *J. Chem. Phys.* **48**, 2094 (1968).
Sc8. Scholten, P. C. and Mysels, K. J. *Trans. Faraday Soc.* **56**, 994 (1960).
Sc9. Schwab, G. M. and Eulitz, G. *Z. Physik. Chem. (Frankfurt)* **55**, 179 (1967).
Sc10. Scott, K. T. and Wassell, L. L. *Proc. Brit. Ceram. Soc.* **7**, 375 (1967).
Se1. Secco, E. A. and Moore, W. J. *J. Chem. Phys.* **23**, 1170 (1955).
Se2. Secco, E. A. and Moore, W. J. *J. Chem. Phys.* **26**, 942 (1957).
Se3. Secco, E. A. *J. Chem. Phys.* **29**, 406 (1958).
Se4. Secco, E. A. *Disc. Faraday Soc.* **28**, 94 (1959).
Se5. Secco, E. A. *Can. J. Chem.* **39**, 1544 (1961).
Se6. Seifert, G. *Z. Physik.* **161**, 132 (1961).
Se7. Seith, W. *Z. Elektrochem.* **39**, 538 (1933).
Se8. Seltzer, M. S. and Wagner, J. B. Jr. *J. Chem. Phys.* **36**, 130 (1962).
Se9. Seltzer, M. S. and Wagner, J. B. Jr. *J. Phys. Chem. Solids* **24**, 1525 (1963).
Se10. Seltzer, M. S. and Wagner, J. B. Jr. *J. Phys. Chem. Solids* **26**, 233 (1965).
Se11. Semenenko, K. N. and Naumova, T. N. *Russ. J. Inorg. Chem.* **9**, 718 (1964).
Sh1. Sheasby, J. S. and Cox, B. *J. Less-Common Metals* **15**, 129 (1968).
Sh2. Sheasby, J. S., Smeltzer, W. W. and Jenkins, A. E. *J. Electrochem. Soc.* **115**, 338 (1968).
Sh3. Sher, A., Solomons, R., Lee, K. and Muller, M. W. *Phys. Rev.* **144**, 593 (1966).
Sh4. Shim, M. T. Thesis, Indiana Univ. (1957).
Sh5. Shim, M. T. and Moore, W. J. *J. Chem. Phys.* **26**, 802 (1957).
Sh6. Shimizu, K. *Rev. Phys. Chem. Japan* **30**, 1 (1960).
Sh7. Shimizu, K. *Rev. Phys. Chem. Japan* **30**, 73 (1960).
Sh8. Shimizu, K. *Rev. Phys. Chem. Japan* **31**, 67 (1962).
Sh9. Short, J. M. and Roy, R. *J. Phys. Chem.* **68**, 3077 (1964).
Si1. Simkovich, G. *J. Phys. Chem. Solids* **24**, 213 (1963).
Si2. Simkovich, G. and Wagner, J. B. Jr. *J. Chem. Phys.* **38**, 1368 (1963).
Si3. Simpson, L. A. and Carter, R. E. *J. Am. Ceram. Soc.* **49**, 139 (1966).
Si4. Sinclair, W. R. and Loomis, T. C. In *Kinetics of High Temp. Processes,* W. D. Kingery. (ed.) (Technology Press of MIT, Cambridge, Mass., 1959), p. 58.
Sl1. Sluss, J. A., Jr. Thesis, Indiana Univ. (1962).
Sm1. Smith, A. W., Meszaros, F. W. and Amata, C. D. *J. Am. Ceram. Soc.* **49**, 240 (1966).
Sm2. Smith, T. *J. Electrochem. Soc.* **112**, 560 (1965).
So1. Sonin, A. S. and Zheludev, I. S. *Kristallografiya* **8**, 57 (1963).
So2. Sonin, A. S. and Zheludev, I. S. *Kristallografiya* **8**, 285 (1963).

So3.   Southgate, P. D. *J. Phys. Chem. Solids* **27**, 1623 (1966).

Sp1.   Spicar, E. Thesis, Stuttgart (1956).

Sp2.   Sproull, R. L., Bever, R. S. and Libowitz, G. G. *Phys. Rev.* **92**, 77 (1953).

St1.   Steele, B. C. H. and Alcock, C. B. *Trans. Met. Soc. AIME* **233**, 1359 (1965).

St2.   Steiger, R. *Chimia* **18**, 306 (1964).

St3.   Steiger, R., Boustany, K. and Boissonnas, Ch. G. *Helv. Chim. Acta* **49**, 787 (1966).

St4.   Stoebe, T. G., Ogurtani, T. O. and Huggins, R. A. *Phys. Rev.* **134**, 963 (1964).

St5.   Stoebe, T. B., Ogurtani, T. O. and Huggins, R. A. *Phys. Stat. Solidi* **12**, 649 (1965).

St6.   Stoebe, T. G. and Huggins, R. A. *J. Mater. Sci.* **1**, 117 (1966).

St7.   Stoebe, T. G. and Pratt, P. L. *Proc. Brit. Ceram. Soc.* **9**, 181 (1967).

St8.   Stone, G. D., Weber, G. R. and Eyring, L., In *Mass Transport in Oxides.* J. B. Wachtman, Jr. and A. D. Franklin (eds.) Nat. Bur. Stand. Spec. Publ. **296**, 179 (1968).

St9.   Strickler, D. W. and Carlson, W. G. *J. Am. Ceram. Soc.* **47**, 122 (1964).

St10.   Strumane, R. and de Batist, R. *Phys. Stat. Solidi* **6**, 817 (1964).

Su1.   Subbarao, E. C., Sutter, P. H. and Hrizo, J. *J. Am. Ceram. Soc.* **48**, 443 (1965).

Su2.   Suchow, L. and Pond, G. R. *J. Am. Chem. Soc.* **75**, 5242 (1953).

Su3.   Surplice, N. A. *Brit. J. Appl. Phys.* **17**, 175 (1966).

Ta1.   Takahashi, T, and Yamamoto, O. *Denki Kagaku* **32**, 610 (1964); **33**, 346 (1965).

Ta2.   Takahashi, T. and Yamamoto, O. *Denki Kagaku* **33**, 733 (1965).

Ta3.   Takahashi, T. and Iwahara, H. *Denki Kagaku* **34**, 254 (1966).

Ta4.   Takahashi, T. and Yamamoto, O. *Electrochim. Acta* **11**, 779 (1966).

Ta5.   Takahashi, T., Yamamoto, O., Tsukada, K. and Baba, A. *Denki Kagaku* **35**, 32 (1967).

Ta6.   Takahashi, T., Kuwabara, K. and Yamamoto, O. *Denki Kagaku* **35**, 682 (1967).

Ta7.   Tallan, N. M., Vest, R. W. and Graham, H. C. *Mater. Sci. Res.* **2**, 33 (1965).

Ta8.   Tallan, N. M. and Vest, R. W. *J. Am. Ceram. Soc.* **49**, 401 (1966).

Ta9.   Tallan, N. M., Tripp, W. C. and Vest, R. W. *J. Am. Ceram. Soc.* **50**, 279 (1967).

Ta10.   Tannhauser, D. S. *J. Phys. Chem. Solids* **5**, 224 (1958).

Ta11.   Tare, V. B. and Schmalzried, H. *Z. Physik. Chem. (Frankfurt)* **43**, 30 (1964).

Ta12.   Taylor, W. H., Daniels, W. B., Royce, B. S. H. and Smoluchowski, R. *J. Phys. Chem. Solids* **27**, 39 (1966).

Te1.   Teltow, J. *Ann. Physik* **5**, 63, 71 (1949).

Th1.   Tharmalingam, K. and Lidiard, A. B. *Phil. Mag.* **4**, 899 (1959).

Th2.   Thompson, B. A. Thesis, Rensselaer Poly. Inst. (1962).

Th3.   Thompson, B. A. and Strong, R. L. *J. Phys. Chem.* **67**, 594 (1963).

Th4.   Thorn, R. J. and Winslow, G. H. *J. Chem. Phys.* **44**, 2822 (1966).

To1.   Tompkins, F. C. and Young, D. A. *Disc. Faraday Soc.* **23**, 202 (1957).

To2.   Torkar, K. and Herzog, G. W. *Monatsh. Chem.* **97**, 765 (1966).

To3.   Torkar, K. and Spath, H. T. *Monatsh. Chem.* **98**, 2382 (1967).

Tr1.   Tretjakov, J. and Schmalzried, H. *Ber. Bunsenges. Phys. Chem.* **69**, 396 (1965).

Tu1.   Tubandt, C., Rindtorff, E. and Jost, W. *Z. Anorg. Allgem. Chem.* **165**, 195 (1927).

Tu2.   Tubandt, C. and Baudouin, M. In *Landolt-Boernstein Physikalisch-Chemische Tabellen,* 5th ed., (Springer-Verlag, Berlin, 1931).

Tu3.   Tubandt, C. and Geiler, J. In *Landolt-Boernstein Physikalisch-Chemische Tabellen*, 5th ed., (Springer-Verlag, Berlin, 1931).
Ue1.   Ueda, A., Asano, Y., Nishimaki, N., Kojima, K. and Ishiguro, M., *Mem. Inst. Sci. Ind. Research, Osaka Univ.* **17**, 89 (1960).
Ur1.   Ure, R. W., Jr., *J. Chem. Phys.* **26**, 1363 (1957).
Ve1.   Verwey, J. F. and Schoonman, J. *Physica* **35**, 386 (1967).
Ve2.   Vest, R. W. and Tallan, N. M. *J. Am. Ceram. Soc.* **48**, 472 (1965).
Ve3.   Vest, R. W. and Tallan, N. M. *J. Appl. Phys.* **36**, 543 (1965).
Vi1.   Vinokurov, I. V., Zonn, Z. N. and Ioffe, V. A. *Inorg. Mater.* **3**, 901 (1967).
Vo1.   Volchenkova, Z. S. and Palguev, S. F. *Tr. Inst. Electrokhim., Akad. Nauk SSSR, Ural'sk. Filial* **5**, 133 (1964).
Wa1.   Wagener, K. *Z. Physik. Chem. (Frankfurt)* **23**, 305 (1960).
Wa2.   Wagner, C. *Z. Physik. Chem.* **B21**, 25 (1933); **B23**, 469 (1933).
Wa3.   Wagner, C. *J. Chem. Phys.* **18**, 62 (1950).
Wa4.   Wagner, C. and Hantlemann, P. *J. Chem. Phys.* **18**, 72 (1950).
Wa5.   Wagner, C. *Proc. Int. Comm. Electrochem. Thermo. and Kinetics*, 7th meeting, 1955 (Butterworth Sci. Publ., London, 1957). p. 361.
Wa6.   Wagner, C. *Z. Elektrochem.* **63**, 1027 (1959).
Wa7.   Wagner, J. B. and Wagner, C. *J. Chem. Phys.* **26**, 1597 (1957).
Wa8.   Wagner, J. B. and Wagner, C. *J. Chem. Phys.* **26**, 1602 (1957).
Wa9.   Wakabayashi, H. *J. Phys. Soc. Japan* **15**, 2000 (1960).
Wa10. Wakabayashi, H. *J. Phys. Soc. Japan* **17**, 292 (1962).
We1.   Weber, M. D. and Friauf, R. J. *J. Phys. Chem. Solids* **30**, 407 (1969).
We2.   Wehefritz, V. *Z. Physik .Chem. (Frankfurt)* **26**, 339 (1960).
We3.   Weil, R. and Lawson, A. W. *J. Chem. Phys.* **41**, 832 (1964).
We4.   Weiss, K. *Z. Physik. Chem. (Frankfurt)* **12**, 68 (1957).
Wh1.   Whitney, E. D. *J. Electrochem. Soc.* **112**, 91 (1965).
Wi1.   Wimmer, J. M., Bidwell, L. R. and Tallan, N. M. *J. Am. Ceram. Soc.* **50**, 198 (1967).
Wi2.   Wirkus, C. D., Berard, M. F. and Wilder, D. R. *J. Am. Ceram. Soc.* **50**, 113 (1967).
Wi3.   Witt, H. *Z. Physik* **134**, 117 (1953).
Wo1.   Woodbury, H. H. *Phys. Rev.* **134**, A492 (1964).
Wo2.   Woodbury, H. H. and Hall, R. B. *Phys. Rev.* **157**, 641 (1967).
Ya1.   Yajima, S., Furuya, H., Hirai, T. *J. Nucl. Mater.* **20**, 162 (1966).
Yo1.   Yokota, I. *J. Phys. Soc. Japan* **8**, 595 (1953).
Yo2.   Yokota, I. and Miyatani, S. *Jap. J. Appl. Phys.* **1**, 144 (1962).
Yo3.   Yokota, I. *J. Phys. Soc. Japan* **21**, 420 (1966).
Za1.   Zakharov, Yu. A. and Savelev, G. G. *Kinetics and Catalysis* **5**, 307 (1964).
Za2.   Zakharov, Yu. A. and Kabanov, A. A. *Russ. J. Phys. Chem.* **38**, 1567 (1964).
Zi1.   Zimen, K. E., Johansson, G. and Hillert, M. *J. Chem. Soc., Suppl.* No. 2, S392 (1949).
Zi2.   Zirkind, P. and Freeman, E. S. *Nature* **199**, 1280 (1963).

Author Index

The numbers in brackets are the reference numbers and those in italic refer to the Reference pages where the references are listed in full. Absence of page number indicates a general reference.

Bell, A. E. *744*
Bell, M. E. 824, *836*
Belle, J. 484(220), *514*, 1125(Be4), 1145 (Ar7, Be3, Be4), *1152*
Bellissent, M. C. 857, 858, 859, 860, 861, *865*, 893, 894, 914, *929*
Bellows, J. C. 82, 99, 106, 112, 120, 124, *125*
Bénard, J. 259, 263, 264, 269, 287, 289, *293*, *294*, *297*, 309, 315, 316, *318*, 1138(La5), 1139(La5), *1158*
Benderskii, 1091, 1093, *1100*
Béniére, F. 252, 259, 260, 261, 262, 263, 264, 265, 266, 267, 268, 270, 274, 275, 279, 280, 282, 283, 284, 286, 287, 288, 289, 290, *292*, *297*, *298*, 312, 313, 316, *317*, 708, 729, *743*, 1138(Be5), *1153*
Béniére, M. 215, 259, 261, 262, 263, 264, 265, 266, 267, 268, 270, 274, 279, 280, 282, 283, 286, 287, 288, 289, 290, *292*, *297*, *298*, 312, 313, 316, *317*
Benton, A. F. 612, *615*
Benveniste, M. 1112(Be8), *1153*
Benz, R. 580, 582, *615*, 631, 632, 645, 646, 649, 650, 652, *674*
Béranger, G. 128, 140, 141, *149*
Berard, M. F. 1107(1), 1143(Be6, Wi2), 1145(Wi2), *1149*, *1153*, *1163*
Bereskhova, G. V. 347, *350*
Bergé, P. 1112(Be7, Be8), 1121(Be9), *1153*
Berger, C. B. 507(292), 508(299), *516*
Bergeron, C. G. 392, *396*
Bergsma, F. 407(7), 408(7), 474(7), 475(202), *509*, *514*
Berkowitz, J. 523, 528, 534, *538*
Bert, J, 728, 738, 740, *744*
Berzins, T. 890, *928*
Besson, H. 1114(Be10), 1122(Be10), *1153*
Besson, J. 330, *344*, 911, *929*, 1129 (Be11, Ro1), 1134(Ro1), 1135(Be11), *1153*, *1160*
Bever, R. S. 1142(Be12, Sp2), *1153*, *1162*
Bhada, R. K. 1042(103), *1049*
Bidwell, L. R. 325, 332, *346*, 576, 578, 581, *615*, *620*, 1135(Wi1), *1163*
Bielen, H. 815, *836*

Bierig, R. W. 112, *126*
Biermann, W. 655, *674*, 1113(Bi2, Bi4), 1115(Bi3), 1121(Bi2, Bi4), 1126(Bi1, Bi2, Bi4), 1127(Bi2, Bi4), 1128(Bi1, Bi3), *1153*
Billard, M. 668, 669, 670, *674*
Billington, J. C. 604, *620*
Biltz, W. 800, *806*, 1117(Bi5, Bi6), *1153*
Binder, H. 1017(60), 1040(60), *1047*
Birchenall, C. E. 1110(54), 1144(Hi1), 1147(Co5, Me3), *1151*, *1154*, *1156*, *1159*
Birks, N. 660, 663, 664, *674*
Biscoe, J. 353, 354, *400*
Bitsianes, G. 611, *616*
Bizouard, M. 286, *294*, 1118(Bi7), 1119(Bi8, Ce2), 1124(Bi8), *1153*
Bjerrum, N. 117, *124*, 494(238), *514*
Blaedel, W. J. 446(144), 448(144, 145, 146), *512*
Blanc, G. 1112(Be8), 1121(Be8), *1153*
Blankeship, F. E. 658, *675*
Bleckmann, P. 632, *676*
Bliton, J. L. 1020(69), *1048*
Block, F. E. 653, 655, *676*
Block, J. 533, 536, *538*
Bloem, J. 819, *836*
Blood, C. M. 1128(De3), 1142(De3), 1148(De3), *1154*
Blount, G. H. 1146(Bl1), *1153*
Blumenthal, R. N. 1011(46), 1019(46), *1047*, 1125(Bl2), *1153*
Bochove, V. 412(43), *510*
Bockris, J. O. M. 509(301), *516*, 583, 607, 608, *615*
Bocquet, J. P. 1135(Bo1, De6), 1144 (Bo1, De6), *1153*, *1154*
Bodine, J. H. 1129(Da1), 1135(Da1), *1154*
Bogdandy, L. von. 611, *615*
Bogren, E. C. 327, 335, 336, 339, *345*, 553, 554, 556, *619*, 679, *698*, 814, 820, *836*, 1134(Pa6), 1135(Pa6), *1160*
Bohnenkamp, K. 582, *621*
Boies, D. B. 416(88), *511*
Boisot, P. 140, 141, *149*
Boissonnas, Ch. G. 1055, *1069*, 1140 (St3), *1162*
Bokiy, G. B. 631, 668, *674*

Darken, L. S. 578, *615*, 685, *697*
Darlak, J. O. 1070, 1076, *1091*
Dasgupta, A. K. 1143(Da2, Pa3), *1154, 1160*
Dasgupta, S. 1129(Da3), *1154*
Dauge, G. 668, 669, 670, *674, 676*
David, D. 127, 128, 130, 137, 140, 141, *149*
Davidge, R. W. 1121(Da4), *1154*
Davidson, N. 93, *124*
Davies, A. E. 607, *615*
Davies, M. O. 337, *344*, 1128(Da5), *1154*
Davis, M. L. 1116(Da6), *1154*
Davis, W. J. 1119(Da7), *1154*
Dawood, R. I. 1117(Da8), *1154*
Dawood, R. J. 347, 348, *350*
Dawson, B. 816, *836*
Dawson, D. G. 401(1), 433(1, 114), *509, 512*
Dawson, D. K. 174, 186, 194, *201, 202*, 268, 269, 270, 286, 288, 290, *292, 293, 298*, 316, *317*, 1113(Da9), 1139(Da10), *1154*
Debiesse, J. 1142(De1), *1154*
De Bruin, H. J. 1128(De3), 1142(De2, De3), 1148(De3), *1154*
Debuigne, J. 1142(De4), 1143(De4), *1154*
Debye, P. P. 83, *125*
De Groot, S. R. 421(101), 425(101), 426(101), *511*, 699, 700, 701, 708, *744*
Dekker, A. J. 33, *76*, 158, 168, *202*
Delahay, P. 868, 870, 882, 890, 891, 918, *928, 929*, 1091, 1099, *1100, 1101*
Delarue, G. 753, *759*
de Levie, R. 1091, 1093, 1096, 1099, *1100*
De Luca, J. P. 392, *396*
Denayer, M. 1125(Na1), *1159*
Denne, W. A. 679, *697*
Dent, L. S. *76*
Deportes, C. 330, *344*, 911, *929*, 1009 (42), *1047*, 1129(Be11, Ro1), 1134 (Ro1), 1135(Be11), *1153, 1160*
Derge, G. 607, 608, 611, *615, 621, 622*
Derge, G. J. 607, *615*
De Rosa, J. A. 965, *987*
De Rossi, M. 977, 981, *986*
Desmarescaux, P. 1135(De6), 1144(De6, De7), *1154*

Desmond, J. E. 971, 974, *987*
Despic, A. 442(134), 493(134, 237), 509 (301), *512, 514, 516*
De Surville, R. 57, *76*
Detry, D. 520, 523, 526, 528, 534, 537, *538*
de Vooys, D. A. 1099, *1101*
Devoto, G. 800, *806*
Devreese, J. 736, *744*
de Vries, W. T. 889, 890, 918, 921, *928, 929*
Diaz, C. 847, 848, 853, *865*
Diaz, C. M. 575, 592, *615*
Di Benedetto, A. T. 506(281), *516*
Dickson, W. R. 607, *615*
Dietzel, A. 994(10), *1046*
Dieumegard, D. 130, 141, *149*
Dismukes, E. D. 607, *615*
Dixon, J. M. 326, 328, 329, *344*, 1000 (33), 1004(33), 1005(33), *1047*, 1129 (Di1), *1154*
Dneprova, V. G. 647, *674*
Dobbs, H. S. 725, 726, *744*, 1123(Ch8), *1154*
Dobner, W. 337, *345*
Dobo, J. 415(73), *511*
Dobos, S. 363, 386, *398*
Dobrovinskaya, E. R. 298
Dobrovinska, O. R. *298*, 1138(Do1), 1139(Do1), *1154*
Dolezalek, F. K. 395, *396*
Dolloff, R. T. 1129(Do2), 1133(Do2), *1154*
Domagala, R. F. 325, *346*
Domange, L. 668, *674*
Dominguez, M. 467(187), *513*
Donnan, F. G. 457(167), *513*
Donner, D. 660, *674*
Doremus, R. H. 352, 361, 370, 371, *396*, 878, *929*
Dornelas, W. 1135(Do3), 1145(Do3), *1154*
Dorokhova, M. L. 372, *397*
Dorst, W. 433(111, 112, 114), 440(127), 441(111), 500(269), *511, 512, 515*
Doty, M. E. 326, *345*, 552, *618*, 995(17), 997(17), 998(17), 1000(17), *1046*, 1129 (Ki1), 1148(Ki1), *1157*
Douglas, D. L. 508(296), *516*, 572, *615*, 1143(Do4, Do5), *1155*

Vetcher, A. A. 626, 631, 632, 642, 643, 644, 649, *676*
Vetcher, D. V. 626, 631, 632, 649, *676*
Vetter, K. J. 834, *838*
Vier, D. T. 1112(Pr1), 1121(Pr1), *1160*
Vierbicky, V. L. 575, 591, *616*
Vikan, M. 1116(Kr2), *1157*
Vineyard, G. H. 254, *295*
Vink, H. J. 706, *745*, 1054, *1069*, 1107 (21), *1150*
Vinokurov, I. V. 1129(Vi1), 1135(Vi1), *1162*
Viškarev, A. F. 559, 575, *618*
Voigt, A. 1117(Bi5), *1153*
Volchenkova, S. 326, 327, 330, 331, 332, *345*, *346*
Volchenkova, Z. S. 994(32), 998(24), 1000(32), 1011(24), *1046*, *1047*, 1135 (Vo1), *1163*
Volger, J. *1151*
Voloshchenko, I. A. 1116(Ma5), *1158*
Volterra, V. 112, 113, *125*
Voropaev, Yu. V. 415(68), *510*
Vorres, K. S. 1020(67), *1048*
Votinova, L. A. 1120(Bo2), 1124(Bo2), *1153*
Vouros, P. 949, *956*, 982, *988*
Vries, K. J. de. 1070, 1071, 1073, *1091*, 1117(De8), 1124(De8), *1154*

W

Wachtman, J. B. Jr. 106, *125*, 1107(56), 1109(50), 1110(54), 1145(St8), 1148 (Cl5), *1150*, *1151*, *1154*, *1162*
Wadsley, A. D. 814, 815, 816, *838*
Wagner, C. 80, *126*, 223, 260, 271, *293*, 299, 300, 310, *318*, 522, *538*, 541, 542, 544, 545, 551, 554, 564, 569, 570, 572, 575, 590, *615*, *618*, *622*, 624, 628, 629, 631, 632, 645, 646, 649, 650, 655, 656, 657, 658, 659, 660, 666, *674*, *675*, *676*, 679, 680, 684, *697*, *698*, 811, 813, 814, 819, 820, *836*, *837*, *838*, 843, 850, 855, *865*, *930*, 939, 953, 954, *956*, 997 (21), 998(21), 1000(29), *1046*, *1047*, 1085, *1091*, 1108(Wa5), Wa6, Wa2), 1115(Wa6), 1122(Wa4), 1128(Wa3, Wa7), 1132(Wa7), 1133(Gu1), 1136

(Wa8), 1144(Ca3), 1145(Ri3), *1151*, *1153*, *1155*, *1160*, *1163*
Wagner, H. 575, 576, 594, 595, *620*
Wagner, J. B. 655, 657, 658, *676*, 820, *838*, 1085, *1091*, 1128(Wa7), 1132 (Wa7), 1136(Wa8), *1163*
Wagner, J. B. Jr. 1117(Se9), 1144(Pr2), 1147(Se8, Se9, Se10, Si2), *1160*, *1161*
Wagner, J. W. 1141(Au3), *1152*
Wagener, K. 1126(Wa1), *1163*
Wakabayashi, H. 1115(Wa9, Wa10), *1163*
Walch, H. 337, *345*, 560, *619*, 1129 (No1), 1130(No1), 1135(No2), *1159*
Wallace, R. A. 486(222), *514*
Warburg, E. 835, *838*
Warkecke, D. 416(82), *511*
Warner, N. A. 610, *614*
Warren, B. E. 78, 353, 354, *400*
Warter, P. J. 1082, *1091*
Wasastjerna, J. A. 62, *78*
Wassell, L. L. 1116(Sc10), *1161*
Watanabe, N. 575, 581, 582, *618*
Watkins, G. D. 86, 101, 104, 106, *126*
Wats, R. E. 443(136), 459(136), *512*
Watson, G. M. 604, *620*, 1128(De3), 1142(De2, De3), 1148(De3), *1154*
Waugh, J. L. T. 816, *837*
Webb, L. E. 583, *620*
Webb, T. H. E. 971, 974, *987*
Weber, F. 815, *836*
Weber, G. R. 1145(St8), *1162*
Weber, H. J. 918, *930*
Weber, M. D. 193, 194, *202*, 271, 272, 273, 274, *295*, 1123(We1), 1139(We1), 1148(We1), *1163*
Weber, M. J. 112, *126*, 1112(Al3), 1121 (Al3), *1152*
Weber, N. 343, *345*, 932, *956*, 970, *988*
Wehefritz, V. 658, 659, *676*, 1130(We2), 1136(We2), 1145(We2), 1148(We2), *1163*
Wehlend, W. 416(82), *511*
Weigt, M. 1146(Go1), *1155*
Weil, R. 1116(We3), 1127(We3), 1128 (We3), 1132(We3), *1163*
Wein, M. 835, *838*
Weininger, J. L. 589, *622*, 719, *745*, 810, *838*, 932, 936, 951, 952, 955, *956*, 964, 966, 976, 978, 980, 982, 984, *988*

Subject Index

A

Activation energy, 552, 667, 703, 814, 1106
Activity,
  coefficients, 590, 644, 870, 884
  of halogen, 821
  of nickel, 582, 852
  of PbO, 584
  of silica, 586
  of silver, 811, 953
  of sulphur, 662, 811
  oxygen, 590, 903
  thermodynamic, 519, 522
Adsorbed layer, 663
Aliovalent impurities, 555
Alloys, 580, 585, 587, 589
Alumina, 589, 603, 815, 827, 909, 1107
Aluminate, 583, 669
Alumino-silicates, 816
Amalgamation, 973
Ammonia gas, 964
Anion,
  free oxygen, 608
  interstitials, 814
  sublattice, 650
Anode, 886, 1031
Anodic,
  depolarization, 695
  oxidation, 868
  oxidation of a metal, 894
  reaction, 967, 973
Autocatalytic effect, 612
Atomic radius, 875

B

Battery,
  aqueous, 810
  solid state, 809, 951
  thin film, 949, 1024

Bessel function, 883
Bessel–Laplace transformation, 892
Boltzmann distribution, 833
Bonds, silver–halogen, 876
Borate melts, 607
Borides, 632, 640
Boron, 670
Bromide,
  lead, 1070, 1077, 1083
  silver, 831, 876, 933
  silver sulphide, 938

C

Calcite, 876
Calcium atoms, 583
Colour centres, 628, 1019
Capacitance, 825, 828, 834, 980
Carbide, 614, 632, 636, 641
Carbon, 614, 670
Carbonate, 604, 756
Catalyst, 905, 989
Cathodic,
  depolarizers, 967, 974
  reaction, 684, 974, 990
Cell,
  concentration, 999
  electrochemical, 623
  electrochemical Knudsen, 519
  galvanic, 632
  hydrogen, 1039
  mixed, 802
  oxygen concentration, 539, 556
  solid state galvanic, 519, 521
  thin-film, 949
Centre F, 738
Ceramic, ion-conductive, 952
Ceria, 1033
Chemical potential,
  of silver, 659, 855
  of sulphur, 660, 663, 855

li

Compound Index